Contents

Introduction

There are as many approaches to studying animal cognition as there are definitions of cognition itself. This diversity is reflected in the essays that follow, to a degree that we believe is unparalleled in any other volume that has been produced on this subject. This diversity is philosophical and methodological, with contributors demonstrating various degrees of acceptance or disdain for terms such as "consciousness" and various degrees of concern for the rigors of laboratory experimentation versus the validity of naturalistic research. The diversity is also apparent in the wide range of species to be found between these covers. Of course there are many chapters on primates, and especially the great apes, reflecting our human-centered interest in our closest relatives, but there are also chapters that touch on cognition in animals as diverse as earthworms, antelopes, dogs, spiders, dolphins, bees, fish, hyenas, snakes, sea lions, prairie dogs, virtual organisms, parrots, rats, ravens, and squirrels, to name but a few. We think we have produced one of the most wide-ranging menageries of scientific studies of animal cognition ever assembled. Yet it is humbling to realize that in terms of the diversity of life itself we cannot claim to have even scratched the surface.

One of our objectives in bringing such a diverse collection of research studies together is to show that whatever the ideological differences, behavioristic psychologists and cognitive ethologists have a lot to learn from each other and from the neurosciences. Yes there are differences of opinion about how to pursue the study of animal cognition, but all sides stand to gain from listening carefully to the concerns of others. Despite the differences, there is in fact a great deal of similarity among the different approaches, for they have, after all, evolved from the same starting point in the theory of evolution by natural selection.

The roots of both comparative ethology and comparative psychology are found in the writings of Charles Darwin, particularly in the *Origin of Species* (1859), *The Descent of Man* (1871),
and *The Expression of the Emotions in Man and Animals* (1872). Consequently, both disciplines are almost inextricably linked to the concept of instinct. Darwin viewed instinct primarily in behavioral terms and considered his ability to explain instinct through natural selection to be one of the most critical tests of his theories. Thus he compared closely related species of bees to explain the evolution of hive building and closely related species of ants to explain the origins of slave making. He also focused on domesticated species to show how human intervention and selection could have formed variants. His prime examples here were different breeds of dogs and pigeons. Soon, zoologists such as C. O. Whitman, E. C. Poulton, Oskar Heinroth, Julian Huxley, and others began to exploit the comparative method to trace the evolution of even the most complex social displays of animals. This work inspired the development of ethology primarily through the research and promotional efforts of the Nobel laureates Konrad Lorenz and Niko Tinbergen.

Although when the *Origin* appeared in 1859 Darwin deliberately discussed neither the evolution nor the behavior of human beings, he did, near the very end of the book, include this intriguing passage:

> In the distant future I see open fields for far more important researches. Psychology will be based on a new foundation, that of the necessary acquirement of each mental power and capacity by gradation. Light will be thrown on the origin of man and his history. (Darwin 1859, p. 488)

Note that this passage considers the psychological aspects of evolution as far more important, in the long run, than the morphological and taxonomic issues to which the *Origin* was largely devoted. Most biologists have ignored the clear evidence, found in Darwin's early notebooks, that while Darwin may have loved natural history, his ultimate agenda was to explain the origin of perhaps the strangest species of all, ourselves. Thus, in his later writings Darwin applied the comparative method to show the possible origins of facial expressions in human beings and speculated about many other aspects of abilities typically considered to be uniquely human, such as conceptual thought, language, loyalty, love, and morality (the latter was to Darwin a social instinct). In this way Darwin sought to show that our differences from other species were not major qualitative leaps, but were based on quantitative change that was due to basic evolutionary processes. Furthermore, these differences in degree, not kind, were not due to some external supernatural intervention or a nonmaterial mind. Darwin thus tried to document that human behavior contained instinctive roots that could be traced to our animal ancestors while at the same time arguing that the conceptual, communicative, intellectual, emotional, social, and moral aspects of our behavior also had roots in the behavior and psychology of other species.

These later writings by Darwin greatly influenced the comparative psychology of the nineteenth century. Many early writers, going back to Aristotle, Pliny, and even earlier, had compared the abilities of people with those of other animals and pointed out the surprising abilities of the latter. However, it was Darwin who systematically set out to show that the gap between humans and other animals was smaller than previously thought, and even more important, how the gap could be bridged by natural selection and sufficient time. Darwin's notebooks from the 40 years preceding *The Descent of Man* showed how influenced he was by the similarities of monkeys and apes to human beings.

Darwin and his protégé in comparative psychology, George John Romanes (see, e.g., Romanes 1883, 1892) often relied on casually collected anecdotes and an uncritical anthropomorphism that troubled more experimentally sophisticated scientists. Soon, the study of animal intelligence and abilities was shaped by the writings and work of C. Lloyd Morgan and E. L. Thorndike (Burghardt 1985a,b; Dewsbury

1984). This focus on parsimonious and fairly simple mechanisms to explain diverse abilities had salutary effects in terms of systematizing quantitative research methodology and interpretation when the field was young (see Boakes 1984). Darwin's cousin, Francis Galton, was among the first to develop quantitative and statistical means to study behavior. Nevertheless, by the 1940s, the study of comparative cognition, especially in psychology in the United States, had narrowed greatly; field and naturalistic research was disparaged and many important problems were ignored. The seminal volume of this period is useful in showing the stated breadth and effective narrowing of comparative psychology (Warden et al. 1935). These authors thoroughly criticized early work that was not based on laboratory experimental paradigms and concluded that most of it was virtually worthless, being contaminated by anecdotal, anthropomorphic, and introspectionist errors.

Furthermore, few biology or zoology departments incorporated behavior as an important element in their scientific training and research. European ethologists, with the primary exception of von Uexküll, von Frisch, and Lorenz, ignored the study of the higher levels of cognition and were especially suspicious of studying consciousness and subjective states. Tinbergen (1951) was adamant on this point, almost certainly because of the lingering vitalism concerning instinct and mind found in so many scientists, including his countryman Birrens de Haan. Concerning play behavior, Tinbergen (1963, p. 413) wrote: "Concepts such as play and learning have not yet been purged completely from their subjectivist, anthropomorphic undertones. Both terms have not yet been satisfactorily defined objectively, and this might well prove impossible...."

However, the seeds of the conceptual and methodological tools necessary to reexplore the complex lives of animals were already in place. Wallace Craig (1918) had shown how to separate the motivational and cognitive aspects of instinct as well as the importance of sequential analysis, primarily in his distinction between appetitive behavior and consummatory acts, a distinction now finally being formally recognized and extended in comparative cognition, as many chapters in this book acknowledge. Von Uexküll (1909/1985) gave us the concepts of Umwelt, Innenwelt, counterworld, sign stimulus, search image, and other useful means for approaching the behavior of other species from their perceptual worlds and not ours. Von Frisch was a brilliant experimenter who, among other accomplishments, showed that fish could hear, honeybees could see colors, and, most amazingly, that bees could communicate information about distant resources, which virtually no other species, except people, could accomplish.

Then, at around the beginning of World War II, other discoveries were being made that would have consequences. One of these was Baerends' demonstration that wasps could assess and hold in memory for a day or more how many caterpillars they needed to provision a nest. This study, presented in a dissertation, was made widely known by Baerends' professor, Niko Tinbergen, in the seminal volume that brought European ethology to English-speaking scientists (Tinbergen 1951). The tidy view that vertebrates, especially mammals, were the most cognitively advanced animals was in effect being challenged by the bees and wasps.

Also around 1940, Donald Griffin, as a student, co-authored a remarkable paper proving that bats used echolocation to navigate in their environment and locate prey. Decades after this work, Griffin finally integrated much ethological work, together with the implications of the abilities uncovered by experimental work on bats and insects, the results of studies on chimpanzee communication, and other diverse information in a remarkable small book, *The Question of Animal Awareness: Evolutionary Continuity of Mental Experience* (Griffin 1976). Griffin advocated a new field, to be called "cognitive ethology" (see Griffin 2001), because he thought that traditional animal learning and comparative psychology were ignoring the really interesting phenomena presented by animals as a result of their narrowly focused laboratory studies of a few selected domesticated species.

Ethology, on the other hand, was falling under the influence of behaviorists and ignoring the cognitive implications of the flexible and diverse accomplishments their studies of animals were uncovering. For Griffin, cognitive ethology was not only the study of the complex natural behaviors of diverse species, it was also the study of taboo subjects such as consciousness and awareness in other species. A cognitive ethology that ignored consciousness, choice, deliberation, planning, intentions, and other mental processes was, for Griffin, not possible.

The Question of Animal Awareness produced "Amens" in some quarters and outrage in others. Experimental psychologists in particular felt that they had developed methods that could answer the important questions without encouraging unbridled speculation that they, like Tinbergen, felt could never be assessed through the methods of normal science. Nonetheless, in psychology a new cognitivism was displacing behaviorism as the dominant approach and already psychologists studying animal learning were incorporating cognitive approaches while still using standard laboratory methods, although they were applied to more cognitive topics such as concept formation, information retrieval, and memory.

The rich history of behaviorism in fact included such concepts put forth by E. C. Tolman in the 1930s as cognitive maps and latent learning, which were eventually applied to in animal learning (e.g., the radial arm maze) as well as cognitive psychology and ethology. By the 1980s the behavioral concepts and methods of Hull and especially Skinner, which were largely codified in the 1940s, were increasingly applied by psychologists to issues of interest to ethologists, behavioral ecologists, and neuroscientists.

Computer scientists, robotics engineers, and modelers also became interested in the problems faced by those working with real animals, and neuroscientists began exploring more naturalistic phenomena, such as face recognition, as well. The results of all these broadened perspectives are on display in this volume. Nevertheless, there remain real differences, many of them high-lighted in this volume. Some authors are willing, even anxious, to enter the cognitive world and conscious life of their animal subjects. Others explicitly or by omission are clearly uncomfort-able with a cognitive study of nonhuman ani-mals that attempts to explore subjective states, intentionality, consciousness, or self-awareness. And, if the science holds, such diversity is both appropriate and defendable.

In addition, philosophers became interested in developing naturalistic approaches to age-old questions about the nature of mind and thereby came to face the challenge that nonhuman ani-mals presented to the traditional anthropocentric philosophy of mind. Ethology, animal learning and communication, and improved understand-ing of the workings of the brain promised some advances in understanding everything from mind–body relationships to the nature of lan-guage, intentionality, ethics, and knowledge it-self. Tom Nagel, Daniel Dennett, Steven Stich, John Searle, Ruth Millikan, Jerry Fodor, Dale Jamieson, Fred Dretske, and many other philos-ophers, despite often remaining wed to tradi-tional philosophical methods of reflection on thought experiments, became fellow travelers with scientists who were studying animal cogni-tion in both the laboratory and the field.

For many years, the primary philosophers to whom experimental psychologists were in-troduced were the positivists and Ludwig Witt-genstein, while ethologists typically ignored all philosophers (except for Lorenz and von Uex-küll, who were very partial to Kant). Evolution-ary biologists were largely influenced by Karl Popper.

These trends changed as both cognitive ethol-ogy and sociobiology became popular in the mid-1970s. Also, at this time the writings of philosophers such as Peter Singer and Tom Regan, among others, raised issues concerning our treatment of animals, increasing the stakes and relevance of obtaining accurate information on the emotional and cognitive lives of other species at the individual level and challenging philosophers to pay more attention to evolu-tionary relationships among species. At the same time, ecological and environmental concerns arose that involved extinction and habitat de-struction. Accurate knowledge of the complex behavior and needs of animals became seen as imperative for their very survival in nature or in captivity (Bekoff 1998, 2002).

Today, all the issues and concerns outlined above swirl over, within, and below seemingly straightforward questions about the cognitive lives of animals. The way we answer such ques-tions and apply the findings may tell us much about ourselves as a species as well as having serious consequences for the other inhabitants of this planet. Thus, the way in which the research issues addressed in this volume are eventually resolved (or not resolved) has enormous political implications at many levels. This is probably something most of those working on cognitive aspects of animal behavior are aware of at some level, but typically do not address in their formal writings.

With the above thoughts as a backdrop, we now can turn to the book at hand. The editors bring considerable and diverse research back-grounds and perspectives to this project. Marc Bekoff was trained in neurobiology and be-havior, later as an ethologist, and has worked on comparative aspects of behavioral develop-ment, social communication, quantitative meth-ods, play, animal ethics, and cognitive ethology for many years, with a particular focus on mam-malian carnivores. In 1978 he and Gordon Burghardt edited a volume on behavioral devel-opment and evolution (Burghardt and Bekoff 1978). Colin Allen was trained as a philosopher and has written extensively on topics in lan-guage, communication, play, and mental evolution. He has conducted fieldwork on the behavior of birds with Marc Bekoff and has studied learning in domestic pigs. Gordon Burghardt was trained in an interdisciplinary biopsychology program and has focused on the comparative behavioral development of squamate reptiles and the complex interactions between genetic and environmental factors, as well as on play, communication, and the intertwined history of ethology and psychology.

The Cognitive Animal contains fifty-seven compact essays dealing with numerous different topics on a wide range of organisms by researchers in many disciplines. The contributors were asked to consider five questions in their essays. These were (1) What are the central research questions in your study of animal cognition? (2) What theoretical or empirical methods have enabled you to address those questions? (3) What, if anything, do your results reveal about the internal psychological states of animals? (4) What future work is suggested by your investigations? (5) What phylogenetic and methodological limits are there to the study of animal cognition? Some authors stuck to these questions more explicitly than others, but generally the essays address these themes. Also, when taken together, the essays reflect evolutionary, ecological, and comparative approaches to the problems at hand. The importance of careful observation, description, and experiments is emphasized; all are important.

Since our contributors are all well-respected researchers, we allowed them considerable latitude in their presentations, realizing that some readers might prefer some of the chapters to provide more citations to the available literature and to be written in a more formal tone, while other readers might prefer the opposite. As editors, we read each essay carefully and often offered organizational, stylistic, conceptual, and historical suggestions, especially those that would aid readers with different perspectives. However, mindful of our own involvement in the issues covered in this book, we encouraged diverse voices and not the tight control or limited coverage found in many edited books.

Because this volume did not begin as part of a conference or symposium, we were able to invite a far larger and broader set of authors, both geographically and disciplinarily, than is typically found in a single book, certainly in a volume on this topic. The result is a book about which we are very excited, a volume that could easily be used in advanced undergraduate and graduate courses. It can serve as a useful and fascinating introduction to modern studies and provides examples for those with a general interest in animal cognition. For the professional, it provides an update to work from approaches other than that in which he or she works. Thus there are chapters by cognitive ethologists, behavioral ecologists, experimental psychologists, behaviorists, philosophers, neuroscientists, developmental psychologists, computer scientists and modelers, field biologists, and others.

The potpourri of topics covered in this book resists an easy or concise summary. We have chosen to keep this introduction brief rather than attempting to provide a beginner's guide, for our best advice to beginners is to just dive in. Each chapter is short enough that it may be read quickly and easily reread when contrasting views are encountered in other chapters; in this way understanding can be progressively deepened. The topics include (but are not limited to) definitions of cognition, the role of anecdotes in the study of animal cognition, naturalizing the study of animal cognition, anthropomorphism, attention, perception, learning, memory, ecology, evolution (including discussions of different levels of selection), communication, reproduction, thinking, consciousness, intentionality, rationality, play, aggression, dominance, predation, parent–young interactions and care giving, the role of models, planning, anticipation, kin selection, cooperation, recognition, assessment of self and others, neuroethology, choice, social knowledge, the role of touch, empathy, social symmetries and asymmetries, and conflict resolution.

Even a cursory glance at the essays makes it obvious that in some cases there are different views on the same or similar topics, an aspect of *The Cognitive Animal* with which we are very pleased. Pluralism is a good route to take, given the state of the art. Behaviorism is alive and well, as are rich cognitivism and more middle-of-the-road perspectives, and all of these approaches need to be given serious attention in future research. We also hope that these chapters will aid in the development of more understanding and reduce the polemics among all those who strive to elucidate the minds of animals.

Given the rich and wide variety of these essays, the task of organizing them in a linear sequence necessarily involved compromises. Two main lines of organization suggested themselves: topical and methodological. We opted for a topical organization because we found that researchers working on similar topics from a variety of different methodological backgrounds have many interesting things to say to each other. Ultimately, people (particularly potential students) are interested in the field of animal behavior because of what animals do, not because of the methods that we use to study them. However, we also believe that the importance of thinking about those methods is highlighted by putting contrasting approaches to similar questions alongside each other.

We also chose to organize the essays along broad rather than narrow topical lines. We found that the chapters could be grouped into four categories under the following headings (with brief explanations):

1. The diversity of cognition (taxonomic, methodological, and theoretical)

2. Concepts and categories (ways in which organisms divide up the world)

3. Communication, language, and meaning (a possible window on animal minds, to borrow Griffin's metaphor)

4. Self and other: the evolution of cognitive cooperators (from self-recognition to social cognition)

Narrower categories would have separated contributions that are clearly relevant to one another. But even within the broad categories we ended up with, there are chapters that might reasonably be paired with others in different categories. For instance, many approaches to animal concepts or to social cognition involve language or communication, and many of the chapters on communication touch upon categorization or concept formation. Furthermore, all the chapters provide a theoretical perspective, but those in the first section tend to be more explicit in raising basic and general theoretical questions. The fact that this organizational task was so difficult, and that there are so many overlapping themes even among researchers with very different approaches, speaks to the interdisciplinary nature of cognitive studies of animals.

We hope that these essays stimulate us all to be more interdisciplinary and to rise to challenges to try to answer, and not to dismiss, difficult questions. Difficult does not mean impossible. Learning more about the cognitive capacities of other animals will inform not only the general topic of animal cognition, but also how we interact with and treat other animals. The degree of development of cognitive skills and assessments of intelligence, along with information about the emotional lives of animals, are being increasingly used to make informed decisions about the use of animals in research and education, and for entertainment and food.

We thank all of our contributors and The MIT Press for working with us with enthusiasm and dedication on this project. We are grateful to Mrs. Rosalie Glenn of the Department of Philosophy at Texas A&M University for her help with printing and mailing. Don Griffin graciously agreed to contribute an afterword. Marc Bekoff thanks the University of Colorado for a sabbatical leave for the academic year 2000–2001. Colin Allen acknowledges support in the form of a Texas A&M Faculty Development Leave during 2000–2001, and a Big 12 Faculty Fellowship in 1999 that enabled a trip to the University of Colorado when this book was conceived.

References

Bekoff, M. (ed.) (1998). *Encyclopedia of Animal Rights and Animal Welfare*. Westport, Conn.: Greenwood.

Bekoff, M. (2002). *Minding Animals: Awareness, Emotions, and Heart*. New York, Oxford University Press.

Boakes, R. (1984). *From Darwin to Behaviorism: Psychology and the Minds of Animals*. Cambridge: Cambridge University Press.

Burghardt, G. M. (1985a). Animal awareness: Current perceptions and historical perspective. *American Psychologist* 40: 905–919.

Burghardt, G. M. (ed.) (1985b). *Foundations of Comparative Ethology*. New York: Van Nostrand Reinhold.

Burghardt, G. M. and Bekoff, M. (eds.) (1978). *The Development of Behavior: Comparative and Evolutionary Aspects*. New York: Garland.

Craig, W. (1918). Appetites and aversions as constituents of instincts. *Biological Bulletin* 34: 91–107.

Darwin, C. (1859). *On the Origin of Species*. London: Murray.

Darwin, C. (1871). *The Descent of Man and Selection in Relation to Sex*. London: Murray.

Darwin, C. (1872). *The Expression of the Emotions in Man and Animals*. London: Murray.

Dewsbury, D. A. (1984). *Comparative Psychology in the Twentieth Century*. Stroudsburg, Pa.: Hutchinson Ross.

Griffin, D. R. (1976). *The Question of Animal Awareness: Evolutionary Continuity of Mental Experience*. New York: Rockefeller University Press.

Griffin, D. R. (2001). *Animal Minds: Beyond Cognition to Consciousness*. 2nd ed. Chicago: University of Chicago Press.

Romanes, G. J. (1883). *Mental Life of Animals*. London: Kegan, Paul, Trench, Trübner.

Romanes, G. J. (1892). *Animal Intelligence*. 5th ed. London: Kegan, Paul, Trench, Trübner.

Tinbergen, N. (1951). *The Study of Instinct*. Oxford: Clarendon Press.

Tinbergen, N. (1963). On aims and methods in ethology. *Zeitschrift für Tierpsychologie* 20: 410–433.

Uexküll, J. von (1909/1985). Environment (Umwelt) and the inner world of animals (C. J. Mellor and D. Gove, trans.). In *The Foundations of Comparative Ethology*, G. M. Burghardt, ed., pp. 222–245. New York: Van Nostrand Reinhold. (Reprinted from J. von Uexküll, 1909, *Umwelt and Innenwelt der Tiere*. Berlin: Jena.)

Warden, C. J., Jenkins, T. N., and Warner, L. H. (1935). *Comparative Psychology: A Comprehensive Treatise*. Vol. 1, *Principles and Methods*. New York: Ronald Press.

Contributors

Michael S. Alfieri
Department of Biology
Mount Allison University
New Brunswick, Canada

Colin Allen
Department of Philosophy
Texas A&M University
College Station, Texas

James R. Anderson
Department of Psychology
University of Stirling
Stirling, Scotland

Russell P. Balda
Department of Biological Sciences
Northern Arizona University
Flagstaff, Arizona

Marc Bekoff
Department of Environmental,
Population, and Organismic Biology
University of Colorado
Boulder, Colorado

Alan B. Bond
Department of Biological Sciences
University of Nebraska
Lincoln, Nebraska

Ruud van den Bos
Animal Welfare Centre
Utrecht University
Utrecht, the Netherlands

Elizabeth M. Brannon
Department of Psychology
Duke University
Raleigh, North Carolina

Gordon M. Burghardt
Department of Psychology
University of Tennessee
Knoxville, Tennessee

John A. Byers
Department of Biological Sciences
University of Idaho
Moscow, Idaho

Richard W. Byrne
Scottish Primate Research Group, School
of Psychology
University of St. Andrews
St. Andrews, Scotland

Dorothy L. Cheney
Department of Psychology and Biology
University of Pennsylvania
Philadephia, Pennsylvania

Robert G. Cook
Department of Psychology
Tufts University
Medford, Massachusetts

Eileen Crist
Center for Interdisciplinary Studies
Virginia Polytechnic Institute
Blacksburg, Virginia

Robert O. Deaner
Department of Biological Anthropology
and Anatomy
Duke University
Raleigh, North Carolina

Michael Domjan
Department of Psychology
University of Texas
Austin, Texas

Lee Alan Dugatkin
Department of Biology
University of Louisville
Louisville, Kentucky

Anne L. Engh
Department of Zoology
Michigan State University
East Lansing, Michigan

Christopher S. Evans
Department of Psychology
Macquarle University
Sydney, Australia

PierFrancesco Ferrari
Human Physiology Institute
University of Parma
Parma, Italy

Leonardo Fogassi
Human Physiology Institute and
Department of Psychology
University of Parma
Parma, Italy

Deborah Forster
Department of Cognitive Science
University of California, San Diego
La Jolla, California

Deborah H. Fouts
Chimpanzee and Human Communication
Institute
Central Washington University
Ellensburg, Washington

Roger S. Fouts
Chimpanzee and Human Communication
Institute
Central Washington University
Ellensburg, Washington

Vittorio Gallese
Human Physiology Institute
University of Parma
Parma, Italy

Gordon G. Gallup, Jr.
Department of Psychology
State University of New York
Albany, New York

Asif A. Ghazanfar
Max Planck Institute for Biological
Cybernetics
Tübingen, Germany

Peter Godfrey-Smith
Department of Philosophy
Stanford University
Stanford, California

James L. Gould
Department of Ecology and Evolutionary
Biology
Princeton University
Princeton, New Jersey

James W. Grau
Department of Psychology
Texas A&M University
College Station, Texas

Donald Griffin
Concord Field Station
Harvard University
Bedford, Massachusetts

Lori Gruen
Department of Philosophy
Wesleyan University
Middletown, Connecticut

Güven Güzeldere
Department of Philosophy
Duke University
Raleigh, North Carolina

Brian Hare
Department of Developmental and
Comparative Psychology
Max Planck Institute for Evolutionary
Anthropology
Leipzig, Germany

Marc D. Hauser
Department of Psychology
Harvard University
Cambridge, Massachusetts

Bernd Heinrich
Department of Biology
University of Vermont
Burlington, Vermont

Louis M. Herman
Kewalo Basin Marine Mammal
Laboratory
University of Hawaii
Honolulu, Hawaii

Kay E. Holekamp
Department of Zoology
Michigan State University
East Lansing, Michigan

Bart B. Houx
Animal Welfare Centre
Utrecht University
Utrecht, The Netherlands

Robert Jackson
Department of Zoology
University of Canterbury
Canterbury, New Zealand

Dale Jamieson
Department of Philosophy
Carleton College
Northfield, Minnesota

Mary Lee A. Jensvold
Chimpanzee and Human Communication
Institute
Central Washington University
Ellensburg, Washington

Alan C. Kamil
Department of Biological Sciences and
Psychology
University of Nebraska
Lincoln, Nebraska

Colleen Reichmuth Kastak
Long Marine Laboratory
University of California, Santa Cruz
Santa Cruz, California

David Kastak
Long Marine Laboratory
University of California, Santa Cruz
Santa Cruz, California

Brian L. Keeley
Department of Philosophy
Pitzer College
Claremont, California

Evelyne Kohler
Human Physiology Institute
University of Parma
Parma, Italy

Bruce MacLennan
Department of Computer Science
University of Tennessee
Knoxville, Tennessee

Tetsuro Matsuzawa
Primate Research Institute
Kyoto University
Kyoto, Japan

Cory T. Miller
Department of Psychology
Harvard University
Cambridge, Massachusetts

Robert W. Mitchell
Department of Psychology
Eastern Kentucky University
Richmond, Kentucky

Eddy Nahmias
Department of Philosophy
Florida State University
Tallahassee, Florida

Don Owings
Department of Psychology
University of California, Davis
Davis, California

Michael J. Owren
Department of Psychology
Cornell University
Ithaca, New York

Sue Taylor Parker
Department of Anthropology
Sonoma State University
Rohnert Park, California

Sergio M. Pellis
Department of Psychology and
Neuroscience
University of Lethbridge
Alberta, Canada

Irene Maxine Pepperberg
MIT Media Lab
Massachusetts Institute of Technology
Cambridge, Massachusetts

Diego Pizzagalli
Department of Psychology
University of Wisconsin
Madison, Wisconsin

Drew Rendall
Department of Psychology and
Neuroscience
University of Lethbridge
Alberta, Canada

Jesús Rivas
Venezuela Nature Tours and National
Geographic
Jamul, California

Herbert L. Roitblat
Dolphin Search, Inc.
Ventura, California

Eric Saidel
Department of Philosophy
The George Washington University
Washington, D.C.

Laurie R. Santos
Department of Psychology
Harvard University
Cambridge, Massachusetts

Ronald J. Schusterman
Long Marine Laboratory
University of California, Santa Cruz
Santa Cruz, California

Robert M. Seyfarth
Department of Psychology and Biology
University of Pennsylvania
Philadephia, Pennsylvania

Sara J. Shettleworth
Department of Psychology
University of Toronto
Toronto, Canada

Daniel J. Shillito
Animal Forest
Charles Towne Landing State Historic
Site
Charleston, South Carolina

Robert W. Shumaker
Smithsonian Institution
National Zoological Park
Washington, D.C.

C. N. Slobodchikoff
Department of Biological Sciences
Northern Arizona University
Flagstaff, Arizona

Barbara Smuts
Department of Psychology
University of Michigan
Ann Arbor, Michigan

Elizabeth S. Spelke
Department of Psychology
Harvard University
Cambridge, Massachusetts

Berry M. Spruijt
Animal Welfare Centre
Utrecht University
Utrecht, The Netherlands

Craig B. Stanford
Department of Anthropology
University of Southern California
Los Angeles, California

Karyl B. Swartz
Department of Psychology
Lehman College, City University of New
York
New York, New York

Charles E. Taylor
Department of Organismic Biology,
Ecology and Evolution
University of California, Los Angeles
Los Angeles, California

Herbert S. Terrace
Department of Psychology
Columbia University
New York, New York

William Timberlake
Department of Psychology and Center for
the Integrative Study of Animal Behavior
Indiana University
Bloomington, Indiana

Michael Tomasello
Department of Developmental and
Comparative Psychology
Max Planck Institute for Evolutionary
Anthropology
Leipzig, Germany

Adrian Treves
Department of Psychology and Zoology
University of Wisconsin
Madison, Wisconsin

Jacques Vauclair
Center for Research in Psychology of
Cognition, Language, and Emotion
University of Provence
Aix-en-Provence, France

Elisabetta Visalberghi
Istituto di Psicologia
Consiglio Nazionale delle Ricerche
Rome, Italy

Edward A. Wasserman
Department of Psychology
University of Iowa
Iowa City, Iowa

Andrew Whiten
Scottish Primate Research Group, School
of Psychology
University of St. Andrews
St. Andrews, Scotland

Stim Wilcox
Department of Biology
State University of New York
Binghamton, New York

Richard Wrangham
Department of Anthropology
Harvard University
Cambridge, Massachusetts

Klaus Zuberbühler
Scottish Primate Research Group, School
of Psychology
University of St. Andrews
St. Andrews, Scotland

I THE DIVERSITY OF COGNITION

1 The Inner Life of Earthworms: Darwin's Argument and Its Implications

Eileen Crist

I have been interested in the scientific knowledge of animal life and in how presuppositions, structures of argumentation, and language express and shape that knowledge. The ways of describing animals have varied in behavioral science according to the interconnected play of underlying assumptions, theoretical frameworks, and the kind of language that has been deemed permissible for portraying their lives. Mind is unavoidably implicated in different descriptions of animal life because the way in which behavior is understood and described always speaks to the question of animal mind, whether deliberately, implicitly, or by omission. This is the major theme of my work *Images of Animals: Anthropomorphism and Animal Mind* (2000). In this essay, I address the connections between assumptions, forms of scientific inquiry, language use, and animal mind by discussing Darwin's argument for the inner life of earthworms.

In his last work, *The Formation of Vegetable Mould, through the Action of Worms with Observations on Their Habits* (1881/1985), Charles Darwin investigated the impact of earthworms on the geological and biotic environment, and devoted part of his study to worm behavior and intelligence. He introduced the latter topic by expressing his wish "to learn how far the worms acted consciously and how much mental power they displayed" (Darwin 1881/1985, p. 3)—a formulation that turned out to be incongruent with most ensuing twentieth-century behavioral science. The question of whether animals act consciously has been regarded as problematic: ontologically problematic in that the existence of conscious action has been subject to doubt regarding many (and sometimes all) animals; epistemologically problematic in that conscious action has been regarded as not lending itself to scientific inquiry; and semantically problematic in that "conscious action" has warranted the pejorative and dismissive label of anthropomor-

phism. That Darwin researched conscious action in animals so genealogically distant from human beings was an anomalous action in the trends that came to govern behavioral research. The fact that he argued that earthworms exhibit intelligence was apparently discomfiting; the silence of behavioral science regarding Darwin's argument for worm intelligence speaks volumes.[1]

Darwin made a bold argument about the inner life of earthworms. I use the expression "inner life" here to capture something more comprehensive than "mental life" or "cognitive ability." These latter terms tend to allude to processes such as thinking, deliberating, or judging, whereas "inner life" includes a subjective viewpoint. In Darwin's portrayal, the inner life of worms is indeed a cognitive world—a world about which worms form judgments. The inner life of worms also includes their subjective world—a world of perception and work that they experience, rather than vacantly sleepwalk through. Darwin delivered both aspects of inner life—cognition and subjective experience. I discuss both these aspects here and after examining Darwin's depiction of the behaviors, intelligence, and experience of worms, I draw some conclusions that are pertinent to the question of animal mind in science today.

The Intelligence of Worms

Earthworms plug the openings of their burrows with leaves and petioles. This behavior was Darwin's main interest, and he began by asking why worms do this, surmising several purposes: to keep the burrows free of water and dirt, to provide protection from predators, and to block cold air currents. He then undertook observations and experiments to examine how earthworms handle leaves and other objects. He found that the pattern of plugging was too regular to be random, yet he also recognized that the pattern was too variable to be strictly instinctive. Darwin was ultimately compelled to admit that earthworms use judgment about the best way to pull leaves into their burrows—that they feel the shape of the leaves prior to grasping them. Darwin described this capacity of judgment based on the tactile sense as showing "some degree of intelligence" (1881/1985, p. 91).

By examining hundreds of leaves used by earthworms, Darwin found that most of the leaves were drawn in by their tips and only about 10 percent were pulled in by their stalks (petioles). He concluded that the pattern of plugging the burrows with leaves was not random. After distinguishing more carefully how worms handled different types of leaves, Darwin found that they drew broad-based leaves in particular almost invariably by their tips—only 4 percent were grasped by their stalks. "The presence of a foot-stalk," he observed, "which might have tempted the worms as a convenient handle, has little or no influence in determining the manner in which [broad-based] leaves are dragged into the burrows" (1881/1985, p. 66). Darwin's explanation was that handling a broad-based leaf by its stalk would be unwieldy, for as the worm pulled the leaf into the burrow, its broad base would encounter the ground abruptly, offering resistance that would be relatively difficult (although not impossible) to overcome (1881/1985, p. 68).

Darwin followed up this line of investigation by examining how worms handled leaves whose tip and base were the same width. The majority of such leaves were still drawn by their tips, but now nearly 30 percent were grasped by their stalks—seven times more than was the case for broad-based leaves. The large discrepancy between broad- and narrow-based leaves drawn in by the foot-stalk, Darwin indicated, was not accidental. The fact that worms "break through their habit of avoiding the foot-stalk" (1881/1985, p. 68) for leaves that are easier to pull into their burrows by the base suggested to him that

worms formed judgments about the different shapes of the leaves and acted at least in part on the basis of these judgments.

The idea that the worms judged the shape of the leaves was further supported by Darwin's observations involving Rhododendron leaves, which have the peculiarity of curling around the midrib shortly after falling to the ground. Examining over two hundred fallen Rhododendron leaves—and before looking at what worms actually did—Darwin figured that the most efficient ratio of drawing them would be two-thirds by their base and one-third by their tip. Turning to his subjects, he found an almost exact match between his estimated base-to-tip ratio and the one executed by worms. "In this case," Darwin concluded, "the worms judged with a considerable degree of correctness how best to draw the withered leaves of this foreign plant into their burrows; notwithstanding that they had to depart from their usual habit of avoiding the foot-stalk" (1881/1985, p. 70).

It is impossible to go into all the details and nuances of Darwin's experiments, which include how worms handled pine leaves, various petioles, and artificial "leaves" that Darwin constructed from paper. His extensive observations led to the explanation, however "improbable" as he tactfully put it, that worms "show some degree of intelligence" (1881/1985, pp. 90–91). Darwin did not define "intelligence" in exact or technical terms,[2] yet its meaning emerges clearly in his meticulous study. For Darwin, the earthworms' ability to judge shape was the most significant indication of intelligence. He directly observed worms feeling the shape of leaves before grasping them. The connection between judgment and intelligence is stated clearly in the summary remarks of his study of worm habits:

If worms are able to judge ... how best to drag [an object] in they must acquire some notion of its general shape. This they probably acquire by touching it in many places with the anterior extremity of their bodies which serves as a tactile organ. It may be well to remember how perfect the sense of touch becomes in a man when born blind and deaf, as are worms. If worms have the power of acquiring some notion, however rude, of the shape of an object and of their burrows, as seems to be the case, they deserve to be called intelligent; for they can act in a manner as would a man under similar conditions. (Darwin 1881/1985, p. 97)

It is interesting that Darwin's reasoning was strengthened through analogy to the human case. He reminded his audience that when a person's tactile sense is the sole sensory modality, it becomes acute and capable of fine discernment. And he suggested that if a person used touch to assess the shape of an object and thereby determine how to manipulate it effectively—as worms did—this same act would indisputably be viewed as intelligent.

Darwin used "intelligence" as a distinct explanatory category, but not one that excluded supplementary accounts. He noted that plugging holes "is no doubt instinctive in worms" (1881/1985, p. 74), for they do not need to learn the behavior. However, instinct could not explain *how* worms actually handled leaves (including the leaves of plants foreign to their habitat), and their behaviors were not "so unvarying or inevitable as most true instincts" (1881/1985, p. 93). In the case of leaves that were just as easily drawn by their stalks as by their apex, the worms' persistence in grasping the apex was explained by their "having acquired the habit" (1881/1985, p. 68). So while Darwin argued that worms exhibit intelligence, he presented a composite view of their performance: Worms possess an inborn drive to plug their burrows; their intelligence consists in acting on the basis of the shapes of objects; yet over time, they acquire habits according to which they tend to behave.

Darwin realized that "worm intelligence" would be an oxymoron for skeptics and even for a commonsense viewpoint: "This will strike everyone as very improbable," he wrote (1881/1985, p. 98). Concern to sustain credibility is reflected in the scrupulous way he gathered, assessed, and presented evidence. Darwin also endeavored to preempt certain objections: He noted that little is known about the nervous systems of "lower animals," implying they might possess more cognitive potential than generally assumed. He included examples of "insect stupidity" (my term) perhaps intended to show that his discovery of worm intelligence did not express a sweeping conception of invertebrate capacities. And finally, Darwin averred that he had been initially dubious about the possibility of intelligence in earthworms, but his a priori doubts were swept aside by observational and experimental results. His discovery of intelligence was unbiased; it was not a romantic or tenuous interpretation imputed to their actions. His insistence on this point was not simply for credibility. "Some degree of intelligence appears," he remarked, "a result which has surprised me more than anything else in regard to worms" (1881/1985, p. 35). Darwin was genuinely taken aback by his discovery.

A World of Experience

While Darwin did not set out to find intelligence in worms, it is also clear that he was open to such a possibility. This openness is visible in his posing the question ("how far the worms acted consciously and how much mental power they displayed"), and in his readiness to accept worm intelligence once other explanations were ruled out as insufficient ["one alternative alone is left, namely, that worms, although standing low in the scale of organization, possess some degree of intelligence" (1881/1985, p. 98)]. Darwin's openness to the possibility of awareness (where it is often offhandedly dismissed by modern *Homo sapiens*) is also evident in his implicit portrayal of earthworm life as a world of experience.

This perspective is vividly discernible in Darwin's depiction of earthworm living quarters. In discussing how worms constructed and inhabited their burrows—the tunnels, openings, and chambers—he used the descriptive terms of architecture and home, thereby accenting the skill and life of worms. Darwin's language contributed two significant dimensions to the portrayal of their inner life: It presented structures constructed by worms as products of work, rather than fortuitous outcomes of passive movement; and it presented worms as inhabitants of spaces that possessed features engineered for utility, comfort, and security.

The burrows Darwin studied were often several feet in depth and their walls were lined with worm castings, deposited initially as "voided earth, still soft and viscid" and then spread out as the worm traveled up and down its hole (1881/1985, p. 111). The consequent thin film of dried-out castings provided the burrow with structural support and protected the worm's body from rough walls. "We thus see that the burrows are not mere excavations," observed Darwin, "but may rather be compared with tunnels lined with cement" (1881/1985, p. 112). The idea of "cemented tunnels," inviting comparison with human labor, suggested a space constructed through (and for) action as opposed to a passively created "mere excavation."

Darwin also examined how worms created basketlike structures, held together with leaves, miscellaneous objects, and castings, that protectively enveloped the mouths of their burrows. Worms that he kept in pots formed these structures using pine needles, fragments of other leaves, glass beads, and bits of tile—formations he described, again inviting analogy with human work, as "plastered with viscid castings" (1881/1985, pp. 112–113). Underscoring the skill involved in its construction, he described one case in detail:

The structures thus formed cohered so well, that I succeeded in removing one with only a little earth adhering to it. It consisted of a slightly curved cylindrical case, the interior of which could be seen through holes in the sides of either end. The pine-leaves had all been drawn in by their bases; and the sharp points of the needles had been pressed into the lining of voided earth. Had this not been effectually done, the sharp points would have prevented the retreat of the worms

into their burrows; and these structures would have resembled traps armed with converging points of wire rendering the ingress of an animal easy and its egress difficult or impossible. The skill shown by these worms is noteworthy and is the more remarkable, as the Scotch pine is not a native of this district. (Darwin, 1881/1985, p. 112)

Darwin noted that while the worms were innately inclined to construct these protective basket structures, the effective manipulation of various objects suggested that the worms' handling of materials per se was not innately obtained, at least not in full.

The admiration Darwin expressed for the worms' "noteworthy and remarkable skill" intimates his admission of what might be called "implicate authorship." The care involved in this particular construction was, Darwin suggested, above and beyond what instinct could explain. The detail of pressing the pointed pine needles into the sides of the cylindrical interior so they could not injure the worm's retreating body was potentially akin to an effective precaution rather than a blindly enacted behavioral pattern, since the pine was not a native tree. By "implicate authorship," I am referring to behaviors that suggest the possibility of an aware agency, for (1) they cannot be completely accounted for by extant concepts or frameworks of behavioral science and (2) they clearly have a rationale or purpose. As Darwin put it for this case, if pressing the needles had "not been effectually done, the sharp points would have prevented the retreat of the worms into their burrows."

According to Darwin, the worms often rested within these baskets, absorbing warmth without exposing their bodies. So in addition to noting their protective function for worms, he observed how the baskets were lived in—how body, habit, and comfort fittingly intersected within their configuration. "Worms often remain ... for a long time close to the mouths of their burrows, apparently for warmth; and the basket-like structures formed of leaves would keep their bodies from coming into close contact with the cold damp earth. That they habitually rested on the pine-leaves, was rendered probable by their clean and almost polished surfaces" (1881/1985, p. 114). Darwin thus saw a dimension of experience in their use. The basket structures were not only potentially deliberately constructed, they were also utilized and lived in, providing places where the worms could rest, find warmth, and be cushioned from the dampness of the earth (and, he indicated elsewhere, be somewhat protected from predators).

Darwin completed his study of earthworm living quarters by examining how burrows "terminate in a little enlargement or chamber" (1881/1985, p. 114). The floors were lined with small stones and seeds and in Darwin's pots, by glass beads and bits of tile carried down by the worms from the surface. During the winter, the worms curled up into balls (singly or in numbers) in their padded chambers, their bodies buffered from the soil. Darwin maintained that "the sole conjecture which I can form why worms line their winter-quarters with little stones and seeds, is to prevent their closely coiled-up bodies from coming into close contact with the surrounding cold soil" (1881/1985, p. 116). With his intimate description of these "chambers" and "winter-quarters," Darwin again presented a view of burrows as experienced abodes.

Darwin's language of lived-in space describes a "phenomenology of mind" through the presentation of an experiential world; it does not focus on cognitive processing, but is partial to embodied, interactive, and material manifestations of awareness in the world (see Abram 1996; Crist 2000). The created landscape of worms' lives, as depicted by Darwin, implies a mind at work, for the cemented tunnels, cylindrical baskets, lined chamber floors, and plugged burrows are designed constructions that cannot be fully comprehended by scientific concepts that eschew mind, in particular by that conceptualization of "instinct" which equates it with thoughtless enterprise. [On the ubiquity of this conception of instinct in behavioral science, see

Griffin (2001/1992).] Indeed, Darwin was cognizant of this particular use of the idea of instinct, writing that "the instincts of even the higher animals are often followed in a senseless or purposeless manner" (1881/1985, p. 95). The actions of worms could be understood as driven by instinct—by an urge unbacked by deliberation or planning—but neither the diversity and the details of construction nor the form of life that the constructions afforded were encompassed by the idea of instinct. There was something more. And this something more was not emphatically asserted by Darwin, but pointed to with restrained awe.

Concluding Remarks

Darwin's portrayal of the inner life of earthworms challenged basic assumptions of science and common sense about what sorts of organisms are capable of intelligence and what are not, and about what sorts of organisms are able to experience life and what are assumed to be little more than animated robots. In admiration of this bold thinker, it is fair to state that without much fanfare, but in a gentle and measured manner, Darwin simply did not abide by these assumptions which are, after all, far from self-evidently true. He found mind—both cognition and subjective experience—where it was presumed not to exist.

How is Darwin's study relevant 120 years later? To return to the introductory comments, his study brilliantly shows that the question of "conscious action" in animals is not inherently problematic: not ontologically problematic because it is not rational to presume, prior to inquiry, that the existence of conscious action is unlikely, even among invertebrates; not epistemologically problematic because once the question of conscious action is allowed to be posed, the scientific imagination finds fascinating ways to address it; and finally, not semantically problematic because writing off "conscious action" as anthropomorphism commits the deeper (in my view) fallacy of anthropocentrism. Such dismissal rests on the presumption of an unbridgeable gap between the ostensibly "highest" of animals (humans) and most other organisms (not to mention worms).

And now to some lingering questions. Did Darwin actually prove the operation of intelligence in worms, or their possession of an aware experiential perspective? Not in any incontestable sense; more important though, he opened the door to such possibilities, engaging in intriguing observations and designing ingenious experiments on the way. Does it matter whether earthworms are intelligent or experience their world? I would submit that what matters is that scientists be allowed and encouraged to pose these questions about worms and other animals. It is hoped that following their cue, commonsense views that are flippantly dismissive of such forms of awareness in the world will be discarded. Why is this desirable? The most significant reason today is the need to awaken and deepen our sense of wonder about the living world. For the erosion of this wonder—encouraged, in part, by the dominance of overly mechanistic models of animal behavior in the twentieth century—is internally connected to the gathering speed of the human onslaught on the natural world, and to its darkest corollary, the sixth extinction.

Notes

1. Exceptions are Yerkes (1912), Ghiselin (1969), Graff (1983), and Gould (1983, 1985).

2. See Ghiselin (1969, p. 201). While Darwin cited George Romanes' criterion of deducing intelligence "only when we see an individual profiting by its own experience" (1881/1985, p. 95), he neither endorsed nor applied it as a stringent criterion. Darwin sent relevant pages of the manuscript about worm intelligence to Romanes, asking him to comment. Romanes replied "that there may be intelligence without self-consciousness" (cited in Graff 1983, p. 11).

References

Abram, D. (1996). *The Spell of the Sensuous*. New York: Vintage Books.

Crist, E. (2000). *Images of Animals: Anthropomorphism and Animal Mind*. Philadelphia, Pa.: Temple University Press.

Darwin, C. (1881/1985). *The Formation of Vegetable Mould, through the Action of Worms with Observations on Their Habits*. Chicago: University of Chicago Press.

Ghiselin, M. (1969). *The Triumph of the Darwinian Method*. Berkeley: University of California Press.

Gould, S. J. (1983). Worm for a century, and all seasons. In *Hen's Teeth and Horse's Toes*, pp. 120–133. New York: W. W. Norton.

Gould, S. J. (1985). Foreword to 1985 edition of Darwin, *The Formation of Vegetable Mould*. Chicago: University of Chicago Press.

Graff, O. (1983). Darwin on earthworms—The contemporary background and what critics thought. In *Earthworm Ecology: From Darwin to Vermiculture*, J. E. Satchell, ed., pp. 5–18. London: Chapman and Hall.

Griffin, D. (2001/1992). *Animal Minds*. Chicago: University of Chicago Press.

Yerkes, R. M. (1912). The intelligence of earthworms. *Journal of Animal Behavior* II: 322–352.

2 Crotalomorphism: A Metaphor for Understanding Anthropomorphism by Omission

Jesús Rivas and Gordon M. Burghardt

The Story of Country Blue

When foreign students come to study at the University of Tennessee, the Center for International Education at the university presents them with a story, paraphrased as follows, to help them understand and deal with their new culture.

People from a country called Blue normally wear blue clothes, blue hats, and blue sunglasses. Houses are blue and so are the cars and streets. Country Blue borders country Yellow where people wear yellow clothes, yellow hats, and yellow sunglasses. Houses as well as cars and street are yellow in country Yellow. These two countries are internally peaceful, but have conflicts with each other. They view the customs and policies of the other country as bizarre and evil. One day, a diplomat from Blue decided to visit Yellow, learn about their customs and traditions, and write an extensive article to his fellow Blue citizens explaining how people in Yellow view the world. He was convinced that they were not evil, they just saw the world in a different way. Therefore, the Blue diplomat put on yellow clothes, a yellow hat, and yellow sunglasses. After three months living in Yellow, the Blue diplomat returned to his country and reported that the citizens in Yellow were not bad, bizarre, or stupid. His article claimed that in country Yellow life was actually very nice and Green!

This tale characterizes someone trying to understand another culture who neglected to consider a basic limitation: his own colored glasses. These he did not, or perhaps could not, remove. In an even more profound way, our human glasses are ingrained in us, and are very hard to remove (if possible at all). Nevertheless, if we are aware of having biased spectacles, we can attempt to address their effects upon us. In order to understand the cognitive accomplishments of a bee or beetle, squid, or chimpanzee, we need to evaluate how they perceive *their* world. In doing so, technology can assist us, but we need to constantly remind ourselves that we are using our human senses and human-based technology, and are processing the information with a human brain.

Like the Poor, Anthropomorphism Will Always Be with Us

Anthropomorphism is defined as the attribution of human properties to nonhuman entities. Such entities can be supernatural (gods) or animate or inanimate nature. The problem with anthropomorphism is that it often leads to the attribution to nonhumans of properties that they do not possess. It is but an extension of the problems facing anyone trying to understand another human culture, as in the Blue/Yellow example, or actually, the experiences of any person other than yourself. The problems inherent in using overt behavior to infer what other people or animals experience was recognized by Romanes and the early comparative psychologists (Burghardt 1985a), but they sought ways of surmounting the problem. After several decades, however, psychologists and ethologists came to regard anthropomorphism as a serious error that must be avoided at any cost. When Griffin's writings, *as a scientist* (Griffin 1978) seemed to be encouraging unfettered and untestable speculation about consciousness and awareness in nonhuman animals, the critical reaction was swift. It reached its zenith in the book by Kennedy (1992), who, nevertheless admitted that the tendency to be anthropomorphic seems endemic to human beings and can never be eliminated.

Some recent attempts look more closely at what anthropomorphism really is and how it operates. Lockwood (1989) argued that not all anthropomorphic attributions were equal. For example, two kinds of anthropomorphism (allegorical and personification) are restricted to nonscientific writing and therefore are not a problem in science. Two others, which Lockwood called "superficial" and "explanatory," have potentially harmful consequences in science, and these were the main culprits for Kennedy. Lockwood also identified a fifth kind of anthropomorphism, applied, which he considered a legitimate strategy that had been used by authorities such as Darwin: the use of the personal perspective to convey what it is like to be another living being.

Anthropomorphism can be further demystified by showing that it is a legitimate and perhaps particularly creative way to do science if it is used to develop hypotheses that can be tested in a rigorous manner (Burghardt 1985a). Critical anthropomorphism was introduced as a way of using various sources of information, including "natural history, our perceptions, intuitions, feelings, careful behavior descriptions, identifying with the animal, optimization models, previous studies and so forth in order to generate ideas that may prove useful in gaining understanding and the ability to predict outcomes of planned (experimental) and unplanned interventions" (Burghardt 1991, p. 73).

Critical anthropomorphism was deliberately modeled after a proposed solution to a centuries-old controversy in science and philosophy: the nature of the external world (Mandelbaum 1964). Here one important contrast is that between direct or "naive" realists who accept that the world is just as it appears to us, and variants of subjectivists, idealists, and solipsists who basically argue that nothing exists but our own minds (or as Edgar Allen Poe wrote, "All that we see or seem is but a dream within a dream").

More germane is the position of skeptics who do not deny the existence of an external world, but assert that our flawed senses show that we can never learn anything about it. Illusions serve the purpose of the latter nicely (see Gregory 2001). Mandelbaum's (1964) solution was to advocate a radical critical realism based on both relevant sensory and neural data and predictive inferences. Furthermore, a process called "transdiction" could be applied to ground "inferences to objects or events which not only have not yet been observed, but which in principle cannot be observed" (Mandelbaum 1964, p. 63). Just as in the study of perception critical realism is the

most scientifically congenial approach, so in the realm of animal minds, critical anthropomorphism is required.

Recently, an edited book (Mitchell et al. 1997), a monograph (Crist 2000), and articles (e.g., Fisher 1990) have been devoted to reviewing and exploring the problems and perils of anthropomorphism and anecdotes in modern studies of animal behavior. Perhaps the default condition of the human mind is anthropomorphic and this condition functions in understanding and dealing with other people (Caporael and Heyes 1997) and animals, especially domesticated and economically valuable ones (Morgan 1894). Perhaps anthropomorphism is only harmful in science when it is unacknowledged, unrecognized, or used as the basis for accepting conclusions by circumventing the need to actually test them.

Anthropomorphism by Omission

Anthropomorphism by omission is the failure to consider that other animals have a different world than ours. We can, without realizing it, attribute human traits to other species by failing to consider that many animals perceive the world in a different manner than do we. Scientists may know this in theory, but if they do not deliberately acknowledge that different species have different perspectives and priorities than we do, they may draw anthropomorphic conclusions that are erroneous.

The idea of studying the private worlds of other animals was pioneered by Jacob von Uexküll (1909/1985), who attempted to bring the latest neural, physiological, and perceptual findings to bear in understanding the behavior of animals by considering both their inner world (Innenwelt) and how they perceived and responded to their environment (Umwelt). A major aspect of this approach was to evaluate differences among species in the salience of biologically relevant perceptual cues (Tinbergen

1951; Burghardt 1985b). The cognitive ethology movement as pioneered by Don Griffin in the mid-1970s explicitly focused on the ways that animals perceive, interpret, and experience the world (e.g., Griffin 1978; Allen and Bekoff 1997; and chapters in this volume). An important component of this approach, though often understated, is to consider the animal being studied as an active participant, with the researcher trying to put him or herself in the animal's situation. Timberlake and Delamater proposed that to understand the behavior of an animal, "Experimenters not only need to put themselves in the subject's shoes, they need to wear them—walk, watch, hear, touch, and act like the subject" (Timberlake and Delamater 1991, p. 39). The power of this concept is shown by the recent demonstration by David Carrier's lab that strapping humans in a weighted suit framed like a theropod carnivorous dinosaur gave plausible insights into the dinosaur's maneuverbility (Stokestad 2001). Even skeptics allow that "Carrier's creative approach to dino motion has given them things to consider" (Stokestad 2001, p. 1572). More recently, Bekoff (2000) has extended this view to advocate a biocentric anthropomorphism.

Although it is true that we will never fully appreciate how another animal experiences the world, by doing our best to accomplish this through applying critical anthropomorphism, including the full range of available scientific data, we will get closer to understanding the life of the animal. Conceptually the task is no different from that of trying to understand another person who may differ from us in age, gender, sensory and motor abilities, personality, temperament, language, health, profession, wealth, status, or a host of other variables. Nevertheless, although we can never obtain access to the full inner life and private experiences of another human being, some people seem to be more successful at generalizing to the situations of other people and are thus considered particularly insightful, empathic, or privy to human nature writ large (e.g., major novelists, playwrights, composers). Others can exploit their knowledge in political, social, and deceptive activities (Burghardt 1997). Partial knowledge is possible and useful even if full knowledge is unobtainable both in practice and in principle.

Our aim in this essay is to document the presence of anthropomorphism by omission and raise awareness about its presence and its detrimental influences on science. If this argument is convincing, then when we analyze the behavior of animals, we not only can but *must* deliberately put ourselves in the animal's shoes; not doing so is potentially and truly anthropomorphic. Thus we are extending the approach advocated by Timberlake and Delamater (1991) one step further. Omitting to put oneself in the animal's shoes often leads to default anthropomorphism or anthropomorphism by omission.

Crotalomorphism—More than a Metaphor?

Consider another story, but one involving the study of other species, species in which, unlike our tendencies with primates and domestic animals, anthropomorphism is not usually considered a serious threat to the work of trained scientists.

A researcher is studying the behavior of a very colorful lizard. When this lizard sees a person, it rapidly changes its color to match its background, just as octopuses are well known to do. The researcher concludes that the change in color is a cryptic response to avoid predation. Just at this time, however, a large female timber rattlesnake (*Crotalus horridus*) that is quietly observing the researcher from some nearby brush is suddenly spotted by the researcher, who is both startled and scared. Rattlesnakes, being pit vipers, can detect patterns of infrared radiation from mammals through the loreal pits situated between the eyes and nostrils. Therefore, when the snake perceives the researcher, she detects it as a very warm animal moving in a much cooler background (not unlike the way the human sees the colorful lizard). When the startled researcher saw the rattlesnake, adrenaline kicked in and the flow of blood to the arms and legs was reduced, along with all other peripheral circulation; this is a normal response to stress. The researcher turned cooler and was therefore less visible to the infrared-detecting "eyes" of the snake. Our clever rattlesnake concludes that the person is trying to escape by matching the cooler background. The drop in peripheral temperature is a cryptic response to predators with heat-sensing organs.

This is an example of crotalomorphism by omission; for although there is evidence that predator stress can lower body temperature (Gabrielsen and Smith 1985), the snake's conclusion would probably be dismissed as erroneous by most human scientists. However, is the snake's conclusion different in any essential way from the conclusion of the human researcher studying lizards? Crotalomorphism illustrates the problem of interpreting the world solely by one species' standards. Together the snake and the scientist are playing out the same game as the people from country Blue and country Yellow!

We are convinced that unwitting anthropomorphism by omission is frequently present in several scientific fields related to animal cognition and we use examples from the literature, including our own work (Burghardt 1998), for we have not been immune ourselves.

A Few Case Histories

Foraging Tactics in Snakes

When a northern water snake (*Nerodia sipedon*) catches a fish, it typically lifts the fish out of the water and takes it to the shore, where it swallows it. Water snakes are generally viewed as not using any specialized technique, other than grab and hold, to subdue or kill their prey, since they lack venom and do not constrict. However, if an anaconda, a large and largely aquatic tropical species, catches a deer and drags it to water where the animal cannot breathe and its capacities to move are reduced, it is often concluded, reasonably we might add, that the anaconda is subduing the deer, not only by constriction, but also by bringing the deer to an environment where it cannot breathe or run.

However, a deer forced into water is no different than a fish carried to land. While recognizing that the deer is being deliberately subdued, we fail to see that the fish is being subdued as well. The former is much more evident to us because we would also be subdued if we were in the position of the deer and not in that of the fish. So, even without directly attributing human traits to the animals, we may fail to consider the traits in which they are different. The fact that most scientists might readily agree when this possibility is pointed out does not invalidate our observation that removing a fish from water is not readily viewed as a functional predatory tactic.

Warning Coloration

Aposematic coloration has been considered a means to warn predators of potential danger and has been the basis of much theoretical work. Yet there is often little consideration of how the presumed predators of aposematically patterned and colored animals actually identify prey, if they can even see the presumed warning cues, or what warning coloration looks like to different predators. This is particularly evident when the predators are invertebrates rather than vertebrates. Even with vertebrates, many may be effectively color blind. A few years ago we came close to making an error in interpreting the responses of garter snakes to warning colors (Terrick et al. 1995) since, although snakes have cones in the retina, the latest research finds that they have no ability to discriminate wavelength (Jacobs et al. 1992). The discrimination we found may have been due, as we noted, to contrast, not wavelength. To other predators, color may be important and our study clearly began with this view.

Drawing conclusions about the aposematic coloration of a species without first asking whether it is aposematic for the species' predators is another example of anthropomorphism by omission. In fact, it would be less anthropomorphic if less research was devoted to studying aposematic coloration and more time researching aposematic scents, aposematic vocalizations, and aposematic textures.

Courtship in *Drosophila*

Another example of the problems of anthropomorphism by omission is seen in a study of the role of sound during the courtship of fruit flies (*Drosophila* spp.) carried out by Boake and Poulsen (1997). Males shake their wings vigorously during courtship, producing sounds, and it was hypothesized that such sounds were an essential part of courtship. An experiment was carried out with wing-clipped males that could not produce the sound. Such males were expected to have reduced mating success. To the authors' surprise, clipped males did better than the controls in one species! The authors then mention that a reviewer (R. Hoy) pointed out that the clipped males had lighter wings (stumps) and thus might vibrate them faster—hence vibration, more than sound, may be the important part of the stimulus. In the world of humans, the shaking of the wings of an insect can only be detected by the sound they make. Owing to the much larger body mass of the researchers, it escaped their attention that the movement of the wings could produce substantial vibratory stimuli for an insect, though not for them.

The Cat and the Mouse

In the field of cognitive ethology, anthropomorphism by omission has also taken its toll. Colin Beer (1997, p. 203) states that "the reach and complexity of connections attaching to ideas in the human case will usually *far exceed* [our italics] what is conceivable for any animal." This respectable and plausible claim is supported by Beer (1997, p. 203) by describing a cat crouching beside a hole down which it has just chased a mouse: "We should be inclined to say that the cat thought there was a mouse down the hole. But consider what thinking that would mean to us: it would mean that there was a furry mammal down the hole, a tetrapod vertebrate, a whiskered rodent, a warm-blooded cheese-eater, and a whole lot more that could not possibly occur to the cat. Only a *small part* [our italics] of the network within which mouseness is nested for us extends into the cat's world."

Beer apparently failed to consider that the world of the cat is different than ours, and that only a small part of that world is obvious to us. The odors left by a chased and stressed mouse might allow the cat to obtain information as to whether the mouse is fat or thin, young or old, male or female, sick or injured. Perhaps it is aware of, and even enjoys, listening to the pitter-patter of the mouse running down the hole or smelling the odors left behind. We can't even begin to imagine the number of things that the cat may be aware of in that moment, such as the taste of the last mouse it caught, the feeling of grabbing and biting a mouse, or the memory of a former encounter when a mouse bit him. Given the salience and importance of rodents to cats, their "thoughts" and private experiences about mice in this situation might be far richer and certainly quite different than ours.

Human and Nonhuman Language

In the literature dealing with comparative communication and language, the superiority of human language as a means of communication is contrasted with "less evolved," "simpler," or "less advanced" systems in other species (cf., Brickerton 1998; Allen and Saidel 1998; Ujhelyi 1996). In addition to the various criticisms that

Figure 2.1
Calvin and Hobbes, by Bill Watterson. Reprinted by permission of Universal Press Syndicate.

can be made of such formulations (e.g., Allen and Saidel 1998), such statements also are prone to committing the error of anthropomorphism by omission, for they often fail to recognize that other species have different worlds than ours (figure 2.1). We often ignore the complex information contained in chemical cues and pheromones since we are so limited chemically.

As another example, consider honey bees studying communication between humans. To them, we would appear to perform rather poorly since we do not, and perhaps cannot, give our partners the location of the closest restaurant by dancing! Claiming that our language is superior risks not being aware of its limitations compared with communication in other species because we are biased in our understanding of it. We are not denying that human language might be, and probably is, generally more complex than other forms of communication in all other animals. Nevertheless, it is uncritically anthropomorphic to begin research on comparative language with this bias, which often goes unrecognized.

Zoos as Products of Anthropomorphism by Design

Zoo exhibits, even the most modern, are often shaped much more by the needs of the human visitors or human caretakers than the animals shown living in supposedly "natural" settings. Naturalistic exhibits reflect human concepts of nature, not necessarily those of the animals exhibited. Indeed, if the exhibit is too effective, the animals may not be readily spotted by visitors, and the exhibits will be changed for people, the paying customers, not for the resident animals. This has happened in gorilla exhibits (Burghardt 1996).

It has actually been argued that the best modern zoos are those that treat the zoo as a theatrical experience for the public, not one oriented to the lifestyle of the captive species (Polakowski 1989). This view seems to be pervasive in action if not in rhetoric in most zoos. An honest and explicit recognition of the anthropomorphic nature of modern zoos would be helpful. It is not enough to have eliminated tea parties for chimpanzees at zoos.

Conservation Planning

Decision makers often develop wildlife management plans that include various ecological benchmarks while ignoring the perceptual capabilities and life histories of the animals that are to be protected. The use of travel corridors to connect patches of habitat has been urged as a management measure to allow species to move between separated natural areas. Metapopulation analysis has identified gene flow across relatively isolated populations as a central issue in conservation biology (Wiens 1996). The spatial scale involved is a critical issue. If the patches are too far apart, then the animals will not be able to find a patch by using their natural navigation method, and gene flow will be interrupted (Wiens 1996).

A well-documented example of how landscape modeling has neglected the perceptual world of animals and their private experiences involved the white-footed mouse, *Peromyscus leucopus* (Lima and Zollner 1996). Agricultural lands that allow some patches of natural vegetation are considered sufficient to allow gene flow from one population to the other. Zollner and Lima (1997) challenged this notion by testing the ability of mice to return to a forest patch from different distances. They found that animals released in agricultural lands would aim straight toward the forest when they were released at distances less than 20 m from it. However, mice released as close as 30 m from a forest orientated randomly, suggesting that these mice cannot discriminate a patch of forest at that distance. Hence the perceptive world of these mice (and not ours) must be considered when we formulate plans about the spatial layout of habitat patches for conservation.

Addressing the Pervasive

We have provided a few examples of anthropomorphism by omission that show that its presence can be detrimental to work in a variety of disciplines. It is not enough to avoid an anthropomorphic vocabulary and claim to be strictly objective. Anthropomorphism is like Satan in the Bible—it comes in many guises and can catch you unawares! Lockwood (1989) pointed out some of the guises. The most easily recognized are not the problem; the conceit that one is immune to them is more often the problem.

If anthropomorphism is a natural tendency of human beings, scientists are not immune; lurking unseen, it can compromise efforts in many areas. By using critical anthropomorphism and trying to wear the animals' "shoes," we can overcome part of our natural bias and obtain a more legitimate understanding of the life of other species and of nature (Rivas and Burghardt 2001). We encourage other researchers to put themselves in the position of their study animals, not only as a novel, complementary approach to their work, but as a required step in conducting good science. As the essays in this volume attest, issues of animal cognitive abilities are prime areas where anthropomorphism by omission may occur, but it is also those researchers working on animal cognitive behavior who are in the best position to discover what is necessary in order to avoid it.

Finally, the view of science embedded in our essay here is one being urged in various quarters. For example, in a recent technical article on the use of statistical inference to interpret data in experimental psychology, the following quotation appeared that puts our message in a broader context:

The selection of hypotheses, their number, their location on the continuum of possible hypotheses, and their prior probabilities depend on the researchers' experience, their theoretical frame of mind, and the state of the field at the time of the study. (Kreuger 2001, p. 19)

Acknowledgments

This work was supported by grants from the National Science Foundation, the University of Tennessee Science Alliance, the National Geographic Society, and other sources of support over many years. We thank Colin Allen and Marc Bekoff for valuable suggestions.

References

Allen, C. and Bekoff, M. (1997). *Species of Mind: The Biology and Philosophy of Cognitive Ethology*. Cambridge, Mass.: MIT Press.

Allen, C. and Saidel, E. (1998). The evolution of reference. In *The Evolution of the Mind*. D. Cummins and C. Allen, eds., pp. 183–203. New York: Oxford University Press.

Beer, C. (1997). Expressions of mind in animal behavior. In *Anthropomorphism, Anecdotes, and Animals*. R. W. Mitchell, N. S. Thompson, and H. L. Miles, eds., pp. 198–209. Albany: State University of New York Press.

Bekoff, M. (2000). Animal emotions: Exploring passionate natures. *Bioscience* 50: 861–870.

Boake, C. R. B. and Poulsen, T. (1997). Correlates versus predictors of courtship success: Courtship song in *Drosophila silvestris* and *D. heteroneura*. *Animal Behaviour* 54: 699–704.

Brickerton, D. (1998). The creation and re-creation of language. In *Handbook of Evolutionary Psychology: Ideas, Issues and Applications*. C. Crawford and D. L. Krebs, eds., pp. 613–634. Hillsdale, N.J.: Lawrence Erlbaum Associates.

Burghardt, G. M. (1985a). Animal awareness: Current perceptions and historical perspective. *American Psychologist* 40: 905–919.

Burghardt, G. M. (1985b). *Foundations of Comparative Ethology*. Van Nostrand Reinhold: New York.

Burghardt, G. M. (1991). Cognitive ethology and critical anthropomorphism: A snake with two heads and hognose snakes that play dead. In *Cognitive Ethology: The Minds of Other Animals*, C. A. Ristau, ed., pp. 53–90. Hillsdale, N.J.: Lawrence Erlbaum Associates.

Burghardt, G. M. (1996). Environmental enrichment or controlled deprivation? In *The Well-being of Animals in Zoo and Aquarium Sponsored Research*, G. M. Burghardt, J. T. Bielitski, J. R. Boyce, and D. O. Schaefer, eds., pp. 91–101. Greenbelt, Md.: Scientists Center for Animal Welfare.

Burghardt, G. M. (1997). Amending Tinbergen: A fifth aim for ethology. In *Anthropomorphism, Anecdotes, and Animals*, R. W. Mitchell, N. S. Thompson, and H. L. Miles, eds., pp. 254–276. Albany: State University of New York Press.

Burghardt, G. M. (1998). Snake stories: From the additive model to ethology's fifth aim. In *Responsible Conduct of Research in Animal Behavior*, L. Hart, ed., pp. 77–95. Oxford: Oxford University Press.

Caporael, L. R. and Heyes, C. M. (1997). Why anthropomorphize? Folk psychology and other stories. In *Anthropomorphism, Anecdotes, and Animals*, R. W. Mitchell, N. S. Thompson, and H. L. Miles, eds., pp. 59–73. Albany: State University of New York Press.

Crist, E. (2000). *Images of Animals: Anthropomorphism and Animal Mind*. Philadelphia, Pa.: Temple University Press.

Fisher, J. A. (1990). The myth of anthropomorphism. In *Interpretation and Explanation in the Study of Animal Behavior*, M. Bekoff and D. Jamison, eds., pp. 96–117. Boulder, Col.: Westview Press.

Gabrielsen, G. W. and Smith, E. N. (1985). Physiological responses associated with feigned death in the American opossum. *Acta Physiologica Scandinavian* 123: 393–398.

Gregory, R. L. (2001). Perceptions of knowledge. *Nature* 410: 21.

Griffin, D. R. (1978). Prospects for a cognitive ethology. *Behavioral and Brain Sciences* 1: 527–538.

Jacobs, G. H., Fenwick, J. A., Crognale, M. A., and Deegan, F. F. III (1992). The allcone retina of the garter snake: Spectral mechanisms and photopigment. *Journal of Comparative Physiology A* 170: 701–707.

Kennedy, J. S. (1992). *The New Anthropomorphism*. New York: Cambridge University Press.

Kreuger, J. (2001). Null hypothesis significance testing: On the survival of a flawed method. *American Psychologist* 56: 16–26.

Lockwood, R. (1989). Anthropomorphism is not a four-letter word. In *Perceptions of Animals in American Culture*, R. J. Hoage, ed., pp. 41–56. Washington, D.C.: Smithsonian Institution Press.

Lima, S. L. and Zollner, P. A. (1996). Towards a behavioral ecology of ecological landscapes. *Trends in Ecology and Evolution* 11: 131–135.

Mandelbaum, M. (1964). *Philosophy, Science, and Sense Perception*. Baltimore, Md.: Johns Hopkins University Press.

Mitchell, R. W., Thompson, N. S., and H. L. Miles (eds.) 1997. *Anthropomorphism, Anecdotes, and Animals*. Albany: State University of New York Press.

Morgan, C. L. (1894). *An Introduction to Comparative Psychology*. London: Scott.

Polakowski, K. J. (1989). A design approach to zoological exhibits: The zoo as theater. *Zoo Biology Supplement* 1: 127–139.

Rivas, J. and Burghardt, G. M. (2001). Understanding sexual size dimorphism in snakes: Wearing the snake's shoes. *Animal Behaviour* 62: F1–F6.

Stokstad, E. (2001). Did saurian predators fold up on turns? *Science* 293: 1572.

Terrick, T. D., Mumme, R. L., and Burghardt, G. M. (1995). Aposematic coloration enhances chemosensory recognition of noxious prey in the garter snake, *Thamnophis radix*. *Animal Behaviour* 49: 857–866.

Tinbergen, N. (1951). *The study of Instinct*. New York: Oxford University Press.

Timberlake, W. and Delamater, A. R. (1991). Humility, science and ethological behaviorism. *Behavior Analyst* 14: 37–41.

Ujhelyi, M. (1996). Is there any intermediate stage between animal communication and language? *Journal of Theoretical Biology* 180: 71–76.

Uexküll, J. von (1985). Environment (Umwelt) and the inner world of animals (C. J. Mellor and D. Gove, trans.). In *The Foundations of Comparative Ethology*, G. M. Burghardt, ed., pp. 222–245. New York: Van Nostrand Reinhold. (Reprinted from von Uexküll, 1909, *Umwelt and Innenwelt der Tiere*. Berlin: Jena.)

Wiens, J. A. (1996). Wildlife in patchy environments: Metapopulations, mosaics and management. In *Metapopulations and Wildlife Conservation*, D. R. McCullough, ed., pp. 53–84. Washington, D.C.: Island Press.

Zollner, P. A. and Lima, S. L. (1997). Landscape-level perceptual abilities in white-footed mice: Perceptual range and the detection of forested habitat. *Oikos* 80: 51–60.

3 The Cognitive Defender: How Ground Squirrels Assess Their Predators

Donald H. Owings

Ground squirrel 09 flagged her fluffed-out tail from side to side as she explored the area around a burrow where we had last seen a rattlesnake. Followed by five of her pups, she then traveled to a burrow where her adult son from a previous year was living. Squirrel 09 left her youngsters there as she began traveling to and fro several times, alternately looking for the snake at the original location and returning to interact with her babies. On some of these excursions, she was accompanied closely by one or more of her pups, who often followed in a tight little "flock." Finally, 09 spotted the snake and approached it very closely, flagging her tail, jumping back, and reapproaching repeatedly. One pup had followed her to the snake, and as the mother continued to deal with the snake, two more pups joined them, and later a fourth. Our concern for the young squirrels intensified as they also began to confront the snake, behaving much as their mother did. Surprisingly, 09 did not step forward to protect her babies when one of them approached the snake. However, she was most consistently the closest squirrel to the snake, and her presence there appeared to keep the pups out of this dangerously close region.

As the snake continued its exploration of the area, 09 led four of her pups back to her son's burrow and all five squirrels went underground there. A little later, a fifth youngster arrived at the same burrow, wobbly, limping, and not using its left forepaw, apparently snake bitten. We never saw this pup again after its mother emerged to lead it into the burrow. Subsequently, 09 reemerged and resumed her search for the snake, which was no longer in view. During this time, some of her pups intermittently accompanied her, perhaps endangering themselves again. After more than an hour of dealing with this snake, mother and pups could no longer locate it. They gradually calmed down and began to feed in preparation for the coming night. But

their problems were not over. The snake harassed this family for three more days, killing a second pup, and blind-siding 09 by delivering a sublethal bite to the side of her face where she had earlier lost her sight.

Exploring What Ground Squirrels Know about Their Predators and How They Know It

My research program has long focused on predatory contexts like this one in order to explore the behavioral abilities of California ground squirrels (*Spermophilus beecheyi*). This rattlesnake episode illustrates the challenges that ground squirrels face in dealing with the problems they encounter in nature. Here the maternal ground squirrel had to juggle at least three different tasks, i.e., keeping herself safe while also protecting her pups and managing the behavior of the rattlesnake. The best way for squirrels to proceed in such situations depends on the details of the threat they face, and they must apply their cognitive systems to the task of uncovering those details and finding a way to proceed, as we will see later. The use of predatory contexts ensures that I am studying behavior in situations that are meaningful to my animals in an evolutionary and ecological sense. In this way, I maximize my chances of discovering behavioral processes that have been most strongly shaped by natural selection and individual experience during development. Toward that goal, I have studied the antipredator activities of ground squirrels as they deal with the variety of predators that have historically been important to them, including not only rattlesnakes but also gopher snakes, badgers, coyotes, bobcats, red-tailed hawks, and golden eagles.

These different classes of predators use different hunting techniques that require different antipredator strategies. Avian predators pose the most immediate threat to squirrels, appearing suddenly and launching rapid aerial attacks. More slowly moving mammalian predators pose threats of intermediate urgency; and very slowly moving, ambush-hunting snakes pose the least immediate threat. Ground squirrels vary their antipredator response to these different threats in ways that reflect a tradeoff between self-preservation and the acquisition of additional information about the predators (Coss and Owings 1985; Owings and Hennessy 1984).

Activities that facilitate assessment require getting close to and maintaining sensory contact with the predator, which increases the squirrel's vulnerability to the predator. Activities that reduce a squirrel's vulnerability, for example by minimizing conspicuousness, often involve staying farther away from the predator, which reduces access to assessment cues. Where danger is most immediate, as with avian predators, squirrels opt for more self-preservation by evading the predator and maintaining a low profile. As a result, their options for assessment are limited. In contrast, where the danger is least immediate, as with snakes, squirrels are able to emphasize assessment more heavily by approaching and interacting with the snake as squirrel 09 did.

Thus squirrels shift the balance among the conflicting demands of assessment and self-preservation in ways that indicate that they know how the immediacy of threat from their various predators differs. The demonstration of such knowledge is the first important step in the biological study of cognitive processes (Kamil 1994).

The next step is to explore *how* these animals know what they know. Our research on this type of question has concentrated on the relationship between ground squirrels and snakes. Throughout much of their range, California ground squirrels are the prey of both Pacific gopher snakes (*Pituophis melanoleucus catenifer*) and northern Pacific rattlesnakes (*Crotalus viridis oreganus*). Venomous rattlesnakes are more dangerous than nonvenomous gopher snakes, a

difference indicated by the fact that these squirrels comprise 69 percent of the diet of rattlesnakes, but only 44 percent of the diet of gopher snakes in the foothills of the central Sierra Nevada of California (Fitch 1948, 1949). Nevertheless, the danger posed by rattlesnakes is moderated by blood proteins that confer resistance to rattlesnake venom in both young and adult California ground squirrels (Biardi et al. 2000; Poran and Coss 1990). Despite this resistance, pups succumb to rattlesnake bites because they cannot neutralize as much venom as adults can. Consequently, rattlesnakes almost exclusively eat pups, and adult squirrels can be quite assertive in defending their pups, risking injury but not death from a rattlesnake bite (Fitch 1949; Poran and Coss 1990).

These squirrel–snake relationships comprise a very useful set of predator–prey systems for studying processes of predator recognition and assessment, for at least two reasons. First, squirrels have had a long time to evolve such cognitive defenses against these snakes; according to the fossil record, the ancestors of modern rattlesnakes and gopher snakes have been potential sources of natural selection in ancestors of California ground squirrels for as long as 10 million years (Coss 1991). Second, squirrel–snake encounters lasting hours or even days (as for squirrel 09) provide abundant time for recognition and assessment of snakes.

Our research on snake recognition and assessment has explored both the details and the development of these processes. We have discovered that California ground squirrels do not require experience with snakes during development in order to recognize them. Pups of just postweaning age, born and reared in the lab without contact with snakes, have impressed us with their sophistication. They distinguish snakes from novel animate objects when tested alone and behave in the complex ways adults do toward snakes, flagging their tails, approaching cautiously, investigating in elongate postures, throwing substrate, and jumping back (Owings and Coss 1977). Apparently pups use both visual and olfactory cues to recognize snakes. Preweanling pups, 40 to 41 days old, become alarmed by the odor of gopher snakes and will cautiously investigate a ruler painted on the wall of the test chamber, a visual pattern resembling a gopher snake or rattlesnake with transverse patterning.

However, these findings do not mean that experience with snakes is unimportant in the development of cognitive antisnake defenses. More subtle aspects of antisnake cognition may require the refinements of learning. For example, some (but not all) populations of California ground squirrels appear to require experience with snakes in order to develop the ability to distinguish rattlesnakes from gopher snakes (Coss et al. 1993). The development of such subtle distinctions may depend in part upon becoming attuned to the defensive sounds that the snakes produce in response to confrontive squirrels. Rattlesnakes rattle when pressed by squirrels, and experimental elimination of the rattling sound leads to attenuation of the antisnake behavior of adult squirrels (Rowe and Owings 1978). Closely related populations of California ground squirrels that differ in their current contact with rattlesnakes also differ in their perception of danger from playbacks of rattling and other sibilant sounds (Rowe et al. 1986), a contrast that may be generated by greater rattlesnake experience in one population than the other.

California ground squirrels are also sensitive to variation in the danger posed by rattlesnakes. For example, large rattlesnakes pose a greater threat than small rattlesnakes because larger rattlesnakes deliver more venom per strike (Kardong 1986) and strike with higher velocity and over longer distances (Rowe and Owings 1990). (Unlike birds and mammals, rattlesnakes grow throughout their lives, so that adult sizes are more variable within species of rattlesnake than within species of birds or mammals.) California ground squirrels differentiate large from small

rattlesnakes in ways that are similar to their discrimination of snake species; they stay farther back from larger rattlers, but monitor them and signal more persistently (Swaisgood et al. 1999a). Body temperature also influences the danger that a rattlesnake poses; rattlesnakes are ectotherms, so that their body temperature varies more than that of endothermic birds and mammals. As a snake heats up, the biochemical processes that support behavior speed up, with the consequence that warmer snakes are more dangerous because they deliver strikes with higher velocity, less hesitance, and greater accuracy (Rowe and Owings 1990).

Thus squirrels need to know about a rattlesnake's size and temperature, but extracting this information is not as straightforward as it might initially seem. Squirrels might assess body size visually, but they often encounter snakes in the darkness of burrows or thick vegetation where visual cues about rattlesnake size are limited (Coss and Owings 1978). The detection and assessment of rattlesnakes is difficult even in well-lighted, open locations because rattlesnakes are so cryptic (Hennessy and Owings 1988; Hersek and Owings 1993). The assessment of snake body temperature can represent an even greater challenge. Simple thermal cues such as ambient temperature can be unreliable predictors of a rattlesnake's body temperature, which can change rapidly as a function of microclimatic conditions (Gannon and Secoy 1985). An assessment of snake size and temperature is further complicated by the possibility that small and cold snakes would benefit from hiding their vulnerability because they are susceptible to attack by aggressive adult squirrels.

The squirrels' solution to this assessment challenge is proactive in two ways; they use very interactive methods to gain access to size and temperature cues, and the cues they use are byproducts, not formalized features of the snakes' defensive signaling. By confronting rattlesnakes, squirrels place snakes on the defensive and induce them to rattle, a sound that leaks clues about both size and temperature. Larger rattle-snakes rattle with higher amplitude and lower dominant frequencies, cues that are available as a by-product of the larger tail-shaker muscles of bigger snakes and the lower resonant frequency of their larger rattles. Warmer snakes rattle with faster click rates, higher amplitudes, and shorter latencies, all features that are by-products of the fact that higher temperatures speed snakes up. In field playback studies, these ground squirrels demonstrated their ability to use both categories of acoustic cues by responding with greater caution to the sounds of larger and warmer snakes (Swaisgood et al. 1999b).

Conceptual Framework, Implications, and Future Directions

The term *cognition* refers most fundamentally to all of those processes by which organisms come to know about their environments. The acquisition of this knowledge (knowing) can take a lot of forms and involve a lot of body parts. Ground squirrels, for example, use more than their sense organs and brains to obtain information about rattlesnakes; they also integrate their forelegs and the rattlesnakes' rattles into a "knowing loop," throwing substrate to elicit the rattling so useful for determining the species, size, and temperature of the snake they are dealing with. From this perspective, cognitive processes are embodied processes, but they are not internal to the organism like brain processes are; they are emergent properties of whole organisms (Fragaszy and Visalberghi 1996; Mason 1986; Michaels and Carello 1981). Such a view of cognition helps us to shed our anthropocentric views of how animals acquire knowledge, and instead to think of cognitive processes as fundamental to all organisms, not as a small set of special abilities confined to "a handful of privileged evolutionary newcomers" (Mason 1986, p. 306).

The processes of acquiring knowledge serve and are guided by equally broadly distributed processes of wanting (Mason 1986; Owings 1994). Wanting refers to psychological processes that involve both motivation and emotion, i.e., processes involving action based on an evaluation of objects and events (e.g., how important a snake is and therefore how much attention, time, and energy should be dedicated to dealing with it). As Mason has noted (1979, p. 225): "From a biological perspective the two great themes in the evolution of behavior are 'wanting' and 'knowing.' ... And surely it will be apparent that these themes are interwoven throughout evolution. Knowledge, for the vast majority of organisms, is pragmatic and utilitarian: Animals are prepared to know what they need to know." It follows that most of an organism's cognitive activities involve "hot" judgments about the personal significance of events, rather than "cold" assessments about how things work in general.

The interrelatedness of wanting and knowing is very evident in the antisnake behavior of California ground squirrels. For example, we have already seen how these squirrels make accurate judgments about the danger posed by rattlesnakes, e.g., staying farther back from larger snakes and monitoring them more consistently. Such assessments are, however, very much dependent on the significance of rattlesnakes to the individual squirrel. In particular, rattlesnakes are more important to maternal females with vulnerable pups than to any other class of adult squirrels. Thus mothers with young pups spend more time dealing with rattlesnakes than do nonmaternal adult females or adult males (Swaisgood et al. 1999a). Mothers are also more discriminating, distinguishing much more clearly both among rattlesnakes of different sizes and among the rattling sounds of snakes of differing temperatures and sizes (Swaisgood 1994). Similarly, all squirrels are sensitive to where they discover a rattlesnake; when a snake is found near their own burrow, squirrels are more confrontational and engage in much more tail flagging than when they detect a snake near someone else's burrow (Swaisgood et al. 1999a).

Like most research, this program has raised as many questions as it has answered. We have not explored the many distinct processes that fall under the heading of wanting as discussed, for example, by Berridge (1996). Similarly, the broad definition of cognition offered here does not imply that all animals function at the same level of cognitive complexity (for discussions of levels of cognition, see Capitanio and Mason 2000; Dennett 1983).

We have some intriguing hints of complexity in this antipredator system. For example, these squirrels behave as though they know (1) *where* their danger from snakes is greatest [e.g., in burrows and other areas where visibility is poor (Coss and Owings 1985; Hersek and Owings 1993)], (2) *how long* snake danger persists [days, because snakes lie in ambush that long (Hersek and Owings 1993)], and (3) *how to* lead rattlesnakes away from the burrows housing their pups (Hennessy and Owings 1988). Future research could explore the levels at which these squirrels function cognitively, focusing especially on how specialized these cognitive mechanisms are for the antisnake context that has been such a strong source of natural selection in this species (see Cheney and Seyfarth 1990 on the domain specificity of cognition). Variation among populations will be especially valuable in such research. Not all populations of these squirrels have consistently been under selection from rattlesnakes or gopher snakes in their recent evolutionary history, and relaxation of selection from snakes is associated with reductions in resistance to rattlesnake venom and changes in the higher-level organization of antisnake behavior (Coss 1999).

Some of the most fascinating future research will explore developmental questions. As we have already indicated, many features of this remarkably complex behavioral system develop without experience with snakes. Such findings beg for detailed studies of cognitive mechanisms, including how these mechanisms change during development and what kinds of inputs are important in their ontogenetic modification. We know that the higher-level organization of antisnake behavior is altered as squirrels mature; pups and adults, for example, use the tail-flagging signal differently. However, these differences do not simply reflect the incompleteness of development in young squirrels. Pups tail flag differently but not less proficiently; they are more skillful than adults in using tail flagging to keep other squirrels near, and they tail flag in ways that distinguish more clearly than adults do between snake-free days and days when snakes have been seen at the site (Hersek and Owings 1994).

Such findings are not unique to the development of tail flagging (Owings 1994; Owings and Loughry 1985). Young animals in general are more than incompletely developed adults; different ages occupy distinct developmental niches, with associated adaptive differences in behavior (for elaboration of this idea, see Galef 1981). Such observations raise very exciting questions. How are corresponding cognitive systems transformed as young animals mature? Do cognitive systems undergo metamorphosis in ways analogous to the metamorphoses that transform caterpillars into butterflies? If so, how do animals navigate the uncertain terrain that lies between stable stages of cognitive functioning? The study of cognitive development from this perspective of "ontogenetic adaptation" (Alberts 1987) is one of the greatest challenges in the biological study of cognition.

References

Alberts, J. R. (1987). Early learning and ontogenetic adaptation. In *Perinatal Development: A Psychological Perspective*, N. A. Krasnegor, E. M. Blass, M. A. Hofer, and W. P. Smotherman, eds., pp. 11–37. Orlando, Fla.: Harcourt Brace Jovanovich.

Berridge, K. C. (1996). Food reward—Brain substrates of wanting and liking. *Neuroscience and Biobehavioral Reviews* 20: 1–25.

Biardi, J., Coss, R. G., and Smith, D. G. (2000). California ground squirrel (*Spermophilus beecheyi*) blood sera inhibits crotalid venom proteolytic activity. *Toxicon* 38: 713–721.

Capitanio, J. P. and Mason, W. A. (2000). Cognitive style: Problem solving by rhesus macaques (*Macaca mulatta*) reared with living or inanimate substitute mothers. *Journal of Comparative Psychology* 114: 115–125.

Cheney, D. L. and Seyfarth, R. M. (1990). *How Monkeys See the World: Inside the Mind of Another Species.* Chicago: University of Chicago Press.

Coss, R. G. (1991). Context and animal behavior: III. The relationship between early development and evolutionary persistence of ground squirrel antisnake behavior. *Ecological Psychology* 3: 277–315.

Coss, R. G. (1999). Effects of relaxed natural selection on the evolution of behavior. In *Geographic Variation in Behavior: Perspectives on Evolutionary Mechanisms,* S. A. Foster and J. A. Endler, eds., pp. 180–208. Oxford: Oxford University Press.

Coss, R. G. and Owings, D. H. (1978). Snake-directed behavior by snake naive and experienced California ground squirrels in a simulated burrow. *Zeitschrift für Tierpsychologie* 48: 421–435.

Coss, R. G. and Owings, D. H. (1985). Restraints on ground squirrel antipredator behavior: Adjustments over multiple time scales. In *Issues in the Ecological Study of Learning,* T. D. Johnston and A. T. Pietrewicz, eds., pp. 167–200. Hillsdale, N.J.: Lawrence Erlbaum Associates.

Coss, R. G., Guse, K. L., Poran, N. S., and Smith, D. G. (1993). Development of antisnake defenses in California ground squirrels (*Spermophilus beecheyi*): II. Microevolutionary effects of relaxed selection from rattlesnakes. *Behaviour* 124: 137–164.

Dennett, D. C. (1983). Intentional systems in cognitive ethology: The "Panglossian paradigm" defended. *Behavioral and Brain Sciences* 6: 343–390.

Fitch, H. S. (1948). Ecology of the California ground squirrel on grazing lands. *American Midland Naturalist* 39: 513–596.

Fitch, H. S. (1949). Study of snake populations in central California. *American Midland Naturalist* 41: 513–579.

Fragaszy, D. M. and Visalberghi, E. (1996). Social learning in monkeys: Primate "primacy" reconsidered. In *Social Learning in Animals: The Roots of Culture,* C. M. Heyes and B. G. Galef, eds., pp. 65–84. New York: Academic Press.

Galef, B. G. (1981). The ecology of weaning: Parasitism and the achievement of independence by altricial mammals. In *Parental Care in Mammals,* D. J. Gubernick and P. H. Klopfer, eds., pp. 211–241. New York: Plenum.

Gannon, V. P. J. and Secoy, D. M. (1985). Seasonal and daily activity patterns in a Canadian population of the prairie rattlesnake, *Crotalus viridis viridis. Canadian Journal of Zoology* 63: 86–91.

Hennessy, D. F. and Owings, D. H. (1988). Rattlesnakes create a context for localizing their search for potential prey. *Ethology* 77: 317–329.

Hersek, M. J. and Owings, D. H. (1993). Tail flagging by adult California ground squirrels: A tonic signal that serves different functions for males and females. *Animal Behaviour* 46: 129–138.

Hersek, M. J. and Owings, D. H. (1994). Tail flagging by young California ground squirrels, *Spermophilus beecheyi*: Age-specific participation in a tonic communicative system. *Animal Behaviour* 48: 803–811.

Kamil, A. C. (1994). A synthetic approach to the study of animal intelligence. In *Behavioral Mechanisms in Evolutionary Ecology,* L. A. Real, ed., pp. 11–45. Chicago: University of Chicago Press.

Kardong, K. V. (1986). Predatory strike behavior of the rattlesnake, *Crotalus viridis oreganus. Journal of Comparative Psychology* 100: 304–314.

Mason, W. A. (1979). Wanting and knowing: A biological perspective on maternal deprivation. In *Origins of the Infant's Social Responsiveness,* E. Thoman, ed., pp. 225–249. Hillsdale, N.J.: Lawrence Erlbaum Associates.

Mason, W. A. (1986). Behavior implies cognition. In *Integrating Scientific Disciplines,* W. Bechtel, ed., pp. 297–307. Dordrecht, The Netherlands: Martinus Nijhoff.

Michaels, C. F. and Carello, C. (1981). *Direct Perception.* Englewood Cliffs, N.J.: Prentice-Hall.

Owings, D. H. (1994). How monkeys feel about the world: A review of *How Monkeys See the World. Language and Communication* 14: 15–30.

Owings, D. H. and Coss, R. G. (1977). Snake mobbing by California ground squirrels: Adaptive variation and ontogeny. *Behaviour* 62: 50–69.

Owings, D. H. and Hennessy, D. F. (1984). The importance of variation in sciurid visual and vocal communication. In *The Biology of Ground-Dwelling Squirrels: Annual Cycles, Behavioral Ecology, and Sociality*, J. A. Murie and G. R. Michener, eds., pp. 169–200. Lincoln: University of Nebraska Press.

Owings, D. H. and Loughry, W. J. (1985). Variation in snake-elicited jump-yipping by black-tailed prairie dogs: Ontogeny and snake-specificity. *Zeitschrift für Tierpsychologie* 70: 177–200.

Poran, N. S. and Coss, R. G. (1990). Development of antisnake defenses in California ground squirrels (*Spermophilus beecheyi*): I. Behavioral and immunological relationships. *Behaviour* 112: 222–245.

Rowe, M. P., Coss, R. G., and Owings, D. H. (1986). Rattlesnake rattles and burrowing owl hisses: A case of acoustic Batesian mimicry. *Ethology* 72: 53–71.

Rowe, M. P. and Owings, D. H. (1978). The meaning of the sound of rattling by rattlesnakes to California ground squirrels. *Behaviour* 66: 252–267.

Rowe, M. P. and Owings, D. H. (1990). Probing, assessment, and management during interactions between ground squirrels and rattlesnakes. Part 1: Risks related to rattlesnake size and body temperature. *Ethology* 86: 237–249.

Swaisgood, R. R. (1994). Assessment of Rattlesnake Dangerousness by California Ground Squirrels. Unpublished Ph.D. dissertation, University of California, Davis.

Swaisgood, R. R., Owings, D. H., and Rowe, M. P. (1999a). Conflict and assessment in a predator–prey system: Ground squirrels versus rattlesnakes. *Animal Behaviour* 57: 1033–1044.

Swaisgood, R. R., Rowe, M. P., and Owings, D. H. (1999b). Assessment of rattlesnake dangerousness by California ground squirrels: Exploitation of cues from rattling sounds. *Animal Behaviour* 57: 1301–1310.

4   Jumping Spider Tricksters: Deceit, Predation, and Cognition

Stim Wilcox and Robert Jackson

In an emerald rainforest of northeastern Australia, a sunbeam pierces the canopy, touches broad green leaves on the way down, and beams onto a lichen-spotted rock surface. In the beam's circle, the slow, careful motions of a brownish jumping spider are illuminated. The jumping spider belongs to the genus *Portia* and it is stalking its prey, a different species of spider sitting in its own web. *Portia* steps cautiously from the rock surface out onto the web and stops. Delicately, *Portia* begins to pluck the web with its palps and legs, making signals that mimic the struggles of a trapped insect. When the prey spider ignores *Portia*'s plucking, *Portia* varies the characteristics of the signals, generating a kaleidoscopic of what appears to be a random selection of signals. Eventually, in response to one of these signals, the prey spider swivels toward *Portia*. Immediately, *Portia* backtracks to that particular signal and repeats it again and again. There being no further response from the prey, *Portia* eventually reverts to broadcasting a kaleidoscope of signals. When the prey spider still moves no farther, *Portia* adopts another ploy.

Now *Portia* slowly and carefully stalks across the web toward the resident spider, intermittently making a variety of signals. From time to time, a soft breeze blows, ruffling the web. The ruffling creates background noise in the web, and *Portia* exploits these moments, during which the resident spider's ability to detect an intruder is impaired, by stalking faster and farther during these periods than when the air is still. Nearing the resident spider, *Portia* makes a signal that elicits from the resident spider a sudden, rapid approach. However, the spider advances very aggressively, and *Portia* scrambles to the edge of the web, then turns around to look over the scene. Soon *Portia* moves away from the web and undertakes a lengthy detour, first going away from the prey and around a large projection on the rock surface, losing sight of prey spider along the way.

About an hour later, *Portia* appears again, but now is positioned above the web on a small overhanging portion of the rock. After anchoring itself to the rock with a silk dragline, *Portia* next slowly lowers itself down though the air, not touching the web at all. Arriving level with the resident spider, *Portia* suddenly swings in, grabs hold of the unsuspecting spider, and sinks its poison-injecting fangs into the hapless victim. So ends another spider-eat-spider episode from the rainforest; it is typical of hundreds that we have witnessed in the field and raises interesting questions about spider cognition. In the discussion to follow, we will return repeatedly to this hunting example.

When we began studying the species of *Portia* about 20 years ago, little was known about the behavior of these unusual tropical members of the spider family Salticidae (jumping spiders or salticids for short). About two dozen species of *Portia* have been described. They are distributed from Australia through the Indonesian and Malaysian island chains into China, the Indian subcontinent, and Africa. The adults tend to be 8–12 mm in body length and live in habitats ranging from low-elevation rainforest to montane pine forests to savannah.

Across all habitats, whenever *Portia* has been studied, it has been shown to specialize on other spiders as prey, invade webs, and practice aggressive mimicry (Jackson and Wilcox 1998). All *Portia* also build their own webs, which they use for capturing both insects and spiders. Web-based behavior is unexpected in a salticid. Typical salticids neither build nor invade webs. Instead, they use their acute eyesight to guide stalk-and-leap sequences on insects carried out on the ground, on tree trunks, and in foliage (Jackson and Pollard 1996). Remarkably, *Portia* also practices away-from-webs stalk-and-leap sequences. Being highly effective at capturing prey in each setting (away from webs, in its own web, and in another spider's web), among salticids *Portia* is a jack of all trades and the master of them all (Jackson and Hallas 1986).

It is the intricate details of web invasion and aggressive mimicry that especially raise questions about cognition. Web-building spiders from families other than salticids have simple eyes and only poor eyesight (Land 1985). Web signals (i.e., the tension and movement patterns of silk threads) can be envisaged as the language of the typical web-building spider (Foelix 1996). When *Portia* enters an alien web, it manipulates the silk, making web signals that deceive and control the behavior of the resident spider. Simply overpowering the resident spider with strength and speed would not appear to be an option for *Portia*. A spider's web is extremely sensitive to encounters with objects of *Portia*'s size, making undetected web entry exceedingly difficult. Making matters worse, the resident spider is also a predator. There is a serious potential for the tables to be turned. Preying on other spiders is a game where *Portia* may pay for mistakes with its life (Jackson and Wilcox 1998).

That *Portia* might use aggressive mimicry to prey effectively on one or a few types of webbuilding spiders would have been an interesting finding, but what we found was unexpected. *Portia* is highly effective at taking almost *any kind* of web-building spider. Not only are virtually all the web builders in its natural habitat taken, but in the laboratory, on first exposure, *Portia* routinely makes effective use of aggressive mimicry to control the behavior of, and prey upon, spiders it would never have encountered in its evolutionary history (Jackson and Wilcox 1993a). One of our initial objectives was to understand the basis for this exceptional flexibility.

Devising a method for studying Portia's signals became a critical requirement. For finegrained detail, this can be done using laser recording technology (Tarsitano et al. 2000). For routine work, however, we devised a homegrown, computerized system for recording, analyzing, and playing back signals (Wilcox and Jackson 1998). We recorded signals with a gal-

vanometer connected by its stylus to the web, coded the very low-frequency signals with a frequency modulation (FM) coder, and stored the coded signals on the soundtrack of a videotape, while simultaneously recording by camera the behavior sequence being observed. Decoded FM signals were converted to digital form when they were input into a laptop computer. The computer was used to analyze the signals for frequency, length, and other characteristics. Signals were played back by making a tiny magnet oscillate. This was achieved by amplifying a signal played by the computer into a coil of magnet wire. When the magnet was glued to an object we wished to vibrate, we played an amplified signal into the coil, which made the magnet oscillate in concert with the electromagnetic waves (Wilcox and Kashinsky 1980).

Unlike the resident spider, *Portia* can see shape and form, and we know from experiments that *Portia* sometimes distinguishes the type of prey it has encountered before web contact, thereby being able to make appropriate adjustments in its mode of approach. For example, there is a Philippine population of *Portia* living in a habitat where an especially dangerous prey spider is common, a spitting spider that is itself a specialist at feeding on salticids. Upon seeing a spitting spider, these Philippine *Portia* (but not *Portia* from other habitats) consistently approach from the rear. Typically this requires a detour. That this inclination to approach from behind is innate (i.e., does not require prior experience with spitting spiders) was shown by testing *Portia* individuals reared in the laboratory (Jackson et al. 1998).

In *Portia*'s signal-making behavior, there is also evidence of adaptation to particular prey species. For example, after contacting the web, the Philippine *Portia* tends to make only faint signals that fail to provoke a full-scale spitting attack (Jackson et al. 1998). In encounters with the females of certain species of prey, the Australia rainforest *Portia* may begin by making signals that simulate the courtship signals of the males of the resident spider (e.g., Jackson and Wilcox 1990). On the whole, if the predatory sequence is short, signal generation may appear more or less stereotyped. However, most predatory sequences are lengthy. It is routine in lengthy sequences for *Portia* eventually to broadcast a kaleidoscope of signals until some particular signal "works," i.e., elicits an appropriate response from *Portia*'s viewpoint. *Portia* then repeats over and over the signal that worked. If it ceases to work, eventually *Portia* may switch back to broadcasting a kaleidoscope. This flexible problem-solving behavior is known as the trial-and-error tactic (Jackson and Wilcox 1993a), and it appears to have a central role in almost all of *Portia*'s signal-making sequences. Thus biases toward particular signals with particular types of prey serve primarily to get a sequence off to a good start, with trial and error being used to finish the job (Jackson and Wilcox 1998).

We have demonstrated experimentally that *Portia* derives signals by trial and error. We successfully encouraged *Portia* to repeat signals we chose at random for reinforcement, where reinforcement might be the spider approaching *Portia* or localized movement of a spider that remained in one place. The coil-and-magnet system gave us control over the prey's behavior and enabled us to provide these kinds of reinforcement (Jackson and Wilcox 1993a).

Portia's trial-and-error tactic might be viewed as at least a rudimentary example of a spider thinking, or more technically, spider cognition. A discussion of animal cognition often seems like a walk through a minefield, there being almost as many definitions of cognition as there have been authors discussing the topic. A more rewarding approach is to apply frameworks that raise questions about cognitive processes. For example, Dukas and Real (1993) based a framework on six cognitive properties, which we list here along with a rough, informal indication of what each means in everyday terms: reception (taking in information), attention (focusing on particular tasks), representation (maintaining a mental image or cognitive map), memory (retaining information), problem solving (deriving pathways to the achievement of goals), and communication language (influencing other individuals by manipulating symbols). This framework directs interest toward understanding processes that underpin cognitive phenomena, and the thorny problem of defining cognition in any rigorous way is sidestepped, the rationale being that once we understand the underlying processes, worrying about a global definition of cognition becomes irrelevant (see Dennett 1991).

Returning to our hunting example, although *Portia* makes use of tactile and chemical cues when hunting, *Portia*'s acute eyesight seems to be the most critical factor in making making predatory decisions that are interesting to discuss in a cognitive framework. *Portia*'s forward-facing anterior medial eyes support a spatial acuity exceeding that known for any other animal of comparable size, rivaling that of much larger animals such as cephalopod mollusks and primates (Land 1985; Harland et al. 1999). The eyes of *Portia* and other salticids are unique and complex evolutionary solutions to the problem of how to see shape and form using drastically fewer receptors than are present in the eyes of cephalopods, birds, and mammals (Land 1974). The human eye, for example, has over 100 million receptors, but the salticid eye has only 10,000 to 100,000 (Land 1985).

We know a great deal about how salticid eyes achieve exceptional acuity, despite their small size, because of extensive research over the past 80 years. The pioneering work of Homann (1928) was significantly extended by Michael Land's work on *Phidippus johnsoni* (Land 1969a, b), a tour de force in small-scale physiological optics, and more recently by the wide-ranging comparative and developmental studies of David Blest (Blest et al. 1990). Salticid research is now at a point where we can begin linking decision-making processes directly to details concerning the information made available by a unique eye (see Harland and Jackson 2000a).

The link between acute vision and cognitive capacities is currently being considered in research on *Portia* (Harland and Jackson 2000b). *Portia's* predatory strategy seems to require especially precise decisions before the spider comes into close proximity to its prey. For this, acute vision would seem to have important inherent advantages. By sight, *Portia* can precisely locate and identify spiders from a distance of 30–40 body lengths away, monitor the spider's orientation and behavior during the course of a predatory sequence, and in general quickly gain critical information for predatory decisions during complex interactions with a dangerous prey (Jackson 1992).

Moving to the next category in the framework of Dukas and Real (1993), the trial-and-error tactic highlights how attention may be critical for *Portia's* success as a predator on other spiders. *Portia's* trial-and-error tactic can be envisaged as at least a rudimentary example of learning (see Staddon 1983), and there has been a tendency to emphasize learning in the literature on animal cognition (see Yoerg 1991). Yet learning in itself may not tell us anything particularly interesting about how animals differ in cognitive capacity because learning of one sort or another appears to be more or less universal within the animal kingdom (Bitterman 1965), and even in single cells (i.e., in single-celled protists and in single cells of multicellular animals; see Staddon 1983).

Portia's trial-and-error tactic may be more interesting in relation to attention. The relevance of attention is apparent whenever one watches *Portia* hunting. *Portia's* attentive ability is especially dramatic when *Portia* is preying on social spiders, where many potential prey are present simultaneously in close proximity. *Portia* singles out one prey spider, sometimes with other potential prey spiders active close by in the same web (Jackson and Wilcox 1993a). Dynamic fine control of the targeted social spider's behavior is achieved by means of the focused flexibility inherent in use of trial and error. Rather than

emphasizing how learning is implied by this tactic, however, we prefer to emphasize the *unlearning* evidenced when *Portia* reverts to the kaleidoscope of signals once a given signal ceases to elicit an appropriate response. Something like this sort of unlearning may be a precondition for much of what interests us in relation to animal cognition, however cognition might be defined.

There is more to *Portia's* flexibility than just switching signals, however. Going back to our hunting example, we saw *Portia* making decisions concerning whether to go out onto the web, whether to undertake a detour, and so forth. Moving onto another spider's web illustrates especially important constraints on *Portia's* strategy. The other spider is a predator, and the web is more than just an arena in which the resident spider normally takes its prey. It is also a critical component of the web-builder's sensory system (Witt 1975).

To enter another spider's web is almost literally to walk right into the spider's primary sensory organ (Jackson and Pollard 1996). The extreme sensitivity of spider webs to movement and weight (Barth 1982) means that for an animal of *Portia's* size, it is probably not a realistic option to walk softly enough in the web to avoid making a signal that is detectable to the resident spider. What *Portia* does instead is usually to precisely control the nature of the signal going to the resident spider (i.e., practice aggressive mimicry). However, there are times when *Portia* does simply walk across the web, which brings us to the opportunistic smokescreen tactic.

In our hunting example, when the wind created background noise in the web (an opportunistic vibratory "smokescreen"), *Portia* moved faster and farther across the web than when the air was still. We have shown experimentally (Wilcox et al. 1996) that wind and other disturbances (e.g., an insect struggling in the web) make large-scale web signals that mask the fainter signals from *Portia's* footsteps. *Portia* is flexible and takes advantage of these opportunities to move rapidly across the web without alerting the resident spider. The opportunistic smokescreen tactic also illustrates interesting levels of attention because this tactic is practiced only when *Portia* is approaching a resident spider and not when it is approaching an ensnared insect or an egg sac of the resident (Wilcox et al. 1996).

In our hunting example, the resident spider responded to *Portia*'s signals and rapidly moved toward *Portia*, but the resident approached too fast and *Portia* moved away. When it was at the edge of the web again, *Portia* looked the situation over and made a decision not to reenter the web. Instead, it chose to plan and undertake a detour. The detour ended with *Portia* positioned better than before, ready for an attack where entering the web would not be necessary. *Portia* chose instead to drop on a line of silk parallel to the web and swing in to capture the resident. Detours similar to this have been observed hundreds of times in nature and the laboratory (Jackson and Wilcox 1993b).

There have been extensive experimental studies of *Portia*'s detouring behavior in the laboratory (Tarsitano and Jackson 1992, 1994, 1997; Tarsitano and Andrew 1999) showing, for example, that *Portia* can choose between correct and incorrect pathways leading to a spider lure, and make detours that initially require moving away from the prey, being out of sight of it, and even bypassing an incorrect pathway on the way to choosing the correct one. Other jumping spiders are also known to take detours (Hill 1979), but *Portia* takes detours more readily, takes longer detours, and seems to be unusual in not needing to maintain visual orientation on the prey spider when conducting the detour (Tarsitano and Andrew 1999).

The utilization of planned detours has especially interesting cognitive implications, suggesting the use of mental maps (representation; see Dyer 1998) and prolonged memory. Planned detours (Tarsitano and Jackson 1997) are also interesting as an example of problem solving. The trial-and-error method was also an example

of problem solving, but planned detours differ because the solution is derived before execution of the behavior. Deriving a solution before execution of behavior comes especially close to what would be called "thinking" in lay terms (see Dennett 1996).

What about the sixth category in the framework of Dukas and Real, communication language? The manipulation of symbols with arbitrarily assigned meanings is inherent to verbal language and surely this is beyond anything achievable by a spider brain. Yet the stringing together of signals during aggressive mimicry sequences is at least remotely suggestive of something akin to verbal language. It is a much more dynamic undertaking than we originally appreciated. Calling *Portia*'s signal-making behavior "aggressive mimicry" raises the question of what *Portia* mimics, but this may be less important than trying to understand how *Portia* achieves fine control of its victim's behavior. The emphasis should perhaps be on how *Portia* takes advantage of biases in the victim's nervous system, adopting a perspective akin to recent ideas about receiver psychology (Guilford and Dawkins 1991) and sensory exploitation (Proctor 1992; Ryan and Rand 1993; Clark and Uetz 1993), but with a greater emphasis on complexity, flexibility, and dynamic interaction between signaler and receiver. Studying *Portia*'s signal-making strategy from this perspective may bring us closer than we initially expected to something like the cognitive implications of verbal language.

References

Barth, F. G. (1982). Spiders and vibratory signals: Sensory reception and behavioral significance. In *Spider Communication: Mechanisms and Ecological Significance*, P. N. Witt and J. S. Rovner, eds., pp. 67–122. Princeton, N.J.: Princeton University Press.

Bitterman, M. E. (1965). Phyletic differences in learning. *American Psychologist* 20: 396–410.

Blest, A. D., O'Carroll, D. C., and Carter, M. (1990). Comparative ultrastructure of Layer I receptor mosaics in principal eyes of jumping spiders: The evolution of regular arrays of light guides. *Cell and Tissue Research* 262: 445–460.

Clark, D. L. and Uetz, G. W. (1993). Signal efficacy and the evolution of male dimorphism in the jumping spider, *Maevia inclemens*. *Proceedings of the National Academy of Science U.S.A.* 90: 1954–1957.

Dennett, D. C. (1991). *Consciousness Explained*. Boston: Little, Brown.

Dennett, D. C. (1996). *Kinds of Minds: Toward an Understanding of Consciousness*. New York: Basic Books.

Dukas, R. and Real, L. A. (1993). Cognition in bees: From stimulus reception to behavioral change. In *Animal Cognition in Nature*, D. R. Papaj and A. C. Lewis, eds., pp. 343–373. New York: Chapman and Hall.

Dyer, F. C. (1998). Cognitive ecology of navigation. In *Cognitive Ecology*, R. Dukas, ed., pp. 201–260. Chicago: University of Chicago Press.

Foelix, R. F. (1996). *Biology of Spiders*. 2nd ed. Oxford: Oxford University Press and Georg Thieme Verlag.

Guilford, T. and Dawkins, M. S. (1991). Receiver psychology and the evolution of animal signals. *Animal Behaviour* 42: 1–14.

Harland, D. P., Jackson, R. R., and Macnab, A. M. (1999). Distances at which jumping spiders distinguish between prey and conspecific rivals. *Journal of Zoology, London* 247: 357–364.

Harland, D. P. and Jackson, R. R. (2000a). Cues by which *Portia fimbriata*, an araneophagic jumping spider, distinguishes jumping spider prey from other prey. *Journal of Experimental Biology* 203: 3485–3494.

Harland, D. P. and Jackson, R. R. (2000b). "Eight-legged cats" and how they see—A review of recent work on jumping spiders. *Cimbebasia* 16: 231–240.

Hill, D. E. (1979). Orientation by jumping spiders of the genus *Phiddipus* (Araneae: Salticidae) during the pursuit of prey. *Behavioral Ecology and Sociobiology* 5: 301–322.

Homann, H. (1928). Die Augen der Araneen. *Zeitschrift für Morphologische Okologie der Tiere* 69: 201–272.

Jackson, R. R. (1992). Eight-legged tricksters: Spiders that specialize in catching other spiders. *Bioscience* 42: 590–598.

Jackson, R. R. and Hallas, S. E. A. (1986). Capture efficiencies of web-building spiders (Araneae, Salticidae): Is the jack-of-all-trades the master of none? *Journal of Zoology, London* 209: 1–7.

Jackson, R. R. and Pollard, S. D. (1996). Predatory behavior of jumping spiders. *Annual Review of Entomology* 41: 287–308.

Jackson, R. R. and Wilcox, R. S. (1990). Aggressive mimicry, predator-specific predatory behavior and predator-recognition in the predator-prey interactions of *Portia fimbriata* and *Euryattus* sp., jumping spiders from Queensland. *Behavioral Ecology and Sociobiology* 26: 111–119.

Jackson, R. R. and Wilcox, R. S. (1993a). Spider flexibly chooses aggressive mimicry signals for different prey by trial and error. *Behaviour* 127: 21–36.

Jackson, R. R. and Wilcox, R. S. (1993b). Observations in nature of detouring behavior by *Portia fimbriata*, a web-invading aggressive mimic jumping spider from Queensland. *Journal of Zoology, London* 230: 135–139.

Jackson, R. R. and Wilcox, R. S. (1998). Spider-eating spiders. *American Scientist* 86: 350–357.

Jackson, R. R., Li, D., Fijn, N., and Barrion, A. (1998). Predator-prey interactions between aggressive-mimic jumping spiders (Salticidae) and araeneophagic spitting spiders (Scytodidae) from the Philippines. *Journal of Insect Behavior* 11: 319–342.

Land, M. F. (1969a). Structure of the retinae of the principal eyes of jumping spiders (Salticidae: Dendryphantinae) in relation to visual optics. *Journal of Experimental Biology* 51: 443–470.

Land, M. F. (1969b). Movements of the retinae of jumping spiders (Salticidae: Dendryphantinae) in response to visual stimuli. *Journal of Experimental Biology* 51: 471–493.

Land, M. F. (1974). A comparison of the visual behaviour of a predatory arthropod with that of a mammal. In *Invertebrate Neurons and Behaviour*, C. A. G. Wiersma, ed., pp. 411–418. Cambridge, Mass.: MIT Press.

Land, M. F. (1985). The morphology and optics of spider eyes. In *Neurobiology of Arachnids*, F. G. Barth, ed., pp. 53–78. Berlin: Springer-Verlag.

Proctor, H. C. (1992). Sensory exploitation and the evolution of male mating behavior: A cladistic test using water mites (Acari: Parasitengona). *Animal Behaviour* 44: 745–752.

Ryan, M. J. and Rand, A. S. (1993). Sexual selection and signal evolution: The ghost of biases past. *Proceedings of the Royal Society of London B* 340: 187–195.

Staddon, J. E. R. (1983). *Adaptive Behavior and Learning*. New York: Cambridge University Press.

Tarsitano, M. S. and Andrew, R. (1999). Scanning and route selection in the jumping spider *Portia labiata*. *Animal Behaviour* 58: 255–265.

Tarsitano, M. S. and Jackson, R. R. (1992). Influence of prey movement on the performance of simple detours by jumping spiders. *Behaviour* 123: 106–120.

Tarsitano, M. S. and Jackson, R. R. (1994). Jumping spiders make predatory detours requiring movement away from prey. *Behaviour* 131: 65–73.

Tarsitano, M. S. and Jackson, R. R. (1997). Araneophagic jumping spiders discriminate between detour routes that do and do not lead to prey. *Animal Behaviour* 53: 257–266.

Tarsitano, M., Jackson, R. R., and Kirchner, W. (2000). Signals and signal choices made by araneophagic jumping spiders while hunting the orb-weaving spiders *Zygiella x-notata* and *Zosis genicularis*. *Ethology* 106: 595–615.

Wilcox, R. S. and R. R. Jackson (1998). Cognitive abilities of araneophagic jumping spiders. In *Animal Cognition in Nature*, I. Pepperberg, A. Kamil, and R. Balda, eds., pp. 411–434. New York: Academic Press.

Wilcox, R. S. and Kashinsky, W. (1980). A computerized method of analyzing and playing back vibratory animal signals. *Behavior Research Methods and Instrumentation* 12: 361–363.

Wilcox, R. S., Jackson, R. R., and Gentile, K. (1996). Spiderweb smokescreens: Spider trickster uses background noise to mask stalking movements. *Animal Behaviour* 51: 313–326.

Witt, P. N. (1975). The web as a means of communication. *Bioscience Communications* 1: 7–23.

Yoerg, S. I. (1991). Ecological frames of mind: The role of cognition in behavioral ecology. *Quarterly Review of Biology* 66: 287–301.

5 The Ungulate Mind

John A. Byers

Bovine: 1. Of, relating to, or resembling a ruminant mammal of the genus Bos, such as an ox, cow, or buffalo. 2. Sluggish, dull, and stolid.
—The American Heritage Dictionary of the English Language, 3rd ed.

As the definition of bovine suggests, many see a cow standing in barnyard muck, its head lowered, a rope of drool hanging from its mouth, and conclude that the space between its ears is filled with bone, or perhaps air. A horse that has traveled the same path many times is likely to shy away in fright when it encounters a newspaper or other new object near the path. These and other common observations support the general view that the ungulates are a fairly dim lot. Clever Hans excepted, no ungulate is or has been the subject in tests of cognitive ability. However, ungulate brains are not conspicuously small (Eisenberg 1981), so we might ask whether there is an underappreciated mental ability in the group. I am going to argue that the ungulates are smarter than previously believed, but that their cognitive abilities are specialized, and most likely are limited to just a few kinds of situations. Like vervet monkeys (Cheney and Seyfarth 1990), ungulates appear to have domain-specific cognitive ability. However, these domains are conspicuously different than those that brought about the intelligence of monkeys and us.

Intelligence and predictive cognitive ability are ecological adaptations. For monkeys and other primates, the relevant aspect of the ecology, that part of the animals' environment that selects for mental ability, is the social environment. Monkeys and apes appear to gain fitness advantages by being able to predict the actions of other group members, and by their ability to use social signals to manipulate the behavior of conspecifics (de Waal 1982). Perhaps because of anthropocentrism, this social intelligence hypothesis, as it is called, has dominated discussion on the evolution of cognition. However, I think that other ecological domains may be the drivers of intelligence in other taxa. Just as other environments may select for sensory abilities that are alien to us (e.g., echolocation in bats, electric field communication in mormyrid fishes, and magnetic field orientation in birds and bees), so other environments may select for cognitive abilities that we may not immediately recognize as such.

In the ungulates, two aspects of ecology are likely to create selection for specialized cognitive ability. These aspects are predation on young and the dynamics of polygynous mating systems. I am going to discuss ungulate cognition from the perspective of my observations on pronghorn antelopes (*Antilocapra americana*) (Byers 1997), but I do not think that pronghorn represent a special case. Many other ungulates live in similar ecological circumstances, have almost identical behavioral traits, and are likely to have similar sets of cognitive traits.

Predation on Young and What Mothers Do about It

Generally, the ungulates avoid being eaten by predators either by being large bodied or fast runners, or both. For many of the fast ungulates, such as pronghorn, all deer, and many species of antelopes, the young are not fast runners when they are born. A specialized strategy called "hiding" has evolved (FitzGibbon 1990; Lent 1974). Hiding represents coordinated behavior of the mother and her young. Shortly after birth, the mother leads the tottering infant away from the birth site, then signals to it to move away. The infant walks a short distance and reclines. Now the incredible part of the hiding strategy begins. The infant remains motionless and refrains from urinating or defecating for 3–4 hours, until the mother returns. Upon the mother's return, the infant sucks in a load of milk that would kill a follower ungulate (Carl and Robbins 1988), and it urinates and defecates into the mother's mouth in response to her licking. The adaptive value of hiding is that it conceals the location of the slow infant from predators. The concealment depends on coordinated behavior of the mother and the infant. The infant must recline and remain motionless, and the mother must somehow not give away the location of the hidden infant.

What does it mean to "not give away" the location of the infant? First the mother must remain sufficiently far from the infant so that her own location is not a valuable search clue to a predator. Second, a mother should not, by her activity, indicate that she is about to return to her infant. Third, the mother should not reveal the location of the infant by looking directly at it more often than would be expected by chance. When Karen Byers and I tested these hypotheses (Byers and Byers 1983), we found that pronghorn mothers were amazingly effective in fulfilling conditions one and two, but were somewhat imperfect in their tendency to look in the infant's direction too much.

Impressive as these aspects of mother performance were, none seemed to demand cognition as an underlying mechanism. However, we also observed that mothers did something even more sophisticated than the activities described (Byers 1997). In the half-hour before returning to the infant, mothers often engaged in what looked startlingly like a search for hidden predators. A mother with an infant hidden midway up a slope might run to the bottom of the slope, look up and down the dry creek bottom, then run to the top of the slope to stare intently for several minutes before returning to the infant. Mothers that acted like this gave the impression that they anticipated the return to the infant and that they were searching for danger in advance. There is certainly plenty of danger in most years on the National Bison Range in Northwestern Montana, my study site; 75–100 percent of each year's crop of fawns succumbs to either coyotes

or golden eagles (Barrett and Miller 1984; Byers 1997).

One spring I observed an incident that strongly reinforced the notion that mothers had some kind of conscious anticipation of returning to the infant. On a rainy, blustery day I watched a mother who was across a ravine, about half a mile away. As she approached her hidden twins, she was suddenly startled by two golden eagles, flying fast and low over the ridge top. The mother ran away from her fawns and stood, craning her head back, to watch the eagles as they circled overhead. Pronghorn hold their heads back like this only when they are looking at golden eagles. The motion is odd looking and unambiguous. The eagles searched for several minutes, then flew away. The mother waited about 30 minutes, then moved toward her fawns. She was only a few meters away from them when the eagles suddenly reappeared, flying across the ridge top about 1 m above the ground. Once again, the mother jumped away and watched the eagles as they again circled overhead. Once again, the eagles did not find the fawns and departed after several minutes. (Incidentally, the failure of the keen-eyed eagles is testimony to perfection in the hiding behavior of the fawns; not even a tiny ear flick occurred while death circled just overhead.)

Now the mother waited for another half-hour before she moved toward her fawns. Just before she reached them, she stopped, then craned her head back and moved it from side to side, as if looking at eagles overhead. However, the eagles were not present. The mother quickly stepped to her fawns and led them out of sight over the ridge top. In this instance it is almost impossible to avoid the conclusion that the mother anticipated her return to the fawns, remembered the eagles as a threat, and thus searched for them before signaling the fawns to move.

Other evidence that pronghorn mothers have a kind of conscious planning comes from my observations of their interactions with coyotes that are actively searching for a hidden fawn. Mothers always seem to know exactly where the fawn is, and they use this knowledge to determine the proper course of action against a searching coyote. As I indicated, the mother is usually far from the fawn (on average, 70 m). Thus when a coyote approaches and begins to search, the optimal response is not simple. The optimal response will prevent the coyote from detecting the fawn while preserving the option of actively defending the fawn, should the coyote detect it.

The course of action that will accomplish these goals depends upon the locations of the mother, fawn, and coyote, and upon the path that the coyote's movements predict. To see this, envision yourself at 12 o'clock, your hidden fawn at 6 o'clock; the coyote now enters at 3 o'clock, trotting toward 6. If you are a pronghorn mother, you will run to the center of the clock, in front of the coyote. You will then flash your big white rump patch and will prance away in a manner designed to cause the coyote to give chase. Now suppose that you and the fawn are positioned again at 12 and 6, but now the coyote enters at 5, trotting toward 10. The coyote is actually closer to the fawn than it would have been in the preceding example, but now a pronghorn mother likely will simply stand and watch the coyote. I have witnessed these types of interactions scores of times, and always the mother displays the ability to extrapolate from the coyote's path, never showing alarm or an attempt to distract or lure unless the coyote is on an interception course. Under intense pressure, with the life of a helpless infant on the line, pronghorn mothers display a level of cool restraint greater than most humans could maintain. They are able to do this because of their superb ability to remember the exact location of a distant spot and to predict whether the path of another animal will intercept that point.

It is instructive to compare this calculated behavior, which relies on planning and anticipation, with that of killdeer parents, which also practice a distraction display. I have provoked many killdeer displays and have observed none of the re-

straint and anticipation that pronghorn mothers show. For killdeer (*Charadrius vociferus*), the distraction display seems to be triggered simply by my approach within a certain distance. No matter what my path, if I reach the minimum approach distance, the parents swing into their loud distraction display. Thus, the responses of killdeer parents to a threat show no sign of conscious planning or intent, but the actions of pronghorn mothers, in a very similar situation, show restraint and apparent calculation that seems to be driven by a kind of conscious planning.

Planning and Anticipation by Males

And now to the guys: Are they the pelvic-brained morons that the proponents of the "testosterone dementia" concept advance? My observations of pronghorn males suggest that they do indeed care about little except copulation, but that they can be impressively clever as they pursue this elusive goal. To show how they operate, I need to explain a little about the pronghorn mating system.

Females come into estrus once a year and within a population, 90 percent of the estruses occur in a 10-day period, usually in mid-September. For about 2 weeks there is a kind of controlled pandemonium in which females move among potential mates, apparently looking for evidence of vigor, while males attempt to hold and hide groups of females (Byers et al. 1994). Each female makes a sampling visit to several males, which have been solitary and site faithful since May. The female groups that males try to control thus are temporary aggregations; individual females move independently.

Each female moves at an increasing rate as she approaches estrus, and she always leaves a male that fails to defend an adequate perimeter around his group. As a female approaches sexual receptivity, she allows a male to advance toward her, then to attempt to mount, then to mount without intromission over a long, gradually building sequence that typically lasts 24–36 hours. Finally, the female braces back against the male when he mounts, and this allows him to probe for intromission. A male ejaculates immediately as soon as he gains intromission.

The most successful males are those that are able to maintain control over large groups of females for many consecutive days during the rut (Byers et al. 1994; Byers 1997). A successful male may begin his morning by moving through his harem, checking each female for signs of estrus, and directing courtship toward those that smell right. If the checking reveals no females in estrus, the male probably will move away from the female group to scent mark and stare into the distance. If he detects another male, he usually announces his presence with a loud "snort-wheeze" vocalization, and he may chase the other male(s) out of sight. Returning to the harem, he is likely to find that females are starting to drift apart and away; he uses mild threatening gestures to move them back together and usually into his special hiding place. This cycle, which represents essentially continuous activity for the male, may be repeated several times between 7:00 and 11:00 A.M.

When one or more females in the harem comes into estrus, the level of activity becomes much more intense. Other males are drawn to the harem, probably by an odor that the females release. The harem male now courts intensely, runs aggressively at approaching males, sprints back to his harem to court, and so on. In these situations, when one or more females is close to accepting a copulation, and a ring of other males is tightening around the group, male stamina and vigor count for a lot, but male tactical sense is equally important.

Tactical sense is needed to assign priority of performance to mutually exclusive but equally important tasks (chasing males, courting females, reassembling and compacting the harem). Tactical sense also is needed to make decisions about the motion vector that will deal most effectively with spatially distributed threats (a ring of males, each at a different distance and on a separate path around the harem). I have witnessed scores of such situations when the harem male was faced with a daunting array of challenges and possibilities, and I have always been impressed by the ability of the males to choose what appears to be the rationally best course of action out of many possible actions. Often, I have watched a male suddenly pause when faced with a difficult choice, and stand motionless for several seconds as a melee began to erupt around him, then abruptly take action. It was difficult for me to avoid the interpretation that the male was in some way *thinking* about what to do next. An alternative interpretation is that the male was waiting for more information before taking action, but with either interpretation, we are left with an animal that appears to be engaged in a kind of conscious planning of activity. On several occasions, I have observed males lose the opportunity to copulate owing to what I saw as a "stupid" decision. A male might persist in chasing a rival far away when the defended female was very close to accepting copulation. However, such observations are very rare, and their rarity demonstrates that pronghorn males are far more than stimulus-response machines.

My tentative conclusions about pronghorn thinking did not arise from a research program that was designed to study cognition. I was interested in observable behavior and its relation to fitness. However, thousands of hours of observation in nature thrust certain observations upon me. My field observations, of course, can only be suggestive. They do not prove that pronghorn ever think about what they are doing. Worse still, it is difficult for me and probably for most researchers to think of the proper experiments that might produce such proof. Observations such as mine, however, do broaden our view of which species out there are thinking, and of why they might be doing so. There are many ungulate species like pronghorn (Estes 1974, 1991; Gaillard et al. 1998) that face the same challenges that I have described here.

Acknowledgments

My research and preparation of this manuscript were supported by the National Science Foundation (grants IBN 9808377 and DEB 0097115) and the National Geographic Society (grant 6396-98). I thank the U.S. Fish and Wildlife Service (National Bison Range) for cooperation and other assistance.

References

Barrett, M. W. and Miller, L. L. W. (1984). Movements, habitat use, and predation on pronghorn fawns in Alberta. *Journal of Wildlife Management* 48: 542–550.

Byers, J. A. (1997). *American Pronghorn. Social Adaptations and the Ghosts of Predators Past*. Chicago: University of Chicago Press.

Byers, J. A. and Byers, K. Z. (1983). Do pronghorn mothers reveal the locations of their hidden fawns? *Behavioral Ecology and Sociobiology* 13: 147–156.

Byers, J. A., Moodie, J. D., and Hall, N. (1994). Pronghorn females choose vigorous mates. *Animal Behaviour* 47: 33–43.

Carl, G. R. and Robbins, C. T. (1988). The energetic cost of predator avoidance in neonatal ungulates: Hiding versus following. *Canadian Journal of Zoology* 66: 239–246.

Cheney, D. L. and Seyfarth, R. M. (1990). *How Monkeys See the World. Inside the Mind of Another Species*. Chicago: University of Chicago Press.

de Waal, F. (1982). *Chimpanzee Politics. Power and Sex among Apes*. New York: Harper and Row.

Eisenberg, J. F. (1981). *The Mammalian Radiations. An Analysis of Trends in Evolution, Adaptation, and Behavior*. Chicago: University of Chicago Press.

Estes, R. D. (1974). Social organization of the African Bovidae. In *The Behavior of Ungulates and Its Relation to Management*. Vol. 1, V. Geist and F. Walther, eds., pp. 166–205. Morges, Switzerland: International Union for the Conservation of Nature.

Estes, R. D. (1991). *The Behavior Guide to African Mammals: Including Hoofed Mammals, Carnivores, Primates*. Berkeley: University of California Press.

FitzGibbon, C. D. (1990). Anti-predator strategies of immature Thomson's gazelles: Hiding and the prone response. *Animal Behaviour* 40: 846–855.

Gaillard, J.-M., Festa-Bianchet, M., and Yoccoz, N. G. (1998). Population dynamics of large herbivores: Variable recruitment with constant adult survival. *Trends in Ecology and Evolution* 13: 58–63.

Lent, P. C. (1974). Mother-infant relationships in ungulates. In *The Behaviour of Ungulates and Its Relation to Management*, V. Geist and F. R. Walther, eds., pp. 14–55. Morges, Switzerland: International Union for the Conservation of Nature.

6 Can Honey Bees Create Cognitive Maps?

James L. Gould

Honey bees (*Apis mellifera*) have attracted the attention of scientists, philosophers, and the world at large for several reasons (Crane 1983; Gould and Gould 1988). For centuries, they were the only source of a sweetener available year-round in much of the world. Beeswax, too, contributed to the economic importance of honey bees; candles of beeswax burn cleaner than tallow candles and do not sag in warm climates. (Indeed, in some parts of Europe taxes were levied as quantities of beeswax.) The economic importance of honey and beeswax led to practical attempts to understand the behavior and social organization of honey bees, with the very tangible goal of improving the ease and efficiency of harvesting these valuable resources.

Another source of interest was the peaceful and apparently efficient social organization of honey bees, as well as their seemingly selfless work ethic. Countless sermons and philosophical essays took inspiration from this paragon of insect socialism.

Finally, and most important, the techniques developed to study honey bees made the details of their behavior and sensory abilities relatively easy to discover. Thus it was that color vision, ultraviolet vision, polarized-light sensitivity, an internal time sense, sun compensation, polarized-light navigation, the use of backup systems in behavior, and a host of other abilities were uncovered first in honey bees (von Frisch 1967; Gould and Gould 1988). Perhaps the most remarkable of the (then) novel abilities of bees was their dance-language system of communication.

Cognition?

Prior to about 1980, honey bees provided perhaps the best example of intricate innate programming to be seen in nature (Gould and Gould 1982). Their dance language—second only to human language in its ability to communicate information—was one example; their remarkable navigational abilities provided another; and the elaborate innate organization of their flower-learning programming was the most complex instance of species-specific learning known (Gould and Gould 1988). In addition, bee learning displayed many apparent similarities to the learning behavior of vertebrates (Bitterman 1996), inviting comparisons with the seemingly mindless conditioning so extensively studied by Behaviorists (Schwartz 1984).

Yet there were hints that innate wiring might not entirely account for honey bee behavior (Lindauer 1961; von Frisch 1967; Griffin 1976, 1984). The decision-making process in swarming, eerie anomalies during training to a food station, and a too-quick ability to grasp learning tasks combined to sow seeds of doubt. However, asking intelligently whether honey bees might have abilities beyond the basics of instinct and conditioning requires criteria for cognition.

Technically, cognition is knowing or knowledge; by this rather generous standard, innate information provides animals with one level of cognition. To most minds, however, cognition implies an ability to step outside the bounds of the innate, including the innate wiring that permits animals to learn through classical and operant conditioning. It means, instead, a capacity to perform mental operations or transformations and thus to plan or make decisions.

This definition may still be too broad, since some mental operations (such as the ability to infer the sun's position from the polarization of a patch of blue sky, or to compensate for the sun's movement) are hardwired in honey bees (Gould and Gould 1988). There is a real danger of a double standard in such criteria. For instance, one common component of certain sorts of human intelligence tests is the ability to recognize a rotated object. It now transpires that both honey bees and bumble bees can do the same (Gould and Gould 1988; Plowright et al. 2001); so can pigeons (Holland and Delius 1983). Because pigeons are faster and more reliable at this task than humans, the usual interpretation is that the ability must be hardwired in them, and thus it is not a cognitive ability (for pigeons). (The experimental conditions necessary to test bees making visual choices do not allow researchers to measure short response times; thus we do not know whether, like pigeons, bees can judge any rotation with equal speed. Humans, in contrast, take longer to analyze larger rotations, leading to the suggestion that rotation matching is a cognitive rather than a native ability in humans.)

Route Planning

Perhaps the least controversial criteria for cognition are (1) planning of a novel response—often a route to a goal—and (2) concept formation. Both abilities, if they exist in bees, qualify as "cognitive maps" (Tolman 1948). As Tolman, who coined the term, envisioned a cognitive map, it was any mental transformation that enabled an animal to formulate a plan or make a cognitive decision. Later workers have sometimes supposed that some sort of literal map needs to be involved, but the original definition is the one used here.

The route-planning issue is slightly problematic with bees in that they are hard to see when flying. One can judge departure bearings and arrival locations and times, and generally infer that if the departure directions and arrival locations are consistent, and the transit time short, then the animals probably flew directly from the release location to the arrival spot. Another potential problem with this experimental strategy is the poor visual resolution of honey bees: about 1–2° real-time vision (roughly equivalent to 20/2000 human vision, a value that exceeds the threshold for legal blindness), and 3–4° for the

landmark memory that presumably is used in solving displacement tasks (Gould 1987). Thus testing must take into account the possibility that the bees may not be able to infer their location unless large and unambiguous landmarks are clearly visible at the release site.

The first test aimed at discovering whether honey bees could plan novel routes involved training foragers to go a feeding station in a forest clearing 150 m from the hive, and then after several days of regular visitation to this station, capturing these foragers on their departure from the hive (Gould 1986). The foragers were then carried in the dark to a site at the edge of an open field 150 m and 60° from both the hive and the training station; the training station, hive, and release point thus formed an equilateral triangle. The release site was near a large tree that stood alone in the field, away from the forest that lined the field. The kidnapped bees were released one at a time and their release bearings recorded. The mean vector of the release bearings was significantly oriented toward the unseen feeding station.

Initial attempts to repeat this experiment met with mixed results. Tests at undescribed sites failed, except for some bees under overcast conditions (Wehner and Menzel 1990). [Under overcast skies, bees depend heavily on landmarks (Dyer and Gould 1981).] It is possible that these disappointing results were a consequence of failing to provide the bees with large and unambiguous landmarks (Gould 1990, 1991). A second set of tests also provided mixed results. At a site along forest edges, where unambiguous landmarks (as seen by honey bees) would be absent, the foragers were disoriented; at a site at the corner of a woodlot, where their position would be unambiguous to bees familiar with the area, the foragers were well oriented (Dyer 1991). The researcher explained the difference in results in terms of the presumed ability of the bees to see the foraging station from the corner release site; given the visual resolution of bees, this explanation seems unlikely (Gould 1991; Gould and Gould 1995).

The most recent pair of tests have provided clear proof for the ability of honey bees to use novel routes (Menzel et al. 1998, 2000). In both sets of experiments, the researchers performed their tests quite near a prominent landmark (a steep, isolated hill). They used a slightly different technique in which the foragers were kidnapped at the training station and moved to the release site. In the first test, the bees were well oriented upon release from a novel location midway between two familiar training stations. The second test sought to see if this result might be an artifact of the intermediate position of the release site. In this experiment, release sites were chosen from around the compass, up to 180° from the hive. The release bearings were nevertheless well oriented toward the hive. As in the original tests (Gould 1986), bees released from much greater distances were not well oriented.

There are two lessons from these tests. The first is the importance of techniques (in this case, the use of proper landmarks) and an understanding of the sensory limitations of the animal being tested. The second is that bees, like many animals, have redundancy in their navigational systems and thus may use one strategy when it works best and another when the primary strategy cannot be used. Thus the researchers performing the last test described here believe (quite reasonably) that route memorization takes precedence over use of a maplike representation of familiar landmarks in guiding flights. In short, animals have backup systems and an ability to choose the one likely to yield the best results under a given set of circumstances.

What we do not know is how the map of nearby landmarks is first created, stored, and used. The inability to track the complete movements of individual honey bees over the many days during which mapping and testing must occur limits our potential knowledge. So too does the poor visual resolution of bees, which makes

experimentation (for instance, adding, moving, or removing large landmarks) difficult. [This sort of manipulation can be done on a local scale near food sources or the hive (Gould 1987), but it is not at all obvious whether the behavior observed in these contexts is relevant to route planning.]

Concept Formation

Concepts are abstractions that make it possible for animals to solve novel choice problems without prior experience of the specific exemplars offered. For instance, pigeons can learn such concepts as tree, fish, or human (Herrnstein 1984). Alex the parrot can identify the color, material, number, and other characteristics of an object or object set without having seen the object(s) before (Pepperberg 1990). The animals in these tests depend instead on an abstract property or (in the case of pigeons, a set of properties of probabilistic value) that is independent of the exemplar.

Preliminary tests showed that honey bees could learn to recognize and distinguish human letters independent of size, color, position, or font (Gould and Gould 1988). Recent work has focused on more specific concept-related questions. In one set of tests, foragers were taught that symmetrical targets offered food while asymmetrical ones did not (Giurfa et al. 1996); in another set they were taught the opposite lesson. By the seventh visit, the bees could chose the correct novel stimulus over the incorrect one.

The learning curve is different from that of more standard tests in which bees are taught that a particular odor, color, or shape is always rewarded. During concept learning there is no evident improvement over chance performance until about the fifth or sixth test, whereas in normal learning there is incremental improvement beginning with the first test. This delay is characteristic of what has been called "learning how to learn," which is interpreted as a kind of "ah-ha" point at which the animal figures out the task (Schwartz 1984).

The main difference is that honey bees are much quicker at deciphering what the experimenter wants than are pigeons and other standard laboratory animals. Another difference is that the researchers testing honey bees chose to interpret their results as indicating an innate sense of symmetry in bees, and thus imply a noncognitive basis for their results (Giurfa et al. 1996). Of course, on the one hand there is good evidence that human infants prefer symmetrical visual stimuli (Grammer and Thornhill 1994) (which would argue that symmetry is not entirely a learned concept for humans either), while on the other we are still left to wonder what sort of mental leap allowed these bees to understand that this particular concept was the one that the experimenters wanted them to key in on.

Another kind of concept formation and use has been demonstrated in honey bees. In this case, the concepts are "same" and "different." The technique, which is well known from conventional laboratory tests, is delayed match-to-sample (Schwartz 1984). The animal is shown a pattern or color and then is later offered a choice between the same pattern or color and a different pattern or color. If the animal learns the concept, it can be shown a novel initial stimulus and then choose the correct same or different stimulus (depending on which concept is being taught) when presented with the choice. After 30–40 training trials, honey bees began to respond to the "same" or "different" stimulus at levels above chance (Giurfa et al. 2001).

Clearly the time has come to try honey bees with more conventional concept-learning tasks. Such experiments must keep in mind their low visual resolution and their many innate biases in approach; for instance, a preference for "busy" targets—figures with high spatial frequency (Gould and Gould 1988). Perhaps the simplest and most objective tests would focus on number: three petals during training, say, of varying color, shape, and position, followed by a choice between a target with a novel combination of three petals versus two- and four-petaled targets. Extension of the training for letter recognition also seems like a powerful way to probe the potential for abstract concept formation in honey bees.

Conclusion

The evidence that honey bees can perform tasks that are considered to require cognitive powers when they are performed by higher vertebrates suggests at least three possibilities. One is that cognition is a capacity that has evolved as needed among animals, independent of size, number of legs, or whether the creature has an external or internal skeleton. As such, cognitive differences among phyla would be quantitative rather than qualitative (Gould and Gould 1994).

Another (not mutually exclusive) possibility suggested by these observations is that behaviors that require cognition in humans may be innate in "lower" species. Thus it could be that map formation and use by bees is hardwired, using the kind of fill-in-the-blanks strategy so evident in their learning (Gould and Towne 1987). In rodents and primates, on the other hand, the ability is genuinely cognitive; that is, it is not a consequence of innate skills.

A third alternative is that the human capacities we commonly label as cognitive have, at least in part, an unappreciated innate basis (Gould and Gould 1994). The animal kingdom is filled with examples of innately directed learning, including no less an achievement than human language (Gould and Marler 1984). This possibility, for which there is considerable suggestive evidence in the form of species-specific "cognitive" abilities (Shettleworth 1998), brings us back to the basic definition of cognition; by the strictest standards, perhaps there is no genuine cognition in any species, our own included. To the extent that cognition is a product of evolution, we should not be surprised if natural selection has provided a rich set of adaptive biases that help shape cognitive performance.

References

Bitterman, M. (1996). Comparative analysis of learning in honeybees. *Animal Learning & Behavior* 24: 123–141.

Crane, E. (1983). *The Archaeology of Beekeeping.* Ithaca, N.Y.: Cornell University Press.

Dyer, F. C. (1991). Bees acquire route-based memories but not cognitive maps in a familiar landscape. *Animal Behaviour* 41: 239–246.

Dyer, F. C. and Gould, J. L. (1981). Honey bee orientation: A backup system for cloudy days. *Science* 214: 1041–1042.

Frisch, K. von (1967). *The Dance Language and Orientation of Bees.* Cambridge, Mass.: Harvard University Press.

Giurfa, M., Eichmann, B., and Menzel, R. (1996). Symmetry perception in an insect. *Nature* 382: 548–461.

Giurfa, M., Zhang, S., Jenett, A., Menzel, R., and Srinivasan, M. V. (2001). A principle of sameness in an insect. *Nature* 410: 930–933.

Gould, J. L. (1986). The locale map of honey bees: Do insects have cognitive maps? *Science* 232: 861–863.

Gould, J. L. (1987). Landmark learning in honey bees. *Animal Behaviour* 35: 26–34.

Gould, J. L. (1988). A mirror-image ambiguity in honey bee visual memory. *Animal Behaviour* 36: 487–492.

Gould, J. L. (1990). Honey bee cognition. *Cognition* 37: 83–103.

Gould, J. L. (1991). The ecology of honey bee learning. In *The Behaviour and Physiology of Bees*, L. J. Goodman and R. C. Fisher, eds., pp. 306–322. Wallingford, UK: CAB International.

Gould, J. L. and Gould, C. G. (1982). The insect mind: Physics or metaphysics? In *Animal Mind—Human Mind*, D. R. Griffin, ed., pp. 269–298. Berlin: Springer-Verlag.

Gould, J. L. and Gould, C. G. (1988). *The Honey Bee.* New York: W. H. Freeman.

Gould, J. L. and Gould, C. G. (1994). *The Animal Mind.* New York: W. H. Freeman.

Gould, J. L. and Gould, C. G. (1995). *The Honey Bee.* Rev. ed. New York: W. H. Freeman.

Gould, J. L. and Marler, P. (1984). Ethology and the natural history of learning. In *The Biology of Learning*, P. Marler and H. Terrace, eds., pp. 47–74. Berlin: Springer-Verlag.

Gould, J. L. and Towne, W. T. (1987). Honey bee learning. *Advances in Insect Physiology* 20: 55–75.

Grammer, K. and Thornhill, R. (1994). Human facial attractiveness and sexual selection: The role of symmetry and averageness. *Journal of Comparative Psychology* 108: 233–242.

Griffin, D. R. (1976). *The Question of Animal Awareness.* New York: Rockefeller University Press.

Griffin, D. R. (1984). *Animal Thinking.* Cambridge, Mass.: Harvard University Press.

Herrnstein, R. J. (1984). Objects, categories, and discriminative stimuli. In *Animal Cognition*, H. L. Roitblat, T. G. Beaver, and H. S. Terrace, eds., Hillside, N.J.: Lawrence Erlbaum Associates.

Holland, V. C. and Delius, J. D. (1983). Rotational invariance in visual pattern recognition by pigeons and humans. *Science* 218: 804–806.

Lindauer, M. (1961). *Communication Among Social Bees.* Cambridge, Mass.: Harvard University Press.

Menzel, R., Brandt, R., Gumbert, A., Komischke, B., and Kunze, J. (2000). Two spatial memories for honeybee navigation. *Proceedings of the Royal Society of London Series B Biological Sciences* 267: 961–968.

Menzel, R., Geiger, K., Möller, U., Joerges, J., and Chittka, L. (1998). Bees travel novel homeward routes by integrating separately acquired memories. *Animal Behaviour* 55: 139–152.

Pepperberg, I. (1990). Cognition in an African Grey parrot. *Journal of Comparative Physiology* 104: 41–52.

Plowright, C. M. S., Lamdry, F., Church, D., Heyding, J., Dupuis-Roy, N., Thivierge, J. P., and Simonds, V. (2001). A change in orientation: Recognition of rotated patterns by bumble bees. *Journal of Insect Behavior* 14: 113–127.

Schwartz, B. (1984). *Psychology of Learning and Behavior.* 2nd ed. New York: W. W. Norton.

Shettleworth, S. J. (1998). *Cognition, Evolution, and Behavior.* Oxford: Oxford University Press.

Tolman, E. C. (1948). Cognitive maps in rats and men. *Psychological Review* 55: 189–208.

Wehner, R. and Menzel, R. (1990). Do insects have cognitive maps? *Annual Review of Neuroscience* 13: 731–733.

7 Raven Consciousness

Bernd Heinrich

In my most recent research I tried to figure out if ravens (*Corvus corax*) can think; that is, if they have the ability to execute the best solution to a simple but at least novel problem without first being programmed to do it (such as by purely hardwired responses or by trial-and-error learning). Before starting this project, I had not given much thought to the idea of trying to collect data on what may or may not be occurring in an animal's mind, largely because I was skeptical of being able to get such results. My intent here is to provide an overview of a research program that spans a range of taxa with which I have had experience, and to provide my assumptions and approaches. The results, conclusions, and steps in the research have been published elsewhere.

Beginning with the Bees

Starting with insects in the 1960s, I tried to solve questions that involved primarily physiology and evolution, such as: Is body temperature regulated, and if so, how and why? Relatively clear answers could be found through long-standard methods of measuring body temperature, blood flow, energy expenditure, heart and breathing rates, heating and cooling rates and so forth, in the context of comparative physiology. However, when trying to solve puzzles of evolution and adaptation, the ultimate reference is the field, where there is no clear boundary between physiology and behavior. The laboratory situation, because it is controlled and thus contrived, allows discrete answers to the most basic, fundamental questions about mechanisms that, like bricks, build the whole animal.

Thus, at one kind of flower, in one kind of weather, under one condition of the colony, a bumblebee might precisely regulate its thoracic temperature to within a degree of 42°C and have a variable abdominal temperature of 25–30°C.

Change any of these parameters, and the bee's thoracic temperature might be 30°C and its abdominal temperature 10°C, or both temperatures might be regulated near 35–40°C (Heinrich 1979b).

In another taxon the data would most likely be radically different, despite similar underlying generalities that apply to both. Details matter profoundly. The complexity that was revealed in insects hinted at a sophistication that seemed unanticipated and surprising, but it ultimately made sense when seen in terms of the larger picture of adaptation (Heinrich 1993).

Not every potentially relevant factor could be measured. For example, it seemed that a bee exhibited something akin to excitement when it found flowers with a high nectar content. Its breathing rate and body temperature shot up immediately; it flew much faster; its flight tone went from a hum to a buzz; it became more selective in flower choice; and it made more frequent foraging trips. The change of behavior clearly and unambiguously registered that the animal could measure food quality, but whether it might know this consciously, as opposed to reflexively, was of no relevance to the questions I asked or felt I could ask. The behavior could be accounted for in terms of rote learning superimposed on innate programming (Heinrich 1976, 1979b; Heinrich et al. 1977). Bumblebees have a relatively open program concerning which flowers to visit and how to manipulate them to most quickly extract either pollen or nectar (Heinrich 1979a), but within a few flower visits they learn to heed specific flower signals and adjust their foraging routes and flower-handling skills accordingly.

The bees' behavior was, after all, predictable, and much like their physiology, the responses served specific functions either in the context of a predictable environment or predictable changes in the environment. They were ideal organisms for demonstrating often highly intricate evolved responses, including specific learning tendencies, to all sorts of environmental contingencies. Although I saw no evidence that their sometimes complex responses could not be accounted for by programming alone, there was, of course, no objective reason to either exclude or accept the possibility that they consciously "knew" what they were doing after they were doing it.

In the whole animal, various responses are integrated and make sense in terms of a larger program. Thus, the energetics of thermoregulation is a component of foraging behavior, because thermoregulation is primarily used for foraging (Heinrich 1979b). In bees, furthermore, the foraging responses of individuals tie in with the colony's economy and cannot be fully understood except through the perspective of the colony's response in the context of a specific environment. For example, honeybee workers communicate the location and quality of potential food sources to hive mates. Bumblebees, who are "equally" social, do not. The difference is that honeybees, which originated in the tropics, are adapted for harvesting from clumped resources, such as flowering trees. Bumblebees, on the other hand, are tundra- or taiga-adapted animals who forage from widely dispersed flowers where communication is of less importance to the hive's economy (i.e., the queen's reproductive output).

Going to the Birds

This is where the ravens came in. Ravens are well known to be solitary and territorial breeders (Boerman and Heinrich 1999). As such, they should have no apparent advantage, like honeybees, in communicating the locations of food bonanzas. However, since I was myself attracted to a ravens' feast by the birds' loud activity, I was impelled to test whether their vocalizations attracted other ravens. Indeed they did. That is, other birds came to playbacks of vocalizations and then also fed; strictly and objectively defined, the food was being shared. To me, whether the food was being shared "willingly" in the sense of "deliberate" recruitment, or whether recruitment resulted "inadvertently" or from the fact that the birds behaved mindlessly (without knowledge of consequences) but, as in the bees, in a way that was adaptive, was at that point not a relevant question. Other questions had to be answered first: (1) Does their vocal activity draw in others? (2) Do those that are drawn in get to feed? (3) Is there an advantage for those whose vocal activity attracts the others to have them come and feed? The psychological underpinnings of their behavior were surely interesting, but they were out of my realm as a behavioral ecologist.

As in the bees, sharing behavior among ravens could evolve by natural selection. For example, there would be some advantage for ravens to share very rare superbonanzas if they all did it. The biggest theoretical hurdle to this idea was that there seemed to be no mechanism for ensuring "honesty" in what would involve altruistic behavior, given that the raven crowds are not likely to be groups of kin or closed flocks of individuals who know each other and would, furthermore, remember favors and be able to play tit-for-tat.

The research that ensued to try to decipher the ravens' sharing behavior was physically demanding, but perhaps the intellectually most rewarding for me so far. I knew that within the birds' overt behavior lay a huge enigma (Heinrich 1989). At the heart of this puzzle was the question of how or why sharing among strangers, or near strangers, could occur on the basis of self-interest. There had to be an immediate advantage for attracting others to the feast. It turned out, of course, that there was: The sharers were juveniles who got access to new, untested, and hence feared food and/or food defended by more dominant adults (Heinrich 1988; Heinrich and Marzluff 1995). Given this advantage, the other and perhaps later even main advantages (such as sharing the risk of not finding food) could be easily added on as "riders." Recruitment and sharing occurred (Heinrich and Marzluff 1991) even in the unlikelihood of any psychological willingness to share (Marzluff and Heinrich 1991; Heinrich et al. 1993) and it occurred with non-kin (Parker et al. 1994), i.e., without kin selection. These data thus closed the loop on the problem I set out to solve.

"Cognition," used in the sense of at least some conscious knowing with resultant purposive actions, then seemed like a possibility to think about. I had not credited bees with knowing or being conscious of the consequences of their waggle dances and thus performing them because they anticipated the positive consequences (i.e., not doing them if the situation were manipulated to cause negative consequences). Why? Largely because this scenario presupposes that they not only get satisfaction from dancing as such, but that they also get a reward from the consequences of their dance, i.e., seeing others rush out of the hive to forage at the source indicated. Not crediting bees with such—to them—probably superfluous powers, I would therefore not waste valuable research time hoping for positive results in trying to test such a scenario. With ravens, on the other hand, there is a difference—a huge difference. Close observation of various pet birds since my childhood has acquainted me with their emotional nature, a nature that is presumably adaptive (by rewarding fitness-enhancing behavior). Might not a raven be emotionally rewarded if it attracts others and makes it easier for them to feed? And might it therefore also not be motivated to recruit because it anticipates the same psychic and hence later material rewards?

I could not and have not eliminated certain aspects of cognition from the mechanism that we have elucidated by which ravens recruit strangers to food bonanzas and share the food. I do not know what the birds intend or are consciously aware of and what behaviors are reactions to stimuli. However, I am thrilled that sharing can be explained without invoking any motive for sharing, because that makes it all the more remarkable and rational. It is much more convincing and elegant to find a mechanism that allows cooperation to occur as each individual attends to its immediate interests without having to invoke purposive logic (which all too easily can be incorrect in the long term since it is subject to faulty or incomplete information). Nevertheless, this in no way precludes conscious involvement, even though the latter is often a detriment to efficient or rational responses (such as in gambling, for example).

The logic of this behavior (i.e., seeing what is out of sight) is always time bound; there are instant or immediate consequences, consequences hours or days later, and potential consequences for breeding. Awareness in terms of consciousness, if it is present, could be used for deliberate planning for almost any time span. But the first and basic premise of logic is that steps can be tried out in the mind (Heinrich 1996, 1999) and mistakes corrected (see Allen and Bekoff 1997) to achieve an anticipated outcome. How far into time consciousness may extend the reality perceived by any one animal is, however, less relevant to me than experimentally determining whether awareness, as opposed to programming, plays any role for any time span at all, in any animals other than ourselves.

Bees cannot rely on conscious planning for the future in storing pollen and honey, etc. (what if some forget the locations?) anymore than any animal can safely rely on having sex solely on the rational basis of trying to produce offspring. The ultimate rewards must be subservient to stronger, move immediate rewards when the intervening steps are long, arduous, and complex.

On the other hand, it may be quite difficult to preprogram a squirrel to choose the best route through unpredictable mazes of branches to a nut. Conscious planning by mentally trying out a number of possible routes would likely be simpler and more reliable. Even jumping spiders appear to be capable of pursuing prey that is out of sight when they invade the webs of other spiders (Jackson and Wilcox 1993, 1998); they use indirect routes and change their tactics as required.

My first intimation that ravens have some sort of awareness of immediate consequences, which is necessary for conscious planning, that would then guide their actions concerned their food-caching behavior (Heinrich and Pepper 1998). Since I had numerous birds in a large outdoor aviary, it was an education to observe their interactions. Bees could, through programming, execute impressive behaviors, but the ravens acted as though they could gauge the results of their actions even before they executed them; they altered their responses moment to moment, depending on what was happening. When some birds went out of their way to bury excess food, others tried to follow them even though the food was carried out of sight in the gular pouch. The followers (if subordinate individuals) acted surreptitiously, and they did not venture near the others' hidden food until the latter had left the area. The cachers (if dominant) in turn either attacked the raiders (but not others) when they came near the caches, or relocated their caches after they had been watched.

Nevertheless, as much as all of this behavior looked as if each bird knew what the other was going to do, it was still possible that the birds did not "know," in the sense of anticipating others' actions, until after they had taught themselves or had learned from experience. Of course, as in our own learning behavior, the birds may become conscious ("knowing") after learning the consequences of specific actions, so the conservative criterion of a test of cognition (knowing without prior learning) was not met. Nonetheless, given my day-to-day observations of the ravens, I eventually wondered if they might know something even though they had not learned it or had not been genetically programmed to know it. In short, I wondered if they could go through behavioral steps in the mind without also committing the body to the same steps first. If so, they could perform the equivalent of trial-and-error learning in their heads, thereby avoiding the commitment of many errors.

It would not have occurred to me to present naive birds with a test involving food dangled from a perch on an almost meter-long string if it were not for my close observations of ravens caching, which sometimes suggested deliberateness and hence potential awareness. The food on a string puzzle (Heinrich 1996, 1999, 2000) was ultimately presented to a series of different ravens that had been reared from nestlings.

Prior to the test, these birds had never seen food or some other object dangled by a string, so that I could examine the details of their behavior on their very first exposure. Could they perform dozens of consecutive steps that had to be executed in a very specific sequence? Could they reach down from their perch, grasp the string, pull it up, lay it on the perch, step onto the string before releasing it with the bill, then apply variable pressure to hold the string fast to the perch while reaching down again and repeating the exact steps several more times in succession? Completing the whole task would require that they get a psychic reward not only from anticipation and/or eating the food, but also from completing each of the proper intermediary steps in a sequence that made their ultimate eating of food more likely. In short, no satisfaction could be gained from proximally unrewarding steps unless the ravens realized (i.e., understood) that what they were doing contributed to their objective (Craig 1918; Timberlake and Silva 1995).

Furthermore, with this test, I could cross check what they knew or did not know by prior training to say, red string, and seeing if they were then conditioned to respond to red string or would preferentially pull up food provided for the first time on a green string. The strings could be crossed to see if the birds' concept of reaching the food was to pull up the string "above food" or "attached to food." I could arrange the string so that they had to pull down on a string to have food come up. I could determine if they knew the food was attached to a string by forcibly chasing them off the perch after they had pulled the food up to see if they would fly off with food that was tied on.

In short, the string test provided opportunities for obtaining a wealth of information where the relative contributions of innate behavior, learning, and cognition could all be at least partially teased apart. Obviously any one behavior contains some aspects of all three, but my main objective was to be as conservative as possible, to see if one could rigorously prove that at least some cognition involving consciousness was involved. The results (Heinrich 1996, 1999, 2000) could not be plausibly explained by the alternatives (random chance, rote learning, or innate programming) as sole explanations for the ravens' problem-solving behavior.

Future work should include other birds, especially other corvids. Other birds with similar body construction should be physically just as capable as ravens are in performing the same task. Can they solve the same puzzle? If not, then why not? Future work will also test whether ravens can keep track of objects that are out of sight, a prerequisite for conscious planning. We already know that ravens routinely keep track of food that others (other ravens and humans) hide. But can they project the path of a moving object that is out of sight (such as a rodent moving through an opaque tube)? By these and other tests, conducted with a variety of taxa, we hope to elucidate one of the perhaps most variable phenomena in the animal kingdom, the ability to solve problems by the application of consciousness, which has eloquently been suggested by the numerous animal studies summarized by Griffin (1998).

References

Allen, C. and Bekoff, M. (1997). *Species of Mind*. Cambridge, Mass.: MIT Press.

Boerman, W. I. and Heinrich, B. (1999). *The Common Raven*. (The Birds of North America series edited by A. Poole.) Washington, D.C.: National Academy of Sciences.

Craig, W. (1918). Appetites and aversions as constituents of instincts. *Biological Bulletin* 34: 91–107.

Griffin, D. R. (1998). From cognition to consciousness. *Animal Cognition* 1: 3–16.

Heinrich, B. (1976). Foraging specializations of individual bumblebees. *Ecological Monographs* 46: 129–133.

Heinrich, B. (1979a). "Majoring" and "minoring" by foraging bumblebees, *Bombus vagans*: An experimental analysis. *Ecology* 60: 245–255.

Heinrich, B. (1979b). *Bumblebee Economics*. Cambridge, Mass.: Harvard University Press.

Heinrich, B. (1988). Winter foraging at carcasses by three sympatric corvids, with emphasis on recruitment by the raven, *Corvus corax*. *Behavioral Ecology and Sociobiology* 23: 141–156.

Heinrich, B. (1989). *Ravens in Winter*. New York: Simon and Schuster.

Heinrich, B. (1993). *The Hot-Blooded Insects: Mechanisms and Evolution of Thermoregulation*. Cambridge, Mass.: Harvard University Press.

Heinrich, B. (1996). An experimental investigation of insight in common ravens, *Corvus corax*. *The Auk* 112: 994–1003.

Heinrich, B. (1999). *Mind of the Raven: Investigations and Adventures with Wolf Birds*. New York: Harper Collins.

Heinrich, B. (2000). Testing insight in ravens. In *The Evolution of Cognition*, C. Heyes and L. Huber, eds., pp. 289–305. Cambridge, Mass.: MIT Press.

Heinrich, B. and Marzluff, J. M. (1991). Do common ravens yell because they want to attract others? *Behavioral Ecology and Sociobiology* 28: 13–21.

Heinrich, B. and Marzluff, J. M. (1995). How ravens share. *American Scientist* 83: 342–349.

Heinrich, B. and Pepper, J. (1998). Influence of competitors on caching behavior in the common raven, *Corvus corax*. *Animal Behaviour* 56: 1083–1090.

Heinrich, B. and Smolker, R. (1998). Play of common ravens (*Corvus corax*). In *Animal Play*, M. Bekoff and J. Byers, eds., pp. 27–44. Cambridge: Cambridge University Press.

Heinrich, B., Mudge, P., and Deringis, P. (1977). A laboratory analysis of flower constancy in foraging bumblebees: *B. ternarius* and *B. terricola*. *Behavioral Ecology and Sociobiology* 2: 247–266.

Heinrich, B., Marzluff, J. M., and Marzluff, C. S. (1993). Ravens are attracted to the appeasement calls of discoverers when they are attacked at defended food. *The Auk* 110: 247–254.

Jackson, R. R. and Wilcox, R. S. (1993). Observations in nature of detouring behaviour by *Portia fimbriatat*, a web-invading aggressive mimic jumping spider from Queensland. *Journal of Ecology, London* 230: 135–139.

Jackson, R. R. and Wilcox, R. S. (1998). Spider-eating spiders. *American Scientist* 86: 350–357.

Marzluff, J. B. and Heinrich, B. (1991). Foraging by common ravens in the presence and absence of territory holders: An experimental analysis of social foraging. *Animal Behaviour* 42: 755–770.

Parker, P. G., Waite, T. A., Heinrich, B., and Marzluff, J. M. (1994). Do common ravens share food bonanzas with kin? DNA fingerprinting evidence. *Animal Behaviour* 48: 1085–1093.

Timberlake, W. and Silva, K. (1995). Appetitive behavior in ethology, psychology and behavior systems. In *Perspectives in Ethology*. Vol. 11, *Behavioral Design*, N. S. Thompson, ed., pp. 211–253. New York: Plenum.

8 Animal Minds, Human Minds

Eric Saidel

Every consideration whatsoever which contributes to my perception of ... any other body cannot but establish even more effectively the nature of my own mind.
—Descartes (1641/1984, p. 33)

What Do Minds Do?

My research in animal cognition is centered on questions about the role(s) representations play in causing animal behavior. I ask if animals represent the world, or if they simply respond to it, and then, for animals that do represent the world, if those representations—specifically the content of those representations—are involved in their behavior. Exactly why I ask these questions, why I take the approach I do toward answering them, and what I think the questions and answers have the potential to reveal about all animal minds—not just the minds of non-human animals—is an involved tale.

Philosophers typically answer questions by asking questions. When asked if animals have minds, or asked about the nature of animal minds, the philosopher will want to start by asking what a mind is, for only when we know what a mind is can we know if animals have minds, or what animal minds are like. But the answer to that question is at best an ending point, not a useful starting point. Consider a similar question.

Is Duchamp's Fountain (a urinal that is like any other factory-made urinal except for the artist's signature) a work of art? In order to answer this question, we might examine paradigm examples of art for common properties and then ask if Duchamp's Fountain shares these properties. However, this could lead us astray by focusing our attention on shared properties that are irrelevant. More important, the examples of art are paradigms not simply because they are clearly works of art, but also because they do not challenge our preconceived notion of what a

work of art is. This method would not permit challenges to that notion to be ratified as art.

Another way to answer the question would be to theorize about what property (or properties) a work of art has that makes it art. However, this approach is too essentialist; it makes an object a work of art if it has one (or more) essential property. It might be the case that an object is art if it has some of a range of properties, none of which are shared by all works of art. Furthermore, our choice of property is going to be determined largely by what we think are the paradigm works of art. For example, if we are strongly influenced by the Old Masters, we might think that having the right subject or meeting certain formal standards is necessary for something to be a work of art. The conclusions we draw are determined by the paradigm works of art with which we started.

Still another way to answer this question would be to ask what it is that a work of art does or is supposed to do. Then things that fulfill that function are works of art. Call this a functional definition of art. Such a definition has the advantage of freeing us from the constraints that are invariably attached to definitions of art that are derived from paradigms. This allows challenges to our preconceived notion of what a work of art is.

The pitfalls we face when trying to define "mind" are similar to those we face in trying to define "art." We might start by looking at what we think of as the paradigm of a mind: the human mind. Then we might claim that an object is a mind if it is sufficiently similar to this paradigm. However, this raises several problems similar to those described above. Any paradigm-driven definition is going to be a slave to the properties shared by the paradigms; it will not be able to encompass things that differ strongly from the paradigms. It may be that things that are wildly different from human minds are not minds, but we should not begin our investigation into the nature of animal minds by assuming this. It should be something we discover as a result of our empirical work rather than a boundary condition that we impose on our work at the outset. Nor should we adopt the second approach, choosing properties that are essential for being a mind. This also binds us to our preconceived notions of what is and is not a mind. So perhaps here too we should adopt a functional definition; something is a mind if it does what a mind does (or if it is supposed to do what a mind is supposed to do).[1]

This is a step in the right direction, but it leaves us with another important question: What do minds do? I am not going to try to answer this question, although I do suggest some answers. [See Godfrey-Smith (1996) for a discussion of the function of minds.] Not only is it as difficult to answer as the question "What is a mind?" it is also, as I argue here, a question with many correct answers. However, it is a more helpful question; it opens the door to fruitful research questions. Once we postulate a trait (an ability) that some minds have, we can ask if particular organisms have that trait. Then we can ask how the organisms accomplish that task. This tells us more about cognition in that animal and more about cognition in general. [See Allen (forthcoming) for a discussion of the importance of focusing on traits.]

Of course, this functional definition is just as mired in our anthropomorphic sense of what a mind is as are the paradigm-based and essential property definitions. Here that is acceptable; our interest in minds is an interest in the nature of things that accomplish tasks similar to those our own minds accomplish. Among the things we want to know are how they manage to accomplish those tasks, what aspects of the mind are essential for accomplishing those tasks, and why minds are organized the way they are. Our interest, that is, is not confined to things that do what our minds do in the way our minds do; we are more broadly interested in things that may

do what our minds do, in any way that they do them. In investigating the minds of other creatures it may be impossible to escape anthropomorphism in our assessment of what minds do, but that anthropomorphism frees us from thinking that only things like our own minds are indeed minds.

Does this mean that all things that do what minds do are themselves minds? If one cognitive function is to gather information from the environment and categorize it so that it can be acted upon at some later date, then are we obliged to say that simple machines that are sensitive to only a few aspects of the environment, but are able to act on that information, have minds? For example, in the late 1980s the MIT Mobile Robot Laboratory designed a robot (dubbed "Herbert") that was able to collect empty soda cans from the tops of tables in a laboratory (Connell 1989; for discussion, see Clark 1997 and Saidel 1999). Herbert accomplishes a task that human beings use their minds to accomplish. Does that mean that Herbert has a mind and uses it when he is collecting empty soda cans?

Rats are apparently able to navigate through complex terrains, even to the extent of quickly finding alternative routes when their preferred route is blocked (Gallistel 1980, pp. 334–354). Again, this is a task that humans accomplish by utilizing various mental abilities. Does it follow that the rats have minds?[2]

This is at best an unhelpful question. Instead of focusing on whether an animal has a mind (something that depends solely on whether they have minds, not on the nature or variety of their cognitive abilities, or the nature of our evidence in support of the proposition that they have minds), we should focus our attention on the evidence and what that evidence reveals about the cognitive processes of that animal. That is, we should worry about the nature of the evidence, what the evidence tells us about the animal we are studying, and what it tells us about mental traits more generally. We can look at the evidence, not as evidence for some abstract organ—

identified with the mind—but as evidence for some simpler trait, one of the many that goes into composing the mind. Thus, in the case of the rats and Herbert, I want to ask if there is evidence that representations of the world are part of the causal nexus of their behavior.

This question is important in part because philosophers have long thought that intentionality—the characteristic that representations have of being about something—is the mark of the mental. More important than that (especially because to look for a sign that something is a mind is to embrace the vague pursuit of minds over the more productive pursuit of cognitive capacities), representation introduces a special kind of distance between the organism and the world. When an organism uses representations, it is not responding to the world, but instead is responding to the way it *pictures* the world. Perhaps the best way to understand this is in contrast to von Uexküll's *Umwelt*. The *Umwelt* of an organism is defined by its perceptual abilities and limitations; an ant detects and removes deceased members of its nest by detecting folic acid, something to which members of other species (e.g., human beings) are blind. The folic acid is part of the ant's *Umwelt*, but not the human's. However, the ant does not need to represent its nestmate as dead, it only needs to respond to the folic acid it perceives.

Organisms that navigate the world without representations are merely responding to the cues to which they are sensitive. Organisms that use representations are responding instead to the way they represent the world. [Von Uexküll's interest in the *Umwelt* of an organism was in the way that the *Umwelt* meshed with the organism's structure to create an internal world shaped by those features of the external world to which the organism was sensitive. As I read von Uexküll, this *Innenwelt* is not quite the same thing as the representational world discussed here. The *Innenwelt* is the way the organism's structure responds to the *Umwelt*, whereas the representational world introduces an even greater distance between an organism and a world, so that

the organism is responding, not to its *Umwelt*, but to its representation of its *Umwelt*; see von Uexküll (1909/1985).] This layer introduces the possibility of error (which means that organisms that use representation must be able somehow to overcome error) and the possibility of overcoming obstacles because an organism that uses representation has the ability to relinquish the means to a goal while remaining directed toward that goal; see Saidel (1998). Thus, by focusing on representations, I am focusing on the cognitive ability some organisms apparently have to interact plastically with their environments, to experiment, and to make mistakes.

Another advantage of focusing on what minds do—in this case, represent—is that it prompts more directed questions. For example, in looking for evidence that an organism uses representations, we might look for evidence that it is able to overcome obstacles. Or we might look for evidence that a behavior that is not productive is not simply triggered by a misleading cue in the environment, such as folic acid painted on a pebble. In addition to focusing our attention on more fruitful questions, the functional approach has the added benefit of reminding us that when we think about cognition, and when we wonder about the nature of animal cognition, we are often not interested in the question abstractly. There are certain aspects of cognition that we find more interesting than others. By focusing on more specific questions, we highlight those questions that are really driving our concerns. And by asking the questions in a more focused way, we ensure that we remember that research into the nature of animal cognition is comparative research. In answering our questions it helps to look at the evidence displayed by members of different species.

Animal Minds, Human Minds

I want to ask about the role that representations play in rats' behavior. If rats do indeed have the ability to find alternative routes to a goal when a preferred route is blocked, that is, if they have the ability to remain directed toward a goal while relinquishing various means to achieving that goal, then it seems likely that representations are causally responsible for their behavior to at least some extent. Contrast this with the robot Herbert's behavior. Herbert is unable to overcome obstacles; he merely moves about his environment, retrieving cans. Having a representation of his environment would not change the way he sweeps the environment for cans.

What might these questions and answers reveal about the nature of the internal psychological states of rats (and of Herbert)? At least this: We learn that the evidence suggests that some of the rats' behavior is guided by representations, rather than being merely a response to the world. That suggests that the rats have internal psychological states. If their behavior were directed by the nature of the world rather than by their representations of the world, no inner cognitive interpretation of the world would be implicated. That means there would be no evidence for inner psychological states. That is the case for Herbert.

However, when we wonder about the nature of an organism's internal psychological states, we are seldom wondering about something as minimal as the presence of a representational layer. Instead we are often wondering about the nature of that layer—the richness of the representations, the degree to which those representations are causally responsible for the organism's behavior, and the structure of those representations. These worries, I suggest, are comparative worries. That is, we may consider the ways in which representations are causally responsible for human behavior and ask if they are causally implicated in rat behavior in the same way. For example, do rats represent pictorially or do their representations have a combinatorial, language-like structure? Or we might compare the behavior of rats with the behavior of another organism that we have reason to believe uses representations and ask how their use of representations is

similar and how it differs. Doing so will help us learn about the minds of both organisms, and it will help us learn about the nature of mind in general.

My interest is more specific; I am curious about the nature of human minds. The answers we find when exploring the role that representations play in the behavior of rats (and other organisms) help us learn in several different ways about human minds: (1) We learn something about the nature of the evidence that representations are active in causing human behavior. This can help us learn when human behavior is properly explained by reference to representational states and when reference to such states goes beyond the evidence. (2) This helps us recognize other ways of achieving behavioral goals (the results toward which behavior is aimed, either by the agent or by evolution) than by using representations, some of which may be part of the human behavioral repertoire. (3) We can also use what we learn about the minds of rats to determine what the structure of human minds might be and why, evolutionarily, human minds might have that structure. Understanding this can help us understand better how human minds work.

This is an example of a more general strategy. By asking specific questions about the specific abilities of other organisms, we are able to learn about the different means that organisms use to achieve similar goals. This sheds light on the means that human beings[3] use to achieve those goals, and it helps us understand the evolutionary history of that particular human cognitive ability. That in turn helps us understand how the human mind is structured and why it is structured that way. And that in turn helps us understand the nature of the human mind.

This, for me, is the grail, but chasing this grail has side benefits as well. Learning about the varied nature of animal minds shows us that a mind is not a simple organ, but is instead a complex mosaic of traits (Allen and Saidel 1998). There are other ways of being a mind than the way that human minds are organized. Recognizing this helps us down from our self-imposed pedestal and lets us see that other animals not only have minds that are different from ours but also that they have mental abilities that human beings lack. And, as is the case with art, recognizing that the world of minds is far greater than we previously supposed can be a wonderful, stimulating revelation.

Notes

1. It would be a mistake to say that only things that do what minds do are minds, for that would mean that malfunctioning minds are not minds. That something has a particular function is a fact about the properties and abilities of that thing, or things of the same type, not a fact about how we may decide to think of that thing. For example, I cannot simply decide by fiat that my toaster oven has the function of warning me when telemarketers are phoning. Thus some object is a mind if it reasonably has as (one of) its function(s) a mental function.

2. One of these questions is about the mental abilities of a nonhuman animal, the other is about the mental abilities of a machine. Some might want to think of these as different areas of research, but I do not see the fruitfulness of such an arbitrary (I think) boundary. In the case of both the nonhuman animal and the machine we ask the same key questions: Does this behavior provide evidence that this thing is using representations? What would such evidence look like?

3. Or other organisms; the comparative strategy I am outlining can be used to learn about the mind of any organism.

References

Allen, C. (forthcoming). Real traits, real functions? In *Functions and Functional Analysis in the Philosophy of Biology and Psychology*, A. Ariew, M. Perlman, and R. Cummins, eds., Oxford: Oxford University Press.

Allen, C. and Saidel, E. (1998). The evolution of reference. In *The Evolution of Mind*, D. Cummins and C. Allen, eds., pp. 183–203. Oxford: Oxford University Press.

Clark, A. (1997). *Being There*. Cambridge, Mass.: MIT Press.

Connell, J. (1989). A Colony Architecture for an Artificial Creature. Technical Report II 5 1, Artificial Intelligence Laboratory, Massachusetts Institute of Technology, Cambridge, Mass.

Descartes, R. (1641/1984) Meditations on first philosophy. In *The Philosophical Writings of Descartes*. Vol. II. (J. Cottingham, R. Stoothoff, and D. Murdoch, trans.) Cambridge: Cambridge University Press.

Gallistel, C. R. (1980). *Organization of Action: A New Synthesis*. Hillsdale, N.J.: Lawrence Erlbaum Associates.

Godfrey-Smith, P. (1996). *Complexity and the Function of Mind in Nature*. Cambridge: Cambridge University Press.

Saidel, E. (1998). Beliefs, desires, and the ability to learn. *American Philosophical Quarterly* 35: 21–37.

Saidel, E. (1999). Critical notice of Andy Clark's *Being There*. *Canadian Journal of Philosophy* 29: 299–318.

Uexküll, J. von (1909/1985). Environment (*Umwelt*) and the inner world of animals (C. J. Mellor and D. Gove, trans.). In *The Foundations of Comparative Ethology*, G. M. Burghardt, ed., pp. 222–245. New York: Van Nostrand Reinhold. (Reprinted from J. von Uexküll, 1909, *Umwelt and Innenwelt der Tiere*. Berlin: Jena.)

9 Comparative Developmental Evolutionary Psychology and Cognitive Ethology: Contrasting but Compatible Research Programs

Sue Taylor Parker

Comparative developmental evolutionary psychology (CDEP) is the name I have given to the kinds of studies my colleagues and I have done (Parker 1990; Parker and Gibson 1990; Parker et al. 1994; Byrne 1995; Russon et al. 1996; Parker et al. 1999). My work in this area has focused on the following questions: What are the patterns of similarities and differences in cognitive abilities among humans, apes, and monkeys? How, when, and in which species have these patterns evolved (Parker and Gibson 1977, 1979)?

In addition, I have focused on the similarities and differences in the developmental extent and timing of these abilities among primate species (Parker 1977). In accord with their developmental focus, my comparative studies and those of my colleagues have used a variety of frameworks from developmental psychology, including Piagetian and neo-Piagetian stages, to compare abilities across cognitive domains.

Also, in accord with their evolutionary focus, studies of this kind have used evolutionary methodologies. I have used cladistic methods to identify cognitive adaptations and to pinpoint their origins; I have used heterochronic concepts to reconstruct evolutionary changes in the extent and timing of cognitive development. These studies have revealed a pattern of terminal addition of new stages of cognitive development in ape and human ancestors as well as a pattern of accelerated rates of cognitive development in humans compared with great apes (Parker 1996; Parker and McKinney 1999).

Beginning in the 1970s and 1980s, comparative developmental evolutionary psychology has grown up coincidentally with, but largely isolated from, the development of a parallel field known as cognitive ethology (CE). This essay compares and contrasts cognitive ethology and comparative developmental evolutionary psychology. It suggests ways each of the two research programs could benefit by adopting elements of the other and how both programs could benefit from strengthening their ties with evolutionary biology. It also briefly contrasts these programs with that of evolutionary psychology (EP).

Both cognitive ethology and comparative developmental evolutionary psychology are based in evolutionary biology, particularly animal behavior. As such, both focus to varying degrees on Tinbergen's (1963) four kinds of problems in the study of animal behavior: proximate causation, ultimate causation or adaptive significance, phylogenetic history, and ontogeny. These in turn map onto various subfields of evolutionary biology, including genetics, physiology, ecology, systematics, and phylogenetics. Table 9.1 summarizes these relationships.

Both CE and CDEP reject behaviorism and embrace folk psychology to address animal mentalities (Jamieson and Bekoff 1996). Researchers in both programs therefore have had to respond to charges of anthropomorphism (Mitchell et al. 1997). They differ primarily in the psychological frameworks and methods they have adopted. Whereas cognitive ethologists use frameworks from cognitive psychology, CDEP researchers use frameworks from developmental psychology.

The origins of cognitive ethology have been traced to Donald Griffin's (1978) papers on animal awareness, which explicitly turned the attention of ethologists and animal behaviorists to questions of animal minds and animal consciousness (Ristau 1991). It is fair to say that Griffin's work was grounded in the concepts of species-specific animal learning (Hinde and Stevenson-Hinde 1973; Roitblat et al. 1984) that emerged out of the ethology-comparative psychology wars in the 1950s and 1960s.

Cognitive ethologists differ from both comparative psychologists and animal behaviorists and from classical ethologists in focusing on animal consciousness, awareness, and intention-

Table 9.1

Treatment of Tinbergen's (1963) four complementary lines of inquiry by cognitive ethology, evolutionary psychology, and comparative developmental evolutionary psychology

Subdisciplinary bases of the four lines of inquiry	Cognitive Ethology	Comparative Developmental Evolutionary Psychology	Evolutionary Psychology
Proximate causation: molecular biology, physiology and anatomy, animal behavior	Yes	Yes	Yes, but weakly so
Ultimate causation or adaptive significance: genetics, behavioral ecology	Yes, strongly so	Yes, but weakly so	Yes
Evolutionary history: comparative paleontology, animal behavior, phylogenetics	Yes, but rarely	Yes, but often poorly	No
Ontogeny: developmental biology, evolutionary developmental biology	Yes, but rarely	Yes	No

ality. Burghardt (1997) has gone so far as to suggest that understanding private experience should be a fifth aim of ethology. Critiques of CE have focused on the difficulties of defining consciousness operationally (Dawkins 1995; Bekoff and Allen 1997).

Cognitive ethology has also been influenced by concepts of information processing from the newly emerging field of cognitive psychology (Newell 1990). Consistent with learning theory and information processing theory, it focuses primarily on species-typical modes and mechanisms of information processing, rather than on development (Hoage and Goldman 1986; Ristau 1991).

Consistent with the discipline's origins in ethology, the most salient feature of cognitive ethological studies is their focus on the behavior of animals in their natural habitats. This explains their strong focus on the ultimate causality or adaptive significance of cognition. The resulting realization that each species is uniquely adapted to its peculiar niche may have discouraged systematic comparative studies of cognition using an integrated framework (Bailey 1986).

In contrast, the origins of CDEP can be traced (Parker 1990) first to the use of models of children's language acquisition by ape language researchers (Gardner and Gardner 1969; Patterson 1980; Miles 1983; Gardner et al. 1989; Savage-Rumbaugh et al. 1989) in the late 1960s and the 1970s and 1980s. Second, it can be traced to subsequent use of stage models of cognitive development in human infants and children (Jolly 1972; Chevalier-Skolnikoff and Poirier 1977; Redshaw 1978; Antinucci 1989; Parker 1977, 1990). CDEP is based in comparative psychology and animal behavior, more specifically that arising from primatology and biological anthropology.

It differs from cognitive ethology, however, in its lesser emphasis on behavior in natural settings. Most CDEP studies done to date have been on captive animals, some in colonies, some cross-fostered, and others in laboratory settings. There are some important exceptions, however, including work on cognition in wild chimpanzees (Boesch and Boesch 1984; Boesch 1991a,b, 1993; Boesch and Boesch-Achermann 2000; Matsuzawa 1994).

CDEP also differs from cognitive ethology in comparing primate cognition in terms of achievements of developmental stages within and across traditional domains of cognition (physical, logical-mathematical, and social) in humans. Physical cognition includes knowledge of objects, space, and causality; logical knowledge includes classification, seriation, and number; social knowledge includes imitation, pretend play, self-awareness, and theory of mind.

Most CDEP studies have focused on the highest levels of species-typical abilities achieved by monkeys or apes relative to those of human children (Chevalier-Skolnikoff and Poirier 1977; Mitchell and Thompson 1986; Whiten and Byrne 1988; Antinucci 1989; Parker and Gibson 1990; Whiten 1991; Boysen and Capaldi 1993; Parker et al. 1994; Russon et al. 1996; Whiten and Byrne 1997; Parker et al. 1999). These comparative studies of the terminal levels of development achieved by related species add a new dimension to comparative psychology, first because they allow systematic comparisons among related human and nonhuman primate species, second because they can be used to reconstruct the evolutionary origins of specific cognitive abilities (Chevalier-Skolnikoff 1976; Parker and Gibson 1977; Parker 1991; Povinelli 1994; Byrne 1995).

Comparative studies that describe the pace and sequence of cognitive development across diverse domains and subdomains further increase the heuristic power of CDEP studies because they allow systematic comparisons of the sequence in which knowledge in various domains develops among related primates and the pace or speed at which it develops. These comparative developmental data also provide material for the reconstruction of patterns of heterochrony in the evolution of cognitive development (Parker 1996).

Heterochrony refers to changes in the pace and/or timing of development in descendant and ancestral species (Gould 1977; McKinney and McNamara 1991, 1997). Its component processes can produce significant changes among related species in a short time, with relatively few mutations. The nature of these changes can be inferred from comparative data on the timing and pace of development in related species. The identification of abilities that are shared among an in-group of closely related sister species such as the great apes, but not by the next most closely related out-group such as lesser apes or Old World monkeys (shared derived character states), provides the basis for this analysis (Brooks and McLennan 1991).

Specifically, comparative developmental data imply that human cognitive development entailed the addition of several new subperiods of cognitive development following divergence from our common ancestor with chimpanzees. These are Piaget's late preoperations, early and late concrete operations, and early formal operations subperiods of cognitive development (Piaget and Inhelder 1969). Second, human development entailed the elaboration and acceleration of late sensorimotor and early preoperations subperiods compared with those of the great apes (Parker 1996; Parker and McKinney 1999). Third, it entailed the realignment of developmental patterns, resulting in more synchronous development across domains (Langer 2000a,b).

In contrast to CDEP researchers, cognitive ethologists, like classical ethologists, have studied a broad range of vertebrate species, With some notable exceptions (e.g., Herzog et al. 1992; Bekoff 1995), however, as with their disciplinary cousins the comparative psychologists, cognitive ethologists have studied distantly related model species such as the white rat, the pigeon, and the rhesus monkey (Beach 1965). These species were selected for convenience of study rather than for clades of closely related sister species, such as the great apes, that share adaptations because of a recent common ancestry (see Parker and McKinney 1999 for references). The breadth and selection of their subjects has generally precluded phylogenetic reconstruction of the evolution of characteristics in related clades (Martins 1996).

Ironically, given its more limited scope, the CDEP focus on a clade of closely related primate species has facilitated efforts to reconstruct cognitive evolution. Likewise, its focus on comparative development has facilitated efforts to reconstruct the evolution of cognitive development in apes and humans through heterochrony (Parker 1996; Parker and McKinney 1999).

Although they are beginning to engage in phylogenetic reconstruction, comparative developmental psychologists often lack training in framing and testing adaptive hypotheses and in reconstructing the evolution of character states. The extension of CDEP training to include life history theory, cladistics, and phylogenetics would greatly aid these efforts (Parker and McKinney 1999).

The boundaries between CE and CDEP—and also between these approaches and those of comparative (CP) and evolutionary psychology—are somewhat fluid. Many of the same topics are investigated by researchers in these four groups, including language and communication, imitation and other forms of social learning, culture, theory of mind, spatial cognition, number, deception, and object concepts. Some researchers, primarily by virtue of their taxonomic focus on primates, cross boundaries of the subfields (e.g., Cheney and Seyfarth 1990; Vauclair 1996; Tomasello and Call 1997). Evolutionary psychologists tend to focus primarily on the ultimate causes or adaptive significance of behaviors, but they also study such proximate factors as facial symmetry and hourglass figures involved in intersexual choice, and they postulate mental modules that mediate these and other behaviors (e.g., Geary 1998). The similarities and differences between CE and CEDP are summarized in table 9.2.

Finally, CE and CEDP researchers have shared experiences of attacks from comparative psychologists. Most notable in the case of CEDP were the ape language wars (Terrace et al. 1979; Sebeok and Rosenthal 1981). This attack effectively limited funding for research in acquisition of language by apes. Similarly, comparative psychologists have harshly criticized CE researchers for their focus on consciousness (Bekoff and Allen 1997). Likewise, tensions between neo-Piagetians and neoinnatists in developmental psychology (Fischer and Bidell 1991) promise to extend to studies of cognition in nonhuman animals as neoinnatist methodologies are adopted by students of primate cognition.

Neoinnatists are cognitive developmental psychologists who reject constructivist models of human development in favor of models of innate organization of cognition. Most of their work is based studies of young infants using a habituation paradigm that infers cognitive abilities from preferential looking patterns (Baillargeon 1987a, b; Spelke et al. 1992). Their work suggests that human infants are born with essentially mature cognitive systems. Consequently, they reject stage or sequence models that would facilitate comparisons of species-typical developmental patterns.

Neoinnatists, like evolutionary psychologists, frequently argue that human cognitive abilities are modular, that is, are more or less discrete and independently evolved (Tooby and Cosmides 1992). This conclusion is contested by biological anthropologists and biologists who use life history models emphasizing that humans are a long-lived, slow-maturing species with a low reproductive rate and a large, slow-developing brain (Martin 1983; Gibson 1990, 1995; Deacon 1997).

Recently some cognitive ethologists and evolutionary psychologists (Hauser 1998) have begun to use perceptual and habituation tests developed for human infants by neoinnatists (Carey and Gelman 1991). These tests allow them to compare human and nonhuman primates, but they are not developmental in the strict sense because they are not part of a stage or sequence model.

Future Prospects

Clearly both CE and CDEP researchers could benefit from adopting certain elements from each

Table 9.2

Contrasts between cognitive ethology and comparative developmental evolutionary psychology

	Cognitive Ethology	Comparative Developmental Evolutionary Psychology
Disciplinary origins	Comparative psychology, ethology/animal behavior, cognitive psychology	Comparative psychology, animal behavior, biological anthropology, developmental psychology
Key concepts	Species-specific learning, consciousness, adaptation	Species-specific developmental stages, adaptation
Topics	Communication, intentionality, consciousness, self-awareness, cognitive maps, number, Social learning	Developmental stages in physical knowledge, logical knowledge, social knowledge, symbolic knowledge, self-awareness
Some key researchers	Griffin, Ristau, Burghardt, Bekoff	The Gardners, Redshaw, Jolly, Miles, Patterson
Methodologies	Observation in wild, model testing, Experimental playback	Observation, Clinical-critical testing, Cross-fostering experiments
Taxa	Distantly related model species of birds and mammals: pigeons, plovers, snakes, bats, rats	Closely related sister species: great apes (in-group) in contrast to monkeys (out-group)
Goals	Identifying species-specific learning abilities; discovering adaptive significance of abilities	Identifying similarities and differences among primates; reconstructing the evolution of cognitive development

other. Cognitive ethologists could benefit from investigating and/or devising comparative developmental models. Although Piagetian models are the only comprehensive models of human cognitive development, investigators of other taxa might base their comparative studies on a detailed longitudinal study of cognitive development in another species, e.g., cetacean or carnivore species.

The Piagetian framework has several advantages for comparative studies of primate cognition. These include its apparently epigenetic nature, which results in an ordinal developmental scale in some domains (Uzgiris and Hunt 1975), and its comprehensiveness across physical, logical, and social domains. Perhaps its greatest advantage lies in its focus on the organization of spontaneous behavior, that is, the sequence, timing, goals, reinforcers, and modalities of behavior (Parker 1977). Finally, its focus on the most complex abilities alerts investigators to the absence of such abilities in related species. For these reasons, it seems desirable to base a comparative framework on the development of the most cognitively complex species in a clade rather than on that of a less cognitively complex species.

Cognitive ethologists could benefit from returning to the classical ethological practice of comparing closely related species and reconstructing the evolution of behavior. By focusing on cognitive development in clades of closely related species, they could generate comparative data that would allow reconstruction of the origins of shared derived character states, and in the case of developmental data, reconstruction of the evolution of developmental patterns.

Comparative developmental evolutionary psychologists, on the other hand, could benefit from extending their scope to include cognitive development in clades of nonprimate mammals and birds. This would greatly expand their knowledge of animal adaptations and provide material for adaptive models based on convergent evolution.

CEDP researchers could benefit from the investigation of systematics and phylogenetics and evolutionary developmental biology. These subfields provide critical tools for reconstructing the evolution of character states and developmental patterns, and for framing and testing adaptational hypotheses.

Both programs, but especially CDEP, would be enhanced by increasing their focus on cognition in wild populations of monkeys and apes (McGrew 1992; McGrew et al. 1996; Boesch and Boesch-Acherman 2000). CDEP researchers need to keep abreast of field studies of nonhuman primates to understand the kinds of ecological contexts in which various species live. This is particularly important for the generation of adaptive hypotheses.

Both CE and CDEP programs could benefit from investigations of paleontological data and paleoenvironments in which the putative ancestors of their study clade existed. In the case of hominids, this should include archeological data on past technologies (Wynn 1989; Gibson and Ingold 1993; Mithen 1996). Likewise, both research programs could benefit from comparative studies of brain development in closely related species—a growing trend in biological anthropology (Gibson and Peterson 1991; Deacon 1997; Parker et al. 2000).

This advice applies even more urgently to evolutionary psychologists, who, despite their interest in understanding the adaptive significance of human behaviors, typically neglect to use the comparative method in reconstructing the evolution of human behavior. The need to close this anomalous gap in their investigations was addressed recently by Marc Hauser in his plenary address to the Human Behavior and Evolution Society (Hauser 2000). There are encouraging signs that this group is beginning to recognize that the failure to use comparative data robs them of the chief tools of evolutionary reconstruction.

References

Antinucci, F. (1989). *Cognitive Structure and Development in Nonhuman Primates.* Hillsdale, N.J.: Lawrence Erlbaum Associates.

Bailey, M. B. (1986). Every animal is the smartest: Intelligence and ecological niche. In *Animal Intelligence: Insights into the Animal Mind*, R. J. Hoage and L. Goldman, eds., pp. 105–114. Washington, D.C.: Smithsonian Institution Press.

Baillargeon, R. (1987a). Object permanence in 3.5- to 4.5-month-old infants. *Developmental Psychology* 23: 655–664.

Baillargeon, R. (1987b). Young infants' reasoning about the physical and spatial properties of a hidden object. *Cognitive Development* 2: 179–220.

Beach, F. (1965). The snark was a boojum. In *Readings in Animal Behavior*, T. E. McGill, ed., pp. 3–14. New York: Holt, Rinehart and Winston.

Bekoff, M. (1995). Play signals as punctuation: The structure of social play in canids. *Behaviour* 132: 419–429.

Bekoff, M. and Allen, C. (1997). Cognitive ethology: Slayers, skeptics, and proponents. In *Anthropomorphism, Anecdotes, and Animals*, R. W. Mitchell, ed., pp. 313–334. Albany: State University of New York Press.

Boesch, C. (1991a). Symbolic communication in wild chimpanzees. *Human Evolution* 6: 81–90.

Boesch, C. (1991b). Teaching among wild chimpanzees. *Animal Behaviour* 41: 530–532.

Boesch, C. (1993). Aspects of transmission of tool use in wild chimpanzees. In *Tools, Language and Cognition in Human Evolution*, K. R. Gibson and T. Ingold, eds., pp. 171–183. Cambridge: Cambridge University Press.

Boesch, C. and Boesch, H. (1984). Mental map in wild chimpanzees: An analysis of hammer transports for nut cracking. *Primates* 25: 160–170.

Boesch, C. and Boesch-Acherman, H. (2000). *The Chimpanzees of the Tai Forest*. Oxford: Oxford University Press.

Boysen, S. T. and Capaldi, E. J. (eds.) (1993). *The Development of Numerical Competence: Animal and Human Models*. Hillsdale, N.J.: Lawrence Erlbaum Associates.

Brooks, D. and McLennan, D. (1991). *Phylogeny, Ecology, and Behavior*. Chicago: University of Chicago Press.

Burghardt, G. (1997). Amending Tinbergen: A fifth aim for ethology. In *Anthropomorphism, Anecdotes, and Animals*, R. W. Mitchell, ed., pp. 254–275. Albany: State University of New York Press.

Byrne, R. W. (1995). *The Thinking Ape: Evolutionary Origins of Intelligence*. Oxford: Oxford University Press.

Carey, S. and Gelman, R. (eds.) (1991). *The Epigenesis of Mind: Essays on Biology and Cognition*. Hillsdale, N.J.: Lawrence Erlbaum Associates.

Cheney, D. and Seyfarth, R. (1990). *How Monkeys See the World*. Chicago: University of Chicago Press.

Chevalier-Skolnikoff, S. (1976). The ontogeny of primate intelligence and its implications for communicative potential: A preliminary report. *Annals of the New York Academy of Sciences* 280: 173–211.

Chevalier-Skolnikoff, S. and Poirier, F. (eds.) (1977). *Primate Biosocial Development*. New York: Garland.

Dawkins, M. S. (1995). *Unraveling Animal Behaviour*. Essex, UK: Longman.

Deacon, T. (1997). *The Symbolic Species*. New York: W. W. Norton.

Fischer, K. W. and Bidell, T. R. (1991). Constraining nativist inferences about cognitive capacities. In *The Epigenesis of Mind: Essays on Biology and Mind*, S. Carey and R. Gelman, eds., pp. 199–235. Hillsdale, N.J.: Lawrence Erlbaum Associates.

Gardner, R. A. and Gardner, B. T. (1969). Teaching sign language to a chimpanzee. *Science* 165: 664–672.

Gardner, R. A., Gardner, B. T., and van Cantfort, T. E. (eds.) (1989). *Teaching Sign Language to Chimpanzees*. Albany: State University of New York Press.

Geary, D. C. (1998). *Male, Female: The Evolution of Human Sex Differences*. Washington, D.C.: American Psychological Association.

Gibson, K. G. and Peterson, A. C. (eds.) (1991). *Brain Maturation and Cognitive Development*. New York: Aldine.

Gibson, K. R. (1990). New perspectives on instincts and intelligence: Brain size and the emergence of hierarchical mental construction skills. In *"Language" and Intelligence in Monkeys and Apes*, S. T. Parker and K. R. Gibson, eds., pp. 97–128. New York: Cambridge University Press.

Gibson, K. R. (1995). Hypermorphosis in hominid brain evolution. Paper presented at the American Association for the Advancement of Science Meeting.

Gibson, K. R. and Ingold, T. (eds.) (1993). *Tools, Language and Cognition in Human Evolution*. Cambridge: Cambridge University Press.

Gould, S. J. (1977). *Ontogeny and Phylogeny*. Cambridge, Mass.: Harvard University Press.

Griffin, D. (1978). Prospects for a cognitive ethology. *Behavioral and Brain Sciences* 1: 527–538.

Hauser, M. (1998). A nonhuman primate's expectations about object motion and destination: The importance of self-propelled movement and animacy. *Developmental Science* 1: 31–37.

Hauser, M. (2000). Why humans are the wrong species in which to study the evolution of human intelligence. Human Behavior and Evolution Meetings.

Herzog, H. A. J., Bowers, B. B., and Burghardt, G. M. (1992). Development of antipredator responses in snakes: V. Species differences in ontogenetic trajectories. *Developmental Psychobiology* 25: 199–211.

Hinde, R. and Stevenson-Hinde, J. (eds.) (1973). *Constraints on Learning*. New York: Academic Press.

Hoage, R. J. and Goldman, L. (eds.) (1986). *Animal Intelligence: Insights into the Animal Mind*. Washington, D.C.: Smithsonian Institution Press.

Jamieson, D. and Bekoff, M. (1996). On aims and methods of cognitive ethology. In *Readings in Animal Cognition*, M. Bekoff and D. Jamieson, eds., pp. 65–78. Cambridge, Mass.: MIT Press.

Jolly, A. (1972). *The Evolution of Primate Behavior*. New York: McGraw-Hill.

Langer, J. (2000a). The descent of cognitive development. *Developmental Science* 3: 361–379 and 385–389.

Langer, J. (2000b). The heterochronic evolution of primate cognitive development. In *Brains, Bodies, and Behavior: The Evolution of Human Development*, S. T.

Parker, J. Langer, and M. McKinney, eds., pp. 215–236. Santa Fe, N.M.: School of American Research.

Martin, R. D. (1983). *Human Brain Evolution in an Ecological Context*. New York: American Museum of Natural History.

Martins, E. P. (ed.) (1996). *Phylogenies and the Comparative Method in Animal Behavior*. New York: Oxford University Press.

Matsuzawa, T. (1994). Field experiments on use of stone tools in the wild. In *Chimpanzee Cultures*, R. Wrangham, W. C. McGrew, F. B. M. de Waal, P. Heltne, and L. A. Marquardt, eds., pp. 351–370. Cambridge, Mass.: Harvard University Press.

McGrew, W. (1992). *Chimpanzee Material Culture*. New York: Cambridge University Press.

McGrew, W., Marchant, L., and Nishida, T. (eds.) (1996). *Great Ape Societies*. Cambridge: Cambridge University Press.

McKinney, M. and McNamara, K. (1991). *Heterochrony: The Evolution of Ontogeny*. New York: Plenum.

McNamara, K. (1997). *The Shapes of Time*. Baltimore, Md.: Johns Hopkins University Press.

Miles, H. L. (1983). Apes and language: The search for communicative competence. In *Language in Primates: Perspectives and Implications*, J. D. Luce and H. T. Wilder, eds., pp. 43–61. New York: Springer-Verlag.

Mitchell, R. and Thompson, N. (eds.) (1986). *Deception: Perspectives on Human and Nonhuman Deceit*. Albany: State University of New York Press.

Mitchell, R. W., Thompson, N., and Miles, H. L. (eds.) (1997). *Anthropomorphism, Anecdotes and Animals*. Albany: State University of New York Press.

Mithen, S. (1996). *The Prehistory of the Mind*. New York: Thames and Hudson.

Newell, A. (1990). *Unified Theories of Cognition*. Cambridge, Mass.: Harvard University Press.

Parker, S. T. (1977). Piaget's sensorimotor series in an infant macaque: A model for comparing unstereotyped behavior and intelligence in human and nonhuman primates. In *Primate Biosocial Development*, S. Chevalier-Skolnikoff and F. Poirier, eds., pp. 43–112. New York: Garland.

Parker, S. T. (1990). The origins of comparative developmental evolutionary studies of primate mental abilities. In *"Language" and Intelligence in Monkeys and Apes*, S. T. Parker and K. R. Gibson, eds., pp. 3–64. New York: Cambridge University Press.

Parker, S. T. (1991). A developmental approach to the origins of self-awareness in great apes and human infants. *Human Evolution* 6: 435–449.

Parker, S. T. (1996). Using cladistic analysis of comparative data to reconstruct the evolution of cognitive development in hominids. In *Phylogenies and the Comparative Method in Animal Behavior*, E. Martins, ed., pp. 361–398. New York: Oxford University Press.

Parker, S. T. and Gibson, K. R. (1977). Object manipulation, tool use, and sensorimotor intelligence as feeding adaptations in cebus monkeys and great apes. *Journal of Human Evolution* 6: 623–641.

Parker, S. T. and Gibson, K. R. (1979). A developmental model for the evolution of language and intelligence in early hominids. *Behavioral and Brain Sciences* 2: 367–408.

Parker, S. T. and Gibson, K. R. (eds.) (1990). *"Language" and Intelligence in Monkeys and Apes*. New York: Cambridge University Press.

Parker, S. T. and McKinney, M. L. (1999). *Origins of Intelligence: The Evolution of Cognitive Development in Monkeys, Apes, and Humans*. Baltimore, Md.: Johns Hopkins University Press.

Parker, S. T., Mitchell, R. W., and Boccia, M. L. (eds.) (1994). *Self-awareness in Animals and Humans*. New York: Cambridge University Press.

Parker, S. T., Miles, H. L., and Mitchell, R. W. (eds.) (1999). *The Mentalities of Gorillas and Orangutans*. Cambridge: Cambridge University Press.

Parker, S. T., Langer, J., and McKinney, M. L. (eds.) (2000). *Brains, Bodies, and Behavior: The Evolution of Human Development*. Santa Fe, N.M.: School of American Research.

Patterson, F. (1980). Innovative use of language by gorilla: A case study. In *Children's Language*, K. Nelson, ed., pp. 497–561. New York: Gardner.

Piaget, J. and Inhelder, B. (1969). *The Psychology of the Child*. New York: Basic Books.

Povinelli, D. (1994). How to create a self-recognizing gorilla (but don't try it on macaques). In *Self-Awareness in Animals and Humans*, S. T. Parker, R. W.

Mitchell, and M. L. Boccia, eds., pp. 291–300. New York: Cambridge University Press.

Redshaw, M. (1978). Cognitive development in human and gorilla infants. *Journal of Human Evolution* 7: 113–141.

Ristau, C. (ed.) (1991). *Cognitive Ethology: The Minds of Other Animals*. Hillsdale, N.J.: Lawrence Erlbaum Associates.

Roitblat, T. G., Bever, T., and Terrace, H. (eds.) (1984). *Animal Cognition*. Hillsdale, N.J.: Lawrence Erlbaum Associates.

Russon, A., Bard, K., and Parker, S. T. (eds.) (1996). *Reaching into Thought: The Minds of Great Apes*. Cambridge: Cambridge University Press.

Savage-Rumbaugh, E. S., Romski, M. A., Hopkins, W. D., and Sevcik, R. (1989). Symbol acquisition and use by *Pan troglodytes*, *Pan paniscus*, and *Homo sapiens*. In *Understanding Chimpanzees*, P. G. Heltne and L. A. Marquardt, eds., pp. 266–295. Cambridge, Mass.: Harvard University Press.

Sebeok, T. A. and Rosenthal, R. (eds.) (1981). *The Clever Hans Phenomenon: Communication with Horses, Whales, Apes, and People*. New York: New York Academy of Sciences.

Spelke, E., Breinlinger, K., Macomber, J., and Jacobson, K. (1992). Origins of knowledge. *Psychological Review* 99: 605–612.

Terrace, H., Petitito, L., Saunders, R. J., and Bever, T. (1979). Can an ape create a sentence? *Science* 206: 891–902.

Tinbergen, N. (1963). On aims and methods of ethology. *Zeitschrift für Tierpsychologie* 20: 410–429.

Tomasello, M. and Call, J. (1997). *Primate Cognition*. New York: Oxford University Press.

Tooby, J. and Cosmides, L. (1992). The psychological foundations of culture. In *The Adapted Mind*, J. Barkow, L. Cosmides, and J. Tooby, eds., pp. 19–136. New York: Oxford University Press.

Uzgiris, I. and Hunt, M. (1975). *Assessment in Infancy: Ordinal Scales of Psychological Development*. Urbana: University of Illinois Press.

Vauclair, J. (1996). *Animal Cognition*. Cambridge, Mass.: Harvard University Press.

Whiten, A. (ed.) (1991). *Natural Theories of Mind: Evolution, Development and Simulation of Everyday Mindreading*. Oxford: Blackwell.

Whiten, A. and Byrne, R. (eds.) (1988). *Machiavellian Intelligence*. Oxford: Oxford University Press.

Whiten, A. and Byrne, R. (eds.) (1997). *Machiavellian Intelligence II: Extensions and Evaluations*. Cambridge: Cambridge University Press.

Wynn, T. (1989). *The Evolution of Spatial Competence*. Urbana: University of Illinois Press.

Cognitive Ethology at the End of Neuroscience

Dale Jamieson

Eliminative Materialism is the thesis that our common-sense conception of psychological phenomena constitutes a radically false theory, a theory so fundamentally defective that both the principles and the ontology of that theory will eventually be displaced, rather than smoothly reduced, by completed neuroscience.
—Paul Churchland (1981, p. 67)

A Short, Simple History

In the beginning, humans were animals. Accounts of the belief systems of aboriginal peoples often emphasize the fact that these peoples viewed themselves as continuous with the rest of nature (Whitt et al. 2001). Animals were worshiped, hunted, and respected. They were also agents with whom one made agreements (Martin 1978), and in some cases even entered into conjugal relationships (Passmore 1974). Of course aboriginal peoples distinguished between those who were members of their own group and those who were outsiders. But in many cases some animals were considered insiders and other humans were treated as outsiders. Thus, for many aboriginal peoples, life was fully lived in interspecies communities.

Then along came humanism. There are many ways of characterizing humanism and dating its arrival. Viewed historically, it was a cultural movement that arose during the Italian Renaissance, although it looked back to the classical world. Protagoras's oft-quoted remark, "man is the measure of all things," however it was originally intended, conveys the spirit of humanism. Humanism can broadly be characterized as "[A]ny philosophy concerned to emphasize human welfare and dignity, and optimistic about the powers of unaided human understanding" (Blackburn 1994, p. 178). On this view, humans are seen as morally distinctive, and the moral difference between humans and other animals is typically thought to rest on a nonmoral categori-

cal distinction—for example that humans are different from other animals in being rational; or that only humans are capable of language, tool use, or some other favored activity.

The rise of humanism and modern science was temporally coincident, and humanism's optimism about human understanding helps to explain this association. Humanism advocated science on the ground that scientific knowledge contributes to human welfare. Humanism thus provided a justification for modern science's inauguration of the large-scale, systematic infliction of pain and death on nonhuman animals in the pursuit of knowledge. Indeed, since there are no elephant Galileos, the very practice of science itself also helped to distinguish humans from other animals. Some might say that humanism was the theory and science was the practice.

The great exemplar of humanism's attitude toward animals was the seventeenth-century thinker, René Descartes. Descartes, who is often regarded as the founder of modern philosophy, also did important work in optics and analytical geometry. He emphasized the importance of reason and exalted humans over other animals.

Descartes was a dualist in at least two respects. First, he taught that humans are composed of two interacting substances: a material substance that is the body and an immaterial substance that is the mind. Second, he was a dualist with respect to the relation between humans and the rest of nature. Humans and other animals are distinct because while nonhuman animals are material substances, humans are essentially immaterial substances associated with material substances. Stated simply, his view was that while humans are minded creatures, nonhuman animals are organic automata who are not harmed when they are subjected to invasive procedures. An unknown contemporary wrote of the scientist followers of Descartes that

[T]hey administered beatings to dogs with perfect indifference; and made fun of those who pitied the creatures as if they felt pain.... They nailed poor animals up on boards by their four paws to vivisect them and see the circulation of the blood which was a great subject of controversy. (Quoted in Rosenfield 1968, p. 54)

Humanism died in 1859 with the publication of Darwin's *On the Origin of Species*, but it has not yet become extinct; it remains a "dead man walking." Various philosophers (e.g., Rachels 1990) have shown how evolutionary theory undermines the human claim to categorical uniqueness, thus rendering views about the moral distinctiveness of humans implausible. Whether we look at humans behaviorally, taxonomically, or genetically, the categorical distinction that is supposed to underwrite the moral difference between humans and other animals does not seem to obtain. Behaviorally, the overwhelming similarities between humans and many other animals are obvious to anyone who bothers to pay close attention (see, e.g., Bekoff 2000). Taxonomically, humans are one of several species of great ape, more closely related to chimpanzees than chimpanzees are to gorillas or orangutans. Genetically, the similarities are overwhelming. Commenting on the publication of the human genome, Svante Pääbo wrote:

The first comparisons will be between the human genome and distantly related genomes such as those of yeast, flies, worms, and mice ... [w]e share much of our genetic scaffold even with very distant relatives. The similarity between humans and other animals will become even more evident when genome sequences from organisms such as the mouse, with whom we share a more recent common ancestor, become available. For these species, both the number of genes and the general structure of the genome are likely to be very similar to ours ... [T]he close similarity of our genome to those of other organisms will make the unity of life more obvious to everyone. No doubt the genomic view of our place in nature will be both a source of humility and a blow to the idea of human uniqueness.

However, the most obvious challenge to the notion of human uniqueness is likely to come from comparisons of genomes of closely related species. We already

know that the overall DNA sequence similarity between humans and chimpanzees is about 99%. When the chimpanzee genome sequence becomes available, we are sure to find that its gene content and organization are very similar (if not identical) to our own. The result is sure to be an even more powerful challenge to the notion of human uniqueness than the comparison of the human genome to those of other mammals. (Pääbo 2001, p. 1219)

For much of the twentieth century behaviorism held sway. It became the "normal science" of university psychology departments and for the most part happily coexisted with prevailing humanist values. Although many behaviorists considered themselves materialists, in some respects their doctrine was oddly unbiological. While they emphasized the importance of learning, they minimized the role of underlying biological structures and seldom attempted evolutionary explanations. The word "evolution" rarely appears in the foundational treatises of the movement; there are six occurrences in the index in Skinner (1953), and none at all in the index in Watson (1930). However, Watson and Skinner did not flinch from the radical antihumanist implications of their theory. Watson, reflecting on his Columbia University lectures of 1912 and the storm of criticism they provoked, wrote that

We believed then, as we do now, that man is an animal different from other animals only in the types of behavior he displays.... Human beings do not want to class themselves with other animals.... The raw fact that you, as a psychologist, if you are to remain scientific, must describe the behavior of man in no other terms than those you would use in describing the behavior of the ox you slaughter, drove and still drives many timid souls away from behaviorism. (Watson 1930, p. v)

The very title of Skinner's book, *Beyond Freedom and Dignity*, indicates the lengths he was willing to go in articulating the antihumanist case.

At least two sources contributed to the decline of behaviorism. One was the development of more biological understandings of behavior in the work of such classical ethologists as Konrad Lorenz and Niko Tinbergen. The other was the cognitive revolution that originated with the work of Noam Chomsky in linguistics, and then migrated into philosophy and psychology. Chomsky's cognitivism was developed in direct response to Skinner's behaviorism. In his devastating review of Skinner's *Verbal Behavior*, Chomsky (1959) showed that behaviorist learning theory, which neglected the innate endowment of organisms, was not powerful enough to explain human linguistic behavior.

Three Questions about Cognitive Ethology

The cognitive turn came late to ethology, dating perhaps from Griffin (1976/1981). Although most of the central concepts and claims in this field are contested, we can start with the simple thought that cognitive ethology proposes that some behavior of some animals can be explained by reference to their cognitive and affective states. Cognitive states are typically understood as representational states produced by natural selection. Representational states are in turn characterized by their semantic content. So, cognitive ethology proposes to explain some animal behavior by reference to semantic content. A simple example of such an explanation is this: Grete (my dog) walks to the door because she wants to go out. Wanting to go out is a representational state that figures in the explanation of her behavior.

There are many different ways of filling in the details of this program. The proffered explanations could be causal or noncausal. They may or may not involve psychobehavioral laws. Content could be wide or narrow. A story must also be told about the relationship between cognitive explanations and those that might be given for the phototropic behavior of plants or the heat-seeking behavior of missiles. However these are questions that I shall put aside. Instead, there are three questions that I wish to explore. The first

two are relatively straightforward: First, what is the relation between cognitive ethology and folk psychological explanation? Second, how can we discover the content of an animal's thought? Finally, I wish to return to the epigraph to this essay and comment on the very large question of whether the rise of neuroscience is a threat to cognitive ethology.

On the first question, it may appear that the future of cognitive ethology is essentially linked to the fate of folk psychology. Certainly part of the intuitive case for cognitive ethology flows from the naturalness of applying folk psychological categories and generalizations to nonhuman animals. Grete, like my mother, sometimes gets jealous, and both Grete and my mother get testy when they are frustrated. Some cognitive ethologists do not shrink from explicitly endorsing such folk psychological explanations of animal behavior. This is apparent in *The Smile of a Dolphin*, a remarkable book in which eminent ethologists, behavioral ecologists, psychologists, sociologists, and anthropologists let their hair down and describe their most memorable encounters with nonhuman animals. The chapter headings say it all: "love"; "fear, aggression, and anger"; "joy and grief"; and "fellow feelings." I myself am not at all chary about applying folk psychological vocabularies and generalizations to both my mother and to Grete. However, the question here is whether cognitive ethology necessarily stands or falls with the tenability of folk psychology. I claim that it does not.

Folk psychology provides one way of providing cognitive explanations, but it is not the only way. Perhaps concepts such as jealousy and frustration will be replaced by ones that more adequately individuate states and explain behavior. Someday more useful generalizations may be found for making behavior intelligible. Cognitive ethology can avail itself of improved cognitive and psychological theories without subverting itself. It is essential to cognitive ethology that its explanations appeal to representational states of organisms, but these states need not be the familiar ones of folk psychology. Thus, the demise of folk psychology does not in itself portend the end of cognitive ethology.

This takes us to the second question of how we can identify the content of animals' cognitive states. Work on this issue has proceeded both from the top down and from the bottom up. Some researchers, including myself (for example in 1998), have insisted that animals think, but have generally avoided serious discussion of what they think. Others have sketched specific approaches for empirically characterizing concepts that might figure in an animal's cognitive economy (e.g., Allen 1999). The first sort of work sometimes seems unconvincing since the best evidence for the claim that an animal is thinking involves some account of the content of its thought. On the other hand, the second kind of work dose not always seem very cognitive. Content that is inferred from fairly crude discrimination experiments and concepts that are straightforwardly reducible to neural states seem rather remote from human cognition. However, in my view, the difficulty in systematically characterizing the content of animals' cognitive states is not so much because there are problems with the various research strategies that have been employed as with the notion of content itself.

The concept of content plays a role in a particular way of conceiving the mind. On this view, a mental state involves a three-place relation between a creature, an attitude, and a content. For example, when Grete (the creature) believes that her treats are in the closet, she has the attitude of believing toward the content, "my treats are in the closet." When the matter is stated this way, it is easy to see why many are skeptical about whether languageless creatures can have cognitive states. If cognitive states are attitudes, and if contents are sentences, as they appear to be in the example given, then some fancy footwork is required to resist this skepticism. This is not the concern I wish to address here, however (see Allen and Bekoff 1997). My point at present is

simply that the notion of content occurs as part of a particular way of looking at the mind.

What I want to suggest is that content ascription is part of a practice used in order to make ourselves and others intelligible. Within this practice, content ascription is a heuristic that is fundamentally interpretative and interest relative. These features are reflected in "the holism of the mental" (Davidson 1999), a feature noted by Quine (1960) and vigorously advocated by Davidson throughout his career. Grete's behavior of walking to the door can be variously explained by mutually adjusting beliefs and desires. If we fix a desire, for example, that Grete wants to urinate, then we can specify a belief (for example, that on the other side of the door is a place in which it is appropriate for her to urinate) that will make the behavior intelligible. But if we fix a different desire (for example, that Grete wants to play with Jethro), then we will have to adjust Grete's beliefs accordingly in order to explain the behavior.

This story about the interactions between contents and attitudes ramifies. While not anything goes, content ascriptions answer to various pragmatic concerns, including those involving other content ascriptions, and not only to what it is known about the organism's body and the world in which it is embedded. For this reason we should not expect content ascriptions to be uniquely determined by empirical observation.

If I am right about this, then assigning content is as much a matter of marshaling conceptual considerations as empirical ones. There will not be a decisive observation or critical experiment that will uniquely determine what an animal "really" thinks. However, it does not follow from this that what an animal thinks is unknowable or a matter of inference or guesswork. For such a skeptical view presupposes that there is some determinate fact of the matter to know, infer, or guess. What I am suggesting is that it is the very nature of content specification, thus cognitive explanation, to be plural and indeterminate and therefore conceptually, not just empirically, grounded. To be blunt, there is no unique fact of the matter about what a nonhuman animal "really" thinks.

This is not a weakness of cognitive ethology. For there is no unique fact of the matter about what you or I "really" think. The same slackness that is at work in content attributions to nonhuman animals is at work in content attributions in humans as well. However, this is obscured by familiarity, deference, and especially language. But linguistic behavior is behavior nonetheless, and the task of the interpreter is not in principle different when faced with my verbalization or Grete's tail wagging (see also Jamieson 1998; Jamieson and Bekoff 1992).

The view that I am urging is not entirely original. Its origins are in Quine and Wittgenstein, and it owes a lot to philosophers who have already been cited. In substance, my view may be closest to Dennett's. The key idea is that we should not expect to find propositional attitudes or their ilk written in the brain or anywhere else. Instead, propositional attitudes are attributed by interpreters who take the "intentional stance" (Dennett 1987). These attributions are ways of keeping track of what the organism is doing, has done, and might do. The propositional attitudes are like a grid projected onto a field. What gives the grid-points their significance is their relation to other grid-points, not their absolute locations in the field. Grid-points and propositional attitudes are means of sorting, classifying, and assessing rather than invariant, sober, descriptions of aspects of the world. On this view we attribute propositional attitudes to humans and nonhumans for the same reason: in order to keep track of behavior.

Since his earliest writings on cognitive ethology, Dennett (1987) has attempted to balance "romantic" interpretations of animal behavior with "killjoy" understandings. In his recent work, the killjoy seems to have gained the upper hand. He has become more scientistic in his views and seems now to believe that there are real biological differences between humans and other animals

that count against attributing propositional attitudes to most nonhumans (1996, p. 132). He has also come to think, as many did before him, that the hyperintensionality afforded by language marks an important difference in the cognitive possibilities of humans and other animals (1996, p. 159). Finally, perhaps motivated in part by the scientism, he seems uncomfortable with the moral uses to which cognitive attributions to nonhumans have been put (1996, p. 161ff).

This is not the place to attempt to thwart Dennett's slide into conventional orthodoxies about nonhuman animals. I will confine myself to two remarks. First, a thoroughgoing interpretivist should not be scientistic (here I side with Davidson against Quine). Scientific statements must also be interpreted, and they are as indeterminate and inscrutable as nonscientific statements. Second (as I have already noted), in principle for an interpreter, the task is the same whether confronted with my verbalizing or Grete's tail wagging. Davidson provides a surprising reason for this (especially in light of his own views with respect to animal minds):

[W]e have erased the boundary between knowing a language and knowing our way around the world generally.... I conclude that there is no such thing as a language, not if a language is anything like what many philosophers and linguists have supposed. (Davidson 1986, p. 446)

This brings us to our final and deepest question. What would be the fate of cognitive ethology in a world in which every behavior yielded to neuroscientific explanation? My own view is that while this would bring an end to cognitive science generally, including cognitive ethology specifically, it would not necessarily put a stop to the productive deployment of cognitive vocabularies. For these vocabularies have a place in everyday discourse, whatever their status as theoretical terms. In particular, they often carry our evaluational attitudes. Cognitive language is closely tied to practices of moral appraisal—of blaming, praising, and so on. Thus, if such language were to outlive its scientific usefulness, it would not necessarily vanish. It might still have other roles to play in everyday life.

Consider an analog from physics. We have been instructed by our epistemological betters that space and time do not exist as independent dimensions with linear structures. Yet this has not led to the abolition of the alarm clock. Even in the face of relativity theory, we speak usefully and responsibly of the sun rising and setting, although no reasonable person thinks that these notions should figure in a scientific conception of the world.

These considerations seem to suggest an important moral for cognitive ethology. If the logic of neuroscientific explanation is quite different from that of cognitive explanation, as I seem to be suggesting, then it will be a dangerous mistake to mix them in an unreflective way. Nevertheless, much of the literature of cognitive ethology does just this. Many scientists seem to go back and forth between neural and cognitive explanations, as if they were working with the same vocabularies, at the same levels of description, employing the same logic of explanation (for examples, see Griffin 1992). They seem to assume that microlevel explanations simply reveal what subserves macrolevel phenomena, while leaving macrolevel phenomena untouched. But that is far from obvious. As Paul Churchland points out in the epigraph, macrolevel phenomena are sometimes displaced, rather than smoothly reduced, by microlevel explanations. This is exactly the concern that many people have about the Human Genome Project, fearing that genetic-level explanations will drive out the language of responsibility.

The suggestion of confusion can be resisted by showing that in fact cognitive and neuroscientific explanations work in the same way. One strategy would be to construe cognitive explanation in a way that is as determinate and mechanical as neuroscientific explanation. The second strategy would be to show that neuroscientific explanation is itself as pluralistic and indeterminate as cognitive explanation. There might be two rea-

sons for thinking this. One reason would be because explanation itself is pragmatic and pluralistic (perhaps this view is implicit in Quine 1960). A second reason would involve claiming that neuroscientific explanation itself appeals to content and thus has the same features as any other content explanation. Patricia Churchland seems to suggest this when she writes:

> It is important ... to emphasize that when neuroscientists do address such questions as how neurons manage to store information, or how cell assemblies do pattern recognition, or how they manage to effect sensorimotor control, they are addressing questions concerning neurodynamics—concerning information and how the brain processes it. In doing so, they are up to their ears in theorizing, and even more shocking, in theorizing about representations and computations. (Patricia Churchland 1986, p. 361)

What I have been suggesting could be summarized by saying that cognitive explanations are appropriate when we are too ignorant to give real (i.e., neural) explanations. I resist this way of putting the point for the reasons suggested in the preceding paragraph, but there is something right about this view. Once we achieved a physical understanding of the occurrence of lightning, we no longer had to appeal to the moods of the gods. However, even if things inexorably move toward micro and mechanical explanations of behavior, and away from macro and functional ones, cognitive ethology will still have performed a great service. Some of its contributions are methodological. It returns scientists to the field; it requires that they watch animals, that they reflect on behavioral similarities and singularities, and so on. However, from a larger cultural perspective, the real contribution of cognitive ethology is that it helps to complete the circle and restore unity to our picture of nature.

Conclusion

I opened this essay with a short, simple history of human attitudes toward animals. Some may quarrel with the history, disagree with my account of the science, or rightly claim that it is all much more complicated than I suggest. What cannot be denied or evaded is that this science has a moral dimension. How we study animals and what we assert about their minds and behavior greatly affects how they are treated, as well as our own view of ourselves. Humanism is dead and its foundation is in tatters, but the full force of this fact has not yet been felt. Cognitive ethology helps us to accept this by showing that the same explanations that apply in one case often apply in the other as well. This is an important scientific lesson, but it also carries deep and profound moral lessons. Indeed, it is because of these moral lessons that some people find this science to be subversive.

Acknowledgments

I thank the editors of this volume, and especially Brian Keeley, for their stimulating comments on an earlier draft.

References

Allen, C. (1999). Animal concepts revisited. *Erkenntnis* 51: 537–544.

Allen, C. and Bekoff, M. (1997). *Species of Mind*. pp. 63ff. Cambridge, Mass.: MIT Press.

Bekoff, M. (2000). *The Smile of a Dolphin: Remarkable Accounts of Animal Emotions*. New York: Discovery Books.

Blackburn, S. (1994). *The Oxford Dictionary of Philosophy*. p. 178. New York: Oxford University Press.

Chomsky, N. (1959). Review of B. F. Skinner's *Verbal Behavior*. *Language* 35: 26–58.

Churchland, Patricia (1986). *Neurophilosophy*. Cambridge, Mass.: MIT Press.

Churchland, Paul (1981). Eliminative materialism and the propositional attitudes. *Journal of Philosophy* 78: 67–90.

Davidson, D. (1986). A nice derangement of epitaphs. In *Truth and Interpretation: Perspectives on the Philosophy of Donald Davidson*, Ernest LePore, ed. New York: Basil Blackwell.

Davidson, D. (1999). The emergence of thought. *Erkenntnis* 51: 7–17.

Dennett, D. C. (1987). *The Intentional Stance*. Cambridge, Mass.: MIT Press.

Dennett, D. C. (1996). *Kinds of Minds*. New York: Basic Books.

Griffin, D. (1976/1981). *The Question of Animal Awareness: Evolutionary Continuity of Mental Experience*. New York: Rockefeller University Press.

Griffin, D. R. (1992). *Animal Minds*. Chicago: University of Chicago Press.

Jamieson, D. (1998). Science, knowledge, and animal minds. *Proceedings of the Aristotelian Society* 98: 79–102.

Jamieson, D. and Bekoff, M. (1993). On aims and methods of cognitive ethology. In *PSA 2*, M. Forbes, D. Hull, and K. Okruhlik, eds., pp. 110–124. Lansing, Mich.: Philosophy of Science Association (1992); reprinted in *Readings in Animal Cognition*, M. Bekoff and D. Jamieson, eds., pp. 65–78. Cambridge, Mass.: MIT Press.

Martin, C. (1978). *Keepers of the Game: Indian–Animal Relationships and the Fur Trade*. Berkeley: University of California Press.

Pääbo, S. (2001). The human genome and our view of ourselves. *Science* 291 (Feb. 16, 2001): 1219–1220.

Passmore, J. (1974). *Man's Responsibility for Nature; Ecological Problems and Western Traditions*. London: Duckworth.

Quine, W. V. O. (1960). *Word and Object*. Cambridge, Mass.: MIT Press.

Rachels, J. (1990). *Created from Animals: The Moral Implications of Darwinism*. New York: Oxford University Press.

Rosenfield, L. (1968). *From Animal Machine to Beast Machine*. p. 54. New York: Octagon Books.

Skinner, B. F. (1953). *Science and Human Behavior*. New York: Macmillan.

Watson, J. B. (1930). *Behaviorism*. p. v. Chicago: University of Chicago Press.

Whitt, L. A., Roberts, M., Norman, W., and Grieves, V. (2001). Indigenous perspectives. In *A Companion to Environmental Philosophy*, D. Jamieson, ed., pp. 3–20. Oxford: Blackwell.

11 Learning and Memory Without a Brain

James W. Grau

Over the past 40 years, the study of animal learning has gone through a radical metamorphosis. Animal learning of the 1960s variety was largely dominated by researchers who took Watson's behaviorist dictates as scripture. They hoped to explain complex behavior in simple stimulus-response (S-R) terms while avoiding reference to unobservable constructs. They were, in retrospect, the killjoys of creativity, always suspicious of anything new and wary of most internal or theoretical constructs. Against this backdrop came the cognitive approach, opening the doors to a new faith that promised greater tolerance for internal constructs.

Today we often take for granted that all higher vertebrates process information in a limited-capacity device (short-term memory or STM) (Wagner et al. 1973), that attention guides the learning process (Mackintosh 1975), and that the organism can deduce the connection between its behavior and an outcome in its environment (Maier and Seligman 1976). Yet 30 to 40 years ago, each of these claims was greeted with skepticism, with many researchers questioning whether explanations of animal behavior required such cognitive constructs.

I began my research career well indoctrinated in the cognitive approach, and I championed its benefits (e.g., Moye et al. 1981; Rescorla et al. 1985; Grau 1987a). However, in recent years, my collaborators have led me down an alternative path that I initially thought led in an incomprehensible direction, a course that sought evidence of learning and memory within a vertebrate that effectively lacks a brain. As we will see, this path led to a new vista, one that has forced me to question my cognitive faith. It seems that many of the behavioral effects I thought were best described in cognitive terms can be observed in the absence of a brain. I now find myself in the unfortunate position of the killjoy, questioning the application of cognitive constructs to infra-human species.

Memory within the Spinal Cord

The starting point for this work was a series of studies that explored how the body (and mind) regulate pain. I had shown that exposure to a mildly painful stimulus engages an inhibitory mechanism that reduces behavioral reactivity to subsequent noxious stimuli, a phenomenon known as antinociception. For example, in rats, exposure to a few brief tail shocks inhibits tail withdrawal from radiant heat (the tail-flick test). This response is mediated by a nociceptive (pain signal) reflex that is organized within the spinal cord; it is a spinal reflex that can be readily elicited after the lower spinal cord has been surgically disconnected from the brain. Exposure to moderate shock appears to inhibit nociceptive reactivity (antinociception) by engaging neural mechanisms within the brain that inhibit spinal nociceptive reflexes through descending pathways. I argued that the memory of the aversive event helped maintain the antinociception after shock exposure. Specifically, I argued that the central representation of the aversive event in short-term memory continues to drive the antinociceptive systems after exposure to shock, producing an antinociception that lasts 10 minutes or more.

Short-term memory in humans is generally envisioned as a kind of rehearsal buffer where information can be temporarily maintained (e.g., Atkinson and Shiffrin 1968). Because it is thought to have a limited capacity, it is subject to distraction; new information can disrupt the rehearsal of items already in STM, causing the memory of them to decay rapidly. Interesting, a distracting stimulus also disrupts memory in other animals, which suggests that they too process information in an STM-like device (Wagner 1981).

Assuming this perspective, I reasoned that if the central representation of an aversive event in STM maintains the activation of the antinociceptive systems, then displacing this memory with a distractor should cause the antinociception to decay more rapidly. As predicted, presentation of a visual distractor (a flashing light) after shock decreased the duration of the subsequent antinociception (figure 11.1A) (Grau 1987a).

To derive predictions from this memory-oriented perspective, I had to assume that the magnitude of antinociception observed depends on the hedonic value of the representation; the more aversive the memory, the greater the antinociception. Given this, suppose rats are exposed to a moderately intense shock (1 mA) followed by a weak shock (0.1 mA) distractor. The weak shock should displace the memory of the strong shock. By introducing a "better end" (Kahneman et al. 1993), I should be able to reduce the magnitude of antinociception observed. As predicted by this memory-based account, subjects that experienced a weak shock after moderate shock exhibited less antinociception (figure 11.1B) (Grau 1987b). Normally, adding shock increases the magnitude of antinociception. Because adding a weak shock had the opposite effect, the finding appeared to provide particularly strong evidence that memorial systems mediate the generation of antinociception.

A few years later, Mary Meagher examined whether incoming nociceptive signals can engage antinociceptive systems within the spinal cord in the absence of the brain (Meagher et al. 1993). To address this issue, she cut the spinal cord of rats at the second thoracic vertebra (T2). This operation produces a condition in which the rat can move about using its front paws, but is paralyzed below its midsection. She found that the moderate shocks used to study the activation of antinociceptive systems in intact subjects had little effect on tail-flick latencies after the spinal cord was transected. However, exposure to severe shocks, which were more intense and longer, caused a dramatic increase in tail-flick latencies.[1]

We assumed that this antinociception reflected an unconditioned response, a passive reaction to the noxious stimulus that was independent of learning and memory, processes everyone knew required a brain. Given this, a shock distractor should not cause the antinociception observed in a spinally transected rat to decay more rapidly. Rather, in this case, more shock should produce greater antinociception. To test this, one group (Int Shk → Nothing) of spinal rats was exposed to three intense shocks that Meagher had shown induce a strong antinociception. Another group (Nothing → Mod Shk) received moderate shock alone, which produces only a weak antinociception in spinalized rats. Because the moderate shock was apparently detected, but has only a weak effect, it was used as the distractor in this study (Grau et al. 1990).

We expected that the distractor would have an additive effect and, if anything, augment the antinociception produced by intense shock. Contrary to our expectations, rats that received the moderate shock distractor after intense shock (Int Shk → Mod Shk) exhibited an antinociception that decayed more rapidly (figure 11.1C), a result that formally mirrors the results obtained in intact rats (figure 11.1B).

Because these findings ran counter to my theory (Grau 1987a), we sought additional evidence. Indeed, this has remained a common feature of all our studies; because we generally doubt the processing capacity of spinal cord neurons, we routinely hold this system to a higher standard. In the present case, this led us to examine the effect of reversing shock order. According to the memory hypothesis, a distractor should be effective only if it is presented after the target event. Presenting a shock distractor prior to a strong shock should have no effect. Again, the results were consistent with our memory-oriented perspective (figure 11.1D). Spinalized rats that experienced the shock distractor before the intense tail shock (Mod Shk → Int Shk) exhibited a robust antinocicep-

A. Intact Rats: Visual Distractor

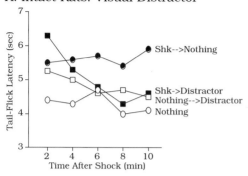

B. Intact Rats: Shock Distractor

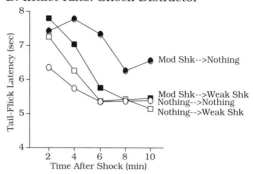

C. Spinal Rats: Shock Distractor

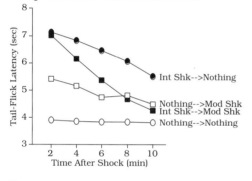

D. Spinal Rats: Temporal Order

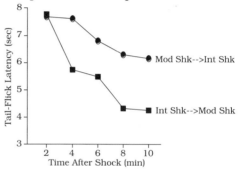

Figure 11.1

A postshock distractor speeds the decay of antinociception in intact and spinal rats. (A) Rats that received shock alone (Shk → Nothing) exhibited longer tail-flick latencies (antinociception) over the 10-minute test period. Presentation of a visual distractor (Shk → Distractor) caused the antinociception to decay more rapidly. (B) Presentation of a weak (0.1 mA) shock distractor after the moderate (1 mA) tail shock (Mod Shk → Weak Shk) also caused the antinociception to decay more rapidly. (C) Exposure to more intense (3 mA) tail shock (Int Shk → Nothing) increased tail-flick latencies in spinally transected rats. Exposure to a moderate (1 mA) tail shock alone had relatively little effect (Nothing → Mod Shk). Presentation of a moderate shock distractor after each intense shock (Int Shk → Mod Shk) caused the antinociception to decay more rapidly. (D) A shock distractor only caused antinociception to decay more rapidly when it was presented after intense shock (Int Shk → Mod Shk); rats that experienced the same amount of shock, but in the opposite temporal order (Mod Shk → Int Shk), exhibited a longer-lasting antinociception. (Adapted from Grau 1987a,b and Grau et al. 1990.)

tion, while rats that received exactly the same amount of shock, but in the opposite order (Int Shk → Mod Shk), did not (Grau et al. 1990).

Pavlovian Conditioning and Attention

Subsequent studies have explored whether nociceptive mechanisms within the spinal cord are sensitive to Pavlovian relations. In intact rats, a conditioned antinociceptive response can be established by pairing a conditioned stimulus (the CS+) with an aversive tail shock (the unconditioned stimulus, or US). After a few pairings, the CS+ generates an antinociception on the tail-flick test relative to another cue (the CS−) that was presented in an explicitly unpaired fashion (figure 11.2A).

To study Pavlovian conditioning within the spinal cord, Juan Salinas used cutaneous electrical stimuli that were applied to the left or right hind legs (figure 11.2B). These served as our CSs, one of which (the CS+) was paired with an intense tail shock 30 times while the other (the CS−) was presented in an unpaired fashion. An hour later, we tested tail-flick latencies during each CS. Subjects exhibited longer tail-flick latencies during the CS+ (conditioned antinociception), and this effect extinguished over the course of testing (Grau et al. 1990).

In intact subjects, preexposure to the CS alone prior to training generally undermines the acquisition of a conditioned response to that CS, a phenomenon known as latent inhibition. A common account of this effect assumes that learning is regulated by attentional mechanisms (e.g., Mackintosh 1975). From this perspective, the repeated presentation of a CS alone decreases its capacity to command attention and thereby undermines the rate of learning. Again, most researchers believe that attention is the province of the brain and, consequently, an explanation couched in attentional terms seemingly would predict that preexposure to the CS alone would have little effect on an intraspinal learning

mechanism. Yet once more our preconceptions proved wrong, for we found that the presentation of the CS alone prior to training undermined conditioning in spinalized rats (Illich et al. 1994).

Another effect that is often described in attentional terms is overshadowing. In an overshadowing experiment, subjects experience a compound cue composed of two elements. These elements are chosen to differ in their salience, or noticeability. One element (B) is very salient while the other (X) is much less salient. A control group is included to show that subjects can learn about the less salient element (X) when it is presented alone and paired with the US. But when X is presented jointly with the more salient cue B, and the two are paired with the US, intact subjects later exhibit a conditioned response to B, but not to X. It appears that the more salient element overshadows the less salient cue. A popular account assumes that subjects naturally attend to the more salient element. Because they fail to attend to X during training, they fail to learn about it.

We reasoned that learning within the spinal cord could be governed by a simple rule that depends solely on the number of CS–US pairings (contiguity). At this level of the nervous system, there may be no cue competition. Certainly, attention does not exist, at least in the usual, cognitive, sense of the term. Paul Illich evaluated these possibilities by training spinalized rats with a compound CS (Illich et al. 1994). Stimulation of one hind leg at the usual intensity served as the X element. A more salient element (B) was provided by stimulating the opposite (contralateral) leg at a higher intensity. As usual, rats that experienced X alone paired with the US exhibited conditioned antinociception relative to a group that experienced X and the US unpaired. More important, rats that experienced X in compound with B during training did not exhibit a conditioned antinociception when X was presented alone. It appears that the concurrent presenta-

A. Intact Rats: Conditioned Antinociception

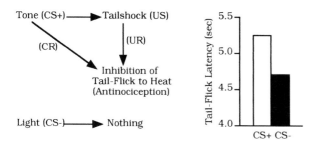

B. Spinal Rats: Conditioned Antinociception

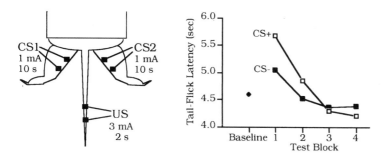

Figure 11.2
Intact (upper panels) and spinalized (lower panels) rats exhibit longer tail-flick latencies during a conditioned stimulus (CS) that has been paired with tail shock (conditioned antinociception). (A) In intact rats, conditioned antinociception can be demonstrated by pairing an auditory cue (a tone) with tail shock (the unconditioned stimulus, US). (B) After a few pairings, rats exhibit longer tail-flick latencies during the paired cue (the CS+) relative to an unpaired cue (the CS−) that was presented an equal number of times. (C) Conditioned antinociception can be demonstrated in spinalized rats by pairing stimulation of one hind leg (the CS) with an intense tail shock (the US). (D) After 30 CS-US pairings, rats exhibited longer tail-flick latencies during the CS+, relative to the CS−. Repeated presentation of the CSs alone during testing caused the CS+/CS− difference to extinguish. (Adapted from Grau et al. 1990.)

tion of B with X undermined the conditioned response produced by X. Neither latent inhibition nor overshadowing necessarily require a brain.

Instrumental Learning and the Cognition of No Control

More recently, we have begun to explore whether neural systems within the spinal cord can encode the relationship between a response and an environmental outcome (reinforcer), a form of learning known as instrumental conditioning. In these studies, Doug Barstow spinalized rats and then exposed them to a response-reinforcer relation using the apparatus illustrated in figure 11.3A. In this apparatus, when a shock is applied to one hind leg, it elicits a flexion response that can be monitored by means of a contact electrode taped to the rat's paw. In the master condition, a leg shock is given whenever the contact electrode touches the underlying salt solution. In this situation, intact rats will quickly learn to maintain the leg in an up (flexed) position, thereby decreasing net shock exposure. What is amazing is that spinalized rats can also learn to perform this response, exhibiting a progressive increase in duration of flexion as a function of training (figure 11.3B) (Grau et al. 1998; Grau and Joynes 2001).

A critic could charge, however, that this reflects an unconditioned response to shock. For example, perhaps shock causes a tetanuslike effect that produces a cumulative increase in muscle tension. To address this possibility, a yoked control was included. Each yoked rat was experimentally coupled to a master rat and received the same shock, but independent of leg position (noncontingent shock). If the change in flexion duration reflects an effect of shock per se, these subjects should also exhibit an increase in flexion duration, but this was not observed (figure 11.3B). Learning, as evidenced by an increase in flexion duration, was observed only when there was a response-reinforcer contin-

gency; remove the contingency and the learning disappears.

To meet the criteria for instrumental learning, we must also show that the experience has an effect that outlasts the environmental contingencies used to produce it, that the consequence of learning can be observed when subjects are later tested under common conditions (Grau et al. 1998). Robin Joynes addressed this issue by testing subjects with response-contingent shock (figure 11.3C). Now all of the subjects had an opportunity to decrease net shock exposure by exhibiting a flexion response. We found that previously trained subjects (master) exhibit some savings and reacquire the instrumental response more rapidly than a control group that had never been shocked (unshocked). Amazingly, yoked rats fail to learn and this is true even if they are tested on the opposite (contralateral) leg (Joynes et al., submitted). It appears that prior exposure to noncontingent shock undermines instrumental behavior, a behavioral deficit reminiscent of the phenomenon of learned helplessness (Maier and Seligman 1976; Overmier and Seligman 1967).

Researchers studying the consequences of noncontingent shock in intact animals have suggested that learned helplessness occurs when the organism develops a cognition of no control (Maier and Seligman 1976; Maier and Jackson 1979). If the cognition of no control mediates the deficit, then behavioral manipulations that alter this cognition should influence the deficit. For example, showing the subject that it can control its environment in some situations might protect it from becoming helpless. Conversely, experiencing a contingent response–outcome relation could reverse the cognition of no control and thereby attenuate the deficit. As predicted, training with contingent shock before (immunization) or after (therapy) inescapable shock attenuates learned helplessness in intact subjects (Seligman et al. 1968, 1975). Notice that the addition of escapable shock training actually increases the

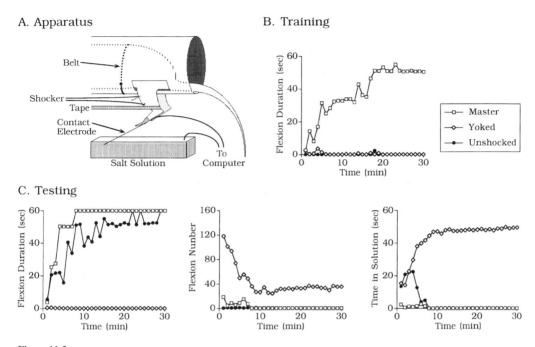

Figure 11.3
(A) The apparatus used to study instrumental learning in spinalized rats. (B) Spinalized rats that received a leg shock whenever the contact electrode touched the saline solution (master) exhibited a progressive increase in flexion duration over the course of the 30-minute training period. Yoked rats that received the same amount of shock independent of leg position did not exhibit an increase in flexion duration. (C) Subjects were then tested under common conditions with response-contingent shock. From flexion duration (left), we can see that rats previously trained with a contingent shock (master) learned more rapidly than a control group that was previously unshocked. Yoked rats repeatedly touched the underlying solution and exhibited the greatest number of flexions (middle). However, this experience with response-contingent shock did not produce an increase in flexion duration (left). As a consequence, yoked rats spent more time contacting the underlying salt solution (right) and received more shock. (Adapted from Grau et al. 1998.)

net exposure to the aversive event. This highlights an essential feature of the phenomenon: It is not the net exposure to shock (aversive stimulation) that is critical, but rather the organism's *perception* of whether the environmental event is controllable or uncontrollable (Maier and Seligman 1976).

Would training with contingent shock affect the development of the behavioral deficit within the spinal cord? To examine this issue, Eric

Crown first trained spinalized rats with contingent shock applied to one hind leg (Crown and Grau 2001). Next, subjects received intermittent tail shock at an intensity and duration known to induce a robust behavioral deficit (Crown et al., submitted). Finally, subjects were tested with contingent shock applied to the contralateral hind leg. As usual, rats that experienced noncontingent tail shock failed to learn, but rats that had experienced contingent shock prior to tail shock acquired the instrumental response at the same rate as the unshocked controls. Evidently training with contingent shock has a protective effect within the spinal cord.

We also examined whether the deficit could be reversed by exposing rats to contingent shock. In this experiment, the deficit was induced by exposing some spinalized rats to intermittent tail shock. Next, a subset of the subjects was given behavioral therapy by training the instrumental response in the presence of naltrexone. Robin Joynes had previously shown that the administration of this opioid antagonist temporarily blocks the expression of the behavioral deficit (Joynes and Grau, in press). Thus, while the drug is present, rats that had previously received inescapable shock were able to learn the instrumental response. Subjects were then tested with response-contingent shock 24 hours later, after the drug had cleared the system. We found that instrumental training in the presence of naltrexone attenuated the behavioral deficit. It seems that instrumental training has a therapeutic effect that helps restore behavioral potential within the spinal cord.

Implications for the Study of Animal Cognition

I admittedly began these experiments as a skeptic, convinced that the spinal cord was largely ineducable. I felt that the behavioral effects I had been studying in intact subjects were best described in terms of cognitive processes. Within a cognitive framework, I had provided evidence that the central representation of an aversive event maintains the activation of antinociceptive systems in intact rats (Grau 1987a,b). To my surprise, the experimental manipulation used to explore this hypothesis generated an identical pattern of results in the absence of the brain (Grau et al. 1990). Similarly, Pavlovian phenomena such as latent inhibition and overshadowing, which are often described in attentional terms, were observed in our spinal preparation (Illich et al. 1994). And experimental designs used to explore the cognition of no control produced remarkably similar results without a brain (Grau et al. 1998).

Our studies suggest that neural systems within the spinal cord are quite a bit smarter than most researchers have assumed. The spinal cord is not a simple conduit for incoming and outgoing neural impulses, but rather it is a complex information-processing system that can learn from experience. [See Patterson and Grau (2001) for other demonstrations of spinal cord plasticity and Patterson (2001) for a review of earlier studies.] It seems that many basic learning phenomena are not localized to particular brain regions, but instead reflect inherent characteristics of neural systems capable of learning, and that the capacity for learning is widely distributed.

What implications does our work have for the study of comparative cognition? Perhaps the most important is a note of caution. While we can readily design experimental procedures that appear formally similar to those used to study a human cognitive capacity (e.g., STM), similar results do not necessarily imply that the same mechanisms are at work. Organisms have most likely evolved many ways to solve environmental puzzles (Joynes and Grau 1996), just as they have evolved many ways to move about. Sensitivity to distraction alone cannot be taken as evidence for STM. Nor can latent inhibition and overshadowing be taken alone as evidence for attention. The same applies for many of the other behavioral tasks used to infer cognitive processing in infrahuman species.

The way in which sensitivity to distraction, cue competition, and the disabling of behavioral potential by noncontingent shock are represented neurally surely varies across different levels of the nervous system. Similar variability must also exist across species. Yet, at a functional or behavioral level, our experimental manipulations often yield similar results. It appears that some operational principles remain surprisingly stable across different neural architectures. Some persons will read this as consistent with the view of the behaviorists and their desire to isolate the general principles of learning. Paradoxically, others will see this as favoring a belief common within cognitive psychology, that the operational principles of an information-processing system can be explored with little attention to the underlying architecture. What we must remember, however, is that nature can write a program for behavior in many ways. Sometimes these programs may call upon cognitive functions, but sometimes they will not. In the latter case, our cognitive description of the behavioral effect is just a metaphor.

Do cognitive systems allow operations that noncognitive systems cannot achieve? Perhaps our studies of learning within the spinal cord can provide some hints, for we remain convinced that the brain (and cognition) adds something to our intellect. At the most obvious level, brain systems are tied to a much more elaborate system for detecting and integrating sensory information. Neural mechanisms in the brain (e.g., the hippocampus) allow the brain to span gaps in time (trace conditioning) (Woodruff-Pak 1993; Clark and Squire 1998), to mediate learning in the absence of a physical stimulus (Holland 1990), and to derive new configural representations (Sutherland and Rudy 1989).

Similarly, whereas the spinal cord may have but one response to an instrumental problem, the brain can select from a range of response alternatives that are tuned to particular environmental situations and temporal-spatial relations (Timberlake and Silva 1995). As a result of these elaborations, brain-mediated behavior often appears more flexible and adaptable, a category of behavior Skinner (1938) distinguished from elicited responses (respondents) by the term *operant learning*. [For a discussion of the relation between instrumental and operant learning, see Grau (2000).]

Researchers rely on operant contingencies when they train an animal to select a stimulus that matches a recently presented novel object (delayed matching-to-sample) or to choose the spatial location they first visited (Blough 1959; Kesner and Novak 1982). Operant behavior of this sort presumably requires a brain. It may also require cognition. Perhaps then it is operant behavior that provides the clearest window into animal cognition. Ironically, our best tool for studying cognition may come from the workbench of Skinner, a researcher who rejected the cognitive approach.

Note

1. People describe our moderate shocks as mildly painful. Severe shocks would presumably elicit intense pain, but because the spinal transection prevents the sensory signal from reaching the brain, spinalized rats do not "consciously" experience pain during severe shock.

Acknowledgment

The research described in this chapter was supported by the National Institute of Mental Health (grants MH54557, MH60157).

References

Atkinson, R. C. and Shiffrin, R. M. (1968). Human memory: A proposed system and its control processes. In *The Psychology of Learning and Motivation*. Vol. 2, K. W. Spence and J. T. Spence, eds., pp. 89–95. New York: Academic Press.

Blough, D. S. (1959). Delayed matching in the pigeon. *Journal of the Experimental Analysis of Behavior* 2: 151–160.

Clark, R. E. and Squire, L. R. (1998). Classical conditioning and brain systems: The role of awareness. *Science* 280: 77–81.

Crown, E. D. and Grau, J. W. (2001). Preserving and restoring behavioral potential within the spinal cord using an instrumental training paradigm. *Journal of Neurophysiology* 86: 845–855.

Crown, E. D., Ferguson, A. R., Joynes, R. L., and Grau, J. W. (submitted). Instrumental learning within the spinal cord: IV. Induction and retention of the behavioral deficit observed after noncontingent shock.

Grau, J. W. (1987a). The central representation of an aversive event maintains the opioid and nonopioid forms of analgesia. *Behavioral Neuroscience* 101: 272–288.

Grau, J. W. (1987b). The variables which control the activation of analgesic systems: Evidence for a memory hypothesis and against the coulometric hypothesis. *Journal of Experimental Psychology: Animal Behavior Processes* 13: 215–225.

Grau, J. W. (2000). Instrumental conditioning. In *The Corsini Encyclopedia of Psychology and Behavioral Science*, 3rd ed., W. E. Craighead and C. B. Nemeroff, eds., pp. 767–769. New York: Wiley.

Grau, J. W. and Joynes, R. L. (2001). Pavlovian and instrumental conditioning within the spinal cord: Methodological issues. In *Spinal Cord Plasticity: Alterations in Reflex Function*, M. M. Patterson and J. W. Grau, eds., pp. 13–54. Boston: Kluwer.

Grau, J. W., Salinas, J. A., Illich, P. A., and Meagher, M. W. (1990). Associative learning and memory for an antinociceptive response in the spinalized rat. *Behavioral Neuroscience* 104: 489–494.

Grau, J. W., Barstow, D. G., and Joynes, R. L. (1998). Instrumental learning within the spinal cord: I. Behavioral properties. *Behavioral Neuroscience* 112: 1366–1386.

Holland, P. C. (1990). Event representation in Pavlovian conditioning: Image and action. *Cognition* 37: 105–131.

Illich, P. A., Salinas, J. A., and Grau, J. W. (1994). Latent inhibition and overshadowing of an antinociceptive response in spinalized rats. *Behavioral and Neural Biology* 62: 140–150.

Joynes, R. L. and Grau, J. W. (1996). Mechanisms of Pavlovian conditioning: The role of protection from habituation in spinal conditioning. *Behavioral Neuroscience* 110: 1375–1387.

Joynes, R. L. and Grau, J. W. (in press). Instrumental learning with the spinal cord: III. Prior exposure to noncontingent shock induces a behavioral deficit that is blocked by an opioid antagonist. *Behavioral Neuroscience*.

Joynes, R. L., Crown, E. D., and Grau, J. W. (submitted). Instrumental learning with the spinal cord: II. Evidence for spinal mediation.

Kahneman, D., Frederickson, B. L., Schreiber, C. A., and Redelmeier, D. A. (1993). When more pain is pre-ferred to less: Adding a better end. *Psychological Science* 4: 401–405.

Kesner, R. P. and Novak, J. M. (1982). Serial position curve in rats: Role of the dorsal hippocampus. *Science* 218: 173–175.

Mackintosh, N. J. (1975). A theory of attention: Variations in the associability of stimuli with reinforcement. *Psychological Review* 82: 276–298.

Maier, S. F. and Jackson, R. L. (1979). Learned helplessness: All of us were right (and wrong): Inescapable shock has multiple effects. In *The Psychology of Learning and Motivation*, G. H. Bower, ed., pp. 155–218. New York: Academic Press.

Maier, S. F. and Seligman, M. E. P. (1976). Learned helplessness: Theory and evidence. *Journal of Experimental Psychology: General* 105: 3–46.

Meagher, M. W., Chen, P., Salinas, J. A., and Grau, J. W. (1993). Activation of the opioid and nonopioid hypoalgesic systems at the level of the brainstem and spinal cord: Does a coulometric relation predict the emergence or form of environmentally-induced hypoalgesia? *Behavioral Neuroscience* 107: 493–505.

Moye, T. B., Coon, D. J., Grau, J. W., and Maier, S. F. (1981). Therapy and immunization of long-term analgesia in rats. *Learning and Motivation* 12: 133–148.

Overmier, J. B. and Seligman, M. E. P. (1967). Effects of inescapable shock upon subsequent escape and avoidance learning. *Journal of Comparative and Physiological Psychology* 63: 28–33.

Patterson, M. M. (2001). Classical conditioning of spinal reflexes: The first seventy years. In *Model Systems and the Neurobiology of Associative Learning*, J. E. Steinmetz, M. A. Gluck, and P. R. Solomon, eds., pp. 1–22. Hillsdale, N.J.: Lawrence Erlbaum Associates.

Patterson, M. M. and Grau, J. W. (2001). *Spinal Cord Plasticity: Alterations in Reflex Function*. Boston: Kluwer.

Rescorla, R. A., Grau, J. W., and Durlach, P. J. (1985). Analysis of the unique cue in configural conditioning. *Journal of Experimental Psychology: Animal Behavior Processes* 11: 356–366.

Seligman, M. E. P., Maier, S. F., and Geer, J. (1968). Alleviation of learned helplessness in the dog. *Journal of Abnormal Psychology* 73: 256–262.

Seligman, M. E. P., Rosellini, R. A., and Kozak, M. (1975). Learned helplessness in the rat: Reversibility, time course, and immunization. *Journal of Comparative of Physiological Psychology* 88: 542–547.

Skinner, B. F. (1938). *The Behavior of Organisms.* Englewood Cliffs, N.J.: Prentice-Hall.

Sutherland, R. J. and Rudy, J. W. (1989). Configural association theory: The role of the hippocampal formation in learning, memory, and amnesia. *Psychobiology* 17: 129–144.

Timberlake, W. and Silva, K. M. (1995). Appetitive behavior in ethology, psychology, and behavior systems. In *Perspectives in Ethology.* Vol. II, *Behavioral Design*, N. S. Thompson, ed., pp. 211–253. New York: Plenum.

Wagner, A. R. (1981). SOP: A model of automatic memory processing in animal behavior. In *Information Processing in Animals: Memory Mechanisms*, N. E. Spear and R. R. Miller, eds., pp. 5–47. Hillsdale, N.J.: Lawrence Erlbaum Associates.

Wagner, A. R., Rudy, J. W., and Whitlow, J. W. (1973). Rehearsal in animal conditioning. *Journal of Experimental Psychology* 97: 407–426.

Woodruff-Pak, D. S. (1993). Eyeblink classical conditioning in H. M.: Delay and trace paradigms. *Behavioral Neuroscience* 107: 911–925.

Michael Domjan

Were it not for sexual behavior, you and I would not be here, nor would many animal species. Given its critical role, it is not surprising that sexual behavior is multiply determined. Hormonal and neuroendocrine processes mediate sexual motivation or readiness to respond to environmental sexual stimuli. These responses are in turn mediated by both preprogrammed instinctive mechanisms as well as learning and memory.

The role of learning and memory in the control of sexual behavior has been recognized for a long time. For example, Craig (1918, p. 100) noted that "the sexual tendency is ... directed, *with much guidance by experience,* toward securing the stimulation required for discharging the sexual reflex" (emphasis added). However, only recently have the precise learning mechanisms involved in sexual behavior been empirically documented. These more analytic empirical studies have also helped to identify a number of cognitive factors involved in sexual conditioning. The evidence suggests that sexual behavior is delicately tuned by learning and memory mechanisms that shape both the stimulus control and the effectiveness of sexual responses.

Conditioning of Sexual Anticipatory Behavior

In many situations, learning is manifest in the development of anticipatory behavior. The proverbial Pavlovian dog salivates in anticipation of the presentation of food. Rats freeze in anticipation of painful stimulation, and pigeons peck a lighted key that signals brief access to grain. Conditioned anticipatory behavior is also readily evident in sexual situations.

Conditioned anticipatory behavior has been examined most extensively in male domesticated quail (*Coturnix japonica*). Male quail learn to approach a localized stimulus (a light, for example) that is presented shortly before each time they receive access to a sexually receptive female. In such situations, the light is referred to as the conditioned stimulus or CS, and access to the female serves as the unconditioned stimulus or US. The development of sexually conditioned approach behavior requires the pairing of the conditioned stimulus with copulatory opportunity (Domjan et al. 1986). Beyond that requirement, however, the learning occurs under a remarkably wide range of circumstances. Conditioned approach behavior develops, for example, even if the conditioned stimulus is presented more than 2 m away from the location of a female that provides copulatory opportunity (Burns and Domjan 2000) and even if contingencies are introduced so that the CS is followed by the US only on those trials when the subject fails to make the conditioned response (Crawford and Domjan 1993).

Under special circumstances, sexual conditioning can also result in conditioned copulatory responses directed toward the conditioned stimulus. Copulation in quail consists of the male grabbing the back of the female's head and/or neck feathers (the grab response), mounting the female's back with both feet (the mount response), and making cloacal thrusts that bring the male's cloaca in contact with the female's (cloacal contact responses). Conditioned grab, mount, and cloacal contact responses occur if the CS is a three-dimensional object that can support copulatory behavior. Figure 12.1 shows an example of such behavior. The conditioned stimulus in this study was a terrycloth object that included a taxidermically prepared head and some neck feathers of a female quail. The pairing of such a CS object with copulatory opportunity results in conditioned grab, mount, and cloacal contact responses if the conditioning trials involve a relatively short interval between the CS and the US (e.g., Akins 2000; Cusato and Domjan 1998).

Figure 12.1
Male quail performing a conditioned grab, mount, and cloacal contact response on a conditioned stimulus object that includes a taxidermically prepared head and partial neck feathers of a female quail.

Sexually conditioned behavior shows many features common to a conditioned responses, including acquisition, extinction, retention, blocking, stimulus discrimination, trace conditioning, second-order conditioning, and conditioned inhibition (Akins and Domjan 1996; Crawford and Domjan 1995, 1996; Domjan et al. 1986; Köksal et al. 1994). To the extent that these features of Pavlovian conditioning reflect cognitive mechanisms, one may assume that cognitive mechanisms also occur in the sexual behavior system.

Stimulus-Response vs. Stimulus-Stimulus Learning

The development of an anticipatory conditioned response suggests that cognitive processes are involved in sexual behavior. However, the complexity of the underlying cognitive mechanisms is not revealed by a simple demonstration of sexual conditioning. In particular, the fact that a male quail responds to a signal for a female does not mean that it is doing so because the signal activates a representation of the female. It may be that the signal or CS automatically activates the conditioned response without activating a representation of the female that serves as the US.

In the Pavlovian conditioning literature, whether a CS activates a conditioned response directly or by first activating a representation of the US is addressed in terms of the distinction between stimulus-response (S-R) and stimulus-stimulus (S-S) learning. An S-R association is one in which the CS automatically elicits the conditioned response, without activating a representation of the US. Thus, an S-R association involves minimal cognitive processing. In contrast, an S-S association is one in which the CS first activates a representation of the US and the conditioned response is a manifestation of this activated representation. Only if sexual conditioning reflected S-S learning could one conclude that the sexual conditioned response reflects cognitive anticipation of the forthcoming sexual encounter.

S-R learning can be distinguished empirically from S-S learning on the basis of manipulations intended to alter the value or attractiveness of the anticipated event or US after conditioning has taken place (e.g., Holland and Rescorla 1975). If the US is food, for example, the attractiveness of the US may be reduced by conditioning an aversion to the food or by making the subjects not hungry. If the conditioned response reflects the cognitive anticipation of food, then procedures that reduce the value or attractiveness of food should reduce the vigor of the conditioned response. In contrast, if the vigor of the conditioned response is not changed by reducing the value or attractiveness of food, then one may conclude that the conditioned response is generated by an S-R mechanism and does not reflect anticipatory cognitions about food.

The available evidence indicates that conditioned sexual behavior is mediated by S-S rather than S-R mechanisms. Thus, conditioned sexual behavior appears to involve the activation of a cognitive representation of the female that is signaled by the CS. In one experiment, for example, Holloway and Domjan (1993) first established a sexually conditioned approach response by pairing a CS with access to a female with whom the males could copulate. The value or attractiveness of the female was then reduced by reducing the sexual motivation of the subjects. This was accomplished by restricting their exposure to light from 16 hours a day to 6 hours a day. Testosterone production (and concomitant sexual motivation) was disrupted by this restricted exposure to light. The reduced sexual motivation (and interest in females) produced a corresponding reduction in the sexually conditioned approach behavior. Conditioned responding was restored when the photoperiod was returned to 16 hours a day.

Subsequent experiments confirmed these findings with direct manipulations of serum testosterone levels and showed that reduced sexual motivation affects sexually conditioned behavior, but does not affect responding conditioned with a food US (Holloway and Domjan 1993). These results indicate that sexually conditioned behavior is mediated by cognitive anticipation of the female that was previously paired with the CS. Alterations in the attractiveness of the female produce corresponding changes in sexually conditioned behavior.

Further evidence of the role of S-S mechanisms in sexual conditioning has been obtained by Hilliard and Domjan (1995), who showed

that sexual satiation also reduces sexual conditioned approach behavior. In a subsequent study, Hilliard et al. (1998) tested the effects of sexual satiation on conditioned responses to a three-dimensional CS object that supported both conditioned approach and conditioned copulatory responses. Sexual satiation reduced both types of conditioned responses, but not to the same extent. Conditioned copulatory responses (mounts and cloacal contacts) were suppressed more by sexual satiation than was conditioned approach behavior.

One interpretation of these results is that S-S mechanisms (or activated representations of a female) have a greater role in conditioned copulatory responses than they do in conditioned approach behavior. Thus, Hilliard et al.'s (1998) findings suggest that cognitive mechanisms may be more important in the control of copulatory or sexual consummatory behavior than they are in the control of approach or appetitive responses that are a precursor to copulation.

Temporal Encoding in the Control of Sexual Anticipatory Behavior

Perception of time is a traditional area of animal cognition (Gibbon and Allan 1984). Most studies of animal timing or temporal coding have been conducted with food and aversive USs. Recent evidence indicates that temporal factors are also important in the control of sexual anticipatory behavior. Furthermore, some of the temporal relations that have been obtained in sexual conditioning are similar to those that have been documented in more conventional learning situations.

Important evidence of the temporal control of sexual anticipatory behavior was first obtained by Akins et al. (1994) in a study of the CS-US interval in sexual conditioning. Male quail received a training procedure in which a CS was presented for either 1 or 20 minutes immediately before access to a sexually receptive female. One conditioning trial was conducted each day, for a total of 15 trials. Control groups received the CS and US in an unpaired fashion. The CS (a foam block) was presented at one end of an experimental chamber that was unusually large (1.2 × 1.8 m). With such a large chamber, approach to the CS could be easily distinguished from nonlocalized locomotor behavior or pacing. Approaching the location of the CS may be considered to be a form of focal search for a potential sexual partner. In contrast, pacing from one side of the experimental chamber to the other may be considered to be akin to a nonlocalized general search response. The short and long CS-US intervals selectively activated these two types of behaviors. Subjects that received the 1-minute CS-US interval came to approach the CS as learning progressed. In contrast, subjects that received the 20-minute CS-US interval developed pacing behavior as the conditioned response. These results were confirmed in a subsequent study by Akins (2000).

These findings are unusual because they contradict the common assumption that learning is disrupted by increasing the CS-US interval. Rather, they indicate that different conditioned responses are activated by short in contrast to long CS-US intervals. Based on extensive analyses of the temporal organization of the behaviors involved in foraging for and eating food, Timberlake (2001) suggested that short and long CS-US intervals activate different response modes. Signals that occur just before an animal encounters food are presumed to activate focal search and food-handling response modes. In contrast, signals that are more remotely related to food are presumed to activate a general search mode. Whether the CS approach and pacing responses identified by Akins et al. in sexual conditioning reflect focal search rather than general search modes is speculative. Nevertheless, the development of different conditioned responses indicates that the subjects encoded the temporal differences between the short and long CS-US intervals.

Further evidence for the importance of temporal factors in sexual anticipatory behavior was recently reported by Burns and Domjan (2001). Unlike Akins et al. (1994), who focused on effects of the CS duration, Burns and Domjan examined how conditioned sexual behavior is influenced by the temporal context in which a given CS duration is presented. In particular, they focused on the ratio of the duration of exposure to the experimental context (C) compared with the duration of the conditioning trial (T) or conditioned stimulus. Studies with both food and aversive USs had shown that conditioned responding is an increasing function of the C/T ratio (see Gallistel and Gibbon 2000 for a review). Burns and Domjan found that sexually conditioned CS approach behavior is also directly related to the C/T ratio. Subjects that are conditioned with a high C/T ratio spend more time near the CS.

In addition to measuring CS-directed conditioned behavior, Burns and Domjan (2001) also measured conditioned behavior directed toward the location of the US (goal tracking). Unlike CS-directed behavior, US-directed conditioned behavior was inversely related to the C/T ratio. The contrasting effects of the C/T ratio on CS-directed and US-directed conditioned behavior are unprecedented in the literature. Furthermore, both findings provide evidence that conditioned behavior reflects encoding of not just the duration of the conditioned stimulus but also the relation between the CS duration and the duration of exposure to the contextual cues in which the CS is embedded.

Modulation of Responding to Sexual Sign Stimuli

As we have seen, depending on temporal factors, sexual anticipation established through Pavlovian conditioning can be manifest in different types of responses elicited by the conditioned stimulus. Another consequence of sexual anticipation is that it enhances responding to the sexual sign stimuli that are provided by a copulation partner. In fact, this modulation of responding to sexual sign stimuli may be of greater functional significance than CS-directed conditioned behavior.

Early evidence that sexual anticipation can enhance responding to sexual sign stimuli was obtained by Zamble et al. (1985) in male rats. In one experiment, for example, the CS was provided by placing the subjects in a plastic tub before exposing them to a receptive female. Sexual conditioning of the CS did not elicit readily identifiable conditioned responses. However, males exposed to the CS just before being allowed to copulate with a female achieved ejaculation significantly faster. Thus, exposure to the CS enhanced the effectiveness of the sexual sign stimuli that were provided by the female copulation partner. Similar facilitation of copulatory behavior in response to a sexual partner has been obtained in studies with male Japanese quail (Domjan et al. 1986).

The increased effectiveness of sign stimuli for sexual behavior engendered by sexual anticipation is clearly illustrated by studies in which only partial sexual sign stimuli are presented. Experiments of this sort have been conducted with male Japanese quail. The sexual behavior of male quail can be triggered by the visual cues of a receptive female, even if those visual cues are provided by a taxidermic model of a female. Partial sign stimuli can be constructed by providing exposure to just the head and neck of a female. Ordinarily, such head and neck cues are not sufficient to elicit sexual behavior on the part of most male quail. However, the limited cues of a female's head can elicit substantial approach and copulatory responses if those sign stimuli are presented in the presence of contextual cues that were previously associated with copulation with a live female.

In the first study of this effect (Domjan et al. 1989), males were alternatively housed in an experimental chamber and in their home cages. One group was allowed to copulate with a female on 15 occasions in the experimental cham-

ber; the other subjects received these copulatory opportunities in their home cage. All of the subjects were then tested with a head and neck model of a female in the experimental chambers. Significantly more grab, mount, and cloacal contact responses were elicited by the female head and neck cues in subjects for whom the experimental chamber had been paired with sexual reinforcement. A subsequent study demonstrated that such enhanced responding to sexual sign stimuli can develop with just one context conditioning trial (Hilliard et al. 1997).

Enhanced responding to the sign stimuli provided by a sexual partner has been also demonstrated in a fish species, the blue gourami (Hollis et al. 1997). A light served as the CS, and conditioning consisted of presenting the light for 10 seconds paired with visual access to a sexual partner. (The CS and US were presented unpaired for a control group.) After 18 daily conditioning trials, the barrier separating the male and female was removed after the CS presentation to enable the fish to engage in their full range of reproductive behavior responses.

The most significant data in the experiment were not provided by what occurred during the CS, but by what occurred when the male and female were allowed to interact after the CS. The results indicated that prior exposure to the CS facilitated the reproductive interactions of the male and female (by decreasing aggressive responses and increasing nesting and copulatory responses) and resulted in significantly more offspring being produced. Related studies with quail have shown that exposure to a sexually conditioned stimulus increases sperm output (Domjan et al. 1998) and results in greater numbers of fertilized eggs (Domjan 2000).

Observational Conditioning of Sexual Behavior

Additional evidence of cognitive modulation of sexual behavior is provided by studies demonstrating observational conditioning of sexual behavior (Köksal and Domjan 1998). In these experiments, a terrycloth object was presented as the CS and paired with the release of a sexually receptive female from a side cage. Male quail in an experimental group were permitted to observe this sequence of events from the other side of a wire mesh barrier on 12 occasions. Conditioning trials were then continued, but now with the observers allowed to interact with the CS object and copulate with the female that was released after the CS presentation. The acquisition of approach to the CS object was measured during this second phase. Males that previously received observational conditioning acquired the conditioned approach response faster than subjects in a control group that observed the CS and US in an unpaired fashion. These results suggest that the observation of a CS paired with the sight of a receptive female is sufficient to establish an association between these events.

Further evidence of observational learning in sexual situations has been obtained by Galef and White (1998), who showed that female quail increase their preference for a male as a result of seeing that male copulate with another female. In contrast, males that observe a female mating with another male show reduced preference for that female (White and Galef 1999a; see also White and Galef 1999b, 2000).

Conclusion

The results of sexual conditioning experiments provide substantial evidence of cognitive mediation of sexual behavior. Sexual anticipatory behavior becomes readily established in male quail and other species through Pavlovian pairings of a signal with sexual reinforcement. Evidence indicates that the conditioned sexual behavior is mediated by activation of a representation of the potential sexual partner. The sexual anticipatory behavior also reflects temporal coding of the duration of the conditioned sexual signal and the relation between the signal's duration and the duration of exposure to the context in which the CS is presented. Sexually conditioned stimuli modulate the effectiveness of the species' typical sign stimuli that are provided by a sexual partner. Further evidence of cognitive mediation is provided by demonstrations of the conditioning of sexual behavior by observation.

Taken together, these findings suggest that sexual behavior is not an automatic outcome of exposure to sexual sign stimuli provided by a potential sexual partner. Rather, the effectiveness of these stimuli is fine tuned by contextual cues, temporal factors, and learning and memory of previous sexual experiences. As Craig (1918) surmised nearly a century ago, sexual behavior is "directed, with much guidance by experience."

Acknowledgment

My research was supported by grant MH39940 from the National Institute of Mental Health.

References

Akins, C. K. (2000). Effects of species-specific cues and the CS-US interval on the topography of the sexually conditioned response. *Learning and Motivation* 31: 211–235.

Akins, C. K. and Domjan, M. (1996). The topography of sexually conditioned behaviour: Effects of a trace interval. *Quarterly Journal of Experimental Psychology* 49B: 346–356.

Akins, C. K., Domjan, M., and Gutiérrez, G. (1994). Topography of sexually conditioned behavior in male Japanese quail (*Coturnix japonica*) depends on the CS-US interval. *Journal of Experimental Psychology: Animal Behavior Processes* 20: 199–209.

Burns, M. and Domjan, M. (2000). Sign tracking in domesticated quail with one trial a day: Generality across CS and US parameters. *Animal Learning & Behavior* 28: 109–119.

Burns, M. and Domjan, M. (2001). Topography of spatially directed conditioned responding: Effects of context and trial duration. *Journal of Experimental Psychology: Animal Behavior Processes* 27: 269–278.

Craig, W. (1918). Appetites and aversions as constituents of instinct. *Biological Bulletin* 34: 91–107.

Crawford, L. L. and Domjan, M. (1993). Sexual approach conditioning: Omission contingency tests. *Animal Learning & Behavior* 21: 42–50.

Crawford, L. L. and Domjan, M. (1995). Second-order sexual conditioning in male Japanese quail (*Coturnix japonica*). *Animal Learning & Behavior* 23: 327–334.

Crawford, L. L. and Domjan, M. (1996). Conditioned inhibition of social approach in male Japanese quail (*Coturnix japonica*) using visual exposure to a female. *Behavioural Processes* 36: 163–169.

Cusato, B. M. and Domjan, M. (1998). Special efficacy of sexual conditioned stimuli that include species typical cues: Tests with a conditioned stimulus preexposure design. *Learning and Motivation* 29: 152–167.

Domjan, M. (2000). New perspectives from a functional approach to Pavlovian conditioning. Presidential address, Division of Behavioral Neuroscience and Comparative Psychology, American Psychological Association, Washington, D.C., August 2000.

Domjan, M., Lyons, R., North, N. C., and Bruell, J. (1986). Sexual Pavlovian conditioned approach behavior in male Japanese quail (*Coturnix coturnix japonica*). *Journal of Comparative Psychology* 100: 413–421.

Domjan, M., Greene, P., and North, N. C. (1989). Contextual conditioning and the control of copulatory behavior by species-specific sign stimuli in male Japanese quail. *Journal of Experimental Psychology: Animal Behavior Processes* 15: 147–153.

Domjan, M., Blesbois, E., and Williams, J. (1998). The adaptive significance of sexual conditioning: Pavlovian control of sperm release. *Psychological Science* 9: 411–415.

Galef, B. G., Jr. and White, D. J. (1998). Mate-choice copying in Japanese quail, *Coturnix coturnix japonica*. *Animal Behaviour* 55: 545–552.

Gallistel, C. R. and Gibbon, J. (2000). Time, rate, and conditioning. *Psychological Review* 107: 289–344.

Gibbon, J. and Allan, L. (eds.) (1984). *Annals of the New York Academy of Sciences*, Vol. 423. *Time and Time Perception*. New York: New York Academy of Sciences.

Hilliard, S. and Domjan, M. (1995). Effects on sexual conditioning of devaluing the US through satiation. *Quarterly Journal of Experimental Psychology* 48B: 84–92.

Hilliard, S., Nguyen, M., and Domjan, M. (1997). One-trial appetitive conditioning in the sexual behavior system. *Psychonomic Bulletin and Review* 4: 237–241.

Hilliard, S., Domjan, M., Nguyen, M., and Cusato, B. (1998). Dissociation of conditioned appetitive and consummatory sexual behavior: Satiation and extinction tests. *Animal Learning & Behavior* 26: 20–33.

Holland, P. C. and Rescorla, R. A. (1975). The effect of two ways of devaluing the unconditioned stimulus after first- and second-order appetitive conditioning. *Journal of Experimental Psychology: Animal Behavior Processes* 1: 355–363.

Hollis, K. L., Pharr, V. L., Dumas, M. J., Britton, G. B., and Field, J. (1997). Classical conditioning provides paternity advantage for territorial male blue gouramis (*Trichogaster trichopterus*). *Journal of Comparative Psychology* 111: 219–225.

Holloway, K. S. and Domjan, M. (1993). Sexual approach conditioning: Tests of unconditioned stimulus devaluation using hormone manipulations. *Journal of Experimental Psychology: Animal Behavior Processes* 19: 47–55.

Köksal, F. and Domjan, M. (1998). Observational conditioning of sexual behavior in the domesticated quail. *Animal Learning & Behavior* 26: 427–432.

Köksal, F., Domjan, M., and Weisman, G. (1994). Blocking of the sexual conditioning of differentially effective conditioned stimulus objects. *Animal Learning & Behavior* 22: 103–111.

Timberlake, W. (2001). Motivational models in behavior systems. In *Handbook of Contemporary Learning Theories*, R. R. Mowrer and S. B. Klein, eds., pp. 155–209. Mahwah, N.J.: Lawrence Erlbaum Associates.

White, D. J. and Galef, B. G., Jr. (1999a). Mate choice copying and conspecific cueing in Japanese quail, *Coturnix coturnix japonica*. *Animal Behaviour* 57: 465–473.

White, D. J. and Galef, B. G., Jr. (1999b). Social effects on mate choices of male Japanese quail, *Coturnix japonica*. *Animal Behaviour* 57: 1005–1012.

White, D. J. and Galef, B. G., Jr. (2000). Differences between the sexes in direction and duration of response to seeing a potential sex partner mate with another. *Animal Behaviour* 59: 1235–1240.

Zamble, E., Hadad, G. M., Mitchell, J. B., and Cutmore, T. R. (1985). Pavlovian conditioning of sexual arousal: First- and second-order effects. *Journal of Experimental Psychology: Animal Behavior Processes* 11: 598–610.

13 Cognition and Emotion in Concert in Human and Nonhuman Animals

Ruud van den Bos, Bart B. Houx, and Berry M. Spruijt

A major question in our research program is how in human and nonhuman animals the balance between positive ("reward") and negative experiences ("punishment," "stress") affects the efficiency of long-term behavior—defined as choosing strategies with the most profitable cost–benefit outcomes—in a "complex" (multiple-choice) environment. The balance between positive and negative experiences results from the continuous integration of such experiences. At least in humans it has been shown that failure of such an integration leads to inefficient long-term behavior in a complex (social) environment (Damasio 1996). In order to address this question, we place experiences in a neurobehavioral model, show how they can be made accessible for experimentation—especially in nonhuman animals—and briefly discuss a method for studying long-term behavioral efficiency.

Motivational Systems and Neurobiology

Emotion and Cognition

The relationships between internal physiological changes on the one hand and behavioral changes on the other in relation to the availability of different commodities in living organisms are described as motivational systems. Commodities are items in the environment that are potentially important to the animal's fitness. Motivational states such as hunger, thirst, and libido arise because of a difference between actual and reference values in an animal's physiological systems, and subsequent behavior—appetitive and consummatory (Craig 1918)—is directed at eliminating this difference.

As reviewed elsewhere, two features of commodities are relevant for an animal's behavior (Spruijt et al. 2001): knowledge of when and where a commodity is available (cognition) and assessment of the incentive or rewarding value of a commodity before and after consumption (emotion).

The "when" component of cognition deals with an animal's capacity to associate stimuli with the arrival of commodities. These mechanisms have been studied in Pavlovian conditioning experiments. The cognitive load increases as the interval between the offset of the cue (conditioned stimulus, CS) and the onset of the arrival of the commodity (unconditioned stimulus, US) increases from zero (delay conditioning) to several seconds or minutes (trace conditioning; Lieberman 2000; Clark and Squire 1998; Wallenstein et al. 1998). The trace-conditioning paradigm measures what others have referred to as beliefs in the context of intentional action or goal-directed behavior (Heyes and Dickinson 1990). The "where" component of cognition deals with an animal's capacity to assess the stimuli that indicate where commodities may be found. These mechanisms have been studied in a number of spatial tasks such as the Olton radial maze and the Morris water maze (Kalat 1998). The cognitive load increases as the stimuli that indicate the commodity's position are progressively less directly "attached" to the commodity itself [the shift from proximal (direct) to distal (configurations of) stimuli (Wallenstein et al. 1998)]. Behavior based on these configurations of stimuli may be conceived of as belief-based behavior, as mentioned earlier.

Incentive value is dependent on an animal's current internal state (the more the animal has been deprived of the commodity, the higher the commodity's incentive value), prior experience with the commodity, and its general properties (for food as a commodity, see Grill and Berridge 1985). Thus incentive value may differ among different commodities and over time for the same commodity.

Berridge and colleagues (Berridge 1996; Berridge and Robinson 1998) have convincingly argued that the system mediating incentive value has two different components with separable functions: the affective component ("liking") and the appetitive component or the disposition to consume the commodity ("wanting"). Both "liking" and "wanting" are theoretical constructs referring to different internal (neuronal or psychological) evaluation processes in the valuation of and behavior toward commodities. "Liking" refers to the immediate appraisal of commodities as pleasurable or not pleasurable (after consumption); "wanting" refers to the disposition to act upon this appraisal on future or new occasions, or in other words, on the activated representation of "the liked" (prior to consumption; see also Spruijt et al. 2001). "Liking" and "wanting" are conceptually similar to what has been described elsewhere as "emotion" and "desire" in the context of intentional action or goal-directed behavior (Heyes and Dickinson 1990; cf. van den Bos 2000).

We (Spruijt et al. 2001) have suggested elsewhere that the role of the system mediating the incentive value may lie in guiding the organism in "doing the best thing," which is captured in Cabanac's maxim: pleasant is useful (Cabanac 1971). In other words, what is good for the individual's fitness in the long run (functional level of analysis of behavior) is directed by an immediate sense of "good" or "bad" (causal level of analysis of behavior). We (Spruijt et al. 2001) have argued that different motivational systems share this reward-mediating system so that conflicts between motivational systems can be settled by using a common currency.

For instance, when selecting whether to eat or to drink, levels of physiological factors representing the metabolic state cannot be directly compared with levels of physiological factors representing the state of bodily fluids. However, if both systems also measure the degree of pleasure they provide, a comparison based on a common currency may be carried out by a structure that is connected to both motivational systems.

We propose that to make this comparison (in fact a comparison of apples and oranges), it is sufficient to compare the degree to which the difference between an actual and a reference value can be reduced, irrespective of the nature of the difference, by a commodity with a specific incentive value. The common currency could be the physiological representation of the reduction, for each motivational system, of the difference between actual and reference values. The system that produces the largest reduction (i.e., the highest concentration of a physiological measure) related to the reward will allow control over behavior to be initiated [this concept is captured in Cabanac's maxim: pleasure is the common currency (Cabanac 1992)]. Thus each motivational system has specific physiological consequences for initiating and terminating activities aimed at a specific goal. Apart from those specific consequences, there are also common consequences that allow a comparison between motivational systems. This comparison can only be made when the level that deals with common consequences has a supervisory position in the hierarchical organization of various motivational systems.

Neurobiology

This hierarchical organization of motivational systems has its counterpart in the organization of the central nervous system. The central nervous system may be conceptualized as a hierarchically organized series of negative feedback loops in which stimuli are processed at different levels and in which each level adds its specific component or programming rule to behavior (see Cools 1985; van den Bos 1997; Powers 1973).

At the lowest levels, the processing of stimuli is more directly related to the physical aspects of the stimuli. The processing is rapid, so that immediate responses are elicited, whereas at increasingly higher levels the processing becomes progressively less directly related to the physical aspects of the stimuli (i.e., more abstract) and less rapid, so that more delayed responses are elicited and facilitated. In other words, the higher levels deal with more general aspects of the programming of behavior whereas the lower levels deal with more concrete aspects. Each level in the hierarchy controls its own aspect in the programming of behavior (Powers 1973).

Motivational systems show such a hierarchical neuronal organization. At lower levels in the central nervous system, specific motor patterns directed at consuming a commodity are regulated, whereas at higher levels, more general aspects of obtaining a commodity and assessing its value are regulated. [See e.g., Grill and Berridge (1985) and the next section for food items; see Spruijt et al. (2001) for a general discussion.] As discussed by Berridge and colleagues, "liking" is mediated by the opioid endorphin (forebrain) system, whereas "wanting" is mediated by the dopamine (forebrain) system (Berridge and Robinson 1998). These systems interact in (inter alia) the ventral tegmental area (VTA). The VTA contains the dopaminergic projections to the ventral striatum, prefrontal cortex, and amygdala, which are activated by the endorphin system (see Spruijt et al. 2001 for a review). This leads to the prediction that "liking" affects "wanting" and not the other way round. The belief cognitive component is at least dependent on the hippocampus for its expression (Clark and Squire 1998; Wallenstein et al. 1998). The hippocampus is an integrating high-level center in the central nervous system (see Spruijt et al. 2001 for a review).

Measuring Emotion, Cognition, and Their Relationship

In order to study the relationship between "liking," "wanting," and "beliefs," independent measures are needed that can be (cor)related at the level of individuals.

Measuring "Liking"

Although the following method concentrates on "liking" in the context of food items, it may be applied to other commodities as well. In the context of food items, "liking" is conceived of as perceived palatability (Berridge and Robinson 1998). Palatability is not a measure of taste quality or intensity, but is a complex central evaluation (Grill and Berridge 1985). In humans, changes in palatability are often measured through verbal reports and are scored on a one-dimensional scale ranging from pleasant to unpleasant (Cabanac 1971), whereas in animals such changes are measured by changes in species-specific behavioral patterns in response to food, so-called "taste reactivity patterns" (TRPs; for a review see Berridge 2000). In general, two kinds of TRPs can be distinguished: ingestive or hedonic patterns and aversive patterns. They have been described in many species, including rats (*Rattus norvegicus*) and various primate species and humans (Berridge 1996, 2000; Grill and Berridge 1985). We have recently described TRPs in domestic cats (*Felis silvestris catus*; van den Bos et al. 2000).

These hedonic and aversive TRPs occur in relation to two different classes of food stimuli or solutions, sweet (glucose being the prime example) and bitter (quinine being the prime example), respectively. They are present early during ontogeny (see Berridge 2000; Grill and Norgren 1978; Grill and Berridge 1985) and do not appear to require structures above the midbrain for their expression (see Grill and Norgren 1978). As such, they appear to represent an immediate and unconditional response to a stimulus (level 0 as defined by Grill and Berridge 1985). Given the appearance of these TRPs as described in different papers (in cats, lip licking as a hedonic pattern; in rats, head shaking as an aversive pattern), it may be argued that they are the product of natural selection operating over many generations either to facilitate the ingestion of food items with a high nutritional value as much as possible or to remove food items with a potentially high mortality risk from the oral cavity or area as much as possible.

The fact that the balance between hedonic and aversive TRPs in response to stimuli may be changed according to the internal state of the organism or previous experiences (for reviews see Berridge 1996, 2000; Berridge and Robinson 1998; Grill and Berridge 1985) shows that these behavioral patterns are also under the control of higher levels in the central nervous system and that these stimuli are also processed at higher levels (levels 1, 2, and 3 according to Grill and Berridge 1985). As such, "palatability" is hierarchically organized.

It is likely that the internal processing at progressively higher levels becomes more unitary, so that at the highest level only a single continuum exists (liking: pleasant-unpleasant) on which the organism will base its behavior (or which feeds into the next system, such as the system dealing with "wanting"), while the direct food-related response at the lowest levels is expressed along two different dimensions. In humans this single continuum is what is experienced and verbally or otherwise reported.

The question now is whether it is possible to take these two different low-level dimensions and integrate them into one high-level dimension. We have reduced these two behavioral dimensions to one hypothetical internal (psychological) dimension by subtracting the sum of the mean durations of the aversive patterns from the sum of the mean durations of the hedonic patterns and labeled this the composite palatability score (CPS; Meijer et al. submitted). We have subsequently explored the use of the CPS in order to measure "liking" in cats. Our results thus far show that the CPS is potentially a valid measure to assess "liking" based on these TRPs.

This procedure leads to the following prediction: There may be individuals in which the im-

mediate responses to food items as reflected by hedonic and aversive TRPs are present (they are directed at ingesting or removing as much of the food as possible) but "liking" as measured by CPS (the internal evaluation of the food's incentive value as reflected by the strength of the hedonic and aversive TRPs) is not; and there may be individuals in which both exist. This refers to the ontogeny of these levels in individuals of the same species or to the phylogeny of these levels between individuals of different species. What is needed therefore is to show that "liking" has consequences for future behavior, and that these consequences may be manipulated independently of the immediate occurrence of the TRPs themselves. As argued earlier, "liking" affects "wanting." By delineating an independent measure for "wanting" and by showing that "liking" (CPS) directly affects "wanting" (measured in whichever way), one step is taken to show ontogenetic and/or phylogenetic processes.

Measuring "Wanting"

If "wanting" reflects the activation of the "liked," then it follows that the more a commodity is liked, the more an individual is willing to work for that commodity, or the more the animal will look forward to the commodity's arrival when it is reliably announced. The former may be measured by operant conditioning using a progressive ratio paradigm, the latter by Pavlovian conditioning, testing during the extinction phase or by using an interval between conditioned and unconditioned stimulus (Spruijt et al. 2001). Combinations of these exist as well (see Wyvell and Berridge 2000). Up to now we have employed mainly the Pavlovian conditioning procedure, although we will apply all three methods in the near future.

In our Pavlovian paradigm, a conditioned stimulus is paired with a commodity (unconditioned stimulus) in which the interval between the CS and US is gradually increased from zero to even 20 minutes in rats, cats, and minks (van den Berg et al. 1999; van den Bos et al., in preparation; von Frijtag et al. 2000; Spruijt et al. 2001). A relatively long interval between CS and US has been used to allow the extensive occurrence of anticipatory behavior. The hypothesis is that the anticipatory behavior seen with different kinds of reward (food, receptive female) has general characteristics in common. We have observed an increase in the number of behavioral transitions (hyperactivity) in the CS-US interval in rats in experiments involving these different rewards (Spruijt et al. 2001). It seems that the major characteristic of this commodity-aspecific anticipatory behavior is the increased frequency of exploratory behavioral patterns, as if short abrupt fragments of these behavioral elements are displayed.

Anticipation is not expressed in the same way in different species under similar conditions because we have recently observed that in cats anticipation is not expressed as an increase in activity in a Pavlovian conditioning procedure, but rather as a decrease using the same paradigm as in rats (van den Bos et al., in preparation; cf. Timberlake and Silva 1995).

"Liking" and "Wanting"

Up to now we have only fully combined these approaches in a Pavlovian paradigm using domestic cats, measuring "liking" by the occurrence of TRPs and "wanting" by the number of behavioral transitions. The data thus far suggest that, as predicted, the more an item is "liked" as measured by the CPS, the more it is "wanted" as measured by the number of behavioral transitions. Owing to the more complex analysis of "food liking" in rats (see Berridge 1996, 2000) we have not managed to fully combine these approaches in rats. However, it was shown recently in rats that, as we predicted, pharmacologically enhancing "wanting" did not increase

"liking" (Wyvell and Berridge 2000). Furthermore, Dickinson and colleagues have shown that rats change their "wanting and belief behavior" (measured by pressing levers in an operant task) according to changes in "liking" (the incentive value of food items) in setups that indicate that they behave by combining these different mental representations (see van den Bos 1997 for a review).

Now the question arises of how we can show that this system affects long-term efficiency in behavior. To show how this might be approached, in the next two sections we discuss an experimental method to change the system's sensitivity (the consequence of social stress in adult rats) and an experimental setup to show how multiple choices may be measured and manipulated (a closed-economy system).

Changing the Sensitivity of the Incentive Value System by Social Stress

In a recent series of experiments we submitted rats (*Rattus norvegicus*) to a social stress paradigm (von Frijtag et al. 2000). Using the Pavlovian conditioning paradigm, it turned out that stressed rats, in contrast to control rats, did not show hyperactive behavior in the interval between the CS and US (5 percent glucose solution; "wanting"), whereas their consumption of the glucose solution ("liking") appeared normal. Accordingly, it would appear that the rats are able to assess the immediate value of glucose— although it should be mentioned that we did not measure TRPs in this study and intake is not a reliable indicator of "liking" (Berridge 1996)— but they are not able to integrate its incentive value in their future behavior because they are not able to recall its value prior to presentation. The effect of stress on anticipatory behavior could be reversed by long-term treatment with the antidepressant drug imipramine (von Frijtag et al., submitted).

Given these results, it is now possible to study these rats in more detail in different kinds of environments that put different loads on their ability to assess the costs and benefits of different strategies. It may be predicted that the more complex the environment is in terms of the number of different choices an animal has to make, the more an animal will fail to behave efficiently. Such a test is described in the next section.

Multiple-Choice Environment: A Closed-Economy System

One approach to measuring long-term behavioral efficiency and the role of cognition and emotion in it in different species is the closed-economy approach that Mason and colleagues (Cooper and Mason 2000; Mason et al. 2001) have used for their studies in mink (*Mustela vison*). In this setup, the animals live for long periods of time in an environment in which they can visit several compartments containing various commodities. Costs can be imposed upon the entrance to these compartments by, for example, attaching weights to entrance doors, and benefits can be increased or decreased by changing the commodity's value. In general, animals will prioritize activities according to the cost–benefit ratios they attach to different commodities (Mason et al. 2001). Thus this system allows the assessment of various features of their behavior toward commodities; for example, whether they differentially appreciate the consumption of commodities ("liking"; cf. method above), whether they look forward to consuming commodities or are willing to work for commodities ("wanting"), and whether they are able to learn the location of and the relationship between cues and the arrival of commodities (cognition). It may be expected that when the "incentive value system" is not functioning properly (e.g., owing to chronic stress), is not yet developed (ontogeny), or is not present (phylogeny), the affected animals will show different priorities in their

behavior than animals in which this system is present and functioning optimally.

Conclusion

Human and nonhuman animals behave efficiently in changing environments. This efficiency of behavior requires that (1) they have knowledge of spatial and temporal relationships in their environment (cognition); (2) they are able to measure and predict the success of the outcome of their actions (emotion); and (3) emotion and cognition are tightly connected. The approach presented here allows us to carefully delineate the system mediating the incentive value in different species and thereby understand its role in long-term efficient behavior. The integration of methods and concepts from neuroscience, psychology, and ethology has led to a fuller understanding of the relationship between an animal and its environment.

References

Berg, C. L. van den, Pijlman, F. T. A., Koning, H. A. M., Diergaarde, L., Van Ree, J. M., and Spruijt, B. M. (1999). Isolation changes the incentive value of sucrose and social behaviour in adult and juvenile rats. *Behavioural Brain Research* 106: 133–142.

Berridge, K. C. (1996). Food reward: Brain substrates of wanting and liking. *Neuroscience and Biobehavioral Reviews* 20: 1–25.

Berridge, K. C. (2000). Measuring hedonic impact in animals and infants: Microstructure of affective taste reactivity patterns. *Neuroscience and Biobehavioral Reviews* 24: 173–198.

Berridge, K. C. and Robinson, T. E. (1998). What is the role of dopamine in reward: Hedonic impact, reward learning, or incentive salience? *Brain Research Reviews* 28: 309–369.

Bos, R. van den (1997). Reflections on the organisation of mind, brain and behaviour. In *Animal Consciousness and Animal Ethics; Perspectives from the Netherlands*, M. Dol, S. Kasanmoentalib, S. Lijmbach, E. Rivas, and R. van den Bos, eds., pp. 144–166. Assen, The Netherlands: Van Gorcum.

Bos, R. van den (2000). General organizational principles of the brain as key to the study of animal consciousness. *Psyche* 6: published online at http://psyche.cs.monash.edu.au/v6/psyche-6-05-vandenbos.html.

Bos, R. van den, Meijer, M. K., and Spruijt, B. M. (2000). Taste reactivity patterns in domestic cats (*Felis silvestris catus*). *Applied Animal Behaviour Science* 69: 149–168.

Bos, R. van den, Meijer, M. K., Van Renselaar, J. P., Van der Harst, J. E., and Spruijt, B. M. (in preparation). Anticipatory behaviour in domestic cats (*Felis silvestris catus*).

Cabanac, M. (1971). Physiological role of pleasure. *Science* 173: 1103–1107.

Cabanac, M. (1992). Pleasure: The common currency. *Journal of Theoretical Biology* 155: 173–200.

Clark, R. E. and Squire, L. R. (1998). Classical conditioning and brain systems: The role of awareness. *Science* 280: 77–81.

Cools, A. R. (1985). Brain and behavior: Hierarchy of feedback systems and control of its input. In *Perspectives in Ethology*. Vol. 6, P. Bateson and P. Klopfer, eds., pp. 109–168. New York: Plenum.

Cooper, J. J. and Mason, G. J. (2000). Increasing costs of access to resources cause re-scheduling of behaviour in American mink (*Mustela vison*): Implications for the assessment of behavioural priorities. *Applied Animal Behaviour Science* 66: 135–151.

Craig, W. (1918). Appetites and aversions as constituents of instincts. *Biological Bulletin of the Marine Biological Laboratory* 34: 91–107.

Damasio, A. R. (1996). The somatic marker hypothesis and the possible functions of the prefrontal cortex. *Philosophical Transactions of the Royal Society of London B* 351: 1413–1420.

Frijtag, J. C. von, Reijmers, L. G. J. E., Harst, J. E. van der, Leus, I. E., Bos, R. van den, and Spruijt, B. M. (2000). Defeat followed by individual housing results in long-term impaired reward- and cognition-related behaviours in rats. *Behavioural Brain Research* 117:137–146.

Frijtag, J. C. von, Bos, R. van den, Spruijt, B. M. (submitted). Imipramine restores the long-term impairment of appetitive behavior in socially stressed rats.

Grill, H. J. and Berridge, K. C. (1985). Taste reactivity as a measure of the neural control of palatability. In *Progress in Psychobiology and Physiological Psychology*, J. M. Sprague and A. N. Epstein, eds., pp. 1–65. Orlando, Fla.: Academic Press.

Grill, H. J. and Norgren, R. (1978). The taste reactivity test. II. Mimetic gustatory responses to stimuli in chronic thalamic and chronic decerebrate rats. *Brain Research* 143: 281–297.

Heyes, C. and Dickinson, A. (1990). The intentionality of animal action. *Mind and Language* 5: 87–104.

Kalat, J. W. (1998). *Biological Psychology*. 6th ed., p. 357. Pacific Grove, Calif.: Brooks/Cole.

Lieberman, D. A. (2000). *Learning. Behavior and Cognition*, 3rd ed., pp. 103–107. Stamford, Conn.: Wadsworth/Thomson Learning.

Mason, G. J., Cooper, J., and Clarebrough, C. (2001). Frustrations of fur-farmed mink. *Nature* 410: 35–36.

Meijer, M. K., Van Renselaar, J. P., and Bos, R. van den (submitted). Measuring "liking" in domestic cats (*Felis silvestris catus*): The composite palatability score.

Powers, W. T. (1973). *Behavior: The Control of Perception*. Chicago: Aldine.

Spruijt, B. M., Bos, van den R., and Pijlman, F. (2001). A concept of welfare based on how the brain evaluates its own activity: Anticipatory behaviour as an indicator for this activity. *Applied Animal Behaviour Science* 72: 145–171.

Timberlake, W. and Silva, K. M. (1995). Appetitive behavior in ethology, psychology, and behavior systems. In *Perspectives in Ethology*. Vol. 11, *Behavioral Design*, N. S. Thompson, ed., pp. 211–255. New York: Plenum.

Wallenstein, G. V., Eichenbaum, H., and Hasselmo, M. E. (1998). The hippocampus as an associator of discontiguous events. *Trends in Neurosciences* 21: 317–323.

Wyvell, C. L. and Berridge, K. C. (2000). Intra-accumbens amphetamine increases the conditioned incentive salience of sucrose reward: Enhancement of reward "wanting" without enhanced "liking" or response reinforcement. *Journal of Neuroscience* 20: 8122–8130.

14 Constructing Animal Cognition

William Timberlake

Cognition refers to a subset of processes that define and operate on the relations between environment and behavior. The current study of animal cognition varies from emphasis on the specialized to the general. Scientists interested in a particular species often focus on complex cognition particular to that species. Thus, students of temperate zone songbirds are interested in song learning and migration. Other scientists are primarily interested in how closely the cognition of nonhuman animals approaches that of humans, as in the case of language (Savage-Rumbaugh et al. 1998). Still others are interested in cognition that is characteristic of a wide range of species, as in the case of scalar timing and conditioning (Gallistel and Gibbon 2000). This essay considers an approach to animal cognition that is compatible with this range of interests, an approach based on constructing the mechanisms, function, and evolution of cognition in one species at a time.

Constructing cognition in this way requires tools and information from a variety of sources. Three sources have roots in the nineteenth century: ethology, learning psychology, and the physiology of perceptual-motor relations. A fourth contributor is a modern version of the art of creating artificial animals, now based in computers and robots (Taylor, chapter 21 in this volume). The final contributor is the ancient human practice of using experience-based knowledge to view the world as though one were, in fact, a particular animal. I will call this practice "theromorphism" (taking the animal's view) to distinguish it from the more common anthropomorphic practice of presuming that the cognition processes of human and nonhuman animals are fundamentally the same, and from the even more common emautomorphic practice of presuming that the cognition processes of other beings, regardless of species, are identical to one's own. In the following discussion I briefly outline what

each source potentially contributes to the study of animal cognition.

Ethology

Ethologists grounded animal cognition in careful observation of the development, control, and vigor of naturally occurring behavior. Influenced by naturalists like von Uexküll, they were also concerned with the animal's view of the world. In an influential paper on the "Umwelt" of an animal, von Uexküll (1934/1957) combined his personal observations with information on the physiology of receptors to create pictures of the sensory world of animals ranging from mollusks to flies and dogs. Ethologists like Tinbergen (1951) created more dynamic scenes by carefully observing naturally occurring sequences of behavior, dividing them into interlocking sets of perceptual-motor units (critical releasing stimuli and species-typical responses). By manipulating characteristics of the releasing stimuli, they explored mechanisms controlling the occurrence and intensity of responses. For example, after carefully illuminating the courtship dance of male and female sticklebacks (a small temperate zone fish), Tinbergen (1951) performed experiments using artificial "models" of males and females to clarify the mechanisms underlying perceptual-motor organization.

Based on both observations and experiments, Tinbergen (1951) summarized the reproductive behavior of male sticklebacks in a hierarchical, motivational model. This model divided the perceptual-motor units into repertoires associated with different motivational states (feeding, migrating, territory defense, courtship, and parental behavior), which were determined by the current stimulus conditions and the previous state. Although he did not extend his modeling efforts beyond this example, other investigators developed motivational systems of fear, aggression, parental behavior, and feeding (see Eibl-Eibesfeldt 1975).

In short, ethology established the importance of careful observation of naturally occurring behavior and showed the value of experimental manipulation of critical stimuli in clarifying the control of perceptual-motor units. Based on observation and experiment, ethologists developed functional models relating stereotyped responses, stimulus filters, and motivational states. Finally, ethologists showed how classical evolutionary comparisons designed to trace phylogenetic descent or environmentally based convergence of morphological characters could also be applied to perceptual-motor units (Lorenz 1950; Tinbergen 1959).

Learning Psychology

Learning psychology defined cognition by using artificial tasks created by experimenters. In reaction to widespread anthropomorphic speculation about the motivations and feelings of animals (e.g., Romanes 1884), early learning researchers aggressively tested and argued for the sufficiency of simple learning explanations for complex tasks. For example, Thorndike (1911) tested the ability of hungry cats to solve latch puzzles to gain access to food. He found that their performance improved trial by trial, based on rewarded repetition, rather than with the suddenness expected from reasoning, observational learning, or general cleverness. He subsequently showed that monkeys solving similar problems also used trial and error. No shrinking violet, Thorndike set an influential precedent by very early summarizing his data in the form of general laws of the effects of reward and punishment in generating efficient new behavior, laws that were presumed to apply to all organisms.

In addition to an abiding interest in general causal laws, learning psychology contributed a set of experimental paradigms (combinations of apparatus, procedure, measures, and species) that provided a "test bench" for establishing functional relations among dependent and independent variables. In some cases these paradigms were used to test predictions of general laws and models, with a strong emphasis on using control groups to isolate the effects of interest. In other cases (notably in Skinnerian and applied psychology) paradigms were used primarily to shape behavior and establish reliable response patterns and relations to stimuli.

In still other cases, tasks were developed to establish the ability of different species to solve complex cognition problems, such as matching one stimulus with another, forming learning sets, discriminating the odd stimulus among three, counting, or reasoning. Initially, the point of this research was to establish a protoevolutionary ranking of species' abilities (Timberlake and Hoffman 1998); however, more recent experimenters, following an analysis of the component skills involved in unique human behaviors, have tested animals separately for each skill (e.g., Pepperberg 1999; Povinelli et al. 1997; Premack 1988; Savage-Rumbaugh et al. 1998).

Physiology of Perceptual-Motor Relations

A good portion of the study of animal cognition during the first half of the twentieth century involved investigation of the anatomy and operating characteristics of the sensory receptors of particular species. When combined with the enumeration of reflexes and the development of learning tasks, considerable knowledge was added about the sensory windows of specific species, including the physical range, sensitivity, and discriminative capabilities of different senses. Based on the study of insects, biologists led by Loeb (1918) proposed a set of simple models of how specific sensorimotor mechanisms controlled orientation and movement. Unfortunately, psychologists studying the sensorimotor control of the orientation and movement of rats in mazes found remarkable interchangeability among different senses in controlling behavior (see Munn 1950). This contrast between simple general models and subsequent causal complexity set a pattern in this area that was repeated several times over the century.

Much of the subsequent data clarifying the sensory and motor worlds of animals has come from combining neurophysiology, mechanics, and the circuitry of sensory receptors with a fine-grained analysis of the behavior controlled by these receptors. Thus the navigational path integration system of desert ants has been shown to be a product of specialized receptors for polarized light and the calculation of optical flow (Wehner and Wehner 1990). Classic work by von Frisch (1965) on communication among foraging bees related characteristics of their dance behavior to the path and energy necessary to find the food source; subsequent work established more of the sensorimotor mechanisms responsible (Dyer 1998; Gould 1998).

The past two decades have witnessed the discovery of remarkable connections between sensory processing of prey cues by predators and their related search and capture behavior. For example, careful work on the visual system of the European toad reveals a clear relation between the firing rate of a class of cell in the optic tectum and the behavioral response of the toad to worms (Ewert 1987). For an extensive summary of other specific examples of the complex and intimate ties between the neurophysiology of the sensorimotor world and behavior, see Carew (2000).

Uttal (1998) recently argued compellingly that it is not possible in principle to reduce cognition processes defined by environmental stimuli and responses to brain circuitry because of the degrees of freedom created by the complexity of brain elements and function. This problem increases in severity the more complex and abstract the cognition under study. The most progress has been made in relating simpler perceptual-motor cognition to physiological mechanisms, especially given the many points of linkage between the environment and behavior that occur in predation (Carew 2000).

Because evolutionary success is not based on a top-down design, we should not expect the nervous system to contain the clear circuitry of a well-designed television set or central processing chip. Instead, we might expect echoes of previous designs and sensorimotor circuits based on unexpected relations involving the environment and activity of the brain. Our increasing technical capacity to peer into brain activity, abetted ultimately (but probably more slowly than hoped) by artful gene knockouts, should facilitate analyses. This knowledge should help rule out implausible assumptions about cognition and its relation to neurophysiology, and suggest more plausible ones, especially in combination with modeling and the multiple ties between environment and behavior.

Computational and Robotic Models

The construction of model animals that move by wind or muscle power is an ancient art. Even model animals based on gears and levers powered by gravity, water, steam, or springs have been around for at least half a millennium. In the last half of the twentieth century, though, scientists began to focus on computers as mimics of the actions of brains. Initially researchers focused on general-purpose artificial intelligence programs designed to solve abstract problems (e.g., Newell and Simon 1958). More recently, researchers have worked on connectionist software models that use layers of neurons and simple learning rules to model sensory processing, categorization, spatial learning, and even language parsing and production (e.g., McClelland and Rumelhart 1986). A limitation of these models is that they are not unlike a brain slice in a dish in their dependence on someone to embed them in an environment.

A second form of model consists of simple, autonomous robots designed with bottom-up (subsumption) rather than top-down architecture (Brooks 1999). A major advantage of working with bottom-up robots is that they typically are designed to function (i.e., survive) in a real environment (such as the Martian landscape or the bottom of the ocean). Thus the builder is forced to include all the processes necessary for survival. There can be no promissory notes that in the future the robot will be made energy efficient or receive sensory organs or motor effectors. As a result, the robot comes closer to mimicking the embodied and situated realities of a living organism in three important ways.

First, there are constraints and tradeoffs involving efficiency and capacity, for example, fineness of discrimination versus speed of reporting, speed of movement versus endurance. Second, because the robot functions in a particular environment, the qualities of that environment can be assumed and used in the robot's cognition. Thus, a functional memory for food locations in an open, flat environment might be achieved by marking the substrate rather than by building a general memory capacity capable of storing the results of triangulating food locations using multiple landmarks. Third, there are potentially powerful advantages to requiring hardware and software to perform multiple functions. The result is that cognition is embedded in the interaction of parts of the robot with each other as well as with the environment. Like real animals, the robot cannot be understood as an isolated brain or slice; its behavior needs to be analyzed within its "selection" environment.

Finally, the bottom-up approach can be combined with genetic algorithms to produce a third form of modeling in which genetic algorithms are applied to either software animals or combinations of software and hardware animals (Beer 1990; Nolfi and Floreano 2001; Yamauchi and Beer 1994). These models provide an important component that has been missing in the study of animal cognition—the possibility of getting at the process of cognitive evolution more directly.

These animals may or may not resemble actual organisms, but it is possible to implement "experiments" to determine environmental and organismic prerequisites for the evolution of communication, or the circumstances conducive to the evolution of more, or of less, dependence on learning. Obviously as we come closer to modeling actual animals, the results of our evolutionary experiments become more relevant to animal cognition.

Constructing Animal Cognition

To this point I have briefly reviewed the kinds of contributions and tools provided by ethology, learning psychology, the physiology of perceptual-motor relations, and computational and robotic models. The next step, combining this information to construct the function and evolution of cognition, raises important questions about the sheer amount of data, its potential incompatibility, and the best way to organize and summarize the data.

Amount of Data

An underappreciated lesson from the successful genome projects of the past several years is that each project focused on laying out the genetic structure of a single species at a time, but the whole species, not just the head genes, or the muscle genes, or the genes on the first two chromosomes. I propose following a similar approach by trying to construct cognition in projects concerned with a single species at a time not just vision, or categorization, or motor capacity, but a functional animal. Once we get the hang of it, the construction of the cognition of different species should go much faster.

Incompatibility of Data

The problem of integrating data from different disciplines may be more apparent than real. The past 20 years of research on the physiology of perceptual-motor relations have focused on naturally occurring behavior that fits well with or was borrowed directly from ethology [see Carew (2000) for classic examples ranging from cricket calls to hunting in toads and barn owls]. Functional motivational systems and perceptual-motor units gleaned from observation and experimentation provide an immediate context to help analyze how the receptor characteristics and neurophysiological pathways relate environmental stimuli to behavior. In turn, neurophysiological analysis has clarified the mechanisms that control both stereotyped and more variable appetitive behavior. Analysis of the mechanisms of perceptual-motor relations also has profited from use of the experimental paradigms provided by learning psychology, while the results provide data about sensory processing that might promote the use of more effective combinations of stimuli, responses, and rewards.

In contrast, there is a history of conflict between ethology and laboratory learning, in part because the former concentrated on naturally occurring behavior and the latter on experimenter-defined behavior in artificial environments. However, three recent developments argue that this separation may be reconcilable. First, more researchers have drawn on the control and hypothesis testing traditions of laboratory learning to clarify the basis of niche-related behaviors, such as the distribution of foraging effort and the mechanisms of food storage and retrieval in birds (see Shettleworth 1998). Second, there is evidence that the presumably artificial paradigms of laboratory learning are based on niche-related mechanisms. In the process of tuning their experimental paradigms to produce reliable, vigorous, and interpretable behavior, it appears that psychologists have inductively made contact with mechanisms of niche-related learning (see Timberlake 2001a). An example is the apparent similarity between laboratory maze paradigms and the observed tendencies of rats to establish and follow trails in natural environments.

A last support for reconciliation lies in the use of motivational systems models similar to that of

Tinbergen (1951) to describe and predict behavior in both natural settings and laboratory paradigms (Timberlake and Lucas 1989; Timberlake and Silva 1995; Timberlake 2001b). A behavior systems model, such as the predatory subsystem of feeding in rats shown in figure 14.1, is based on the combination of observational data from free behavior circumstances and experimental data from laboratory paradigms. Behavioral observations provide the initial basis for the organization and components of the system. Reading across columns under each heading in the figure, the rightmost column represents actions, such as track (visual tracking at a distance). The next column to the left relates these actions to modules (learned or unlearned perceptual-motor units), such as chase (small moving objects). The next column to the left organizes modules in repertoires within modes (such as general search). In the leftmost column, modes are related to functional subsystems, such as predatory (behavior), and systems, such as feeding (level not shown).

Naturally occurring sequences of behavior can be generated by tracing actions (and related modules and modes) from top to bottom of the diagram, with oscillation and retracing when the behavior of the animal locates stimulus support for alternative modules or is unsuccessful in locating stimuli that maintain the present mode or lead to the next. The animal begins by expressing general search mode actions controlled by learned and unlearned perceptual-motor modules. In typical environments, these actions lead to circumstances that produce a shift to actions characteristic of the repertoire of perceptual-motor modules related to the focal search mode, which in turn leads to handling and consuming actions related to still another repertoire of perceptual motor modules.

It is important to note that laboratory procedures such as Pavlovian conditioning can be very useful in testing and clarifying such a model Timberlake (1994, 2001b). For example, the procedure of presenting an artificial moving prey stimulus that predicts food to different rodent

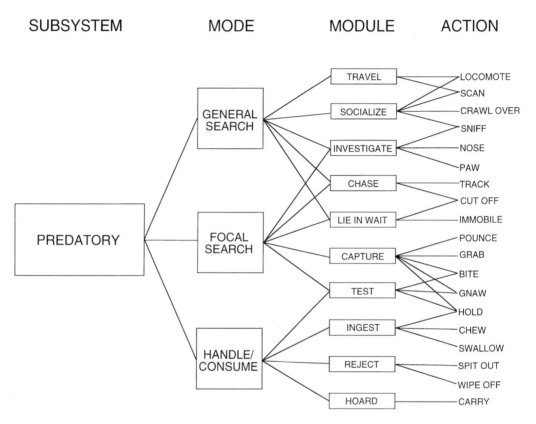

Figure 14.1
Proposed predatory subsystem of the feeding behavior system of the rat (*Rattus norvegicus*) consisting of modes, modules, and actions. The figure shows three search modes (general, focal, and handling/consummatory) and their overlapping repertories of learned and unlearned perceptual-motor modules. Actions controlled by system components in conjunction with the environment are shown on the far right. The environment can affect actions directly through local stimulus support and indirectly through effects at other levels of the system.

species with an appropriate delay reveals the organization of their general and focal search repertories related to predation (Timberlake et al. 1982; Timberlake and Washburne 1989). Finally, such a systems approach to modeling is general, not being limited to either feeding (Fanselow 1994; Domjan 1994) or rats (Domjan 1994; Hogan 1994).

Summarizing and Organizing Data

An efficient way to summarize and organize data is to construct a model representing the knowledge already assembled that can facilitate further thought and experiment, but this is not a trivial task. It appears, though, that humans have an attribute that can help in this process, namely, the ability to use their experience to integrate information about an animal's sensory physiology, behavioral organization, and learning to understand and predict its behavior, namely the theromorphism I mentioned in the introduction. Conversations with fishermen, hunters, and trackers reveal species and even individual specificity in their knowledge and predictions of animal behavior; this implies that they have a model of the animal developed from their experience. Listening to experienced observers of primates suggests that they develop similar models. Such implicit models appear to allow an observer to predict behavior by virtually placing him or herself in the position of a specific animal, not as a human, but as the animal. These models can be made more comprehensive and testable by giving them reality using computation and robotics. These models become more powerful as they include explicit knowledge of the animal's sensorimotor capacities and organization, motivational states, and learning possibilities.

Using Human Cognition as a Standard

This approach can be applied to any species, including humans. However, our tendency to use humans as a standard carries with it drawbacks, such as a tendency to focus on specialized processes that define us rather than cognition more likely to be shared broadly with other animals. There is also a tendency not to treat comparisons with humans using the same criteria as comparisons involving other species, namely, phylogenetic descent and environmentally based convergence and divergence. As in the case of any species, it would help to have carefully prepared motivational system models for humans that included relations with the behavioral physiology of sensorimotor relations. We could also use greater clarity about how human learning fits with niche-related mechanisms (e.g., Cosmides and Tooby 1987), and decrease our resistance to modeling humans in a bottom-up fashion.

Perhaps most important, using humans as a standard for studying other species makes it difficult not to invoke thoughts and feelings as direct causes of behavior, thereby stopping our inquiry short of the data needed to construct working models of cognition. In the approach described here, thoughts and feelings are phenomena to be explained, not the basis of explanation. To be sure, if there is no other information, anthropomorphic and emautomorphic inferences can be useful shortcuts to predicting human behavior. That is undoubtedly why these practices are a nearly ineradicable part of the human social toolkit (Beer 1992). But by continuing to invoke such human-centered explanations of the behavior of all nonhuman animals, we engage at best in a hopeful speciesism.

Acknowledgments

Preparation of this manuscript was supported by the National Science Foundation (grant IBN 17175) and the James McKeen Cattell Foundation. Thanks to Allen Neuringer and the editors for helpful feedback, and to Oregon Health Sciences University and Reed College for hosting me.

References

Beer, C. G. (1992). Conceptual issues in cognitive ethology. In *Advances in the Study of Behavior*. Vol. 21, P. J. B. Slater, J. S. Rosenblatt, C. Beer, and M. Milinski, eds., pp. 69–110. San Diego, Calif.: Academic Press.

Beer, R. D. (1990). *Intelligence as Adaptive Behavior: An Experiment in Computational Neuroethology*. New York: Academic Press.

Brooks, R. A. (1999). *Cambrian Intelligence: The Early History of the New AI*. Cambridge, Mass.: MIT Press.

Carew, T. J. (2000). *Behavioral Neurobiology: The Cellular Organization of Natural Behavior*. Sunderland, Mass.: Sinauer Associates.

Cosmides, L. and Tooby, J. (1987). From evolution to behavior: Evolutionary psychology as the missing link. In *The Latest on the Best*, J. Dupré, ed., pp. 279–306. Cambridge, Mass.: MIT Press.

Domjan, M. (1994). Formulation of a behavior system for sexual conditioning. *Psychonomic Bulletin and Review* 1: 421–428

Dyer, F. C. (1998). Cognitive ecology of navigation. In *Cognitive Ecology: The Evolutionary Ecology of Information Processing and Decision Making*, R. Dukas, ed., pp. 201–260. Chicago: University of Chicago Press.

Eibl-Eibesfeldt, I. (1975). *Ethology: The Biology of Behavior*. 2nd ed. New York: Holt, Rinehart and Winston.

Ewert, J. P. (1987). Neuroethology of releasing mechanisms: Prey-catching in toads. *Behavioral and Brain Sciences* 10: 337–368.

Fanselow, M. S. (1994). Neural organization of the defensive behavior system responsible for fear. *Psychonomic Bulletin and Review* 1: 429–438.

Frisch, K. von (1965). *Dance Language and Orientation in Bees*. New York: Springer-Verlag.

Gallistel, C. R. and Gibbon, J. (2000). Time, rate, and conditioning. *Psychological Review* 107: 289–344.

Gould, J. L. (1998). Sensory bases of navigation. *Current Biology* 8: R731–R738.

Hogan, J. A. (1994). Structure and development of behavior systems. *Psychonomic Bulletin and Review* 1: 439–450.

Loeb, J. (1918). *Forced Movements, Tropisms, and Animal Conduct*. Philadelphia, Pa.: J. B. Lippincott.

Lorenz, K. (1950). The comparative method in studying innate behavior patterns. *Symposia of the Society for Experimental Biology, IV*. Cambridge: Cambridge University Press.

McClelland, J. L. and Rumelhart, D. (1986). *Parallel Distributed Processing* vol. 2: Psychological and Biological Models: *A Handbook of Models, Programs, and Exercises*. Cambridge, Mass.: MIT Press.

Munn, N. R. (1950). *Handbook of Psychological Research on the Rat*. New York: Houghton-Mifflin.

Newell, A. and Simon, H. A. (1958). Elements of a theory of human problem solving. *Psychological Review* 65: 151–166.

Nolfi, S. and Floreano, D. (2001). *Evolutionary Robotics: The Biology, Intelligence and Technology of Self-organizing Machines*. Cambridge, Mass.: MIT Press.

Pepperberg, I. M. (1999). *The Alex Studies: Cognitive and Communicative Abilities of Grey Parrots*. Cambridge Mass.: Harvard University Press.

Povinelli, D. J., Reaux, J. E., Bierschwale, D. T., Allain, A. D., and Simon, B. B. (1997). Exploitation of pointing as a referential gesture in young children, but not adolescent chimpanzees. *Cognition* 12: 327–365.

Premack, D. (1988). Minds with and without language. In *Thought Without Language, a Fyssen Foundation Symposium*, L. Weiskrantz, ed., pp. 46–65. New York: Clarendon Press.

Romanes, G. J. (1884). *Mental Evolution in Animals*. New York: Appleton.

Savage-Rumbaugh, S., Shanker, S. G., and Taylor, T. J. (1998). *Apes, Language, and the Human Mind*. New York: Oxford University Press.

Shettleworth, S. J. (1998). *Cognition, Evolution, and Behavior*. New York: Oxford University Press.

Thorndike, E. L. (1911). *Animal Intelligence*. New York: Macmillan.

Timberlake, W. (1994). Behavior systems, associationism, and Pavlovian conditioning. *Psychonomic Bulletin and Review* 1: 405–420.

Timberlake, W. (2001a). Integrating niche-related and general process approaches in the study of learning. *Behavioural Processes* 54: 79–94.

Timberlake, W. (2001b). Motivational modes in behavior systems. In *Handbook of Contemporary Learning Theories*, R. R. Mowrer and S. B. Klein, eds., pp. 155–209. Hillsdale, N.J.: Lawrence Erlbaum Associates.

Timberlake, W. and Hoffman, C. M. (1998). Comparative analyses of learning. In *Comparative Psychology: A Casebook*, G. Greenberg and M. Haraway, eds., pp. 553–564. New York: Garland.

Timberlake, W. and Lucas, G. A. (1989). Behavior systems and learning: From misbehavior to general principles. In *Contemporary Learning Theories: Instrumental Conditioning Theory and the Impact of Biological Constraints on Learning*, S. B. Klein and R. R. Mowrer, eds., pp. 237–275. Hillsdale, N.J.: Lawrence Erlbaum Associates.

Timberlake, W. and Silva, K. M. (1995). Appetitive behavior in ethology, psychology, and behavior systems. In *Perspectives in Ethology*, N. Thompson, ed., pp. 211–253. New York: Plenum.

Timberlake, W. and Washburne, D. L. (1989). Feeding ecology and laboratory predatory behavior toward live and artificial moving prey in seven rodent species. *Animal Learning & Behavior* 17: 1–10.

Timberlake, W., Wahl, G., and King, D. (1982). Stimulus and response contingencies in the misbehavior of rats. *Journal of Experimental Psychology: Animal Behavior Processes* 8: 62–85.

Tinbergen, N. (1951). *The Study of Instinct*. Oxford: Clarendon Press.

Tinbergen, N. (1959). Comparative studies of the behaviour of gulls (Laridae): A progress report. *Behaviour* 15: 1–70.

Uexküll, J. von (1934/1957). The world of animals and men. In *Instinctive Behavior: The Development of a Modern Concept* (C. H. Schiller, ed. and trans.) pp. 5–81. New York: International Universities Press. (Translated from J. von Uexküll, 1934, *Streifzuge durch die Umwelten von Tieren und Menschen*. Berlin: Springer.)

Uttal, W. R. (1998). *Toward a New Behaviorism: The Case Against Perceptual Reductionism*. Mahwah, N.J.: Lawrence Erlbaum Associates.

Wehner, R. and Wehner, S. (1990). Insect navigation: Use of maps or Ariadne's thread? *Ethology, Ecology, and Evolution* 2: 27–48.

Yamauchi, B. M. and Beer, R. D. (1994). Sequential behavior and learning in evolved dynamical neural networks. *Adaptive Behavior* 2: 219–246.

15 Genetics, Plasticity, and the Evolution of Cognitive Processes

Gordon M. Burghardt

In this volume the emphasis has been on the diverse cognitive abilities of animals; the various ways that they can be studied both in the field and in captivity; and on theoretical issues as to the number, kinds, mechanisms, and comparative distribution of such abilities. It is now generally accepted that cognitive abilities have evolved just as have other characteristics of animals, such as anatomical structures. If natural selection produces animals with abilities to cope with problems in some areas of their lives better than in others, we need to focus on the different abilities animals possess and the contexts in which they are expressed. Moreover, for cognitive abilities to evolve, individuals in a population must differ in their cognitive abilities (or in the processes underlying them) and these differences must have adaptive consequences. Rarely, however, has work on these issues been carried out using modern methods of quantitative genetics in nondomesticated species, and yet such work may be particularly useful.

The evolution of cognitive abilities can be considered a subset of the evolution of plasticity in behavior. Behavioral plasticity has recently become a major concern in evolutionary biology (e.g., Via et al. 1995), but psychological interest in the issue goes back at least as far as the "organic selection" promulgated in the late nineteenth century (Belew and Mitchell 1996). In this effect, selection acts on plasticity itself. At the time, this mechanism was viewed as an elegant means of dealing with the then still-potent Larmarckian views of many scientists who could not envision how natural selection for either instincts or structures could explain the diversity found in nature without the inheritance of acquired traits.

Romanes (1883) and other theorists such as Herbert Spencer developed the neo-Lamarckian lapsed intelligence theory. This theory aimed to explain how instinctive or hardwired behavior could have evolved by postulating that initially such behavior was experience dependent or learned, then was subsequently encoded in the hereditary material and transmitted as inherited "instinct." In essence, the noncognitive evolved from the cognitive in terms of behavior, turning the typical evolutionary scenario on its head. Although the major thrust of organic selection was independently discovered by several eminent scientists (i.e., Baldwin, Poulton, and C. Lloyd Morgan) about 1896, today it is often simply called "the Baldwin effect" (see Belew and Mitchell 1996). Waddington's theory of genetic assimilation (Waddington 1953) was a much later model in this vein, and other more recent ones have been collected in Belew and Mitchell (1996).

However, attempts to study the Baldwin effect empirically, rather than through simulations and models, has proven difficult. One reason for the difficulty is that measuring a gene–environment interaction through behavioral changes in natural populations is difficult (Plomin and Hershberger 1991), although knowledge of such interactions is critical to how populations adapt to changing circumstances both phenotypically and genotypically. In many species dietary selection is a major arena for the operation of plasticity, and it is known that variation in both genetics and dietary experience plays an important role in responses to food (e.g., Burghardt 1993).

There are many reasons for the lack of research on the genetic bases of individual differences in cognitive abilities. First, comparative psychology was historically focused on species differences in "intelligence" and on ranking animals along some continuum. Here the search was for a key method that could produce a reliable indicator of intelligence across species; it assumed that there was such a single indicator if we could only discover and measure it.

Second, genetic studies require large numbers of animals that have been raised and tested identically, and this is hard to accomplish with dogs, monkeys, cats, pigeons, and other typical laboratory animals. Small samples of animals of diverse genetic backgrounds were tested intensively and individual differences dismissed as noise or the effects of pretraining experiences, if they were noted at all. Species that could be reared and tested in large numbers were laboratory mice and rats, and these were typically strains or breeds quite highly inbred, so genetic differences were minimal. Crosses between breeds and selective breeding did show a genetic basis, as demonstrated in the pioneering studies on dogs by Scott and Fuller (1965), but factors such as temperament and sensory function seemed to preclude a measure of "pure" intelligence and so such work fell into disfavor.

Recently, however, studies on the role of specific genes in behavior, including learning, exploration, and motivation, by adding and eliminating specific genes in inbred animals, have renewed interest in the genetic basis of individual differences in behavior within a population (Tang et al. 1999). Quantitative trait locus (QTL) methods are also being used to specify where genetic influences on certain traits are located on chromosomes. Such studies appear to hold promise for the genetic study of specific phenotypic attributes, but so far are limited to highly domesticated species or those for which extensive gene sequencing and pedigree data are available.

In my laboratory, we have focused on two biologically important areas where experience may play an important role—obtaining food and avoiding becoming food. We wanted to do this using fairly natural problems that an animal might need to solve in order to survive. We also wanted to use a species in which postnatal parental care and influence are minimal, and thus from birth individuals have to solve problems on their own or perish. It was also important to be able to obtain large numbers of animals from the

same population so that natural levels of genetic variation could be studied; ample evidence shows that geographical variation in behavior can be considerable (Foster and Endler 1999).

My work over many years on the behavior of snakes, particularly newborn or newly hatched snakes, led me to decide that snakes would be an ideal group to study for the quantitative genetics of basic cognitive processes. Ectothermic reptiles are not generally considered ideal subjects for the study of animal cognition, although there is a long history of their use in learning and discrimination studies (Burghardt 1977). Snakes, however, have a rather dismal success record in traditional learning paradigms in the old literature, primarily because their perceptual world and effector operating space are so alien to ours (Rivas and Burghardt 2001). In addition, it is because of the highly precocial and seemingly hardwired behavior of young snakes that their ability to learn has been neglected. Yet, neonatal snakes do satisfy the advantages listed above in terms of genetic studies. Furthermore, although their behavior is highly constrained by the absence of limbs and, for most species, a great reliance on chemosensory cues, snakes are surprisingly diverse and successful animals (Greene 1997).

While no one would claim they are as cognitively complex as mammalian carnivores and primates, snakes do possess a series of behavioral adaptations that are only beginning to be studied experimentally. For example, the death feigning of hognose snakes is triggered by the presence of a "predator," and recovery is more rapid if the predator, though present, is not gazing directly at the snake (Burghardt 1991). Many young snakes have tail tips with white or yellow markings that are used to lure unsuspecting prey; for example, hatchling Malayan pit vipers, *Calloselasma rhodistoma*, would raise their tail tips and slowly wiggle them in the presence of insectivorous lizards (G. M. Burghardt, Jones, and Schwartz unpublished observations). They would not do so in the presence of mice, although they would eat both kinds of prey. The diversity and complexity of such tail luring behavior has been recently reviewed (Reiserer in press). The use of the tail, not only as a lure, but also as a distraction display to cause a potential prey item, such as a rodent, to momentarily turn toward a rustling leaf and away from the snake's head approaching from a different direction, has been documented experimentally (Mullin 1999). There are also many anecdotal accounts of other tactics used by snakes in locating the nests of birds and turtles, homing, and so on (e.g., Mori et al. 1999). More traditional tests of snake learning, such as spatial tasks, are also being reported (e.g., Holtzman et al. 1999).

Particularly amenable to quantitative analysis are behavior patterns used repeatedly, such as the responses made by neonatal snakes to simulated predators, and the behavior involved in identifying, capturing, subduing, and swallowing prey. For many years we have studied the behavior of neonatal, young, and adult snakes, primarily natricine snakes in the family Colubridae. In North America, this family consists of "highly derived and recently evolved snakes" (Holman 2000). Natricines include the common water (*Nerodia*) and garter snakes (*Thamnophis*) of North America and their many East Asian relatives. All the New World natricines are viviparous and often have large litters, which facilitates genetic studies.

Avoiding Predation

Neonatal natricine snakes lack parental care and are typically small and vulnerable to predation. Natural selection should thus foster deployment of whatever cognitive ability a small snake might possess to the task of survival. Natricine snakes exhibit several antipredator behavior patterns, such as flight (including reversing direction), threatening (body or neck flattening, hissing, striking, slow-motion tongue flicking), and biting. When contacted by a predator, they may

writhe, wrap around the interloper, emit and rub a vile musk on the predator, or defecate. They may also hide their head and wave their tail, seemingly to direct the predator to their least vulnerable part. Both adult and neonatal snakes may have species-typical repertoires (Scudder and Burghardt 1983; Herzog and Burghardt 1986; Bowers et al. 1993). Within a population, the behavior of individual neonatal snakes may also differ at birth, and these temperament or personality differences appear to be heritable (Arnold and Bennett 1984; Brodie and Garland 1993) and can be remarkably stable (Herzog and Burghardt 1988).

The defensive behavior of snakes is not fixed at birth, however. There are changes that may occur over the first weeks of life, and these ontogenetic trajectories also vary by species (Herzog et al. 1992). Furthermore, experience does affect these responses. We discovered that littermates that were handled differently because they were used in different studies were more or less defensive than those given minimal handling (Herzog and Burghardt 1988). Young snakes tested with repeated but controlled contact stimulation for a short period on one day were much more defensive 2 weeks later, suggesting considerable retention if the appropriate test is utilized (Herzog 1990).

However, there was also evidence of genetic involvement in the response to repeated stimulation. Both short-term and long-term habituation were documented in garter snakes and we found differences in habituation processes both within and between species (Herzog et al. 1989). When short-term habituation trials were carried out in young snakes at day 1 and day 60, the slopes of the curves were remarkably consistent and, even more remarkable, the shapes of the curves were themselves consistent (Bowers 1992; B. B. Bowers and G. M. Burghardt, unpublished observations). All this evidence suggests that the way snakes process experience is under genetic control, down to the shape of their learning curves. Such studies, however, do not help eluci-date the nature of the gene–environment inter-actions involved. To begin to study this issue, we turned to aspects of foraging behavior.

Finding Prey

Just as with antipredator behavior, neonatal natricine snakes have species-typical innate responses to the chemical stimuli of prey at birth as well as precocial prey capturing abilities (see the reviews in Burghardt 1993; Burghardt and Schwartz 1999). The latter also show plasticity in early ontogeny (Krause and Burghardt 2001), but only the responses to food chemicals will be discussed here.

The most popular groups for the study of congenital responses to food cues are neonatal garter snakes of the North American natricine genus *Thamnophis*. Natural diets can include earthworms, fish, frogs, salamanders, leeches, slugs, small mammals, and birds. Newborn, in-gestively naive members of several species of this genus respond to aqueous surface extracts of prey with increased tongue flicking and open-mouthed strikes. The quantitative genetics of these and other behavioral characteristics are well established (Brodie and Garland 1993). Be-fore our work, however, all quantitative genetic estimates were based on the initial responses of snakes at birth. Such calculations may under-estimate both the rate of microevolution (Resnick et al. 1997) and the role of experience and onto-genetic processes in evolutionary events. For ex-ample, in humans, the heritability of behavioral and morphological measures increases over years (Petrill et al. 1998).

Snakes have to make many decisions when foraging, and one of the most critical is whether to attack a prey animal when one is encountered. Many studies have shown that postnatal dietary experience can alter the chemosensory-elicited tongue flicking and prey attack responses of young garter snakes (Burghardt 1993). Gener-ally the direction of change is toward increased responsivity to the diet fed, or avoidance if the diet is toxic. However, the amount of change, especially in relative preferences, seems to vary with individuals, species, prey type, and age of the individual. This led me to consider whether variation in how snakes respond to prey experi-ence is itself genetically controlled.

We studied the common garter snake, *Tham-nophis sirtalis*, the most wide-ranging and suc-cessful snake species in North America. This success may be due to the fact that the species is a prey generalist, feeding on almost every prey type recorded for any of the more than thirty garter snake species. The species is also known for its ability to invade areas with introduced food resources, such as fish hatcheries, where snakes living largely on earthworms shift to the new diet quite quickly and do better on it than on the natural diet (Gregory and Nelson 1991). To assess genetic effects on chemosensory prey preferences and their modification, neonatal snakes from 17 litters in a single population ($n = 79$) were tested at birth and after 6 weeks on the same diet (details in Burghardt et al. 2000).

Pregnant female eastern garter snakes, *Tham-nophis sirtalis*, captured over a few days in small field on Beaver Island in upper Lake Michigan, were moved to the laboratory. Beaver Island contains great numbers of garter snakes and their diet is almost exclusively earthworms (*Lum-bricus terrestris*), while ingestion of fish has never been recorded. Snakes from all litters were reared from birth exclusively on mosquito fish (*Gambusia affinis*). After birth, the animals were measured and then tested on their responses to mosquito fish and earthworm aqueous prey chemicals, along with control stimuli (water), presented on cotton swabs using standardized methods (see review in Burghardt 1993). Each trial took a maximum of 30 seconds.

Following the completion of four rounds of testing, all snakes were given twelve meals of live mosquito fish twice a week. All the snakes were again measured and the chemoreceptive tests repeated in the same format as before. The chemoreceptive responses were scored using a standard procedure. The scores for water and fish and worm replicates before the feeding experience were combined, as were those for the responses after dietary experience. Relative preference scores were derived by subtracting worm scores from fish scores both before and after the feeding experience. Fish and worm change scores were derived by subtracting early (prefeeding) from late (postfeeding) results for each prey stimulus. The change in relative preference was calculated by subtracting the early relative preference scores from the late relative preference scores. Heritabilities were calculated using maximum likelihood methods with sibship scores, and only significant values are reported below.

Both initially and after the fish diet experience, the snakes responded much more to prey extracts than to water. Initially fish and worm extracts were responded to almost equally, but after the feeding experience, only the response to fish increased significantly. In contrast to previous reports on other species or populations (Brodie and Garland 1993), there was no significant heritability of initial responses to either fish or earthworm extracts. However, after the fish diet experience, snakes retested in exactly the same manner now evinced a significant heritability of response to both fish (0.32) and worms (0.50). That this was not simply a general increase in responsivity to all extracts is shown by the fact that the change in the fish response was heritable (0.23), but not that to worms (0.16). Furthermore, the relative preference (fish score – worm score) was not heritable before the fish experience (0.07), but was heritable after this experience (0.53). In addition, the change in the relative prey preference was itself heritable (0.26).

These results clearly show that experience with fish was processed differently by the individual snakes and that the experiential effect was heritable. Thus, the ability to utilize experience in

altering perceptual responses to prey is under genetic influence in a natural population existing on a narrow diet. This appears to be the first such demonstration in a natural population.

This heritable plasticity in chemosensory prey preference had other effects that support the adaptive nature of such plasticity. Over the course of the experiment, the snakes grew significantly in mass and length. It is interesting that across all snakes there were significant positive correlations between growth measures and initial worm and fish chemosensory preferences (the results are reported below only for mass; body length results were very similar). After their dietary experience, however, only the correlation with a fish preference remained significant (fish: $r_s = 0.31$; worm: $r_s = 0.13$). On the other hand, the change in worm ($r_s = -0.23$), but not fish ($r_s = 0.12$) chemosensory preference was negatively related to the growth rate. Relative fish preference (fish – worm scores) after the snakes fed on fish was significantly related to growth ($r_s = 0.25$), but the same measure before feeding had no significant predictive power ($r_s = -0.10$). The change in this relative fish preference from before to after dietary experience was significant for growth in mass ($r_s = 0.26$).

Since all the snakes ate virtually the same number of fish, these data suggest that the sensory bias for the experiential effect on prey preference is positively related to the ability of the animals to utilize the diet effectively, which itself differs among individuals in a population. Since snakes with higher chemoreceptive responses to worms grew less well than those snakes that developed a chemosensory bias for fish, a strong unexpected link between perception, experience, diet quality, and growth is supported (Lyman-Henley and Burghardt 1995).

Future Directions

The study of genetic contributions to cognition is no simple matter. For example, multiple paternity occurs in this species (McCracken et al. 1999), and this would increase somewhat the heritabilities reported here. A litter can thus be composed of full and half siblings. Use of this trait and molecular genetic methods for identifying full and half siblings within a litter allows comparisons of two full sibships produced by the same mother at the same time and with the same maternal traits, such as body size, nutrition, stress, and body temperature. The application of this method to some morphological and behavioral traits in neonatal common garter snakes resulted in a better resolution of the genetic contribution through a comparison of paternal and maternal genetic contributions plus environmental effects (King et al. 2001). This method has yet to be applied to the chemosensory preference changes described earlier.

If the ability to profit by experience is an adaptive trait, with both costs and benefits depending on the context, then determining individual differences in such plasticity requires quantitative study in naturalistic contexts. Animals face many recurring problems and among the most important of these are avoiding predators and implementing the series of decisions needed to locate, capture, and ingest prey or other food resources. All animals face similar problems.

Genetic differences clearly underlie the plasticity of snake behavior in response to variable and ecologically relevant experiences. Furthermore, the heritable plasticity to shift behavior to favor novel prey types is related to the ability of snakes to metabolically utilize the relevant resource, in this case fish. This finding demonstrates that the traits underlying behavioral plasticity and cognitive processing are profoundly integrated into the entire biology of the species. If this is true in snakes, the importance of studying genetic and individual differences along with cognitive ethology may be even greater, although much more difficult, with non-domesticated mammals and especially primates.

Scattered throughout much of the primate literature is ample evidence that great individual differences in behavior, temperament, social prowess, and cognition exist. Often great claims are made on the basis of the accomplishments of a particular "star" subject: chimpanzee, dog, monkey, or parrot (Bekoff 1998). However, until the underlying basis and consequences of such differences, *and their variation*, are understood, it is unlikely that a detailed analysis of the evolution of cognitive abilities will be possible.

Acknowledgments

This work was supported by grants from the National Science Foundation, the University of Tennessee Science Alliance, and other sources of support over many years. The primary experiment described above would have been impossible to complete without the help of James Gillingham, Donna Layne, and Lyle Konigsberg. The editors and John Placyk made numerous useful comments on an earlier version of this chapter.

References

Arnold, S. J. and Bennett, A. F. (1984). Behavioral variation in natural populations. III. Antipredator displays in the garter snake *Thamnophis radix*. *Animal Behaviour* 32: 108–1118.

Bekoff, M. (1998). Deep ethology, animal rights, and the great ape/animal project: Resisting speciesism and expanding the community of equals. *Journal of Agricultural and Environmental Ethics* 10: 269–296.

Belew, R. K. and Mitchell, M. (eds.) (1996). *Adaptive Individuals in Evolving Populations*. Reading, Mass.: Addison Wesley.

Bowers, B. B. (1992). Habituation of Antipredator Behaviors and Responses to Chemical Prey in Four Species of Garter Snakes. Unpublished Ph. D. dissertation University of Tennessee, Knoxville.

Bowers, B. B., Bledsoe, A. E., and Burghardt, G. M. (1993). Responses to escalating predatory threat in garter and ribbon snakes (*Thamnophis*). *Journal of Comparative Psychology* 107: 25–33.

Brodie, E. D. III and Garland, T., Jr. (1993). Quantitative genetics of snake populations. In *Snakes: Ecology and Behavior*, R. A. Seigel and J. T. Collins, eds., pp. 315–362. New York: McGraw-Hill.

Burghardt, G. M. (1977). Learning processes in reptiles. In *The Biology of Reptilia*, Vol. 7, *Ecology and Behavior*, C. G. D. Tinkle, ed., pp. 555–681. New York: Academic Press.

Burghardt, G. M. (1991). Cognitive ethology and critical anthropomorphism: A snake with two heads and hognose snakes that play dead. In *Cognitive Ethology: the Minds of Other Animals*, C. A. Ristau, ed., pp. 53–90. Hillsdale, N.J.: Lawrence Erlbaum Associates.

Burghardt, G. M. (1993). The comparative imperative: Genetics and ontogeny of chemoreceptive prey responses in natricine snakes. *Brain, Behavior and Evolution* 41: 138–146.

Burghardt, G. M. and Greene, H. W. (1988). Predator simulation and duration of death feigning in neonate hognose snakes. *Animal Behaviour* 36: 1842–1844.

Burghardt, G. M. and Krause, M. (1999). Plasticity of foraging behavior in garter snakes (*Thamnophis sirtalis*) reared on different diets. *Journal of Comparative Psychology* 113: 277–285.

Burghardt, G. M. and Schwartz, J. M. (1999). Geographic variations on methodological themes from comparative ethology: A natricine snake perspective. In *Geographic Variation in Behavior: An Evolutionary Perspective*, S. A. Foster and J. A. Endler, eds., pp. 69–94. Oxford: Oxford University Press.

Burghardt, G. M., Layne, D. G., and Konigsberg, L. (2000). The genetics of dietary experience in a restricted natural population. *Psychological Science* 11: 69–72.

Foster, S. A. and Endler, J. A. (eds.) (1999). *Geographic Variation in Behavior: An Evolutionary Perspective*. Oxford: Oxford University Press.

Greene, H. W. (1997). *Snakes: The Evolution of Mystery in Nature*. Berkeley: University of California Press.

Gregory, P. T. and Nelson, K. J. (1991). Predation on fish and intersite variation in the diet of common garter snakes, *Thamnophis sirtalis*, on Vancouver Island. *Canadian Journal of Zoology* 69: 988–994.

Herzog, H. A., Jr. (1990). Experiential modification of defensive behaviors in garter snakes, *Thamnophis sirtalis*. *Journal of Comparative Psychology* 104: 334–339.

Herzog, H. A., Jr. and Burghardt, G. M. (1986). The development of antipredator responses in snakes: I. Defensive and open-field behaviors in newborns and adults of three species of garter snakes (*Thamnophis melanogaster, T. sirtalis, T. butleri*). *Journal of Comparative Psychology* 100: 372–379.

Herzog, H. A., Jr. and Burghardt, G. M. (1988). Development of antipredator responses in snakes. III. Stability of individual and litter differences over the first year of life. *Ethology* 77: 250–258.

Herzog, H. A., Jr., Bowers, B. B., and Burghardt, G. M. (1989). Development of antipredator responses in snakes. IV. Interspecific and intraspecific differences in habituation of defensive behavior. *Developmental Psychobiology* 22: 489–508.

Herzog, H. A., Jr., Bowers, B. B., and Burghardt, G. M. (1992). Development of antipredator responses in snakes. V. Species differences in ontogenetic trajectories. *Developmental Psychobiology* 25: 199–211.

Holman, J. A. (2000). *Fossil Snakes of North America*. Bloomington: Indiana University Press.

Holtzman, D. A., Harris, T. W., Aranguren, G., and Bostock, E. (1999). Spatial learning and memory of an escape task by young corn snakes, *Elaphe guttata guttata*. *Animal Behaviour* 57: 51–60.

King, R. B., Milstead, W. B., Gibbs, H. L., Prosser, M. R., Burghardt, G. M., and McCracken, G. F. (2001). Application of microsatellite DNA markers to discriminate between maternal and genetic effects on scalation and behavior in multiply-sired garter snake litters. *Canadian Journal of Zoology* 79: 121–128.

Krause, M. A. and Burghardt, G. M. (2001). Neonatal plasticity and adult foraging behavior in garter snakes (*Thamnophis sirtalis*) from two nearby, but ecologically dissimilar, habitats. *Herpetological Monographs*.

Lyman-Henley, L. P. and Burghardt, G. M. (1995). Diet, litter and sex effects on chemical prey preference, growth and site selection in two sympatric species of *Thamnophis*. *Herpetological Monographs* 9: 140–160.

McCracken, G. F., Burghardt, G. M., and Houts, S. E. (1999). Microsatellite markers and multiple paternity in the garter snake *Thamnophis sirtalis*. *Molecular Ecology* 8: 1475–1479.

Mori, A., Ota, H., and Kamezaki, N. (1999). Foraging on sea turtle nesting beaches: Flexible foraging tactics by *Dinodon semicarinatum* (Serpentes: Colubridae). In *Tropical Island Herpetofauna: Origin, Current Di-*

versity, and Conservation, H. Ota, ed., pp. 99–128. Amsterdam: Elsevier.

Mullin, S. J. (1999). Caudal distraction by rat snakes (Colubridae, *Elaphe*): A novel behavior used when capturing mammalian prey. *Great Basin Naturalist*, 59, 361–367.

Petrill, S. A., Plomin, R., Berg, S., Johansson, B., Pederson, N. L., Ahern, F., and McClearn, G. E. (1998). The genetic and environmental relationship between general and specific cognitive abilities in twins age 80 and older. *Psychological Science* 9: 183–189.

Plomin, R. and Hershberger, S. (1991). Genotype–environment interaction. In *Conceptualization and Measurement of Organism–Environment Interaction*, T. D. Wachs and R. Plomin, eds., pp. 29–43. Washington, D.C.: American Psychological Association.

Reiserer, R. (in press). Stimulus control and other feeding responses: Visual perception in vipers. In *Biology of the Pit Vipers*, G. W. Schuett, M. Hoggren, and H. W. Greene, eds. Traverse City, Mich.: Biological Sciences Press.

Resnick, D. N., Shaw, F. H., Rodd, F. H., and Shaw, R. G. (1997). Evaluation of the rate of evolution in natural populations of guppies (*Poecilia reticulata*). *Science* 275: 1934–1937.

Rivas, J. and Burghardt, G. M. (2001). Understanding sexual size dimorphism in snakes: Wearing the snake's shoes. *Animal Behaviour* 62: F1–F6.

Romanes, G. J. (1883). *Mental Life of Animals*. London: Kegan, Paul, Trench, Tröbner.

Scott, J. P. and Fuller, J. L. (1965). *Genetics and the Social Behavior of the Dog*. Chicago: University of Chicago Press.

Scudder, R. M. and Burghardt, G. M. (1983). A comparative study of defensive behavior in three sympatric species of water snakes. *Zeitschrift für Tierpsychologie* 63: 17–26.

Tang, Y.-P., Shimizu, E., Dube, G. R., Rampon, C., Kerchner, G. A., Zhuo, M., Liu, G., and Tslen, J. Z. (1999). Genetic enhancement of learning and memory in mice. *Nature* 401: 63–69.

Via, S., Gomulkiewicz, R., DeJong, G., Scheiner, S. M., Schlichting, C. D., and van Tiederen, P. H. (1995). Adaptive phenotypic plasticity: Consensus and controversy. *Trends in Ecology and Evolution* 10: 212–214.

Waddington, C. H. (1953). Genetic assimilation of an acquired character. *Evolution* 7: 118.

16 Spatial Behavior, Food Storing, and the Modular Mind

Sara J. Shettleworth

Behavior, together with the brain and cognitive processes that underlie it, is an evolved adaptation like eyes, teeth, wings, fins, and feathers. Questions about how behavior functions in the natural environment and how it evolved have long been prominent in ethology, but for much of its history the study of the psychological mechanisms underlying behavior, including cognition, has been remarkably abiological (Plotkin 1997). Currently, however, scientists studying all aspects of animal mind and behavior are converging on an integrated approach in which the interrelationships among ecology, brain, and behavior across a whole range of species are seen as key to understanding cognition, how it works, what it does for animals in nature, and how it evolved (examples are Balda et al. 1998; Dukas 1998; Hauser 1996; Shettleworth 1998). This essay briefly reviews aspects of one research program taking this approach—spatial memory in birds that store food. To introduce this animal-centered, as opposed to anthropocentric (Shettleworth 1998; Staddon 1989), approach to cognition, I begin with a few remarks on the evolution and organization of spatial behavior.

Self-propelled travel is a fundamental feature of animal life. The oldest fossil evidence of behavior is the tracks and burrows of primitive bottom-dwelling organisms (Raff 1996). Their movements may have been random and undirected, but eventually animals evolved senses for detecting distant objects and connections between sensation and movement that permitted them to approach or avoid things important for survival and reproduction. Even the simplest spatial orientation involves detection and recognition of some correlate of a goal, as when a male moth's antennal receptors detect a species-specific female pheromone and activate searching flight. Orientation toward places of refuge or reliable food sources specific to an individual's own environment may require learning and re-membering responses to otherwise neutral cues so the animal can get there from a distance.

It is a long way from primitive organisms wriggling and slithering through the mud to cognitive maps, which we come to in a moment. Such creatures and their simple behavior are mentioned to emphasize that as we navigate the gap of computational and neural complexity separating them from rats or human beings, we do not necessarily find a clear divide between the primitive and uninteresting, on the one hand, and the cognitively and computationally demanding, on the other. To develop a general comparative, evolutionary approach to the mind, it is essential to abandon hard-and-fast distinctions between cognitive and other mind–brain processes that mediate between sensory input from the environment and behavior.

It is also essential to adopt a modular view of cognition as opposed to assuming some single entity such as learning ability or intelligence that all species possess to some degree. Modularity is a fundamental feature of biological structure (see Raff 1996, chapter 10), the brain included (Barton and Harvey 2000). Cognitive scientists tend to speak of modules, if not always precisely in Fodor's (1983) sense, at least with reference to computationally distinct mechanisms (Coltheart 1999). Similarly, behavioral neuroscientists and neuropsychologists refer to memory systems, distinct areas of the brain that do distinct tasks or store distinct kinds of memories (Nadel 1992). And learning theorists speak of adaptive specializations of learning (Rozin and Kalat 1971), which have some of the same features as memory systems (Sherry and Schacter 1987) or modules (Gallistel 1999).The divisibility of brain and cognition into analytically distinct and somewhat independent subunits identified by all of these terms is well illustrated by spatial behavior.

Accurate spatial orientation can be accomplished by any of a variety of distinct mechanisms (Gallistel 1990; Shettleworth 1998). When intact animals find their way in the real world, more than one of these mechanisms may be at work, and the way in which such modules interact is an important topic of current research. An animal active during daylight can see the global visual panorama as well as landmarks like rocks and bushes near its nest, potential mates, or profitable food sources. Smells, sights, and sounds emanating directly from such places may serve as beacons, drawing the animal directly to them. The global geometry of visible space may be perceived and used independently of the features that make it up (Cheng 1986). Internal cues generated by an animal's own movement potentially allow it to keep track of where it is relative to a known starting point (dead reckoning or path integration; Biegler 2000). In well-traveled parts of its territory, stereotyped motor routines may carry it from place to place (Stamps 1995).

These different kinds of cues demand different implicit mental computations. For example, approaching a goal identified by a beacon is a simple hill-climbing process, whereas locating a place identified by one or more distal landmarks depends on a process like vector addition (Cheng 1989) or visual template matching (Cartwright and Collett 1987). Triggering and executing a stored motor routine is another thing again. Thus the various kinds of spatial information might be processed in dedicated mental modules, or memory systems (Gallistel 1990; Shettleworth 1998). Cognitive mapping (O'Keefe and Nadel 1978; Tolman 1948) refers to a unified allocentric (or earth-centered) representation of space that goes beyond, or perhaps integrates, the capacities of such separable orientation mechanisms to permit novel shortcuts between known locations. However, the term is often used more loosely to indicate, for instance, the use of configurations of distal cues as opposed to beacons (Pearce et al. 1998) or specific responses (Tolman 1948; see also Gallistel 1990). Moreover, once the subtle and sophisticated orientation possible with well-

specified spatial modules is taken into account, little, if any, evidence remains for cognitive mapping in the strict sense in any nonhuman species (Bennett 1996; Shettleworth 1998; but see Gould, chapter 6 in this volume).

Modularity in spatial cognition is delineated by testing animals in artificial environments that offer limited kinds of information. A comparison of species is another approach to defining cognitive modules. Just as there are species-characteristic adaptations of beaks, eyes, or feet, we might expect species-characteristic adaptations of specific aspects of behavior, brain, and cognition. The way in which studying such adaptations can help to illuminate the modular organization and evolution of cognitive processes is well illustrated by studies of natural history, brain, and memory in birds that store food. Here I emphasize work in my laboratory in which black-capped chickadees (*Poecile atricapillus*), birds that store food, were compared with nonstoring dark-eyed juncos (*Junco hyemalis*). Similar research on corvids is discussed by Balda and Kamil (chapter 17 in this volume). This research has been thoroughly reviewed elsewhere (Sherry and Duff 1996; Shettleworth 1995; Shettleworth and Hampton 1998); the focus here is on a series of experiments on memory for locations and colors and its connection with ideas about cognitive modularity.

As I write, the ground outside is deep in snow and the temperature has not risen above freezing for several weeks. The nights, when small passerine birds cannot forage, are 14 hours long. Yet black-capped chickadees spend the winter here in southern Ontario and throughout the higher latitudes of North America. The selective pressure is intense on a 10-g animal with high metabolism. A single day in which a bird does not accumulate enough energy to survive the frigid night is all it takes to eliminate all future breeding opportunities. However, the abundance of chickadees testifies to the fact that many of them do survive, and they do it with the aid of adaptively specialized behavior, memory, and brain.

Black-capped chickadees, like most other chickadees and tits (Paridae), nuthatches (Sittidae), and many corvids (Balda and Kamil, chapter 17 in this volume), store food in scattered locations and later retrieve and consume it. Most early naturalists who saw birds storing food in the wild (e.g., Haftorn 1956) assumed that animals with such tiny brains could not remember the locations of all the individual items they stored. Rather, storing of food must function for the good of the group by moving seeds from locations that become inaccessible under snow to places like the undersides of branches, where any conspecific would encounter them during normal foraging. However, food storing can evolve under individual selection provided that the animals that invest time and energy in hoarding are more likely to retrieve their stores than lazy conspecifics that do not hoard but only pilfer the hoards of others (Andersson and Krebs 1978).

Nonmemory mechanisms such as idiosyncratic foraging site preferences could, and probably do, promote a bird's successful retrieval of its own stores. Nevertheless, the memory of individual hoarding sites clearly plays a role (Brodin 1994). This raises the question of whether, along with the specialized behaviors of storing and retrieving food, food-storing species have some sort of specialization of memory, say in accuracy, capacity, or durability. This is inherently a comparative question, ideally answered by comparing a number of food-storing species with species that store little or not at all, but are otherwise as similar as possible to the food storers in phylogeny and in perceptual or motivational factors that could affect their performance in tests of memory. Methodological nightmares await anyone seeking a clear answer to such a question (Kamil 1988; Shettleworth 1998; Shettleworth and Hampton 1998). It is clearly not enough to compare only two species, one that stores food and one that does not.

Nevertheless, that is the approach we have taken in my laboratory, but the comparative base has then been broadened by other researchers testing other species similarly.

If food-storing species have better memory in any sense than species that do not store, it is most likely spatial memory as opposed to memory for other features of the world, such as patterns and colors of objects. This is especially true of the Clark's nutcrackers and other very long-term storers described by Balda and Kamil (chapter 17 in this volume), but in a snowy climate, local appearances can change drastically in even a few hours. In a series of experiments begun by David Brodbeck (Brodbeck 1994; Brodbeck and Shettleworth 1995), we asked whether chickadees are more likely to remember or use spatial than nonspatial information on the location of food. So we could also test nonstoring species, we devised a task that captures important features of food storing without requiring the birds to store food. Chickadees and juncos were allowed to return after a few minutes or hours to food that they had encountered briefly once before. The food was a peanut hidden in a brightly decorated block of wood on the wall of a large aviary. Four new feeders in four new locations were used on every trial. When the birds returned directly to the baited feeder on a high proportion of trials, they were tested to see what they remembered by swapping the formerly baited feeder with another one of the feeders on occasional unrewarded tests, thus dissociating location from pattern and color cues. Chickadees nearly always went first to the former location of the baited feeder, even though it was now occupied by a feeder that looked entirely different. In contrast, juncos chose about 50:50 between that feeder and the formerly baited feeder in its new location (Brodbeck 1994). Because there were four feeders to choose from and none of them had food in the tests, the birds continued to search after their first choice. Birds of both species were very likely to visit the spatially correct feeder second if they had not visited it first, and vice versa.

Thus both chickadees and juncos seem to remember both location and color, but they differ in their relative weightings of these features. We obtained this same species difference in an operant delayed matching-to-sample task, in which the birds pecked at colored shapes on a computer touch screen (Brodbeck and Shettleworth 1995). Chickadees also remember location better than color when the features are presented separately, whereas juncos do not (Shettleworth and Westwood 2002). In an important test of the generality of these findings, Clayton and Krebs (1994) obtained similar results in parallel experiments with European species of tits, which are closely related to our black-capped chickadees, as well as two corvid species (but see Healy 1995). Similarly, although Clark's nutcrackers excel at operant spatial nonmatching-to-sample, they are no better than other corvids in operant tests of color memory (Olson et al. 1995; but see Gould-Beierle 2000).

Just as in mammals, in birds the hippocampus is important for spatial memory. Chickadees with hippocampal lesions still store food, but cannot remember where they put it (Sherry and Vaccarino 1989). Hippocampal lesions degrade the performance of both chickadees and juncos on spatial delayed matching-to-sample, but leave the same individuals' color matching unaffected (Hampton and Shettleworth 1996). Consistent with the idea that food storing involves the evolution of a modular cognitive capacity, in food-storing species of birds, the hippocampus is larger relative to body size and telencephalon volume (most of the rest of the brain) than in nonstoring species (Krebs et al. 1989; Sherry et al. 1989). These findings have stimulated studies of how the volume as well as the detailed structure of the hippocampus changes as food storing changes both seasonally (Smulders et al. 2000) and developmentally (Clayton 1995b). The hippocampal enlargement that occurs as chickadees

store more food during the autumn is accompanied by increased neurogenesis (see Smulders et al. 2000). Experience storing food and/or using spatial memory contributes to the development of the relatively enlarged hippocampus of food storers, but the effects of experience do not explain all of the species differences (Clayton 1995a). In terms of this essay's theme of modularity and cognitive evolution, these findings mean that the specialized behavior of storing food rather than eating it immediately, a species-specific behavioral module that appears prefunctionally, helps to shape the specialization of a part of the brain which, in turn, presumably plays a role in helping that behavior function adaptively by facilitating memory for the food's location. These discoveries in food-storing birds suggest that any animals with lifestyles that make extraordinary demands on spatial cognition should also have specializations of the hippocampus and spatial cognition (Sherry et al. 1992). This hypothesis has inspired studies of rodents and of several other species of birds, and most have produced results consistent with it. Included are studies in which males and females of the same species differ, perhaps only seasonally, in space use and concomitantly in hippocampal volume (see Sherry and Duff 1996).

The development of food storing deserves special attention in the context of some themes in this book. Food storing has been cited (Griffin 1984) as a behavior suggesting that animals have conscious foresight. In storing food that will be used later, chickadees certainly look as if they are behaving in a consciously planful manner. However, the way in which storing behavior develops and later can be modified, in parids at least, indicates that storing is better described as a hardwired behavior that the birds perform relatively independently of current need and anticipation of future consequences. Fledglings begin storing avidly at around 40 days of age, even when they have ample food and efforts are made to deprive them of storable items and suitable substrates (see Clayton 1995b). The items chosen and the expertise with which they are inserted into suitable sites change with experience, as does the selection of sites by adults (Hampton and Sherry 1994). However, parids store food even under circumstances where they seem unlikely to be anticipating retrieving it. For instance, in our laboratory some birds persist indefinitely in storing peanuts in places where they drop out of reach. These observations do not necessarily rule out conscious planning. However, they do make it clear that the cognitive and brain mechanisms underlying a fascinating natural behavior like storing and retrieving can be studied while remaining at best agnostic about the nature of the animals' possible awareness. The study of food storing and indeed the broader study of animal spatial cognition is but one example of how studying the mechanisms by which animals behave adaptively in their worlds illuminates general questions about the evolution of mind.

References

Andersson, M. and Krebs, J. (1978). On the evolution of hoarding behaviour. *Animal Behaviour* 26: 707–711.

Balda, R. P., Pepperberg, I. M., and Kamil, A. C. (eds.) (1998). *Animal Cognition in Nature.* San Diego: Academic Press.

Barton, R. A. and Harvey, P. H. (2000). Mosaic evolution of brain structure in mammals. *Nature* 405: 1055–1058.

Bennett, A. T. D. (1996). Do animals have cognitive maps? *Journal of Experimental Biology* 199: 219–224.

Biegler, R. (2000). Possible uses of path integration in animal navigation. *Animal Learning & Behavior* 28: 257–277.

Brodbeck, D. R. (1994). Memory for spatial and local cues: A comparison of a storing and a nonstoring species. *Animal Learning & Behavior* 22: 119–133.

Brodbeck, D. R. and Shettleworth, S. J. (1995). Matching location and color of a compound stimulus: Comparison of a food-storing and a non-storing bird species. *Journal of Experimental Psychology: Animal Behavior Processes* 21: 64–77.

Brodin, A. (1994). The disappearance of caches that have been stored by naturally foraging willow tits. *Animal Behaviour* 47: 730–732.

Cartwright, B. A. and Collett, T. S. (1987). Landmark maps for honeybees. *Biological Cybernetics* 57: 85–93.

Cheng, K. (1986). A purely geometric module in the rat's spatial representation. *Cognition* 23: 149–178.

Cheng, K. (1989). The vector sum model of pigeon landmark use. *Journal of Experimental Psychology: Animal Behavior Processes* 15: 366–375.

Clayton, N. S. (1995a). Development of memory and the hippocampus: Comparison of food-storing and non-storing birds on a one-trial associative memory task. *Journal of Neuroscience* 15: 2796–2807.

Clayton, N. S. (1995b). The neuroethological development of food-storing memory: A case of use it, or lose it! *Behavioral Brain Research* 70: 95–102.

Clayton, N. S. and Krebs, J. R. (1994). Memory for spatial and object-specific cues in food-storing and non-storing birds. *Journal of Comparative Physiology A* 174: 371–379.

Coltheart, M. (1999). Modularity and cognition. *Trends in Cognitive Sciences* 3: 115–120.

Dukas, R. (ed.) (1998). *Cognitive Ecology.* Chicago: University of Chicago Press.

Fodor, J. A. (1983). *The Modularity of Mind.* Cambridge, Mass.: MIT Press.

Gallistel, C. R. (1990). *The Organization of Learning.* Cambridge, Mass.: MIT Press.

Gallistel, C. R. (1999). The replacement of general-purpose learning models with adaptively specialized learning modules. In *The Cognitive Neurosciences*, M. Gazziniga, ed., pp. 1179–1191. Cambridge, Mass.: MIT Press.

Gould-Beierle, K. (2000). A comparison of four corvid species in a working and reference memory task using a radial maze. *Journal of Comparative Psychology* 114: 347–356.

Griffin, D. S. (1984). *Animal Thinking.* pp. 69–71. Cambridge Mass.: Harvard University Press.

Haftorn, S. (1956). Contribution to the food biology of tits especially about storing of surplus food. Part IV. A comparative analysis of *Parus atricapillus* L., *P. crista-*

tus L. and P. *ater* L. *Det Kgl Norske Videnskabers Selskabs Skrifter* 1956 Nr. 4: 1–54.

Hampton, R. R. and Sherry, D. F. (1994). The effects of cache loss on choice of cache sites in black-capped chickadees. *Behavioural Ecology* 5: 44–50.

Hampton, R. R. and Shettleworth, S. J. (1996). Hippocampal lesions impair memory for location but not color in passerine birds. *Behavioral Neuroscience* 110: 831–835.

Hauser, M. (1996). *The Evolution of Communication.* Cambridge, Mass.: MIT Press.

Healy, S. D. (1995). Memory for objects and positions: Delayed-non-matching-to-sample in storing and non-storing tits. *Quarterly Journal of Experimental Psychology* 48B: 179–191.

Kamil, A. C. (1988). A synthetic approach to the study of animal intelligence. In *Comparative Perspectives on Modern Psychology. Nebraska Symposium on Motivation*, Vol. 35, D. W. Leger, ed., pp. 230–257. Lincoln: University of Nebraska Press.

Krebs, J. R., Sherry, D. F., Healy, S. D., Perry, V. H., and Vaccarino, A. L. (1989). Hippocampal specialization of food-storing birds. *Proceedings of the National Academy of Sciences, U.S.A.* 86: 1388–1392.

Nadel, L. (1992). Multiple memory systems: What and why. *Journal of Cognitive Neuroscience* 4: 179–188.

O'Keefe, J. and Nadel, L. (1978). *The Hippocampus as a Cognitive Map.* Oxford: Clarendon Press.

Olson, D. J., Kamil, A. C., Balda, R. P., and Nims, P. J. (1995). Performance of four seed-caching corvid species in operant tests of nonspatial and spatial memory. *Journal of Comparative Psychology* 109: 173–181.

Pearce, J. M., Roberts, A. D. L., and Good, M. (1998). Hippocampal lesions disrupt navigation based on cognitive maps but not heading vectors. *Nature* 396: 75–77.

Plotkin, H. (1997). *Evolution in Mind.* London: Penguin Press.

Raff, R. A. (1996). *The Shape of Life.* p. 87. Chicago: University of Chicago Press.

Rozin, P. and Kalat, J. W. (1971). Specific hungers and poison avoidance as adaptive specializations of learning. *Psychological Review* 78: 459–486.

Sherry, D. F. and Duff, S. J. (1996). Behavioural and neural bases of orientation in food-storing birds. *Journal of Experimental Biology* 199: 165–172.

Sherry, D. F. and Schacter, D. L. (1987). The evolution of multiple memory systems. *Psychological Review* 94: 439–454.

Sherry, D. and Vaccarino, A. L. (1989). Hippocampus and memory for food caches in black-capped chickadees. *Behavioral Neuroscience* 103: 308–318.

Sherry, D., Vaccarino, A. L., Buckenham, K., and Herz, R. S. (1989). The hippocampal complex of food-storing birds. *Brain, Behavior and Evolution* 34: 308–317.

Sherry, D. F., Jacobs, L. F., and Gaulin, S. J. C. (1992). Spatial memory and adaptive specialization of the hippocampus. *Trends in Neurosciences* 15: 298–303.

Shettleworth, S. J. (1995). Comparative studies of memory in food storing birds: From the field to the Skinner box. In *Behavioral Brain Research in Naturalistic and Semi-Naturalistic Settings*, E. Alleva, A. Fasolo, H. P. Lipp, L. Nadel, and L. Ricceri, eds., pp. 159–192. Dordrecht, The Netherlands: Kluwer.

Shettleworth, S. J. (1998). *Cognition, Evolution, and Behavior.* New York: Oxford University Press.

Shettleworth, S. J. and Hampton, R. H. (1998). Adaptive specializations of spatial cognition in food storing birds? Approaches to testing a comparative hypothesis. In *Animal Cognition in Nature*, R. P. Balda, I. M. Pepperberg, and A. C. Kamil, eds., pp. 65–98. San Diego: Academic Press.

Shettleworth, S. J. and Westwood, R. P. (2002). Divided attention, memory, and spatial discrimination in food-storing and non-storing birds, black-capped chickadees (*Poecile atricapillus*) and dark-eyed juncos (*Junco hyemalis*). *Journal of Experimental Psychology: Animal Behavior Processes*, in press.

Smulders, T. V., Shiflett, M. W., Sperling, A. J., and DeVoogd, T. J. (2000). Seasonal changes in neuron numbers in the hippocampal formation of a food-hoarding bird. *Journal of Neurobiology* 44: 414–422.

Staddon, J. E. R. (1989). The tyranny of anthropocentrism. *Perspectives in Ethology* 8: 123–135.

Stamps, J. A. (1995). Motor learning and the value of familiar space. *American Naturalist* 146: 41–58.

Tolman, E. C. (1948). Cognitive maps in rats and men. *Psychological Review* 55: 189–208.

Russell P. Balda and Alan C. Kamil

Research Questions

The central research questions that have guided our studies since 1981 combine issues and techniques from both comparative psychology and avian ecology. Most of our questions originate from the cognitive implications of extensive field studies on the natural history, ecology, and behavior of seed-caching corvids. Because our questions have evolved as our studies progressed, we have chosen to give a historical perspective outlining the progression of our ideas and questions (see Shettleworth, chapter 16 in this volume for a description of a similar program with seed-caching tits and chickadees).

Research Paradigm

Our research program began by examining the amazing spatial memory system of the Clark's nutcracker (*Nucifraga columbiana*). A single nutcracker buries up to 33,000 food items in thousands of different subterranean sites and retrieves them months later with a high degree of accuracy. These birds are highly adapted for this behavior because they possess a strong, sharp bill for opening cones, extracting seeds, and burying them in the substrate; a sublingual pouch (Bock et al. 1973) for carrying large numbers of seeds (up to 90 pinyon pine, *Pinus edulis*, seeds); and strong wings for carrying seeds great distances (up to 22 km). Birds have been observed digging up seeds in the field with seemingly uncanny accuracy (Vander Wall and Balda 1977, 1981; Vander Wall and Hutchins 1983). Although this behavior occurs regularly in the field, field conditions do not allow the design of studies to address the questions of how nutcrackers are able to locate their stored food.

Studies of the cognitive mechanisms involved in cache recovery required the development of a research plan using controlled laboratory experiments and captive birds. Fortunately, nutcrackers are quite willing to cache and recover seeds in laboratory settings and do so with a high degree of accuracy, both in a sandy floor indoors (Balda 1980; Balda and Turek 1984) or out of doors (Vander Wall 1982), as well as in a room with a raised floor containing sand-filled cups as potential cache sites (Kamil and Balda 1985). The ability to study caching and cache recovery under controlled laboratory conditions allowed us to test hypotheses on how the nutcrackers find their caches.

For example, because we were able to control when and where the birds cached, we were able to rule out odor, marking the site, list learning, and site preferences (Kamil and Balda 1985). We also learned that these birds remember some cache sites better than others and recover food from the better-remembered sites first, and with greater accuracy (Kamil and Balda 1990a). Birds sometimes revisit cache sites after they have recovered the seeds. On these revisits they treat the cache site differently than when they previously emptied it (Kamil et al. 1993). These birds also showed a long retention interval for cache memory, recovering caches with high levels of accuracy up to 9 months after creating them (Balda and Kamil 1992).

The results of several studies showed that nutcrackers were using visual landmarks for accurate cache recovery (Vander Wall 1982; Balda and Turek 1984). Data obtained by using a clock-shift technique, popular in studies of migratory birds and homing pigeons, suggest that seed-caching corvids may use the sun as a compass under some circumstances (Wiltschko et al. 2000). Thus we successfully brought a behavior prominent in the field into the laboratory, where we could examine it in great detail. From these studies we concluded that nutcrackers were using a spatial memory system to recover their caches and that this system was of long duration and robust (Kamil and Balda 1990b).

Comparative Studies

On the slopes of the San Francisco Peaks in Northern Arizona, five species of corvids cache and recover seeds. These species differ in their degree of dependence on their seed caches to survive winter, as well as in their adaptations for this behavior. The Clark's nutcracker is the most highly specialized of these species and lives at the highest elevations, where winters are harsh and alternative foods are very scarce. At mid-elevations, moderately specialized Steller's jays (*Cyanocitta stelleri*) and pinyon jays (*Gymnorhinus cyanocephalus*) coexist and also cache pine seeds when they are available. Both species have a relatively sharp bill for extracting seeds and an expandible esophagus for carrying pine seeds. A pinyon jay may cache up to 26,000 pine seeds when cone crops are abundant (Balda 1987). At lower elevations, the less specialized western scrub jays (*Aphelocoma californica*) and Mexican jays (*A. ultramarina*) cache and recover seeds much less intensely than the birds at higher elevations. Scrub and Mexican jays possess no morphological adaptations for the harvest, transport, caching, and recovery of seeds, and the lower elevations where they live have mild winters and a year-round supply of arthropods, seeds, and berries.

These differences in natural history raise a compelling question about evolution and cognition. Are these differences in morphological adaptations for food caching, and the concomitant dependence on cached food, associated with differences in the spatial cognitive abilities of these species? Are species that have the highest level of dependence on cached food also better at finding their caches? In a comparative test with three of these species, we found that although all three performed above chance, nutcrackers and pinyon jays recovered their caches more accurately than western scrub jays (Balda and Kamil 1989).

Specificity of Spatial Memory

This difference in the accuracy of cache recovery raises an interesting issue that is important in understanding the evolution of cognitive abilities: Is this spatial memory restricted to remembering where food has been stored or is it more general? Natural selection selects for outcomes, not mechanisms. Thus it could be that the nutcrackers' ability is highly specific. On the other hand, selection could have operated to sharpen already existing spatial cognitive abilities, in which case nutcrackers should perform quite well on a variety of spatial tasks. Therefore, we embarked on a series of comparative studies using different procedures to test spatial memory. These included two- and three-dimensional open-room analogues of the radial maze (Kamil et al. 1994; Balda et al. 1997) and operant nonmatching-to-sample tests. These studies all involved spatial memory, but not the recovery of food previously cached. The results of these studies were consistent with our hypothesis. The species most dependent on seed caches for winter survival performed at higher levels. If dependence on stored food has selected for improved spatial cognitive abilities, it has done so in a way that is not completely domain specific.

However, there is some specificity. Olson et al. (1995) tested three species in an operant nonmatching-to-sample test. In one experiment, the birds were required to remember a spatial location; in another, a color. As in many other experiments, the most seed-dependent species performed best in the spatial test, showing much longer retention intervals. However, this difference disappeared completely during the color test. This suggests a modularity for spatial cognition.

Social Cognition

More recently, we have become interested in extending our natural history-based analysis of cognition to another domain. Primatologists (e.g., Humphrey 1976) have developed a hypothesis about the evolution of intelligence based on the cognitive demands of sociality. Animals that live in large, stable social groups must be able to assess the consequences of their behaviors, classify other animals as members of various groups and coalitions, and recognize and remember traits of many other individuals. Success within the group will be improved if an individual possesses a rich internal representation of the group that will allow it to adjust to the fluid nature of the group. These cognitive demands will necessarily increase as group size increases and can be expected to be greatest for those animals living in larger, well-structured groups. Although the social complexity hypothesis has been considered primarily for primates, its logic is general and its implications can be tested with any appropriately chosen taxon (Balda et al. 1996).

The species we have worked with vary considerably in sociality. Pinyon jays are possibly the most social bird in North America, living in permanent groups of up to 400 individuals. Many of these individuals never leave their natal flock. Mexican jays are also highly social, living in relatively stable groups of 12–18 individuals, where helping at the nest is especially prominent. The Clark's nutcracker and western scrub jay, in contrast, live in family units or pairs year round. Young of the year disperse before the next breeding season. These differences in social living led us to hypothesize that pinyon jays and Mexican jays should be able to solve more complex cognitive tests than the less social nutcracker and western scrub jay. If this hypothesis is correct, the ordering or gradient of the four species will be different from that along the dependency gradient; that is, nutcrackers and scrub jays should perform poorly whereas pinyon jays and Mexican jays should demonstrate superior cognitive skills.

The selection of appropriate tasks for testing this hypothesis presented a challenge. We began with some comparative studies of observational learning. In one series of experiments, nutcrackers, pinyon jays, and Mexican jays watched a conspecific make caches in a room with many open holes for caching in the floor. The observer bird could view all areas of the floor. Later the observer was allowed to attempt to recover the caches it had observed being made. While pinyon jays and Mexican jays recovered caches with an accuracy above chance levels, nutcrackers did not perform above chance (Bednekoff and Balda 1996a,b).

In another experiment, pinyon jays learning a novel task were facilitated by being able to observe a conspecific performing the same task, but nutcrackers were not so facilitated (Templeton et al. 1999). In a third experiment (Bond et al. MS), pinyon jays performed better than western scrub jays in an operant test of transitive inference. Further studies of the ability to identify conspecifics and to form equivalence sets are planned for the near future.

Methodologies

We have combined two methodologies. First, we have used a classical biological comparative method, comparing closely related species that differ in their adaptations. This necessarily involves using natural history as a clue for asking relevant questions about relevant species. Principles of Darwinian evolution are central for understanding the dynamics of biological systems. The results of our studies led us to an important conclusion concerning the evolution of cognition. The accuracy of locating seed caches is an adaptive trait and as such is shaped by the actions of natural selection.

Cognitive abilities are a part of the adaptive arsenal that consists of a collective suite of characters that allows for swift and efficient harvest, transport, and caching and then accurate recovery of the cache at a later time. As such, cognitive traits can be viewed as playing a role in the biological success of an organism, much as morphological and physiological traits do. Species differences reflect, in part, differences in selective pressures among the species that are due to differences in their ecologies. Thus, cognitive processes have evolved and must be viewed as biological processes, not only because they have their roots in neurophysiology, but also because they are biologically significant as adaptations that contribute to the biological success of the organisms so endowed.

Second, we have dealt with the learning–performance distinction that bedevils comparative studies of cognition by using multiple behavioral test procedures. That is, if one species performs differently from another in any specific test, this could be due to differences in how well the test situation is suited to each species rather than to differences in ability. However, when substantially different tasks (such as cache recovery and operant tests) produce similar patterns across species, the likelihood that the behavioral differences are due to real species differences in ability grows. These varied tests provide converging operations, an approach first outlined by Kamil (1988).

Internal States

Our experiments are designed to inform us about the knowledge that our birds possess about their physical and social environments. We do not generally think of this knowledge in terms of internal psychological states because the meaning of this term is unclear. We take the term *internal psychological states* to be equivalent, at least for some people, to internal subjective states (e.g., awareness or consciousness), and we have never found it useful to speculate about the subjective states of our birds.

Future Work

Cache Location

Although it is well established that Clark's nutcrackers (and other seed-caching birds) remember their cache sites, we know relatively little about exactly what they remember. To put it another way, how does a nutcracker know when it is at a cache site? We know that they use the position of landmarks to locate caches (e.g., Balda and Turek 1984). Kamil et al. (1999) compared the body orientations used during caching with those used during recovery. We found that nutcrackers perform just as accurately when they use different orientations as when they use the same orientation. This suggests that they remember each site separately. Recently, Kamil and Cheng (2001) suggested that when the distances between the landmarks and the goal are relatively great, as they often are in nature, the birds remember the directional relationship between the goal and each of several landmarks. We are beginning a series of experiments to test this hypothesis.

Selection for Spatial Cognition

The results of comparative studies such as those reviewed here that indicate that spatial cognition is correlated with dependence on cached food are consistent with the hypothesis that in food-storing, scatter-hoarding species, natural selection has favored spatial abilities. We are currently attempting a more direct test of the hypothesized link between cognitive abilities and biological fitness in natural populations by obtaining measures of spatial abilities and measuring reproductive success in a wild population of pinyon jays.

Social Cognition

Our initial results of the social cognition hypothesis (outlined earlier) are quite exciting. We hope that we and other investigators will be able to expand this work in two ways. First, many more taxa need to be studied, so that we have many independent comparisons, each testing the hypothesis (Felsenstein 1985; Kamil 1988). Second, more tests of social cognition need to be developed and used in this effort.

From Limits to Opportunities

Yesterday's limits are often today's opportunities, and we are reluctant to set limits on the study of animal cognition. However, the study of the adaptive nature of animal cognition and its evolution is much more difficult and challenging than the study of morphological and physiological traits for a number of reasons. Animals are not necessarily programmed to maximize the performance of cognitive behaviors with the same degree of certainty that they often exhibit in morphological and physiological experiments. Cognitive behaviors are often more subtle, are governed by very complex and involved neural circuitry that is not obvious, and are not often performed with maximum intensity. As Humphrey (1976) once pointed out, we would learn little from watching Albert Einstein through a pair of binoculars!

References

Balda, R. P. (1987). Avian impacts on pinyon-juniper woodlands. In *Proceedings of the Pinyon-Juniper Conference*, R. L. Everett, ed., pp. 525–533. Reno, Nev.: USDA Forest Service General Technical Report, INT-215.

Balda, R. P. (1980). Recovery of cached seeds by a captive *Nucifraga caryocatactes. Zeitschrift für Tierpsychologie* 52: 331–346.

Balda, R. P. and Kamil, A. C. (1989). A comparative study of cache recovery by three corvid species. *Animal Behaviour* 38: 486–495.

Balda, R. P. and Kamil, A. C. (1992). Long-term spatial memory in Clark's nutcracker, *Nucifraga columbiana. Animal Behaviour* 44: 761–769.

Balda, R. P. and Turek, R. J. (1984). The cache-recovery system as an example of memory capabilities in Clark's nutcracker. In *Animal Cognition*, H. L. Roitblat, T. G. Bever, and H. S. Terrace, eds., pp. 513–532. Hillsdale, N.J.: Lawrence Erlbaum Associates.

Balda, R. P., Kamil, A. C., and Bednekoff, P. A. (1996). Predicting cognitive capacity from natural history. In *Current Ornithology,* Vol. 13, V. Nolan and E. D. Ketterson, eds., pp. 333–366. New York: Plenum.

Balda, R. P., Kamil, A. C., Bednekoff, P. A., and Hile, A. G. (1997). Species differences in spatial memory on a three-dimensional task. *Ethology* 103: 47–55.

Bednekoff, P. A. and Balda, R. P. (1996a). Social caching and observational spatial memory in pinyon jays. *Behaviour* 133: 807–826.

Bednekoff, P. A. and Balda, R. P. (1966b). Observational spatial memory in Clark's nutcrackers and Mexican jays. *Animal Behaviour* 52: 833–839.

Bock, W. J., Balda, R. P., and Vander Wall, S. B. (1973). Morphology of the sublingual pouch and tongue musculature in Clark's nutcrackers. *The Auk* 90: 491–519.

Bond, A. B., Kamil, A. C., and Balda, R. P. (MS). Social complexity predicts cognitive differences between two corvid species.

Felsenstein, J. (1985). Phylogenies and the comparative method. *American Naturalist* 125: 1–15.

Humphrey, N. K. (1976). The social function of intellect. In *Growing Points in Ethology*, P. P. G. Bateson and R. H. Hinde, eds., pp. 307–317. Cambridge: Cambridge University Press.

Kamil, A. C. (1988). Synthetic approach to the study of animal intelligence. In *Comparative Perspectives in Modern Psychology: Nebraska Symposium on Motivation*, Vol. 35, D. W. Leger, ed., pp. 230–357. Lincoln: University of Nebraska Press.

Kamil, A. C. and Balda, R. P. (1985). Cache recovery and spatial memory in Clark's nutcrackers (*Nucifraga columbiana*). *Journal of Experimental Psychology: Animal Behavior Processes* 11: 95–111.

Kamil, A. C. and Balda, R. P. (1990a). Differential memory for different cache sites by Clark's nutcrackers (*Nucifraga columbiana*). *Journal of Experimental Psychology: Animal Behavior Processes* 16: 162–168.

Kamil, A. C. and Balda, R. P. (1990b). Spatial memory in seed-caching corvids. *Psychology of Learning and Motivation* 26: 1–25.

Kamil, A. C. and Cheng, K. (2001). Way-finding and landmarks: The multiple-bearings hypothesis. *Journal of Experimental Biology* 204: 103–113.

Kamil, A. C., Balda, R. P., Olson, D. J., and Good, S. (1993). Revisits to emptied cache sites by Clark's nutcrackers (*Nucifraga columbiana*): A puzzle revisited. *Animal Behaviour* 45: 241–252.

Kamil, A. C., Balda, R. P., and Olson, D. J. (1994). Performance of four seed-caching corvids in the radial-arm analog. *Journal of Comparative Psychology* 108: 385–393.

Kamil, A. C., Balda, R. P., and Good, S. (1999). Patterns of movement and orientation during caching and recovery by Clark's nutcrackers (*Nucifraga columbiana*). *Animal Behaviour* 57: 1327–1335.

Olson, D. J., Kamil, A. C., Balda, R. P., and Nims, P. J. (1995). Performance of four seed-caching corvid species in operant tests of nonspatial and spatial memory. *Journal of Comparative Psychology* 109: 173–181.

Templeton, J. J., Kamil, A. C., and Balda, R. P. (1999). Sociality and social learning in two species of corvids: The pinyon jay (*Gymnorhinus cyanocephalus*) and the Clark's nutcracker (*Nucifraga columbiana*). *Journal of Comparative Psychology* 113: 450–455.

Vander Wall, S. B. (1982). An experimental analysis of cache recovery in Clark's nutcracker. *Animal Behaviour* 30: 84–94.

Vander Wall, S. B. and Balda, R. P. (1977). Coadaptations of the Clark's nutcracker and the pinon pine for efficient seed harvest and dispersal. *Ecological Monographs* 47: 89–111.

Vander Wall, S. B. and Balda, R. P. (1981). Ecology and evolution of food-storage behavior in conifer-seed-caching corvids. *Zeitschrift für Tierpsychologie* 56: 217–242.

Vander Wall, S. B. and Hutchins, H. E. (1983). Dependence of Clark's nutcracker (*Nucifraga columbiana*) on conifer seeds during the postfledgling period. *Canadian Field Naturalist* 97: 208–214.

Wiltschko, W., Balda, R. P., Jahnel, M., and Wiltschko, R. (2000). Sun compass orientation in seed-caching corvids: Its role in spatial memory. *Animal Cognition* 2: 215–221.

18 Environmental Complexity, Signal Detection, and the Evolution of Cognition

Peter Godfrey-Smith

Basic Principles

What are animal minds, including human minds, *for*? Although some versions of this question are teleological in a sense that has no place within an evolutionary world view, other versions of the question can be coherently asked. We can ask: Is it possible to make a general statement about the kinds of selective pressures and advantages that have been responsible for the evolution of cognitive mechanisms? Why has the expensive and delicate biological machinery underlying mental life evolved?

I suggest that the way to approach this question is not only to stress the continuities between human mental capacities and cognition in non-human animals, but also to recognize continuities between cognition and a wider class of "protocognitive" mechanisms. Cognitive mechanisms are mechanisms for behavioral control. And behavioral control mechanisms comprise one subset of a larger class of mechanisms that have the function of enabling organisms to adapt to changing problems and opportunities in their environments.

My work attempts to defend and develop tools for exploring a view of the mind based on an evolutionary perspective of this kind. This involves a combination of philosophical argumentation, commentary on empirical research, and some modeling. The chief goals of this work are foundational; the aim is to formulate general principles that unify diverse projects of empirical work on simple forms of cognition, and make explicit the connections between this empirical work and philosophical questions about the place of mind in nature. It is also hoped that discussions of this kind might sometimes help those engaged in the empirical research.

In this essay I sketch the basic framework used and some simple mathematical models that illustrate this framework. A range of empirical examples are discussed in Godfrey-Smith (2002).

I begin with the following principle:

Environmental complexity thesis (ECT): The function of cognition (and of a range of proto-cognitive capacities) is to enable an agent to deal with environmental complexity.

Each of the key terms in the ECT requires clarification (see also Godfrey-Smith 1996). "Function" is understood here in a strong sense; the function of a trait or structure is the effect or capacity it has which has been responsible for its success under a regime of natural selection (see Allen et al. 1998). Cognition, as I said earlier, is understood very broadly. We can think of cognition as a biological toolkit used to control behavior; a collection of capacities which, in combination, allow organisms to achieve various kinds of adaptive coordination between their actions and the world. This toolkit typically includes the capacities for perception, internal representation, memory, learning, decision making, and the production of behavior.

As the term "toolkit" suggests, we need not expect to find some single set of tools across all the organisms with cognitive capacities; different organisms have different collections of behavior control devices, according to their circumstances and evolutionary history. Furthermore, the list I gave of some core elements of the toolkit (perception, internal representation of the world, memory, learning) should not be seen as describing a set of recognizable and distinct "modules" found in the same form in all cognitive systems that have them. Rather, this is a set of capacities realized in different ways in different organisms, capacities that shade into each other and off into other, noncognitive parts of the biological machinery.

There is no sharp line between "real" cognition and a range of processes that we can call "protocognitive." By any normal standard, plants and bacteria (for example) do not have minds and do not exhibit cognition. But plants and bacteria do exhibit some capacities for flexible responses to environmental conditions, using environmental cues to control development and metabolism. Bacteria, for example, can modify their metabolism in order to take advantage of changes in the local food supply. In the case of many plants, a wide range of developmental processes are sensitive to the details of their local access to light. In plants, there are also some rapid and reversible changes that merit the title of "behavior." [See also Silvertown and Gordon (1989) although I think they might err on the side of including too much in the category of plant behavior.] Cognition shades off into other kinds of biological processes, and there is no point in trying to draw an absolute line between them.

The simplest biological capacities that we might consider protocognitive are cases of adaptive flexibility in behavior, development, or metabolism that are controlled by a fixed response to a simple environmental cue. This category includes adaptive phenotypic plasticity of all kinds (Schlichting and Pigliucci 1998). As we add different types of flexibility of response and different kinds of inner processing of the output of perceptual mechanisms, we reach clearer and clearer cases of cognition. However, there is no single path that takes us from the simplest cases to the most elaborate. There are various ways of adding sophistication to mechanisms of behavioral control, ways that will be useful to different organisms according to their circumstances. The ability to expand or contract the range of stimuli coupled to a given response is one important sophistication (Sterelny 2001). The ability to learn through mechanisms such as classical conditioning, reinforcement, and imitation is another. Yet another is the ability to construct a "cognitive map" of spatial structures in the environment (Roberts 1998). It is an error to try to describe a single hierarchy of cognitive skills, from the simplest to the most complex.

The ECT asserts a functional connection between this broad sense of cognition and environmental complexity. But what is environmental complexity? I suggest that the most useful concept of complexity here is a simple one. Complexity is *heterogeneity*. Complexity is variety, diversity, doing a lot of different things or having the capacity to occupy a lot of different states.

There are many different kinds of complexity; as with skills, it is not just unnecessary, but positively mistaken, to try to devise a single scale to order all environments from the least to the most complex. Any environment will be heterogeneous in some respects and homogeneous in others. Whether a particular type of complexity is relevant to an organism will depend on what the organism is like—on the organism's size, physiology, needs, and habits. The heterogeneity properties of environments are objective, organism-independent properties, but among the countless ways in which an environment is structured and patterned, only some will be relevant to any given organism. [See Levins (1968) for a classic discussion of these issues.] According to the ECT, the point of acquiring complex systems for behavioral control is to enable an organism to deal with heterogeneity in the challenges the environment presents and in the opportunities it offers.

Environmental complexity itself has many sources. It should not be thought that the only kind of patterns that constitute environmental complexity are those that are causally independent of the activities of the organisms under consideration. On the contrary, in many cases the environmental complexity that organisms must deal with is either a causal product of, or is constituted by, the activities of other organisms within the same population. Then we have a situation that can exhibit feedback, or a coupling of organism and environment (Lewontin 1985; Laland et al. 2000). The most graphic examples in this class—and perhaps the most important examples for hominid cognitive evolution—are those that involve social interactions among individuals within a population. In these cases, the behavioral complexity of other organisms comprises part of the environmental complexity that each individual must deal with. Many hypotheses stressing the role of social or "Machiavellian" intelligence in hominid evolution (Byrne and Whiten 1988) are examples of the ECT.

Some of the original reasoning behind the social intelligence hypothesis illustrates ways in which discussion of general principles like the ECT might help guide empirical work. Nicholas Humphrey and other early advocates of the social intelligence hypothesis were motivated in part by the idea that human and other primate cognitive capacities are too complex for their evolution to be explained in terms of the demands associated with tasks like foraging and predator avoidance (Humphrey 1976). Problems of social living were seen as more likely to have the kind of complexity that would propel the evolution of high-powered mental machinery. This reasoning can be seen as making an implicit appeal to a principle about the overall function of cognition in dealing with environmental complexity. Explicit discussion of this kind of principle might lead to other insights about the kinds of selection pressures responsible for different aspects of the evolution of cognition. Can we make generalizations about the kinds of environmental complexity that select for imitative learning as opposed to simpler kinds of associative learning, for example, or planning intelligence as opposed to simpler kinds of goal-directed behavior (Sterelny 2001)?

Models

Cognitive mechanisms provide a way to track the differences between environmental situations and adjust behavior accordingly. But when is it worth trying to track what is going on, rather than just producing a fixed behavior and hoping for the best? Intuitively, we might expect the value of a flexible approach to depend on the relationship between a number of factors. Are the kinds of environmental heterogeneity that an organism confronts of such a kind that different responses are *worth* producing? Does the organism have some sufficiently reliable way to *track* which state of the world it is confronting or is likely to confront? Are the *costs* of cognitive mechanisms and information processing outweighed by the advantages of flexibility?

These issues can be investigated and illuminated with some simple mathematical models. The models make a huge number of idealizations, but they do give us a way of moving beyond vague arguments and intuitions about when tracking the world is likely to be adaptive. The first model I discuss was proposed independently by Nancy Moran (1992) and Elliot Sober (1994), and resembles various earlier discussions, such as that by Lively (1986). More complex models that explore similar themes have been developed by David Stephens (1989, 1991).

This first model is the most simple, general, and "ground-floor" model of the adaptive advantages of plasticity or flexibility. The model describes optimal phenotype or strategy, but does not address the dynamics of evolution itself. So it would have to be augmented to function as a genuine evolutionary model. The model applies not only to behavioral flexibility but to phenotypic plasticity in general.

We assume there are two alternative states of the world, S_1 and S_2. These states are encountered by the organism with the probabilities P and $(1 - P)$, respectively. The organism has available two phenotypic states, or behaviors, C_1 and C_2. The payoff for producing behavior C_i when the world is in state S_j is V_{ij}. When the world is in S_i, the behavior with the highest payoff is C_i. The behaviors, C_1 and C_2, must be distinguished from the *strategies* available to the organism. The three possible strategies are

All-1: always produce C_1

All-2: always produce C_2

Flex: produce C_1 or C_2, depending on the state of an environmental cue

The point of the model is to describe the situations in which Flex is the best strategy.

The cue used by the flexible strategy is the state of an environmental variable that provides some information about whether the world is in S_1 or S_2. Given that the world is in some particular state, the reliability properties of the flexible strategy based on the cue can be expressed in a matrix of the probabilities that a specific response will be made:

Response likelihoods, $\Pr(C_i \mid S_j)$

	S_1	S_2
C_1	a_1	$(1 - a_2)$
C_2	$(1 - a_1)$	a_2

Here are the expected payoffs of the three strategies:

$$E(\text{All-1}) = PV_{11} + (1 - P)V_{12} \tag{1}$$

$$E(\text{All-2}) = PV_{21} + (1 - P)V_{22} \tag{2}$$

$$E(\text{Flex}) = P[a_1 V_{11} + (1 - a_1)V_{21}]$$
$$+ (1 - P)[(1 - a_2)V_{12} + a_2 V_{22}] \tag{3}$$

This simple model includes an assumption that some persons may see as biasing the case in favor of plasticity. We assume that plasticity comes for "free"; we are not imposing additional costs for setting up and maintaining the mechanisms that make plasticity possible.

Before we ask about the conditions under which Flex is the best strategy, we can determine first when one inflexible strategy is better than the other:

$$P(V_{11} - V_{21}) > (1 - P)(V_{22} - V_{12}) \tag{4}$$

In signal detection theory (discussed later), the quantity $(V_{11} - V_{21})$ is known as the *importance* of S_1, and $(V_{22} - V_{12})$ is the importance of S_2. The importance of a state of the world is the size

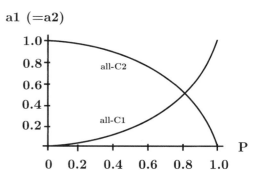

Figure 18.1
Flexible strategy. Minimal acceptable level of a_1 and a_2 for various values of P.

of the difference between the payoffs associated with suitable and unsuitable responses to that state. Sober (1994) introduces a useful term here. The expected importance of a state of the world is its importance multiplied by its probability. So according to formula (4), All-1 is better than All-2 if S_1 has a higher expected importance than S_2.

So if you are going to always do the same thing, you should do the thing suited to the state of the world with the higher expected importance. When will the flexible strategy be better than both inflexible options? Flex is better if and only if:

$$a_2/(1 - a_1) > (1 - P)(V_{22} - V_{12})/P(V_{11} - V_{21})$$
$$> a_1/(1 - a_2) \tag{5}$$

An example is given in figure 18.1. For simplicity, I assume that $a_1 = a_2$. The payoffs used are as follows: $V_{11} = 3$, $V_{12} = 1$, $V_{21} = 1$, $V_{22} = 10$. The figure graphs the minimum acceptable level of a_1 and a_2 for various values of P. The area above both lines is the only area where flexibility is favored over both inflexible options.

In this model, we aim only to describe the behavioral profile of a well-adapted individual. The model does not describe the internal psychological structure of the agent. However, Christopher

Stephens (in press) has extended the model in a psychologistic way. Stephens accepts that the behavioral profile of a well-adapted organism is described by the model given here, but adds that an actual organism can also be psychologically described using a matrix of subjective probabilities and subjective utilities. These are the internal causes of the organism's behavior. Clearly, if an organism's subjective probabilities and subjective utilities exactly match the objective probabilities and the real fitnesses, the organism will do well. But the consequences of inaccurate subjective probabilities and pathological utilities can be complicated because combinations of "bad" subjective probabilities and utilities can interact and sometimes result in an adaptive behavioral profile.

Stephens' extension of the model is interesting because it connects the biological modeling of the evolution of adaptive behavior with a well-studied model of individual psychological processing, the Bayesian model. In other work (Godfrey-Smith 1996) I have connected the simple model of plasticity with another modeling tradition, signal detection theory.

Signal detection theory (Green and Swets 1966) augments the model described here by giving a finer-grained description of how organisms use cues to achieve adaptive flexibility. The model described here looks at relationships between flexible and inflexible ways of dealing with environments while assuming that the cue used by a flexible strategy is a "given," a fixed constraint within which the organism optimizes. Signal detection theory enables us to ask, given the general nature of an organism's physical connections to the world, what is the best cue available for guiding its behavior with respect to a particular problem?

Suppose that the organism must fashion its "cue" by monitoring the concentration of some environmental chemical, which varies in a continuous way. Two likelihood functions describe the relation between the concentration of the chemical and the alternative states of the world.

These functions, $F(X \mid S_1)$ and $F(X \mid S_2)$, contain information about how likely a particular value of X is, given a particular state of the world. Let us suppose that both these functions are normal distributions and that the mean of $F(X \mid S_2)$ is higher than that of $F(X \mid S_1)$. The likelihood ratio function, $lr(X)$, is the ratio between the values of the functions. That is, $lr(X) = F(X \mid S_2)/F(X \mid S_1)$. In the case described here, $lr(X)$ will be a continuously increasing function. The likelihood ratio of an observation gives us a way of describing the evidential quality of that observation.

Let us suppose that because of the payoff matrix and the value of P, the best inflexible strategy is to always produce C_1. Which observed values of X are sufficiently good evidence for S_2, that it is worth producing C_2 instead?

Signal detection theory can determine a threshold value of X, such that if this value or a higher one is observed, then this observation is worth acting on (the expected payoff from producing C_2 given this observation is higher than the expected payoff from producing C_1). That threshold in turn gives us the optimal way for an agent to treat levels of X as a cue directing behavior. The threshold is described initially in terms of the minimal acceptable likelihood ratio of an observation. This can be used to give us in turn the threshold value of the observable variable X. The best threshold is the value of X so that:

$$F(X \mid S_2)/F(X \mid S_1)$$
$$> P(V_{11} - V_{21})/(1 - P)(V_{22} - V_{12}) \qquad (6)$$

The kind of assessment being done using the signal detection model is very similar to that seen in the simple Moran–Sober model of plasticity. Again the crucial relationship is that between the reliability properties of what is observed and the expected importances of the alternative states of the world. These in turn depend on the costs of different kinds of errors and the benefits of different kinds of correct decisions. However, signal detection theory gives us a finer-grained way of describing the use of environmental cues than the other model does because it does not simply take a cue with its reliability properties as a given, but instead describes how an organism might use a continuously varying environmental variable as a cue.

We can also add to a model of this kind various ways of representing the costs of plasticity. I supposed that the likelihood functions $F(X \mid S_1)$ and $F(X \mid S_2)$ were normal distributions with different means. Clearly, the likelihood ratio (which measures evidential quality) will depend not just on the means but also on the variances of these distributions. The smaller the overlap of the distributions, the more that can be learned from observing X. We might suppose that the organism makes its observation by a sampling process of some kind and that the likelihood functions $F(X \mid S_1)$ and $F(X \mid S_2)$ apply to a sample of a given size. If the organism were to observe a larger sample, it could in effect observe a signal with associated likelihood functions that have smaller variances and hence less overlap. But such sampling might have costs. In that case, the organism with optimal cognitive mechanisms will strike a balance between the costs of sampling and the benefits of observing a more informative signal, just as a scientist does when determining how large a sample is needed to answer a statistical question.

These optimality models are obviously extremely simple and in many ways unrealistic. They are no substitute for full evolutionary models or for empirical descriptions of how animals deal with their environments via psychological mechanisms. However, the simplicity of the models gives them the ability to illustrate some themes in a clear and stark way. They give us a ground-floor understanding of what it means for an organism to make use of an environmental cue in controlling its behavior, and how the costs of various kinds of flexibility must be traded off against the benefits. As such, they

are small pieces of a future theory of mind that stresses the role of cognition and protocognition in controlling behavior to deal with environmental complexity.

References

Allen, C., Bekoff, M., and Lauder, G. (eds.) (1998). *Nature's Purposes: Analyses of Function and Design in Biology*. Cambridge Mass.: MIT Press.

Byrne, R. W. and Whiten, A. (eds.) (1988). *Machiavellian Intelligence: Social Expertise and the Evolution of Intellect in Monkeys, Apes and Humans*. Oxford: Clarendon Press.

Godfrey-Smith, P. (1996). *Complexity and the Function of Mind in Nature*. Cambridge: Cambridge University Press.

Godfrey-Smith, P. (2002). Environmental complexity and the evolution of cognition. In *The Evolution of Intelligence*, R. Sternberg and J. Kaufman, eds., pp. 223–249. Hillsdale, N.J.: Lawrence Erlbaum Associates.

Green, D. M. and Swets, J. A. (1966). *Signal Detection and Psychophysics*. New York: Wiley.

Humphrey, N. (1976). The social function of intellect. In *Growing Points in Ethology*, P. P. G. Bateson and R. A. Hinde, eds., pp. 303–317. Cambridge: Cambridge University Press. Reprinted in Byrne and Whiten (1988).

Laland, K., Odling-Smee, J., and Feldman, M. (2000). Niche construction, biological evolution and cultural change. *Behavioral and Brain Sciences* 23: 131–175.

Levins, R. (1968). *Evolution in Changing Environments*. Princeton, N.J.: Princeton University Press.

Lewontin, R. C. (1985). The organism as the subject and object of evolution. In *The Dialectical Biologist*, R. Levins and R. Lewontin, eds., pp. 85–106. Cambridge, Mass.: Harvard University Press.

Lively, C. (1986). Canalization versus developmental conversion in a spatially variable environment. *American Naturalist* 128: 561–572.

Moran, N. (1992). The evolutionary maintenance of alternative phenotypes. *American Naturalist* 139: 971–989.

Roberts, W. A. (1998). *Principles of Animal Cognition*. Chapter 7. Boston: McGraw-Hill.

Schlichting, C. and M. Pigliucci (1998). *Phenotypic Evolution: A Reaction Norm Perspective*. Sunderland, Mass.: Sinauer Associates.

Silvertown, J. and D. Gordon (1989). A Framework for plant behavior. *Annual Review of Ecology and Systematics* 20: 349–366.

Sober, E. (1994). The adaptive advantage of learning and a priori prejudice. In *From a Biological Point of View*, E. Sober, ed., pp. 50–69. Cambridge: Cambridge University Press.

Stephens, C. (in press). When is it selectively advantageous to have true beliefs? Sandwiching the better safe than sorry argument. *Philosophical Studies*.

Stephens, D. (1989). Variance and the value of information. *American Naturalist* 134: 128–140.

Stephens, D. (1991). Change, regularity, and value in the evolution of animal learning. *Behavioral Ecology* 2: 77–89.

Sterelny, K. (2001). *The Evolution of Agency, and Other Essays*. Cambridge: Cambridge University Press.

19 Cognition as an Independent Variable: Virtual Ecology

Alan C. Kamil and Alan B. Bond

On close examination, human cultural artifacts bear the unmistakable impress of the structure of the human mind; our tools, habitations, and methods of communication have been molded to suit the strengths and limitations of the human cognitive system (Norman 1988). It has not commonly been emphasized, however, that similar shaping processes have taken place over the course of biological evolution in response to the cognitive features of other, nonhuman species (Bonner 1980; von Frisch 1974).

Cognitive influences are particularly evident in the modifications of color patterns and behavior of prey species that take advantage of biases and constraints in the perceptual systems of their principal predators. For example, avoidance learning by predators contributes to the evolution of aposematic, or warning, coloration in many distasteful or poisonous species (Guilford 1990; Schuler and Roper 1992); Batesian mimicry (Bates 1862), in which palatable prey evolve to imitate the appearance of aposematic species, takes advantage of the predator's tendency to generalize stimuli (Oaten et al. 1975). But perhaps the most striking illustration of the effects of predator cognition on prey appearance is the large number of species of cryptically colored insects that are polymorphic, with a single species occurring naturally in a variety of disparate forms.

Cryptic prey polymorphism is common among grasshoppers, leafhoppers, and walking-sticks, but it is particularly characteristic of lepidoptera (Poulton 1890). Many moths have evolved cryptic coloration to avoid bird predation while they rest on tree trunks during the daytime, and polymorphism is pervasive among these species. In North America, roughly 45 percent of the noctuid moths in the genus *Catocala* are polymorphic, with some species occurring in as many as nine different morphs (Barnes and McDonnough 1918). In other branches of the same family, the degree of morphological variation can be even more extreme. Adults of the army cutworm, *Euxoa auxiliaris*, are almost continuously variable in appearance.

In 1890, Poulton remarked on the high frequency of polymorphism among cryptic insects and formulated a remarkably perceptive explanation for the phenomenon. He said that in polymorphic species, "the foes have a wider range of objects for which they may mistake the moths, and the search must occupy more time, for equivalent results, than in the case of other species which are not polymorphic" (Poulton 1890, p. 47). His implications are, first, that polymorphism is an adaptive response to the foraging behavior of the predator and by extension, of the cognitive processes that determine successful visual search. Second, the selective advantage of polymorphism results from the fact that it is harder and more time-consuming to search for several things simultaneously than to search for only one.

The cognitive process involved appears to be a transitory increase in a predator's ability to detect cryptic prey when items of a similar appearance are encountered in rapid succession (Pietrewicz and Kamil 1979; Bond and Riley 1991; Reid and Shettleworth 1992), a phenomenon that has been termed "hunting by searching image" (Tinbergen 1960). As a result of the shift in detectability, visual predators tend to search for only a limited number of prey types at any moment in time, focusing on the most common prey available and effectively overlooking the others (Tinbergen 1960; Bond 1983; Bond and Kamil 1999).

The ecological consequence of this perceptual bias is known as apostatic selection (Clarke 1962, 1969), and it has been suggested to be the primary mechanism for maintaining stable prey polymorphism. If predators tend to search most effectively for prey types they have recently found, then the more common any given prey type is, the more heavily it will be preyed upon. Thus as a prey type becomes more common, predation on it increases while the predation pressure on rarer types declines. Common morphs experience relatively higher predation rates and decline in numbers, while rarer ones become more common. The outcome should be a stable configuration of prey types with a much higher degree of morphological diversity than would have been the case in the absence of predatory cognitive biases.

Apostatic selection is an elegant theory, but until recently, empirical support has been only fragmentary and indirect, primarily because experimental evolutionary ecology is an exceedingly difficult undertaking. When predators and prey are brought into the laboratory, the simplified environment interferes with normal population cycles (Murdoch 1969; Murdoch and Oaten 1975). And when evolutionary effects are sought in the field, limited experimental control reduces one's ability to make causal inferences. Although there are many documented instances of apparently stable polymorphisms and a fair amount of field experimentation indicating that predators tend to feed preferentially on more common morphs (Clarke 1969; Allen 1988; Cooper 1984; Cooper and Allen 1994), it has proven impossible to isolate the role of predation in the production and maintenance of prey polymorphism. What has been needed is a method for studying the detection of cryptic prey that allows the isolation of the many variables that can affect the decisions of a predator, including recent experience, and then allows predation to feed back onto the prey population. We have developed such a technique, which we call "virtual ecology," a novel paradigm that combines populations of artificial prey organisms with the foraging behavior of real predators.

Our methods derive from an established experimental system. In North America, noctuid moths are commonly preyed on during the daytime by blue jays (*Cyanocitta cristata*), which are the only avian predator that seems able to break the crypsis and find these insects while they are at rest on tree trunks. In a series of experiments in the 1970s, Pietrewicz and Kamil (1977, 1979) were able to show that jays in operant chambers exhibited the same impressive detection abilities when they were required to locate cryptic moths in slide images, and that the parameters of their search for moth images in the laboratory provided a satisfactory emulation of natural foraging behavior. In particular, (1) when the crypticity was increased, the birds were both slower and less accurate at detecting the moths; (2) responses to the moth images were substantially faster than those to the displays without moths, indicating an exhaustive, self-terminating search; and (3) the birds showed better detection after a run of a single type of moth than during random presentations of differing moth types, which has come to be considered a criterion for the use of searching images (Blough 1991; Bond and Riley 1991; Reid and Shettleworth 1992; Langley 1996; Kono et al. 1998; Bond and Kamil 1999).

For our purposes, photographs of moths on tree trunks were not sufficient. To explore evolutionary issues, we needed standardized backgrounds and much greater control of the features of the prey stimuli. So we undertook to convert this natural predator–prey system into something that was more amenable to digital manipulation, while retaining the essential features of the interaction. Our compromise was to take photographs of *Catocala* moths, render them in gray scale, and reduce them to a 16×16 pixel image, producing virtual moths (for a similar approach, see Plaisted and Mackintosh 1995). For the backgrounds, we reverse engineered the evolutionary process, generating fractal backgrounds based on the distribution of gray levels in the moths, which allowed us to titrate the difficulty of the detection task by manipulating the generating distributions (figure 19.1).

As in the earlier experiments using photographs, the moths were presented one at a time to blue jays in an operant chamber. In each trial

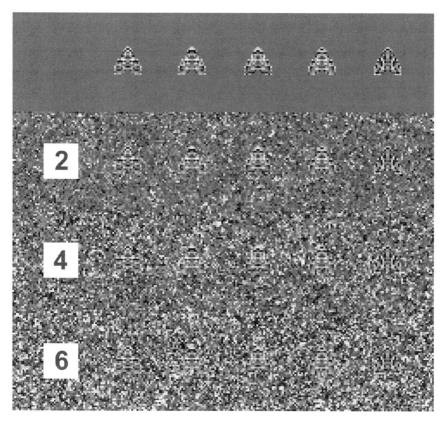

Figure 19.1
The five types of digital moths used in the virtual ecology experiments (moths 1 to 5, from left to right), presented against backgrounds of three levels of crypticity to illustrate the difficulty of the detection task. When projected on a computer monitor, the moths were about 6 mm high.

there either was or was not one moth image embedded in one of two fields of cryptic background on a computer monitor. If the bird correctly detected a moth, it pecked it and was rewarded with a food pellet. If the bird did not find a moth, it pecked a central green circle, in which case the next trial began immediately. The bird was never informed if it overlooked a moth, and if it pecked an area of background with no moth present, the time to the next trial was substantially delayed. Searching image experiments

using this methodology showed results that were equivalent to those produced using photographs (Bond and Kamil 1999), which encouraged us to generate additional novel forms with similar attributes and to develop populations of digital moths that would interact with the real jays in a dynamic virtual ecology.

Our first virtual ecosystem was a moth population of 240 individuals distributed among three fixed, disparate morphs. The population was exposed to predation by six blue jays. Over the

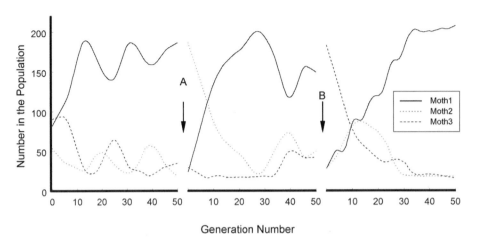

Figure 19.2
Population numbers of three morphs in three successive replications of the virtual ecology procedure. At point A, the second replication began; at point B, the third replication began. Curves were smoothed with weighted least squares using an 8-generation window

course of each daily session, each moth was presented once to one of the predators. Those moths that were detected were considered killed and were removed from the population; those that were overlooked survived and were proportionately cloned to create a new set of 240 individuals for the following day. Each day thus constituted a generation, and the makeup of the population at any given time was a direct reflection of the previous history of predation. We conducted three replications of this selection process, each running for 50 generations, but beginning with different starting conditions. In each case the relative numbers of the three moth types rapidly converged on approximately the same equilibrium values (figure 19.2). Analysis across replications of the rates of detection of each morph demonstrated that this equilibrium was the result of the predicted negative feedback between prey numbers and the probability of detection by the jays. When a moth type became more common, it became more likely to be detected and this in turn made its numbers decrease. Our results thus clearly implicated apostatic selection and demonstrated directly that such selection by visual predators can stabilize a preexisting polymorphism (Bond and Kamil 1998).

We then went on to test the effects of jay predation on the establishment of a novel prey type. If a new morph occurs in a population through immigration or mutation, its initial abundance will be low. Since local predators will have no experience with the new type, density-dependent selection should favor an increase in a new morph. We tested this by introducing a novel, fourth morph. Initially, the jays failed to detect it and it increased in abundance, briefly becoming the most common type in the population. Eventually, however, the jays learned to detect the new morph, and its numbers declined to a new equilibrium level (figure 19.3). This was not, however, the only possible outcome. When we repeated the novel prey introduction with a fifth morph that was extremely cryptic, only two of the six jays learned to detect it, and the novel

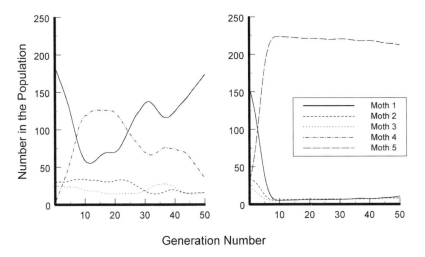

Figure 19.3
Population numbers of four morphs in the last two replications of the virtual ecology procedure. Curves were smoothed with weighted least squares using an 8-generation window. Replication 4, shown in the left panel, included moths 1 through 3 and moth 4; replication 5, shown in the right panel, included moths 1 through 3 and moth 5.

morph quickly came to dominate the population, essentially driving the other types to extinction (figure 19.3) (Bond and Kamil 1998). This result demonstrated that the selective effects of predation can be as powerful in virtual ecology as in the real world.

These results encouraged us to undertake an exploration of the role of predation in the origin of prey polymorphism, testing whether the process of predatory search, in and of itself, will tend to select for color polymorphism in an initially monomorphic population. This is a much more complex issue, since it requires a virtual genetics. We have developed an algorithm loosely based on current understanding of the control wing pattern development in lepidoptera, including (1) loci that code for individual patches of pattern elements, (2) loci that produce global changes in wing brightness or contrast without modifying pattern elements, and (3) linkage mechanisms that protect favorable genetic com-

binations from being lost during recombination (Robinson 1971; Nijhout 1991, 1996; Brakefield et al. 1996).

We are currently conducting experiments in which jays hunt a population in which individual moths are defined by their genotypes, and moths that survive exposure to the predators have a significantly higher probability of being allowed to breed and contribute to the next day's generation. To track the evolution of prey crypticity, we developed a crypticity index based on the correspondence between the phenotype and the background in the distribution of color and size of pixel patches (Endler 1984). The first of these studies has been completed recently (Bond and Kamil 2002) and directional selection did occur. The virtual moth populations evolved to become more cryptic and more phenotypically diverse than under appropriate control conditions. This outcome clearly establishes that our virtual ecology is at least a sufficient emulation of a real predator–prey system. But the real power of this technique lies in its amenability to testing through simulation. Because the selective process operates on virtual prey items under software control, the role of the predator can be subsumed, for purposes of hypothesis testing, by simulations that possess varying degrees of cognitive competence. In this fashion, we have been able to focus on the specific aspects of the jay's behavior that influence the distribution of prey phenotypes.

Conclusions

The significance of a study of animal cognition is usually viewed in terms of understanding the underlying neurophysiological and behavioral mechanisms or of interpreting the adaptive significance and evolutionary history of an organism's cognitive capabilities. However, as our experiments demonstrate, an animal's cognitive characteristics themselves can have direct evolutionary consequences. These are not limited to the detection of cryptic prey. For example, in sexual selection, the preferences expressed in mate choice are clearly a function of the cognitive characteristics of the animals doing the selecting (Endler and Basolo 1998). Thus we believe that students of animal cognition can make an essential contribution to understanding the natural world by participating in the study of cognition as an independent variable in evolutionary processes.

References

Allen, J. A. (1988). Frequency-dependent selection by predators. *Philosophical Transactions of the Royal Society of London, B* 319: 485–503.

Barnes, W. and McDunnough, J. (1918). *Illustrations of the North American Species of the Genus Catocala.* Memoirs of the American Museum of Natural History, Vol. 3, Part 1. New York: American Museum of Natural History.

Bates, H. W. (1862). Contributions to an insect fauna of the Amazon valley (*Lepidoptera*: Heliconidae). *Transactions of the Linnean Society, London* 23: 495–566.

Blough, P. M. (1991). Selective attention and search images in pigeons. *Journal of Experimental Psychology: Animal Behavior Processes* 17: 292–298.

Bond, A. B. (1983). Visual search and selection of natural stimuli in the pigeon: The attention threshold hypothesis. *Journal of Experimental Psychology: Animal Behavior Processes* 9: 292–306.

Bond, A. B. and Kamil, A. C. (1998). Apostatic selection by blue jays produces balanced polymorphism in virtual prey. *Nature* 395: 594–596.

Bond, A. B. and Kamil, A. C. (1999). Searching image in blue jays: Facilitation and interference in sequential priming. *Animal Learning & Behavior* 27: 461–471.

Bond, A. B. and Kamil, A. C. (2002). Visual predators select for crypticity and polymorphism in virtual prey. *Nature* 415: 609–614.

Bond, A. B. and Riley, D. A. (1991). Searching image in the pigeon: A test of three hypothetical mechanisms. *Ethology* 87: 203–224.

Bonner, J. T. (1980). *The Evolution of Culture in Animals.* Princeton, N.J.: Princeton University Press.

Brakefield, P. M., Gates, J., Keyes, D., Kesbeke, F., Wijngaarden, P. J., Monteiro, A., French, V., and Carroll, S. B. (1996). Development, plasticity and evolution of butterfly eyespot patterns. *Nature* 384: 236–242.

Clarke, B. C. (1962). Balanced polymorphism and the diversity of sympatric species. In *Taxonomy and Geography*, D. Nichols, ed., pp. 47–70. Oxford: Systematics Association.

Clarke, B. C. (1969). The evidence for apostatic selection. *Heredity* 24: 347–352.

Cooper, J. M. (1984). Apostatic selection on prey that match the background. *Biological Journal of the Linnean Society* 23: 221–228.

Cooper, J. M. and Allen, J. A. (1994). Selection by wild birds on artificial dimorphic prey on varied backgrounds. *Biological Journal of the Linnean Society* 51: 433–446.

Endler, J. A. (1984). Progressive background matching in moths, and a quantitative measure of crypsis. *Biological Journal of the Linnean Society* 22: 187–231.

Endler, J. A. and Basolo, A. L. (1998). Sensory ecology, receiver biases and sexual selection. *Trends in Ecology and Evolution* 13: 415–420.

Guilford, T. (1990). The evolution of aposematism. In *Insect Defenses: Adaptive Mechanisms and Strategies of Prey and Predators*, D. L. Evans and J. O. Schmidt, eds., pp. 23–61. Albany: State University of New York Press.

Kono, H., Reid, P. J., and Kamil, A. C. (1998). The effect of background cuing on prey detection. *Animal Behaviour* 56: 963–972.

Langley, C. M. (1996). Search images: Selective attention to specific visual features. *Journal of Experimental Psychology: Animal Behavior Processes* 22: 152–163.

Murdoch, W. W. (1969). Switching in general predators: Experiments on predator specificity and stability of prey populations. *Ecological Monographs* 39: 335–354.

Murdoch, W. W. and Oaten, A. (1975). Predation and population stability. *Advances in Ecological Research* 9: 1–131.

Nijhout, H. F. (1991). *The Development and Evolution of Butterfly Wing Patterns.* Washington, D.C.: Smithsonian Institution.

Nijhout, H. F. (1996). Focus on butterfly eyespot development. *Nature* 384: 209–210.

Norman, D. A. (1988). *The Psychology of Everyday Things.* New York: Basic Books.

Oaten, A., Pearce, E. M., and Smyth, M. E. B. (1975). Batesian mimicry and signal detection theory. *Bulletin of Mathematical Biology* 37: 367–387.

Pietrewicz, A. T. and Kamil, A. C. (1977). Visual detection of cryptic prey by blue jays (*Cyanocitta cristata*). *Science* 195: 580–582.

Pietrewicz, A. T. and Kamil, A. C. (1979). Search image formation in the blue jay (*Cyanocitta cristata*). *Science* 204: 1332–1333.

Plaisted, K. C. and Mackintosh, N. J. (1995). Visual search for cryptic stimuli in pigeons: Implications for the search image and search rate hypotheses. *Animal Behaviour* 50: 1219–1232.

Poulton, E. B. (1890). *The Colours of Animals.* New York: Appleton.

Reid, P. J. and Shettleworth, S. J. (1992). Detection of cryptic prey: Search image or search rate? *Journal of Experimental Psychology: Animal Behavior Processes* 18: 273–286.

Robinson, R. (1971). *Lepidopteran Genetics.* Oxford: Pergamon.

Schuler, W. and Roper, T. J. (1992). Responses to warning coloration in avian predators. *Advances in the Study of Behavior* 21: 111–146.

Tinbergen, L. (1960). The natural control of insects in pine woods I. Factors influencing the intensity of predation by songbirds. *Archives Neerlandaises de Zoologie* 13: 265–343.

von Frisch, K. (1974). *Animal Architecture.* New York: Harcourt Brace Jovanovich.

20

Synthetic Ethology: A New Tool for Investigating Animal Cognition

Bruce MacLennan

Goals

Synthetic ethology is based on several methodological commitments. First, it is based on the conviction that the investigation of cognition should look at behavior and the mechanisms underlying that behavior in the agent's environment of evolutionary adaptiveness. Second, this investigation should extend over structural scales from the neurological mechanisms underlying behavior, through individual agents, to the behavior of populations, and over time scales ranging from neurological processes, through an agent's actions, to that of evolutionary processes. Obviously, such a wide range of scales is difficult to encompass in investigations of natural systems. Third is the observation that the discovery of deep scientific laws (especially quantitative ones) requires the sort of control of variables that can be achieved only in an artificial experimental setup.

Thus we are faced with conflicting demands. On the one hand, we need precise experimental control. On the other, ecological validity dictates that agents be studied in their environment of evolutionary adaptiveness, where there are innumerable variables that are not amenable to independent control. Synthetic ethology intends to reconcile these conflicting requirements by constructing a synthetic world in which the phenomena of interest may be investigated. Because the world is synthetic, it can be much simpler than the natural world and thereby permit more careful experimental control. However, although the world is synthetic and simple, it is nevertheless complete in that the agents exist, live, and evolve in it.

The original motivation for synthetic ethology came from one of the central problems in cognitive science: the nature of intentionality, the property that causes mental states to be about something. We felt that an understanding of intentionality would have to encompass both the underlying mechanisms of intentional states and the social-evolutionary structures that lead to the creation of shared meaning. Our analysis of intentionality concluded that something is intrinsically meaningful to an agent when it is potentially relevant to that agent or to its group in its environment of evolutionary adaptedness (MacLennan 1992). Therefore intentionality must be studied in an evolutionary context.

We began our investigation with communication, since it involves both intentionality and shared meaning. We show in this essay how synthetic ethology permits the investigation of signals that are inherently meaningful to the signalers, as opposed to those to which we, as observers, attribute meaning.

Methods

The agents that populate our synthetic worlds can be modeled in many different ways; in particular there are a variety of ways of governing their behavior, including simulated neural networks and rule-based representations. In the experiments described here, an agent's behavior was controlled by a set of stimulus-response rules (64 rules in these experiments). These rules were determined by an agent's (simulated) genetic string, but they could be modified by a simple learning mechanism (described later).

Since our goal is to investigate the synthetic agents in their environment of evolutionary adaptedness, they must evolve. Our world includes a simplified form of simulated evolution, which proceeds as follows: Periodically two agents are chosen to breed, the probability of which is proportional to their "fitness" (as described later). The genetic strings of the two parents are mixed so that each of the offspring's genes comes randomly from either one or the other of the parents. In addition, there is a small probability of a gene being mutated. The resulting genetic string is used to create the stimulus-response rules for the single offspring, which is added to the population. In order to maintain a constant population size (100 in these experiments), one agent was chosen to "die" (i.e., to be removed from the population), the probability of dying being inversely related to "fitness."

We illustrate here the sort of experimental control permitted by synthetic ethology. Because we have complete control over the experimental setup and the course of evolution, we may begin with genetically identical populations and observe their evolution under different experimental conditions. If something interesting is observed in the course of an experiment, we may rerun the exact course of the evolution of the population to that point, and then make additional observations or experimental interventions to investigate the phenomena. Finally, whenever any interesting phenomena are observed, there can be no fundamental mystery, for all the mechanisms are transparent. If some agent exhibits interesting behavior, its entire mechanism is available for investigation. There can be no "ghost in the machine."

In synthetic ethology there is no requirement to model the natural world, as long as the synthetic world retains the essential characteristics of the natural world. That is, although determinate laws govern the evolution of our experimental populations, we are able to decide our world's "physical laws," which determine whether an agent "lives" or "dies," and which select agents for reproduction. The goal, of course, is to create synthetic worlds that are like the natural world in relevant ways, but are much simpler to study. The following experiment illustrates what can be accomplished.

Demonstrating the Evolution of Communication

Methods

Our first series of experiments investigated whether it was even possible for genuine, mean-ingful communication to evolve in an artificial system. We decided to construct the simplest possible system that could be expected to lead to real communication.

Although there are many purposes for which an agent might be expected to communicate, we decided to focus on cooperation. Our reasoning was that communication could be expected to evolve in a context in which some agents have information that other agents could use to facilitate cooperation. Thus we gave each agent a local environment that could be sensed by that agent but by no other. It can be thought of as the situation in an animal's immediate vicinity, but to keep the model as simple as possible, we limited the local environments to be in a small number of discrete states (eight in these experiments).

To make the state of one agent B's local environment relevant to another agent A, we arranged that they could cooperate only if A performed an action suitable for B's environment. To make this cooperation as simple as possible, we made our agents capable of producing an action from the same set as the local environment states. Thus A could cooperate with B only by producing the same item that was in B's local environment, which A could not sense directly.

To select for cooperation, we simply measured the number of times, over a specified period, that each agent was involved in successful cooperation. The probability of an agent reproducing was made proportional to this rate of cooperation, and its probability of dying was inversely related to the rate in a simple way. Thus we placed selective pressure on cooperation, but not directly on communication; indeed, limited cooperation can be achieved by random action (which has a 1:8 chance of succeeding).

Our experiments implemented only microevolution, so our agents were unable to evolve new sensor or effector organs. We gave our agents organs that might be used for communication, but we did not construct the agents to use them in this or any other way.

Again, simplicity was our principal aim. We equipped our synthetic world with a simple global environment, shared by all the agents, which could be in one of a few discrete states (eight in these experiments). The agents had the physical capability of sensing and modifying this global environment. Specifically, the state of the global environment is part of the stimulus to which an agent reacts, and the response can be either a new state for the global environment or an attempt to cooperate.

To test the potential effects of communication on cooperative behavior, we implemented a mechanism for making communication impossible. When communication was being suppressed, we periodically randomized the state of the global environment. This allowed us to measure the effect of apparent communication on the fitness (rate of cooperation) of the population, since genuine communication is defined in terms of its effect on the fitness of the communicators (Burghardt 1970).

We also investigated a very simple form of single-case learning, which could be enabled or disabled. When it was enabled, learning took place when an agent attempted to cooperate but failed. In this case, the stimulus-response rule used was changed to what would have been correct in this situation (although there was no guarantee that it would be the correct response in the future).

We initialized our population with 100 individuals containing random genetic strings. Thus the stimulus-response rules governing their behavior, which were determined by their genomes, were also initially random.

Results

To be able to measure the effect of communication on the fitness of a population, we quantified the fitness by the number of successful cooperations per unit of time, which we called the "degree of coordination" of the population. (The unit of time was a "breeding cycle," in which one individual died and one was born.) Because there

was considerable random variation in the degree of coordination, the time series was smoothed by a moving average. Linear regression was used to establish the rate at which the degree of coordination (fitness) increased or decreased. Details can be found in MacLennan (1990, 1992) and MacLennan and Burghardt (1993).

The baseline for comparison was determined by suppressing all possible communication, as previously described. In this case the degree of coordination stayed near 6.25 cooperations per unit of time, the level the analysis predicted would occur in the absence of communication. Linear regression showed a slight upward trend in the degree of coordination (the reason for which is discussed in the papers cited).

On the other hand, when communication was not suppressed, we found that the degree of coordination increased 26 times faster than when communication was suppressed. Over an interval of 5000 breeding cycles, the degree of coordination reached 10.28 cooperations per unit of time, which is 60 percent higher than the 6.25 achieved when communication was suppressed (figure 20.1). When the agents were permitted to learn from their mistakes, fitness increased at 3.82 times the rate found when learning was disabled, and at approximately 100 times the rate that occurred when communication was suppressed.

As would be expected for experiments of this kind, there is considerable experimental variation from run to run. Nevertheless, the results we have described are typical over more than a hundred experiments. Therefore we can conclude that genuine, meaningful communication is taking place, for it is significantly enhancing the fitness of the population. Furthermore, since communication evolves in our population when it is not suppressed, we may investigate genuine communication in its environment of evolutionary adaptedness.

Since it is genuine communication, the signals passed among the agents are meaningful to them, but not necessarily to us as observers. That is, we have a situation opposite from that of artificial intelligence, in which the computer

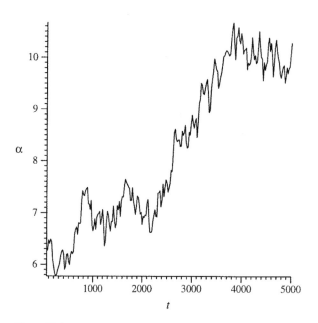

Figure 20.1
Increase in degree of coordination that is due to evolution of communication.

manipulates symbols that are meaningful to us but meaningless to it (or, more precisely, they have only a derived meaning that is dependent on the meaning we attribute). Here we are in the same situation as in natural ethology; we are faced with apparently meaningful communication and must discover its meaning *for the communicators.*

Even in these simple experiments, signals and their interpretation are complex functions of the total situation. The signal emitted by an agent may depend on both its local environment and the shared global environment. Furthermore, an agent's interpretation (use) of a signal may be (and typically is) influenced by its own local environment.

Nevertheless, we would expect that over time a simple meaning would emerge for the signals; that is, that there would be a one-to-one correspondence between signals and local environment states. To determine if this was occurring, we compiled a co-occurrence table, which recorded the number of times particular pairings of a signal (global environment state) and meaning (local environment state) occurred in successful cooperations.

If no communication is taking place, one would expect all signal–meaning combinations to be about equally likely, and that is what we found when communication was suppressed, and at the beginning of the simulations when it was not. However, when communication was not suppressed, the co-occurrence tables became more structured as the "language" self-organized.

We quantified the organization of the co-occurrence tables in a number of different ways, including entropy, a measure of disorder (lower numbers represent greater organization). With our experimental design, the maximum entropy, when all signal–meaning pairs are equally likely, is 6 bits, but when there is a one-to-one symbol–meaning correspondence, the entropy is 3 bits. When communication was suppressed, we observed an entropy of 4.95 bits, which shows that it was not completely disordered. However, when communication was not suppressed, the entropy decreased (after 5000 breeding cycles) to 3.87, representing a much higher degree of organization.

Visual inspection of the evolved co-occurrence tables showed a number of cases in which, almost always, a particular signal corresponded to a particular meaning and vice versa. However, we also observed cases of ambiguity, in which a signal was more or less equally likely to correspond to two or more meanings, and cases of synonymy, in which two or more signals were about equally likely to correspond to a particular meaning. These cases could result from individual agents using ambiguous or synonymous symbols, or from two or more competing "dialects" in the population, but Noble and Cliff (1996) have evidence supporting the former hypothesis.

The observations described here can be called "behavioral" and are analogous to those made in natural ethology. However, synthetic ethology affords additional possibilities, for the structure of the agents is completely transparent. At any time we may "dissect" the agents and analyze their behavioral programs (see MacLennan 1990 for examples). Thus we may relate the mechanisms of behavior to their manifestation in the population.

Brief Overview of Other Experiments

We have been interested in whether a population would evolve to use sequences of symbols for communication if there was a selective advantage in doing so. To explore this possibility, my students and I have conducted a number of experiments similar to those already described; details may be found in MacLennan (2001) and in the references cited there. In these experiments

the agents evolved the ability to communicate with pairs of symbols displaying a very rudimentary "syntax," but the results have been less interesting than we expected. One explanation may be that the very simple behavioral model we used (finite-state machines) is too weak for the sequential perception and control required for more complex communication. Animals, in contrast, have rich, highly structured perceptual-motor systems, which evolution can recruit for communication. Future experiments might need to use more complex models of agents, as well as a more structured environment about which they might communicate.

Discussion

Of necessity, our discussion of related and future work and its implications must be brief. Noble and Cliff (1996) have replicated our earliest studies and extended them in a number of informative ways. A somewhat different approach can be found in Werner and Dyer (1992), who demonstrated the evolution of communication by making it necessary for effective reproduction. Steels (1997a, b) has conducted fundamental studies on the emergence of meaningful symbols.

In discussing related work, it may be worth making a few remarks about the connection between synthetic ethology and a related discipline, artificial life. First, it must be stressed that there is substantial overlap between them, so that the difference is at most one of emphasis.

Artificial life studies artificial systems that are significantly "lively" in some sense. Some investigators are attempting to create systems that are literally alive, while others are content with systems that faithfully imitate life. Synthetic ethology differs from this discipline in that the agents need not be alive in either of these senses, although they may be. Certainly we make no claim that the agents described in this essay are alive in any literal sense.

Current experiments in synthetic ethology are too simple to exhibit psychological states, but future ones may be able to do so; even the current experiments exhibit genuine intentionality. Synthetic ethology indicates how psychological states may be made accessible to scientific investigation.

We have claimed that our agents (although they are not conscious or even alive) exhibit genuine intentionality. The point is certainly arguable and depends on our analysis of intentionality. Nevertheless, all subtleties aside, we claim that the signals are inherently meaningful to the agents because the agents' continued persistence as organized systems depends on their use of the signals.

Are these synthetic worlds and agents so alien that results will not be seen as relevant to nature? In particular, we have argued that we can use abstract, ad hoc selection rules (since the "laws of physics" are under our control), but it can be objected that selection should be more naturalistic (e.g., Werner and Dyer 1992). Certainly this is an important issue, and in the long run we want to explore ever richer synthetic worlds, but introducing gratuitous complexity would defeat the goals of synthetic ethology.

One of the advantages of synthetic ethology is that we can make our worlds as simple as possible, as long as they include the phenomena of interest. On the other hand, we must construct these worlds from scratch; they are not given to us. This becomes a challenge as we begin to investigate phenomena that require larger populations of more complex agents acting in more complex environments. The simulation of such worlds requires ever more powerful computers. Therefore synthetic ethologists must strike a delicate balance between the sophistication of the synthetic world and the resources required to implement it. Indeed, as we move in the direction of greater complexity, synthetic ethology will face some of the same problems as natural ethology. Nevertheless, by affording greater control and an alternative to natural life, it will remain a worthwhile approach.

References

Burghardt, G. M. (1970). Defining communication. In *Communication by Chemical Symbols*, J. W. Johnston, Jr., D. G. Moulton, and A. Turk, eds., pp. 5–18. New York: Appleton-Century-Crofts.

MacLennan, B. J. (1990). *Evolution of Communication in a Population of Simple Machines*. Technical Report No. CS-90-99. Knoxville, Tenn.: Computer Science Department, University of Tennessee.

MacLennan, B. J. (1992). Synthetic ethology: An approach to the study of communication. In *Artificial Life II: The Second Workshop on the Synthesis and Simulation of Living Systems*, C. G. Langton, C. Taylor, J. D. Farmer, and S. Rasmussen, eds., pp. 631–658. Redwood City, Calif.: Addison Wesley.

MacLennan, B. J. (2001). The emergence of communication through synthetic evolution. In *Advances in Evolutionary Synthesis of Neural Systems*, V. Honavar, M. Patel, and K. Balakrishnan, eds., pp. 65–90. Cambridge, Mass.: MIT Press.

MacLennan, B. J. and Burghardt, G. M. (1993). Synthetic ethology and the evolution of cooperative communication. *Adaptive Behavior* 2: 161–188.

Noble, J. and Cliff, D. (1996). On simulating the evolution of communication. In *From Animals to Animats 4: Proceedings of the Fourth International Conference on Simulation of Adaptive Behavior*, P. Maes, M. Mataric, J.-A. Meyer, J. Pollack, and S. W. Wilson, eds., pp. 608–617. Cambridge, Mass.: MIT Press.

Steels, L. (1997a). Constructing and sharing perceptual distinctions. In *Proceedings of the European Conference on Machine Learning*, M. van Someran and G. Widmer, eds., pp. 4–13. Berlin: Springer-Verlag.

Steels, L. (1997b). The synthetic modeling of language origins. *Evolution of Communication* 1: 1–34.

Werner, G. M. and Dyer, M. G. (1992). Evolution of communication in artificial organisms. In *Artificial Life II: The Second Workshop on the Synthesis and Simulation of Living Systems*, C. G. Langton, C. Taylor, J. D. Farmer, and S. Rasmussen, eds., pp. 659–687. Redwood City, Calif.: Addison Wesley.

21

From Cognition in Animals to Cognition in Superorganisms

Charles E. Taylor

Research Questions

My research in cognition has not been directed toward animal cognition per se. Rather, I have been intrigued by the apparent inability of modern science, as I understand it, to explain subjective experience. For this discussion I will refer to it as "the problem of mind." While they are not the same, there is a broad overlap between this problem and those of animal cognition.

My research on the problem of mind has been motivated by beliefs, in no way original with me, that:

• Far from consciousness being explained by anyone, I do not think we even know what an adequate explanation would look like, an opinion articulated especially well by Nagel (1986).

• Probably some animals experience subjective feelings in more or less the same way that we do, although there is a gradation.

• At present, however, we do not know how to objectively identify whether others, animal or human, experience subjective feelings or what those feelings are like, more or less for reasons discussed by Erwin Schrödinger (1967) as "objectivation."

• The study of artificial intelligence (AI) and robotics has contributed a great deal to our understanding of the problem, but certainly has not solved or explained it.

However, it does seem to me that:

• While we are not now able to construct an artificial system that can obviously experience subjective feeling, we may still be able to evolve systems that do, using techniques that have come to be known as evolutionary computation (Mitchell and Taylor 1999).

In recent years, largely as a result of work by Rodney Brooks and his students, much of which is collected in Brooks (1999), I have come to believe further that:

• For purposes of understanding cognition, especially animal cognition, and probably subjective experience, organisms are best viewed as collections of sensors, effectors, and processors of limited abilities surrounding and located throughout the organism. These collections communicate primarily with other sensors, effectors, or processes that are mostly nearby, that mostly have a limited bandwidth, and that function as an ensemble.

This means that cognition and subjective experience are to be understood as the collective behavior of these limited agents acting as one. If so, then one of the most important challenges facing us, but one we can probably address in concrete terms, will be to understand how such an ensemble can function as some sort of superorganism. In the remainder of this essay I attempt to explain how and why I arrived at this opinion, and at the end briefly address some of the promises and limitations it offers.

Past Research

Initially I approached the problem of mind by looking at animals that could reasonably be assumed to experience rudimentary subjectivity. We used mutant strains of *Drosophila melanogaster*, exploring how learning or memory-retention mutations affected habitat selection (Taylor 1987). These insects were chosen because much is known about their biology and because they are easy to breed and evolve. I soon abandoned this approach, however, because it seemed difficult for me to relate to their sensory experiences; they were setting the boundary conditions, not I. It was too difficult to broach the subjective-objective problem mentioned earlier.

Artificial systems, however, can be made more or less complex, as desired. In the early 1970s there was enthusiasm that the "traditional" approach to artificial intelligence would provide insights to the problems of cognition. By "traditional" I mean the approach exemplified by Newell and Simon (1972) in their General Problem Solver (GPS) program, for example. Very simply, a cognitive system is viewed as having a snapshot of a state space encoded by symbols. There are rules by which these symbols may be manipulated, and cognition consists of searching through a sequence of manipulations of these symbols so that they will achieve the desired configuration of state space in a suitably economical fashion. According to this view, problem solving by animals consists of obtaining information about their environment (describable by predicate calculus) and then manipulating this representation to obtain a desired state of affairs, again describable by predicate calculus. The animal would then follow that sequence. Cognition is the deliberation about which chain of actions is most appropriate.

There is a school of thought in evolutionary biology that views the environment as posing problems that animals need to solve and that sees evolution as guiding them in doing so. Pursuing this approach, I developed a computer system in which objects inherited certain propositions at the time they were created. Other propositions could be obtained from the environment. The objects would then act on these propositions using the rules of symbolic logic to see if they could solve problems posed by the environment. Those who did so could reproduce, those who were unable to arrive at those solutions perished without leaving offspring. The system was written in Lisp and worked only modestly well, so I did not publish it. Since then, others, especially John Koza (see Fogel 1998; Mitchell and Taylor 1999), have developed powerful Lisp-based systems for evolutionary computation termed *genetic programming*. Koza et al. (1992) were able to obtain some insights about (possible) reason-

ing by anolis lizards when the lizards choose to pursue or ignore insects while foraging. Audrey Cramer, who started her research in my laboratory as a graduate student, subsequently explored how similar models of cognition might explain how vervet monkeys choose a route in their search for food (Cramer and Gallistel 1997).

In the early 1980s I began working with David Jefferson on a system he had developed that embodies what I feel characterizes the nature of organisms and life. We have articulated this in several papers and the work is summarized in Taylor and Jefferson (1995). Most important is the view that life is a property of *processes*, rather than of *objects*. Living beings are self-organizing processes (or even an ensemble of processes) that interact with the world around them, taking in some things and expelling others, possibly reproducing other such processes, and eventually "dying," in that they are no longer capable of self-perpetuation. Sentiments about process versus object that influenced my thinking include those of Whitehead (1926) and Birch and Cobb (1981), although I have been unable to reconcile certain significant differences between us. In any event, it seems likely to me that mind as we know it can be described only by "living" processes or objects embodying such processes. If so, then my prior work with problem solving had fundamental shortcomings.

Rob Collins, a student of Jefferson's, used connection machines—computers with tens of thousands of processors, each with its own memory and processes—to evolve populations of processes that seemed lifelike in that they could reproduce, move, die, and evolve as they navigated a virtual world (Jefferson et al. 1992). Such systems could also learn, not simply evolve—a combination that has great adaptive potential (Belew and Mitchell 1996).

Jefferson and I became acquainted with the theories of Rodney Brooks, a Massachusetts Institute of Technology (MIT) exponent of "nouvelle AI." In a series of papers with titles like "Intelligence without reason" and "Elephants don't play chess" (collected in Brooks 1999), he argued that intelligent systems need to be (1) embodied, not just simulated; (2) situated in the real world, not simply in a demonstration setting; and that (3) a subsumption architecture is the best way to control such systems. Brooks's views were heavily influenced by certain ethologists, especially von Uexküll (1921). In this view, "traditional AI" is incapable of capturing animal intelligence. Among other reasons, the representation of natural environments by traditional AI models in a way that is sufficiently rich for intelligent behavior requires an unfeasible number of logical manipulations. Brooks suggested that the best way for an animal to represent its environment is to use the environment to represent itself.

To test such theories, one cannot use manipulation of disembodied symbols, but generally needs to use robots in rich environments. These robots have typically consisted of large numbers of distributed sensors, effectors, and processors, each with modest abilities. None of these processors could be said to characterize the environment in any meaningful way by itself, yet collectively these systems are capable of complex coordinated actions such as climbing up stairs or collecting empty soda cans from desk tops. Furthermore, watching robots play tag or attempt to fetch an object comes much closer to meeting the criteria that are now used to infer subjective feeling, described in the introduction, than watching computers that simply manipulate symbols. Hence we thought it worth exploring whether evolving robots might contribute to studying the problem of mind.

At that time few robots were commercially available and supported. So two students in my laboratory, Kourosh Nafisi and Orazio Miglino, constructed a small, fairly robust robot out of Lego blocks using a small controller board we obtained from the MIT artificial intelligence laboratory (Miglino et al. 1994). It resembled an early version of the Lego Technics kits that are now sold in toy stores. It was controlled by neu-

ral nets, and populations of such nets could be made to evolve by a variant of genetic algorithms (see Mitchell and Taylor 1999). Miglino's robot evolved fairly complex neural networks and the associated behavior needed to navigate a simple environment. These robots were a hybrid combining simulated neural nets, similar to those evolved by Collins (Jefferson et al. 1992), and real measurements of occasional physically realized rules. We were unable to physically test all of the robots that needed to be tested each generation because the robots were only plastic and were not sufficiently robust to withstand hours of continuous use; nor was it possible to provide power to them for fully automatic testing. To my knowledge, these were the first hybrid simulated–actual robots that were created by evolutionary computation. Since that time, very robust robots have been developed and ways have been found to provide constant power to them (Nolfi and Floreano 2001). One of these, the Khepera robot, has become the workhorse of evolutionary robotics, although even these versions are still very simple and require an environment so artificial that they can be regarded only as embodied, not situated in the real world (R. A. Brooks, personal communication).

As we attempted to reconcile these studies with those of Brooks, it became evident that a critical feature of intelligent behavior by real organisms is the ability for large numbers of sensors, processors, and actuators to function as a unified ensemble. A robot is surrounded and studded with many such elements communicating together. None of the elements has a complete model of the environment—certainly not one that could be naturally represented by propositional calculus as the earlier AI models required—but each element has some ability to digest and process the information it obtains locally and from sensors and processors. In a living organism, however, cognition seems to be almost a collective consciousness derived from the interaction of numerous parts functioning as a "superorganism."

Current Research

If we are to evolve such a "collective individual" as a robot, we will need to pay much more attention to just what is abstracted from the environment and communicated to other parts of the organism. A collection of sensors and processors indiscriminately pouring their electrons into a bus is not likely to achieve our goal. Instead, each sensor and processor needs to compress the experience that learning or evolution has judged to be important and transmit only that information. It is especially challenging to make the experiences of different sensors and processors mutually intelligible.

With a computational linguist, Edward Stabler, and a student in my laboratory, Tracy Teal, I have begun to explore how such extraction and compression might be achieved. Our first work has been exclusively with symbolic systems (Teal et al. 1999; Teal and Taylor 2000). We are now exploring how to relate such symbols to actual experiences (e.g., in Wee et al. 2001). [Several years ago I cooperated with Takaya Arita on a study in which populations of neural nets learned to acquire a common lexicon, even when they were not all able to observe the same object (Arita and Taylor 1996).]

If this approach is more or less correct, then the challenge ahead is to learn how distributed systems operating in the real world are able to combine and adapt to construct a unified whole —possibly to the extent of making a "whole iguana" (Dennett 1978). This will necessarily involve both mechanical and electrical engineering and computer science, with plenty of theory and real-world constructions. Probably linguistics and evolutionary theory will provide needed insights.

Phylogenetic Appropriateness

The overriding question or test of this approach is one of phylogenetic appropriateness. Is in silico life going to be sufficiently similar to in vivo life, in the way that much in vitro life has proven to be? Will these simpler and manipulable constructions of intelligent behavior really lead to an understanding of mind and cognition? Or will they be merely a diversion, albeit one that is enthralling and no doubt commercially useful, but still just a diversion, from our understanding of the deeper issues?

Although they are poorly understood, subjective and physical events are undeniably part of the same phenomena. "The world is given to me only once, not one existing and one perceived" (Schrödinger 1967, p. 137). Putting aside the problem of how subjective experience is to be recognized at all, some physical systems must be capable of subjective experience and some (we believe) must not, for reasons that are not known. We are currently unable to judge just which system might become capable and which will not. If a system is capable of becoming subjective by incremental changes, and if our criteria for selection are appropriate, then it should be possible to evolve systems that have subjective experiences. But what if the system is simply not capable of the physical interactions necessary to support subjective experience? A liter of oxygen, though unquestionably a physical entity, is capable of becoming solid only under the most unusual of circumstances, if at all. It might be that the silicon chips of an artificial construct are similarly not capable of the physical processes that must underlie subjective experiences. Maybe the necessary interactions occur only between carbon, hydrogen, and oxygen. If that is the case, then computers and robots as now constructed will be phylogenetically inappropriate vehicles for studying the problem of mind.

Or maybe they will have a different sort of subjective experience. Extrapolating from the imaginative and forceful arguments of von Uexküll (1934/1957), we can presume that the "experience" of such creations will be quite different from our own, and probably quite varied from creation to creation. There is every reason to expect that these creations will provide us with an enormously useful test-bed for studying the relation between cognition and subjective experience; perhaps they may even provide evidence for other forms of conscious experience, called for by Nagel (1986).

In the past few decades we have learned a lot from efforts to create computers and robots that behave in ways that might seem to require subjectivity. One need only peruse the robots on display at http://www.androidworld.com/ to be impressed by their variety and accomplishments. Most emphasis has been on human subjectivity, but there has been progress from attempts to explore intelligent behavior and subjective experience at other levels, including that of lizards. There is no reason to believe this progress will stop, and there is plenty of reason for optimism. Whether this category of model is truly phylogenetically appropriate or not, only time will tell.

References

Arita, T. and Taylor, C. (1996). A simple model for the evolution of communication. In *Evolutionary Programming: Proceedings of the Fifth Annual Conference of Evolutionary Programming*, L. Fogel, P. Angeline, and T. Bäck, eds., pp. 405–410. Cambridge, Mass.: MIT Press.

Belew, R. K. and Mitchell, M. (1996). *Adaptive Individuals in Evolving Populations: Models and Algorithms*. Reading, Mass.: Addison Wesley.

Birch, C. and Cobb, J. B., Jr. (1981). *The Liberation of Life*. Cambridge: Cambridge University Press.

Brooks, R. A. (1999). *Cambrian Intelligence: The Early History of the New AI*. Cambridge, Mass.: MIT Press.

Cramer, A. E. and Gallistel, C. R. (1997). Vervet monkeys as travelling salesmen. *Nature* 387: 464.

Dennett, D. C. (1978). Why not the whole iguana? *Behavioral and Brain Sciences* 1: 103–104.

Fogel, D. B. (1998). *Evolutionary Computation; The Fossil Record*. New York: Institute of Electronics and Electrical Engineers.

Jefferson, D., Collins, R., Cooper, C., Dyer, M., Flowers, M., Korf, R., Taylor, C., and Wang, A. (1992). Evolution as a theme in artificial life: The genesys/tracker system. In *Artificial Life II*, C. G. Langton, C. Taylor, J. D. Farmer, and S. Rasmussen, eds., pp. 549–578. Reading, Mass.: Addison Wesley.

Koza, J. R., Rice, J. P., and Roughgarden, J. (1992). Evolution of food foraging strategies for the Caribbean anolis lizard using genetic programming. *Adaptive Behavior* 1: 47–74.

Miglino, O., Nafasi, K., and Taylor, C. E. (1994). Selection for wandering behavior in a small robot. *Artificial Life* 2: 101–116.

Mitchell, M. and Taylor, C. E. (1999). Evolutionary computation: An overview. *Annual Review of Ecology and Systematics* 30: 593–616.

Nagel, T. (1986). *The View from Nowhere*. New York: Oxford University Press.

Newell, A. and Simon, H. A. (1972). *Human Problem Solving*. Englewood Cliffs, N.J.: Prentice-Hall.

Nolfi, S. and Floreano, D. (2001). *Evolutionary Robotics: The Biology, Intelligence and Technology of Self-Organizing Machines*. Cambridge, Mass.: MIT Press.

Schrödinger, E. (1967). *What Is Life? Mind and Matter*. Cambridge: Cambridge University Press.

Taylor, C. E. (1987). Habitat selection within species of *Drosophila*: A review of experimental findings. *Evolutionary Ecology* 1: 389–400.

Taylor, C. E. and Jefferson, D. R. (1995). Artificial life as a tool for biological inquiry. In *Artificial Life: An Overview*, C. G. Langton, ed., pp. 1–13. Cambridge, Mass.: MIT Press.

Teal, T. and Taylor, C. E. (2000). Effects of compression on language evolution. *Artificial Life* 6: 129–143.

Teal, T., Albro, D., Stabler, E., and Taylor, C. (1999). Compression and adaptation. In *Advances in Artificial Life: Fifth European Conference on Artificial Life*, D. Flareans, J.-D. Nicoud, and F. Mordaa, eds., pp. 709–719. Heidelberg: Springer.

Uexküll, J. von (1921). *Umwelt und Innenwelt der Tiere*. Berlin: Jena.

Uexküll, J. von (1934/1957). A stroll through the worlds of animals and men. In *Instinctive Behavior: The Development of a Modern Concept*, C. H. Schiller, ed. (1957), pp. 5–80. New York: International Universities Press.

Wee, K., Collier, T., Kobele, G., Stabler, E., and Taylor, C. (2001). Natural language interface to an intrusion detection system. In *Proceedings of the International Conference on Control, Automation and Systems,* in press.

Whitehead, A. N. (1926). *Science in the Modern World*. New York, NY: Macmillan.

22 Consort Turnovers as Distributed Cognition in Olive Baboons: A Systems Approach to Mind

Deborah Forster

In a section on psychological components of tactics used by males in competition for mates, Smuts provides the following anecdote from her field observations of olive baboons (*Papio anubis*):

Early in the morning, Dante is in consort with Andromeda. Three older males, Alex, Sherlock, and Zim, are following and harassing Dante. Their movements are so perfectly synchronized that they take on an almost dance-like quality. Sherlock and Zim stand side by side facing Dante and, in unison, they rapidly and repeatedly threaten Dante with raised brows and then glance at Alex, 20 m away, soliciting his aid. Alex lopes over to them, places one arm around Sherlock's shoulder, and all three pant-grunt at Dante in an antiphonal chorus. In one smooth motion Zim lip-smacks, touches Alex's rear, looks at him, grunts at Sherlock, and then circles around to the other side of Dante. When he is opposite Alex and Sherlock, he resumes threatening Dante and, at precisely the same instant, they do the same. Alex embraces Sherlock and, together, they circle Dante and join Zim. All three stand in contact and swivel as a unit to face Dante, who avoids them. Dante appears increasingly tense. He repeatedly interposes his body between Andromeda and the other males and then herds her away by shoving her from behind. Each time he pushes her, Andromeda squeaks in protest. She too seems tense, glancing back and forth between Dante and the other three males. A few minutes later, a fifth male, Boz, appears on the hillside above the consort pair. Alex, Sherlock, and Zim immediately solicit Boz's aid against Dante. Boz runs towards them, and at the same time the other three once again move toward Dante. Dante and Andromeda break away from each other and run in opposite directions. Zim and Sherlock chase Dante while Boz and Alex run after Andromeda. Alex reaches her first, and she stops running and lets him copulate with her. A new consortship is formed (focal consorts sample, 2 July 1983). (Smuts 1985, 153–154)

The dynamics leading to a new consortship (referred to as a consort turnover, or CTO, event) serve Smuts as the backdrop for a discussion of individual baboons as psychological beings motivated by sophisticated goals and emotions. The complex dance of activity is transformed into an argument about confidence, tension, frustration, and the ability (of a male baboon) to manipulate the emotion of others (Smuts 1985). In this essay I revisit such social dynamics and propose that alternative links between patterns of activity and their cognitive implications are possible, perhaps even necessary if we are to resolve current issues regarding the nature, development, and evolution of cognition in primates. The general framework of distributed cognition (DCog) I present here was developed by Hutchins (1995) in a human setting. Later, in collaboration with Strum and Hutchins, I adapted this framework to the study of nonhuman primates (Forster et al. 1995; Strum et al. 1997; Strum and Forster 2001).

Research Questions and the Big Picture

Students of animal behavior are, most broadly, interested in the constraints and underlying mechanisms that organize the behavioral patterns we observe. Animal cognition researchers are, more specifically, concerned with the role cognition plays in organizing and controlling behavior, with an understanding that not all behavior is necessarily cognitive and not all cognitive processes necessarily produce observable behavior. There is a more or less common ground in cognitive science, which states that the "stuff" of cognition is representational in nature, intimately related to the brain, and is (or should be) amenable to computational models (for a general review see Gardner 1983; alternatively, see Varela et al. 1991). The popular stance among cognitive scientists studying behavior is that cognition mediates between "experience" (a gloss here for "things" that happen to organisms) and behavior (certain kinds of activities that certain organisms produce; see chapters 7 and 8 in Millikan 1993). Cognitive scientists argue that trying to explain the relationship between experience and behavior without addressing cognitive (i.e., representational) structures and processes will ultimately prove inadequate.

At the core of most controversies in cognitive science are the interpretive variations on the term *representation*. Articulating the arguments is beyond the scope of this essay, yet the DCog theory presented here turns on the notion of cognition as a process of propagating and transforming representational states across representational media (Hutchins 1995). I resist making claims about representational states and media in the world of olive baboons in the wild at this point and instead count on a conceptual link between representational and informational processes (Hutchins 1995).

The outcome of cognitive research is (wistfully) expected to reveal something about the internal psychological states of animals. Moreover, the explicit assumption in much of the literature is that cognition is restricted to what goes on inside the head of individuals; thick boundaries are drawn around the cognitive unit of analysis at the surface of the skull. Historically, the notions of a mind–brain identity and of the brain as an input/output device ("black box") for information processing got reinvigorated with the advent of computers and their mechanical and functional appeal as a model of the mind (for critiques, see Gardner 1983; Hutchins 1995; and Varela et al. 1991).

The reformulation of the cognitive science mission as the search for internal psychological states technically leaves open, not only the straightforward possibility of internal nonpsychological states, but also the possibility of external/interpersonal/supraindividual psychological states. The latter points to the realm of phenomena we are more interested in pursuing in our own research. In other words, we subscribe to the notion that although a lot of cognition happens inside the head of individuals, representational states have a tendency to "leak"

across the internal–external boundary and the permeability of this boundary has much to do with, and much to teach us about, the nature, development, and evolution of cognition. On a continuum from cognition being only inside the head to cognition being only in the external world (e.g., Gibson 1979), distributed cognition of the Hutchins (1995) variety sits somewhere in the middle, with representational structures and processes that span the internal–external boundaries organizing behavior.

Cognition research on social behavior remains strikingly fixated on internal psychological states. The social function of intellect (Humphrey 1976), or SFI, hypothesis, more recognizable as Machiavellian intelligence (Byrne and Whiten 1988, but see Strum et al. 1997), links social behavior and cognition and has its contemporary roots in field observations of nonhuman primates (e.g., Jolly 1966; Humphrey 1976; review in Byrne and Whiten 1997), although historically the "discovery" of social complexity in primates was distinct from, and slightly preceded, the "rediscovery" of the nonhuman animal mind (Strum and Forster 2001; Strum and Fedigan 1999). SFI is an evolutionary argument stating that an increase in social complexity acted as a driving force in the evolution of higher cognition in primates. It thus encompasses the cognitive abilities of individuals as well as the evolution of cognition in primates, levels often confounded in behavioral explanations.

Social complexity was pursued during a period in which the behaviorist tradition and its moratorium on invoking mental states still haunted field researchers, finding its way into the scientific literature through behavioral ecology. This move supported the sociobiological (i.e., selfish gene) approach to explaining behavior, which inadvertently allowed a profusion of cognitive metaphors (tactics, strategies, goals, etc.) into evolutionary explanations (natural selection itself could be talked about metaphorically in intentional terms). Authors added disclaimers to texts to avoid accusations of attributing any real mentality to their subjects (e.g., Dunbar 1988), but by the time cognitive ethology (Griffin 1976; Ristau 1991) gained recognition, an easy slippage between metaphorical and real cognition permeated the literature.

Social complexity remains largely an intuitive concept and only rarely receives operational treatments (Strum et al. 1997). Indeed, the impact of SFI to date has been mainly in determining the presence or absence of specific behavioral patterns (tactical deception, imitation, cultural traditions, teaching, etc.) which supposedly indicate the presence or absence of particular in-the-head abilities (such as symbolic reference, theory of mind, order of intentionality). Moreover, these patterns and abilities are usually defined in very human-centric terms (Caro and Hauser 1992; King 1994). The evidence then fuels speculative arguments about the extent of primate social complexity, the depth of primate understanding, and of domain specificity (Byrne and Whiten 1997; Tomasello and Call 1997), among others. Even though the (social) environment presents the context for these behavioral patterns (i.e., the problem space), the locus of cognitive resources for behavioral organization and control (the solution space?) is still "seen" and sought after mostly inside the head (but see Tomasello 1999).

Our orientation remains linked in spirit to the SFI hypothesis and explores the relation of social complexity to the nature, development, and evolution of cognition. We believe this requires a more thorough examination of how social complexity plays out in these contexts (see discussion in Strum et al. 1997). Since social complexity is a relational concept, we direct our search toward relational aspects of cognitive phenomena, and are drawn to relational perspectives on cognition (e.g., Vygotsky 1978; Rogoff 1990; Lave and Wenger 1991; Fogel 1993; Cole 1996; Nunez 1996). We study socially distributed behavioral phenomena that have cognitive properties: joint decision making, distributed information processing, polyadic conflict resolution, etc.— phenomena that by their temporal pace, variability of execution, and developmental paths

rule out the possibility that they are organized and controlled merely by genetic and/or maturational processes.

The exploration of these issues through research on sexual consort dynamics in baboons was undertaken for reasons that go beyond the fit to the criteria just mentioned. Historically, data on consort dynamics were also central to earlier formulations in behavioral ecology and cognitive ethology (exemplified by the opening anecdote), as well as discussions of tactical deception and other aspects of Machiavellian intelligence. Here I present research organized around dynamics like those described in Smuts's anecdote (see also Forster and Strum 1994; Strum et al. 1997). Briefly, in a female baboon, sexual consortships form in the days prior to ovulation, during which a male partner tries to monopolize access to the receptive female. Male followers and other troop members may actively participate in these efforts, especially in the dynamics that lead up to a switch in male consort partners, the consort turnover event. Here I revisit CTO events as the choice system or unit of cognitive analysis.

Theoretical and Empirical Methods

Although the current perspective is made possible only by standing on the shoulders of (a long line of) giants (see discussions and references in Strum et al. 1997; Strum and Forster 2001), I merely point here to the convergence of two "trailheads" that guide us most directly on our journey to linking social complexity and cognition: Hinde's (1983, 1987) relationship perspective on social structure and Hutchins' (1995) distributed perspective on cognition. While they do not directly overlap, both paradigms have in common an emphasis on development, dialectical relations between levels of organization in complex systems, and the employment of nonreductionist explanatory frameworks. Also, both emphasize the sociocultural context of behavior.

Hinde (1987) describes levels of social complexity (individuals ↔ interactions ↔ relationships ↔ social structure ↔ sociocultural environment) and shows how each level has properties that are not relevant to the ones below it. For example, properties of interactions, such as the level of coordination or synchronicity, are not relevant to descriptions of individual activity; the frequency of interactions over time is relevant to descriptions of relationships; and so on. Hinde's relationship perspective on social structure and its cumulative buildup from interactional data is directly relevant to our questions about social complexity. By taking the CTO event as a focal type of interaction in the context of sexual consorts, we begin to see patterns that lead to revisiting other levels of social complexity as well.

Hutchins' (1995) DCog theory suggests treating the CTO event as the unit of cognitive analysis. A cognitive system is defined by a regularly produced outcome, its boundaries flexibly dependent on an ability to identify meaningful regularities in the activity of the system as a whole. The outcome in this case is a switch in male consort partners (to the question, "What does the cognitive system do?" the answer is that the system decides who will be the next male consort partner). This is a profound analytic step since it allows us to proceed without having to first attribute goals or plans to individual participants. Identification of the switch in male consort partners as the defining outcome of a CTO event guides criteria for where to draw the boundaries of the unit of analysis, making sure that no element that is necessary to explain the outcome is left out. Although the focus here is on socially distributed processes, clearly other elements in the environment are relevant and at times are the defining factors (see a pattern of CTO events labeled "sleeping near the enemy," or SNE, in Forster and Strum 1994; Strum et al. 1997). In sexual consorts in general, the consort pair can be considered the nucleus of the system, yet a consort party may include other participants (see introductory anecdote). The system is not dependent solely on the identity of the participants and will maintain its character in the face of changing "personnel"; in this sense the description of it resembles a script with slots for prescribed roles.

While a system and its boundaries are defined by a clearly produced outcome, it is probed and studied as a process. An examination of the regularities of CTO events (Forster, in preparation) suggests a system that moves between four easily recognizable states: stable configuration (SCN), disruption (DSR), negotiation (NGT), and new configuration (NCN). All CTO events move between the four states, although the order and frequency of their expression in any given case may vary. There is a finite set of possible "flavors" for each transition to a system state; for example, there are nine ways to initiate disruption of an SCN. Significantly, these states are properties of the system and may differ from the properties of any individual participant or even of any dyad. A system state is achieved by a variety of individual activities and configurations, and conversely, an individual or dyadic activity such as grooming can "serve" a variety of system states, a many-to-many mapping, making it impossible to simply reduce system states to the activity of the elements from which they emerge. This redescription of CTO events has implications for explaining the cognitive behavior of individuals. Rather than relying on the outcome to characterize an individual (e.g., winner, loser), we can explore regularities in the way participants move through the various states, obtaining a richer cognitive profile.

This notion of process (of things changing continuously) is pursued over multiple time scales and tracked on several levels of behavioral organization: the system as a whole, the activity of individual participants, and changes in smaller behavioral units in the system (e.g., head movements). The heart of a DCog analysis is in tracking the transformation of and coordination among bits of structure on the various levels throughout the process. However, arriving at this form of description is no easy feat. It requires capturing the various levels simultaneously while preserving the ability to track each level and element separately (so that the relations between them can be empirically examined rather than assumed or extrapolated). Through repeated viewing at variable speeds, video footage of CTO events allows independent tracking of multiple agents over multiple time scales.

An examination of relational dynamics would be relevant for any discipline studying complex systems. Yet, from a traditional cognitive perspective, the most desirable aim is to give a representational account of system dynamics, which requires tracking cognitively significant elements. In human settings, the identification of representational structures is a relatively easy (or at least a less controversial) task compared with one using nonhuman primates in observational studies. What counts as representational in the world of olive baboons in the midst of sexual consort dynamics is not a settled issue and perhaps is best tested experimentally. Here I use a conceptual link between informational activities and representational processes. Representational processes are undoubtedly involved in the unfolding of CTO events, yet this early in the analysis I prefer a more conservative assumption, namely, that certain activities such as head movements are indicative of the management of attentional resources. A comparison of head movement patterns to patterns of other elements in the system (e.g., leg movements), suggests that tracking head movement dynamics is indeed relevant to "higher" cognition (on the order of decision making as opposed to that of motor control).

Internal Psychological States from a Systems Perspective

Reframing CTO dynamics in a systems perspective brings into sharp relief the ways social complexity plays out in the context of long-standing cognitive issues such as processing information, making plans and goals, and the development of skill, providing opportunities for alternative interpretations.

Head Movements Relative to Body Orientation as Information Processing

Head movements relative to body orientation (HMBO) in olive baboons were tracked separately for each participant and then represented in continuous fashion relative to two other levels of behavioral organization: activity patterns of (whole) individual participants and overall patterns of systems states and transitions (SCN, DSR, NGT, NCN) (Forster in preparation). I explored these relations across levels of behavioral organization as well as across individuals participating in the overall system, examining the coordination of movements among them. HMBO profiles of individual participants suggest polyadic monitoring and divided attention (showing as frequent side-to-side head movements while traveling in a third direction), especially before and after activity changes and around certain state transitions.

When HMBO profiles of participants are laid out in parallel, it is apparent that the combined patterns match system states and transitions in ways that are not reflected in the patterns of any of the participants individually or even of any dyadic combination. This is an analytic representation of dynamics like those conveyed in the opening anecdote. Video footage of such events makes it possible to detect the subtleties of coordination among participants. These patterns challenge assumptions about communication made in cognitive interpretations, namely, that communication can be described as a sequential exchange of signals between a sender and receiver. They also force us to reconsider the meaning of actions (i.e., their effect on the outcome). Rather than intrinsic to any particular instantiation, meaning is more likely the emergent result of the coordinated interactive unfolding of behavior distributed over the system as a whole.

Plans, Goals, and Situated Action

The discrepancies between individual and system-level properties are not resolved in any straightforward manner. In the SNE pattern of CTOs (Forster and Strum 1994), the regularities that make up a pattern (and make it identifiable as such to human observers) are distributed among the participants and other elements in their environment. This suggests a process that unfolds with equal regularity even if no single individual has an internal representational structure that matches the event as a whole. Individuals may coordinate with the system and may even anticipate the outcome and yet not have the big picture at their mental disposal (Strum et al. 1997). Instead, they may be responding to the constraints of the unfolding situation, hence "situated action" (Suchman 1987). That cognitive work in such a setting seems to get offloaded onto the environment is at times interpreted as a move to take cognition out of the head. It should rather be seen as a move to make more accurate specifications about what *has* to go inside. More significantly, it suggests how new cognitive structures and processes may arise (Hutchins 1995).

Development of Social Skill

A factor that is not necessary yet is often part of CTO systems is the presence of a young subadult male on the periphery of the consort party. Like other follower males, he synchronizes his activity with that of the consort pair, but refrains from interacting with the other participants. Only rarely will a situation unfold (usually during a prolonged NGT state) in which access to the consort female goes unchallenged by other males or is even initiated by the female. It is difficult to explain the presence of this male in terms of the odds of copulating relative to a constant monitoring effort. From a DCog perspective, however, it suggests that the larger system provides the constraints and the structure for young males

to safely experience regularities and learn how to coordinate with them. This baboon analog to human legitimate peripheral participation, or LPP (Lave and Wenger 1991), is possible only during a phase in which the subadult is too small and too familiar to present any real challenge to the other males. Therefore the social context can be seen as not only defining the cognitive challenge but also as the location of cognitive resources for learning and problem solving. [For a similar LPP interpretation of agonistic buffering, see Strum and Forster (2001).]

Future Research Directions

Subcategories and Partial Systems

With systemic descriptions of CTO events, it is possible to identify subtle variations such as the SNE CTO we mentioned earlier. Behavioral patterns that share only some of the characteristics of CTO systems, such as attempt turnovers (ATOs), are also available for scrutiny. ATOs lack the definitive switch in male partners and did not figure in our cognitive analysis of CTOs. Armed with descriptive regularities of system states, we can now compare CTOs and ATOs and gain insight into the factors that influence one outcome over another.

Life Cycle Transitions

Probing CTOs as a process provided us with a richer cognitive profile of individual participation. By the same token, it provides us with the potential for finding richer developmental paths. Rather than characterizing development as a change in some measure of performance, we can explore how individual participation in CTO events changes over life cycle transitions (subadult to adult female, newcomer to long-term resident male, etc.). The transition from a young subadult male who is a legitimate peripheral participant to a fully active consort male partner or follower may address the cognitive transformation from novice to expert.

Other Contexts and Scales

This framework is easily applied to other contexts. We can examine the relational dynamics of other interactions that occur with regularity in baboon social life (e.g., agonistic buffering, Strum and Forster 2001) and adapt the DCog framework to larger-scale situations (e.g., troop movement, intertroop encounters, mobbing). We emphasized polyadic interactions in our research in order to move beyond the restrictions of imposing dyadic sequential descriptions on communicative dynamics, but there is no reason this framework cannot be applied to dyadic interactions (which often occur in greeting and in grooming). A comparison across contexts provides an alternative route for addressing issues of domain specificity and modularity.

Phylogenetic and Methodological Limits

The study of primate cognition, like that of every long-lived social species, leaves phylogenetic arguments largely in the realm of speculation (for our contribution, see Strum and Forster 2001). Population genetics may offer a measure of how distant common ancestors are, so that comparative studies of behavioral patterns across populations, species, and taxa may provide useful landmarks in the speculative landscape. Adding to the ambiguity created by scrutinizing a relatively small number of successive generations is the complexity of the life cycle itself. A comparative approach can do much to set constraints on the range of patterns a phylogenetic story needs to contend with, but a broadening of the horizons of primate-centered cognition research is crucial to its success (Cords 1997; Rowel 2000). The breadth of research described in this volume is a welcome effort.

The observational methods used in our approach leave ambiguous distinctions that a well-designed experimental setting may be able to control for. Experimental settings in which the results of observational studies are further explored have already proven valuable (e.g., Cheney and Seyfarth 1990; Seyfarth and Cheney, chapter 46 in this volume; Tomasello 1999). Nevertheless, the need to deal with context, process, and relational dynamics limits the kinds of systems that can be studied practically. Phenomena in which system-level qualitative changes are easily identified are needed so that the questions of transitions and the relations among levels can be rigorously pursued. Decades of field research on primates, especially baboons, have yielded a rich source of such phenomena that can now be revisited through the lens of systems thinking and dialectical relations.

Conclusion

I have tried to show how choices made at the outset of the cognitive science revolution (Gardner 1983) have influenced and limited the agenda of cognition research in primates. By relaxing the constraints on locating cognitive resources for behavioral organization, and by adopting a relational dynamic view of the nature and development of cognition, we are able to respond more successfully to issues that have stumped us in the past (e.g., how does social complexity affect the individual? how is it linked to cognition?) as well as to ask a whole host of new questions (e.g., how do individuals acquire the functional abilities necessary to establish and maintain coordination with system-level regularities of, say, a CTO event?). The adaptation of DCog theory and other relational perspectives to the study of nonhuman primates opens up new arenas of exploration. These are many advantages in using the primacy of the system as the basis for cognitive analysis:

• The analysis is not completely dependent on the attribution of goal states to one or more individuals.

- Meaning is not intrinsic to behavioral acts, but rather is determined by their distributed and negotiated effect on the system.
- Behavioral patterns of individuals need not get organized solely by internal mental plans.
- Richer descriptions are available for participation profiles in polyadic interactions.
- Systems become not only context and settings for cognitive problems but also resources and learning environments for solving them.
- Alternative scenarios are plausible for how individuals acquire the functional abilities to establish and maintain coordination with system-level dynamics (situated learning by LPP).

Much research is still needed to flesh out the relations between the cognitive processes that emerge in interactive distributed systems and the cognitive work done by individuals. Most significant perhaps is the need to integrate into this picture a developmental perspective that spans the life cycle. No less important though are methodologies that will help reveal representational features in the relational worlds of nonhuman animals.

Acknowledgments

I would like to thank Shirley Strum and Ed Hutchins for inspiring me with their research frameworks and collaborating with me to produce a hybrid in exemplary DCog fashion. Thanks to Barb Holder, Shirley Strum, and the editors of this volume for helpful comments and suggestions.

References

Byrne, R. W. and Whiten, A. (eds.) (1988). *Machiavellian Intelligence*. Oxford: Clarendon Press.

Byrne, R. W. and Whiten, A. (1997). Machiavellian intelligence. In *Machiavellian Intelligence II: Extensions and Evaluations*, A. Whiten and R. W. Byrne, eds., pp. 1–23. Cambridge: Cambridge University Press.

Caro, T. M. and Hauser, M. D. (1992). Is there teaching in nonhuman animals? *Quarterly Review of Biology* 67: 151–174.

Cheney, D. and Seyfarth, R. (1990). *How Monkeys See the World*. Chicago: University of Chicago Press.

Cole, M. (1996). *Cultural Psychology: The Once and Future Discipline*. Cambridge, Mass.: Harvard University Press.

Cords, M. (1997). Friendships, alliances, reciprocity and repair. In *Machiavellian Intelligence II: Extensions and Evaluations*, A. Whiten and R. W. Byrne, eds., pp. 24–49. Cambridge: Cambridge University Press.

Dunbar, R. I. M. (1988). *Primate Social Systems*. pp. 26–27. Ithaca, N.Y.: Cornell University Press.

Fogel, A. (1993). *Developing Through Relationships: Origins of Communication, Self, and Culture*. Chicago: University of Chicago Press.

Forster, D. (in preparation). From outcome to process: A systems perspective on sexual consort turnovers in baboons.

Forster, D. and Strum, S. C. (1994). Sleeping near the enemy: Patterns of sexual competition in baboons. In *Current Primatology*. Vol. 2. *Social Development, Learning and Behavior*, B. T. J. Roeder, J. Anderson, and N. Herrenschmidt, eds., pp. 19–24. Strasbourg: Louis Pasteur University.

Forster, D., Hutchins, E., and Strum, S. C. (1995). Essence and boundaries: Ontogeny of distributed social skill in baboons. Paper presented at the 25th Annual Symposium of the Piaget Society.

Gardner, H. (1983). *The Mind's New Science: A History of the Cognitive Revolution*. New York: Basic Books.

Gibson, J. J. (1979). *The Ecological Approach to Visual Perception*. Boston: Houghton Mifflin.

Griffin, D. (1976). *The Question of Animal Awareness: Evolutionary Continuity of Mental Experience*. New York: Rockefeller University Press.

Hinde, R. A. (1983). *Primate Social Relationships: An Integrated Approach*. Oxford: Blackwell.

Hinde, R. A. (1987). *Individuals, Relationships and Culture: Links Between Ethology and the Social Sciences*. Cambridge: Cambridge University Press.

Humphrey, N. (1976). The social function of intellect. In *Growing Points in Ethology*, P. P. G. Bateson and R. A. Hinde, eds., pp. 303–317. Cambridge: Cambridge University Press.

Hutchins, E. (1995). *Cognition in the Wild*. Cambridge, Mass.: MIT Press.

Jolly, A. (1966). Lemur social behaviour and primate intelligence. *Science* 153: 501–506.

King, B. J. (1994). *The Information Continuum: Evolution of Social Information Transfer in Monkeys, Apes, and Hominids*. Santa Fe, N.M.: School of American Research.

Lave, J. and Wenger, E. (1991). *Situated Learning*. New York: Cambridge University Press.

Millikan, R. G. (1993). *White Queen Psychology and Other Essays for Alice*. Cambridge, Mass.: MIT Press.

Nunez, R. (1996). Ecological naturalism: Conscious experience as a supra-individual biological (SIB) phenomenon. *Consciousness Research Abstracts, Toward a Science of Consciousness*: 178.

Ristau, C. A. (ed.) (1991). *Cognitive Ethology: The Minds of Other Animals*. Hillsdale, N.J.: Lawrence Erlbaum Associates.

Rogoff, B. (1990). *Apprenticeship in Thinking: Cognitive Development in Social Context*. New York: Oxford University Press.

Rowell, T. (2000). A few peculiar primates. In *Primate Encounters: Models of Science, Gender, and Society*, S. C. Strum and L. M. Fedigan, eds., pp. 57–70. Chicago: University of Chicago Press.

Smuts, B. (1985). *Sex and Friendship in Baboons*. pp. 153–155. New York: Aldine.

Strum, S. C. and Fedigan, L. (1999). Theory, method and gender: What changed our views of primate society. In *The New Physical Anthropology: Science, Humanism and Critical Reflection*, S. C. Strum, D. Lindburg, and D. Hamburg, eds., pp. 67–106. Englewoods Cliffs, N.J.: Prentice-Hall.

Strum, S. C. and Forster, D. (2001). Nonmaterial artifacts: A distributed approach to mind. In *In the Mind's Eye: Multidisciplinary Approaches to the Evolution of Human Cognition*, A. Nowell, ed., pp. 63–82. Ann Arbor, Mich.: International Monographs in Prehistory.

Strum, S. C., Forster, D., and Hutchins, E. (1997). Why Machiavellian intelligence may not be Machiavellian. In *Machiavellian Intelligence II: Extensions and Evaluations*, A. Whiten and R. Byrne, eds., pp. 50–85. Cambridge: Cambridge University Press.

Suchman, L. (1987). *Plans and Situated Actions: The Problems of Human–Machine Communication*. New York: Cambridge University Press.

Tomasello, M. (1999). *The Cultural Origins of Human Cognition*. Cambridge, Mass.: Harvard University Press.

Tomasello, M. and Call, J. (1997). *Primate Cognition*. Oxford: Oxford University Press.

Varela, F. J., Thompson, E., and Rosch, E. (1991). *The Embodied Mind: Cognitive Science and Human Experience*. Cambridge, Mass.: MIT Press.

Vygotsky, L. (1978). *Mind in Society: The Development of Higher Psychological Processes*. Cambridge, Mass.: Harvard University Press.

II CONCEPTS AND CATEGORIES

23 General Signs

Edward A. Wasserman

This essay contends that contemporary conditioning methods provide animals with the very "general signs" that John Locke said animals lacked in disclosing their thinking to us. The button pressing of pigeons can reveal mental imagery, visual illusions, and abstract concepts, if we are smart enough to teach these responses to animals and to interpret their actions carefully.

The interrelation between thought and language has been a central and enduring issue in philosophy and psychology. Of particular interest has been the possibility that nonhuman animals think. Although René Descartes and John Locke disagreed on other matters of animal intelligence (Wilson 1995), they did agree that animals are incapable of thinking or any other advanced cognitive feats.

Descartes denied animal thought because animals do not speak. "[T]he reason why animals do not speak as we do is not that they lack the organs but that they have no thoughts" (Descartes 1646/1970, p. 207). Locke, too, believed that animals were incapable of ideation or abstraction because they did not use words to divulge the operation of those advanced cognitive skills.

I think, I may be positive ... that the power of Abstracting is not at all in them; and ... the having of general *Ideas*, is that which puts a perfect distinction betwixt Man and Brutes; and is an Excellency which the Faculties of Brutes do by no means attain to. For it is evident, we observe no foot-steps in them, of making use of general signs for universal Ideas; from which we have reason to imagine that they have not the faculty of abstracting, or making general *Ideas*, since they have no use of Words, or any other general Signs. (Locke, 1690/1975, pp. 159–160)

Putting aside the question of whether nonhuman animals can be taught a human language (Gardner and Gardner 1984; Pepperberg 1981), my essay concerns Locke's claim that animals lack any other general signs through which we might learn about their intelligence. Locke considered the matter closed; otherwise, he would at least have left open the possibility that new means might be devised that could divulge animal thought. However, the matter is not closed. In fact, the creative application of the trusted methods of classical and operant conditioning (Wasserman and Miller 1997) has flung the door wide open to a fresh, empirical understanding of animal cognition and perception (Wasserman 1993, 1997). These methods provide animals with quite general signs by means of which they can disclose their cognitions and perceptions to us.

Contrary to Locke's unsubstantiated denial, the use of these conditioning methods suggests that animals are capable of advanced cognition and abstraction. Further research suggests that animals experience many of the same perceptual illusions and mental images that we do. All of this scientific work forces us to reconsider the stale convention that thought is impossible without language.

Setting the Agenda

The Darwinian revolution raised the possibility that there was mental continuity between human and nonhuman animals: that "the difference in mind between man and the higher animals, great as it is, certainly is one of degree and not of kind" (Darwin 1871/1920, p. 128).

Charles Darwin and the early comparative psychologists adduced support for this provocative hypothesis with numerous anecdotes of animal genius. But that initial anecdotal evidence did not pass muster; it was of dubious veracity and replicability. The call thus came for more controlled and repeatable evidence. Answering these calls were the fledgling behaviorists, led by John Watson (1913).

For decades, critics (e.g., Griffin 1992) have falsely characterized behaviorism as a dry and conservative school of psychological science. But behaviorism simply insisted on clear empirical evidence of the same psychological processes that had until then been studied only introspectively, through the inward examination of one's own thoughts and feelings.

Watson himself harbored ambitious aspirations for this new branch of psychological science. He was convinced that the full gamut of mental functioning—including perception, memory, imagination, judgment, reasoning, and conception—would eventually yield to objective behavioral analysis: "Psychology as behavior will, after all, have to neglect but few of the really essential problems with which psychology as an introspective science now concerns itself. In all probability even this residue of problems may be phrased in such a way that refined methods in behavior (which certainly must come) will lead to their solution" (Wetson 1913, p. 177). Watson's words were prophetic. Thanks to the innovative behavioral methods of Ivan Pavlov and B. F. Skinner, we can now study an unprecedented range of perceptual and cognitive processes in nonhuman animals—the very mental processes that theorists have for centuries believed animals lacked or that appeared to fall outside the ken of psychological science.

It is interesting that the use of behavioral tools—especially arbitrary conditioned responses —to investigate perception and cognition in nonhuman animals was advocated by Edwin Boring, the esteemed historian of experimental psychology and a student of America's best-known introspectionist, Edward Titchener. Boring (1953, p. 183) disagreed with the common view, "that you learn about human consciousness by direct observation of it in introspection, but that animal consciousness is known only indirectly by analogical inference." Instead, he championed Max Meyer's (1921) psychology of "the other one": "an argument that your own personal consciousness is not material for science, being particular and not general, and that psychology always studies other organisms —other people, other animals" (Boring 1953, p. 183). Boring (1953, p. 183) believed that "both the animal's conduct and man's words are introspection if they are taken as meaning something about the subject's consciousness."

Boring's ideas represent what modern theorists (Zuriff 1985) call "methodological behaviorism" rather than Skinner's radical behaviorism. Methodological behaviorism pursues Watson's ambition of deciphering mental processes with behavioral rather than introspective methods, whereas radical behaviorism emphasizes the study of behavior in its own right, irrespective of its relevance to mental or neural processes (Skinner 1938). I am not able to engage this debate here, having discussed it earlier (Wasserman 1981, 1982, 1983).

Proving the Point

It is one thing to suggest that we can use arbitrary conditioned responses to tell us about perception and cognition in nonhuman animals; it is quite another thing to accomplish that feat. In the next three sections, I document how the use of operant conditioning methods has disclosed some remarkable examples of perception and cognition in the pigeon. I then return to Locke's notion of general signs and relate these examples to that intriguing notion.

Abstraction in Pigeons

Despite his certainty that nonhuman animals are incapable of abstraction and conceptualization, Locke may have been premature in his assessment. Mounting behavioral evidence suggests that nonhuman animals—even pigeons—can learn abstract concepts (Allen and Hauser 1991; Delius 1994; Thompson 1995).

In our own initial experimental investigation of abstraction in pigeons, we (Wasserman et al.

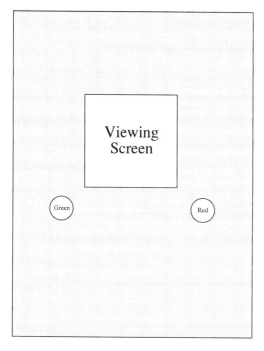

Figure 23.1
The pigeon's stimulus-response panel, including the viewing screen and the two differently colored choice buttons.

1995) first used food reinforcement to teach feral pigeons to peck a red button when they saw any of 16 4 × 4 displays that contained 16 copies of the same visual icon and to peck a green button when they saw any of 16 other 4 × 4 displays that contained 1 copy of 16 different visual icons. Figure 23.1 provides a bird's-eye view of the stimulus-response panel of the Skinner box. The top half of figure 23.2 illustrates 2 of the 16 same displays and 2 of the 16 different displays in the training set of displays (set 1) for half of the pigeons. The 16-icon arrays were first shown to the pigeon for several seconds and then the choice buttons were illuminated. A single choice response was permitted. If the correct button was chosen, then food was given; if the incorrect

button was chosen, then no food was given. Daily sessions consisted of dozens of these same and different training trials.

After the pigeons had reached a high level of discrimination accuracy (exceeding 80 percent correct choices; the chance score was 50 percent correct), in testing sessions we showed the birds 16 new same displays and 16 new different displays, which contained icons that they had never seen before. The bottom half of figure 23.2 illustrates 2 of the 16 same displays and 2 of the 16 different displays in the testing set of displays (set 2) for these pigeons. The other half of the pigeons were first trained with stimuli from set 2 and later tested with stimuli from set 1.

Across both groups of birds, accuracy averaged 83 percent correct responses to the same and different displays from the training set; accuracy averaged 71 percent correct responses to the same and different displays from the testing set. These high levels of accurate responses to both familiar and novel displays are consistent with the pigeons' having learned a general same-different concept. So too are the even stronger results of a replication (Young and Wasserman 1997) of this first project, in which the pigeons' accuracy of choice averaged 93 percent correct responses to stimuli from the training set and 79 percent correct to stimuli from the testing set. Finally, pigeons successfully learned and generalized a same-different discrimination with lists of successively presented icons (Young et al. 1999), thereby proving that low-level perceptual mechanisms such as texture discrimination cannot explain the pigeon's conceptual behavior.

The Ponzo Illusion in Pigeons

Consider the two drawings in the right half of figure 23.3. Do the two horizontal lines appear to be equally long in each? No. The horizontal line in the upper drawing looks longer than the one in the lower drawing. However, the two horizontal lines are actually the same length. The

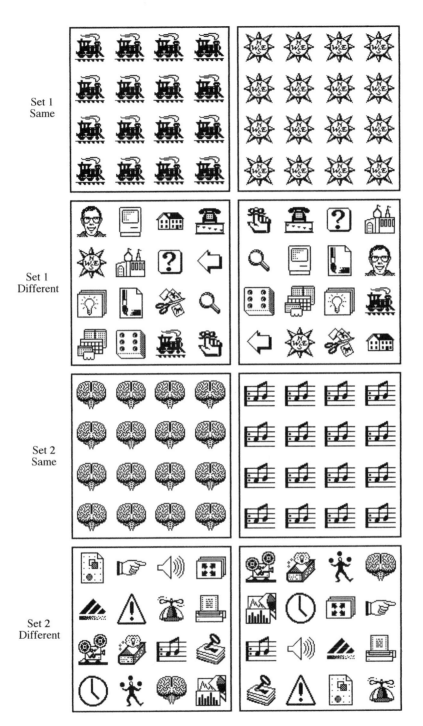

Figure 23.2
Sample same and different arrays from sets 1 and 2 of computer icons. Each set contained 16 icons.

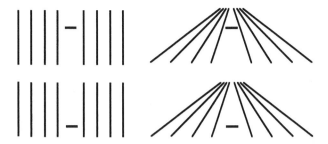

Figure 23.3
Left: Same-sized horizontal lines appearing in the parallel vertical line arrays used by Fujita et al. (1991) during original discrimination training. During this training, short and long lines appeared at various heights in the parallel vertical line field, as in the top and bottom portions of this figure. Right: Same-sized horizontal lines appearing in the nonparallel line arrays used by Fujita et al. (1991) during testing for the Ponzo illusion. During testing, the same-sized lines appeared at various heights in the nonparallel line field, as in the top and bottom portions of this figure. To our eyes, the horizontal lines in the left half of the figure appear to be of the same length, whereas the horizontal lines in the right half of the figure appear to be of different lengths; the top line appears to be longer than the bottom line. The results of this experiment suggest that pigeons also experience the same visual illusion. These drawings were adapted from those depicted in figure 3 of Fujita et al. (1991).

apparent difference in the lengths of the two lines is called the "Ponzo illusion."

This and dozens of other instances of nonveridical or illusory perception are not isolated curiosities; rather, they betray the fundamental and imperfect operating characteristics of our sensory systems. An understanding of the functioning of these sensory systems requires the elucidation of such perceptual illusions.

Of course, most subjects in investigations of illusory perception are adult human beings. However, nonhuman animals can also be studied in this type of research (Blough and Blough 1977). Instead of using verbal behaviors to disclose illusory perception, nonverbal behaviors are conditioned through contingencies of reinforcement. A case study of illusory perception in animals discovered that when pigeons viewed the stimuli in the right half of figure 23.3, they too reported that the upper horizontal line looked longer (Fujita et al. 1991, 1993).

The first stage of the pigeons' training established two different pecking responses to two different sets of stimuli. Through differential food reinforcement, the birds learned to peck one ("long") button when one out of a set of relatively long horizontal lines was positioned at different heights between parallel vertical lines and to peck a second ("short") button when one out of a set of relatively short horizontal lines was similarly positioned between the same parallel vertical lines (see the drawings in the left half of figure 23.3). Then in testing, the birds saw a novel horizontal line of intermediate length that was positioned between nonparallel lines (see the drawings in the right half of figure 23.3). When the novel line was positioned high in the display near the apex of the converging lines, the pigeons were more prone to peck the "long" button than they were when the test line was positioned low in the display, far from the apex of the converging lines.

This pattern of pigeons' pecking responses closely accords with the verbal responses of human beings who also report that the lines are of different length in analogous experiments. In this sense, we can say that the pigeons experience the Ponzo illusion in the same way as human beings.

We can also say that when different members of the same or different species respond in accord with such illusory perceptions, we gain considerable confidence in the basic nature of visual processing, irrespective of experiential or genetic influences. We then expect that the neural mechanisms of visual perception may be quite general indeed.

Mental Imagery in Pigeons

Rilling and Neiworth (1987) studied mental imagery in animals. Here the challenge was to retain the methodological rigor of behaviorism while simultaneously conducting psychologically valid research on a process that most theorists presume to be uniquely human.

Calling on classic human research by Shepard and Cooper (1982), Rilling and Neiworth's experimental response to this daunting challenge was to have a pigeon view a moving clock hand on a computer monitor for a short time, for the clock hand to disappear briefly, and for the pigeon to report whether the location of the clock hand when it reappeared at test was in the proper spot given its earlier position and trajectory. Two buttons were provided: one that was to be pecked when the clock hand appeared at the proper spot and a second that was to be pecked when the clock hand appeared at an improper spot; this could be either a spot that was too near the disappearance point or a spot that was too far from the disappearance point. Through differential reinforcement, Rilling and Neiworth taught pigeons entirely nonverbal behaviors for describing the projected motion of an absent stimulus—a task that can quite reasonably be said to require mental imagery.

The data provided strong empirical support for the possibility that pigeons can properly project the trajectory of a moving stimulus, despite its absence for a period of time. The pigeons not only made correct reports on trials with familiar starting points, trajectories, and stopping points, but they also did so on trials with new starting points, trajectories, and stopping points, thereby attesting to the generality of their imagery knowledge; the pigeons had acquired a concept of "motion," if you will.

Button Pecks as General Signs

There are two buttons. One or the other must be pecked to indicate: (1) that several visual items are the same as or different from one another, (2) whether a horizontal line is relatively short or long, or (3) whether the clock hand that has just reappeared is in the correct or the incorrect position given the hand's location and trajectory prior to its disappearance.

Except for the training procedures themselves, nothing should incline a pigeon to make the correct responses in each of these three experimental examples; the functional significance of these particular button assignments for the receipt of food is completely arbitrary. So I would argue that pigeons' button pecks are truly general signs.

What might Locke say about my analysis of button pecks as general signs of animal perception and cognition? I will not presume to answer this question for the reader. What I will do is to underscore Locke's own distrust of language for communicating mental states and processes: "[I]t is easy to perceive, what imperfection there is in Language, and how the very nature of Words, makes it almost unavoidable, for many of them to be doubtful and uncertain in their significations" (Locke 1690/1975, pp. 475–476).

If not through words, then just how might we best understand the nature and processes of perception and cognition? I believe that carefully devised behavioral studies—of both humans and animals—can and will yield the very kind of verifiable data that can make what was once a purely philosophical undertaking a truly natural science of mind. Pursuit of this path will continue to yield new secrets of adaptive behavior and cognition (Wasserman 1993).

Behavioral science may have its limits, but those limits do not prohibit the systematic study of cognition and perception. The fact that we have been able to make unprecedented progress in investigating these processes in as "intellectually challenged" a beast as the pigeon testifies to the potential of this behavioral program. It also testifies to the continuity of mental processes in human and nonhuman animals—a strong rebuttal of a cornerstone of Cartesian philosophy.

Finally, it will come as little surprise to readers that the perceptual and cognitive processes on which I have focused in this essay—mental imagery, visual illusions, and abstract concepts—have long been held to be central to an understanding of human mental function. We know no mind better than our own. So at least part of the agenda of the study of comparative cognition has a decidedly anthropocentric character (Wasserman 1997). Nevertheless, to the degree that organisms as phylogenetically, ecologically, and physiologically different as humans and pigeons see and think alike, we gain considerable confidence in the adaptive value of advanced perception and cognition. We also place further trust in the worth of general signs as valid markers of these processes. These and other considerations place research in comparative cognition squarely at the center of evolutionary biology.

Acknowledgments

The author thanks Daniel Dennett, Fred Dretske, and Jon Ringen for their helpful comments on earlier drafts of this paper. Some of the research that is described in this paper was supported by research grant IBN 99-04569 from the National Science Foundation.

References

Allen, C. and Hauser, M. D. (1991). Concept attribution in nonhuman animals: Theoretical and methodological problems in ascribing complex mental processes. *Philosophy of Science* 58: 221–240.

Blough, D. and Blough, P. (1977). Animal psychophysics. In *Handbook of Operant Behavior*, W. K. Honig and J. E. R. Staddon, eds., pp. 514–539. Englewood Cliffs, N.J.: Prentice-Hall.

Boring, E. G. (1953). A history of introspection. *Psychological Bulletin* 50: 169–189.

Darwin, C. (1871/1920). *The Descent of Man; and Selection in Relation to Sex*. (2nd ed.) New York: Appleton. (Originally written in 1871.)

Delius, J. D. (1994). Comparative cognition of identity. In *International Perspectives on Psychological Science*. Vol. 1, P. Bertelson, P. Eelen, and G. d'Ydewalle, eds., pp. 25–40. Hillsdale, N.J.: Lawrence Erlbaum Associates.

Descartes, R. (1646/1970). *Descartes's Philosophical Letters* (A. Kenny, ed. and trans.). Oxford: Clarendon Press. (Originally written in 1646.)

Fujita, K., Blough, D. S., and Blough, P. M. (1991). Pigeons see the Ponzo illusion. *Animal Learning & Behavior* 19, 283–293.

Fujita, K., Blough, D. S., and Blough, P. M. (1993). Effects of the inclination of context lines on the perception of the Ponzo illusion by pigeons. *Animal Learning & Behavior* 21: 29–34.

Gardner, R. A. and Gardner, B. T. (1984). A vocabulary test for chimpanzees (*Pan troglodytes*). *Journal of Comparative Psychology* 98: 381–404.

Griffin, D. R. (1992). *Animal Minds*. Chicago: University of Chicago Press.

Locke, J. (1690/1975). *Essay Concerning Human Understanding*, P. H. Nidditch, ed. Oxford: Clarendon Press. (Originally written in 1690.)

Meyer, M. (1921). *Psychology of the Other One*. Columbia, Mo: Missouri Book Company.

Pepperberg, I. M. (1981). Functional vocalizations by an African grey parrot (*Psittacus erithacus*). *Zeitschrift für Tierpsychologie* 55: 139–160.

Rilling, M. E. and Neiworth, J. J. (1987). Theoretical and methodological considerations for the study of imagery in animals. *Learning and Motivation* 18: 57–79.

Shepard, R. N. and Cooper, L. A. (1982). *Mental Images and Their Transformations*. Cambridge, Mass.: MIT Press.

Skinner, B. F. (1938). *The Behavior of Organisms*. New York: Appleton-Century-Crofts.

Thompson, R. K. R. (1995). Natural and relational concepts in animals. In *Comparative Approaches to Cognitive Science*, H. L. Roitblat and J. A. Meyer, eds., pp. 175–224. Cambridge, Mass.: MIT Press.

Wasserman, E. A. (1981). Comparative psychology returns: A review of Hulse, Fowler, and Honig's *Cognitive Processes in Animal Behavior*. *Journal of the Experimental Analysis of Behavior* 35: 243–257.

Wasserman, E. A. (1982). Further remarks on the role of cognition in the comparative analysis of behavior. *Journal of the Experimental Analysis of Behavior* 38: 211–216.

Wasserman, E. A. (1983). Is cognitive psychology behavioral? *Psychological Record* 33: 6–11.

Wasserman, E. A. (1993). Comparative cognition: Beginning the second century of the study of animal intelligence. *Psychological Bulletin* 113: 211–228.

Wasserman, E. A. (1997). Animal cognition: Past, present, and future. *Journal of Experimental Psychology: Animal Behavior Processes* 23: 123–135.

Wasserman, E. A. and Miller, R. R. (1997). What's elementary about associative learning? *Annual Review of Psychology* 48: 573–607.

Wasserman, E. A., Hugart, J. A., and Kirkpatrick-Steger, K. (1995). Pigeons show same-different conceptualization after training with complex visual stimuli. *Journal of Experimental Psychology: Animal Behavior Processes* 21: 248–252.

Watson, J. B. (1913). Psychology as the behaviorist views it. *Psychological Review* 20: 158–177.

Wilson, M. D. (1995). Animal ideas. *Proceedings of the American Philosophical Association* 69: 7–25.

Young, M. E. and Wasserman, E. A. (1997). Entropy detection by pigeons: Response to mixed visual displays after same-different discrimination training. *Journal of Experimental Psychology: Animal Behavior Processes* 23: 157–170.

Young, M. E., Wasserman, E. A., Hilfers, M. A., and Dalrymple R. (1999). The pigeon's variability discrimination with lists of successively presented visual stimuli. *Journal of Experimental Psychology: Animal Behavior Processes* 25: 475–490.

Zuriff, G. E. (1985). *Behaviorism: A Conceptual Reconstruction*. New York: Columbia University Press.

24 The Cognitive Dolphin

Herbert L. Roitblat

Although we may not be able to say definitively what it is like to be a dolphin, there is a good deal that we can know about its perceptual and cognitive system. My work, along with my colleagues and students, has been dedicated to discovering what kinds of representations animals have and how those representations underlie its behavior. The highlight of this research is our work on dolphin biosonar echolocation. Most of this work involves the Atlantic bottlenosed dolphin (*Tursiops truncatus*), although we have on occasion studied other species as well. Like bats, dolphins obtain information about the identity, location, and characteristics of objects in their world by actively interrogating them using their unique biological sonar, which is highly adapted to their aquatic environment.

Although their use of biological sonar is called "echolocation," dolphins use their sonar for far more than just determining how far away objects are. Their biosonar abilities far exceed those of any man-made system. Dolphins can detect and discriminate targets in highly cluttered and noisy environments (Au 1993). One outstanding example of their keen sonar capabilities is their ability to sonically detect, dig out, and feed on fish and small eels buried up to 45 cm beneath the sandy seabed (Rossbach and Herzing 1997). Using echolocation, dolphins can identify many characteristics of submerged objects, including size, structure, shape, and material composition. For example, they can detect the presence of small (7.6 cm diameter) stainless steel spheres at distances of up to 113 m. They can discriminate among aluminum, copper, and brass circular targets, and among circles, squares, and triangular targets covered with neoprene (see Au 1993).

Bottlenosed dolphin biological sonar uses very broadband high-frequency clicks of about 50 μs that emerge from their rounded forehead or melon as a highly directional sound beam with 3 dB (half-power) beam widths of about 10° in both the vertical and horizontal planes (Au et al. 1986). Their echolocation clicks have a peak energy at frequencies ranging from 40 to 130 kHz with source levels of 220 dB re: 1 μPa at 1 m (Au 1993). Dolphin hearing extends to frequencies as high as 150 kHz, which is 8–10 times higher than human hearing limits. They generate their clicks deep within their heads by passing air through a nasal structure called "monkey lips" because of its appearance. The sound travels through the water in a narrow conelike beam and reflects off objects in that beam. The sound is picked up in the dolphin's jaw and conducted to the animal's inner ear, where it is transduced into neural signals for processing by the rest of the brain.

The time between successive clicks depends on the distance between the animal and the target it is scanning. The average time between emitted clicks in a train is typically 15–22 ms longer than the time required for the click to travel through the water to the target and return as an echo (Morozov et al. 1972; Penner 1988).

Although both bats and dolphins use echolocation, the characteristics of the medium in which their signals are emitted, the mechanisms by which the signals are produced, the type of signals, and the neurological apparatus they use to processes those signals differ substantially. Bat biosonar is adapted for use in air, whereas dolphin biosonar is adapted for use underwater. Bat biosonar signals are relatively long in duration (up to several milliseconds), and contain both narrow-band constant-frequency and frequency-modulated components depending on the species (Bellwood 1988; Fenton 1988; Suthers 1988). By contrast, the dolphin echolocation signal is very broadband and, as indicated, extremely short. Echoes typically range in duration from 50 to 200 μs.

Dolphin echolocation is one of the most sophisticated cognitive processes that have been studied. When a dolphin uses its biological sonar to recognize objects, its brain performs the equivalent of some extraordinarily complex computations. These computations transform one-dimensional sound waves arriving at each of the dolphin's two ears into representations of objects and their features in the dolphin's environment. The process by which this transformation occurs is the focus of our interest.

In the preceding paragraph I asserted, perhaps too boldly, that the dolphin transforms the echo's one-dimensional sound waves into a three-dimensional representation of objects, but that is not the only way that dolphins could use their biosonar to recognize objects. It is conceivable that the dolphin does not solve the object-recognition problem per se, but rather solves a listening problem. Rather than using its sonar to determine the characteristics of objects, it could recognize them by detecting the object's characteristic sonic signature. In this hypothesis, a tuna is recognized, not by its structure, but by the sound signature or profile of its echo. In vision, this would be the equivalent of saying that we recognize an object, not by its perceived structure, but by the characteristic pattern it projects on our retina.

Although such a model is conceivable, it seems unlikely from an ecological perspective. Sensory systems evolved in response to real ecological problems, so it seems reasonable to suppose that they actually do provide ecologically relevant information. Treating echolocation as a listening problem would allow the animal to distinguish one group of objects from another, but it would not provide a very reliable basis for dealing with novel objects. More critically, it would make object permanence into a really difficult problem, for example, because of the strong dependence of the echo characteristics on the angle from which the object is ensonified, called "aspect dependence."

The echo returned by an object depends very strongly on the angle from which it is "viewed." Aspect dependence is also a property of visual perception and in the present context it is prob-

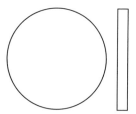

Figure 24.1
A sketch of a coin from the front and from the side.

ably easier to explain in that way. Consider the two objects depicted in figure 24.1, which is a sketch of a coin from the front and from the side. The scene that is projected onto the viewer's retina depends on the angle from which the object is viewed. No features are common between these two views, yet under appropriate circumstances, people can easily tell that it is the same object. We do not get the impression that an object has disappeared, to be replaced by a different one when our viewing angle changes; rather, we tend to perceive the object rather than its projection on the retina—an example of object permanence. Every viewing angle and every viewing distance would project a different pattern on the retina, so it is difficult to imagine how there could be characteristic patterns by which the object could be visually recognized. Visual object constancy over changes in position and angle does not seem to depend on the existence of specific invariant properties in the retinal image per se, but rather seems to be computed from changes in the retinal image.

Sound is also dependent on the angle at which an object is ensonified. Even a small change in angle can have a profound impact on the structure of the echo, especially when there are sharp discontinuities in the object's structure (e.g., corners or edges). Furthermore, unlike vision, sound in the water often penetrates the object so that there are reflections not only from the front surface of the object but also from the back surface. The internal properties of the object also affect the echo. For example, a dolphin can distinguish

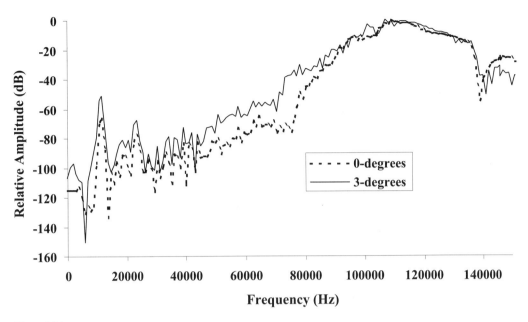

Figure 24.2
Echo spectra from a complex object ensonified at either 0 or 3° relative to the perpendicular.

between identical containers that are filled with different liquids (Roitblat et al. 1993). These physical characteristics all affect the structure of the echo as it is received by the dolphin, but there does not seem to be any simple function that can be used to predict the structure of the echo from one angle given its structure at another angle (the same difficulty exists for vision). The dolphin must use other, presumably cognitive, functions to recognize the constancy in an object viewed at different angles. We can describe some of the characteristics of these cognitive functions.

Figure 24.2 shows example echo spectra from a complex constructed object at two different angles. This object was constructed of a large carriage bolt with several nuts attached. It was ensonified by an artificial dolphin click from a distance of about 1 m. As it rotates, it presents different faces to the sound beam and hence reflects the incident sound in different frequency-

and angle-dependent directions. Notice how even a small rotation of 3° can have a substantial effect on the spectrum of the echo. This is the difference one would expect between the spectrum arriving at the two ears if a dolphin were about 2.7 m away from this complex object, with an interaural distance of about 14 cm. As the dolphin approached the object, the angle between the incident sound beam and the dolphin's two ears would increase and the corresponding differences between the two spectra would also increase. Presumably the dolphin could use these differences as a function of distance and angle to recover many features of the object being echolocated. In vision, Ullman (1979) used similar ideas to show how one could derive shape from motion cues. Each angle and each distance constrains the possible shapes that could be producing the acoustic or visual cues. As a result, one can identify the properties of an object with a high degree of certainty when enough different samples have been obtained. This sequential sampling method is consistent with what we know of dolphin echolocation. [See Möller and Schnitzler (1999) for a similar analysis in bats.]

Dolphins appear to combine information from multiple echoes when recognizing objects (Roitblat et al. 1990). In the laboratory, they emit on average between five and a hundred echolocation clicks per trial, even when they are recognizing familiar objects. Their performance indicates that they combine information from multiple echoes, rather than simply clicking until they get a particular kind of echo (Roitblat et al. 1991).

The dolphin ear provides an excellent frequency analysis of the echo. Because of the physical properties of the dolphin's cochlea (or any other mammalian ear, for that matter), each point along the basilar membrane resonates at a characteristic frequency. Vibrations received at the oval window of the cochlea excite a traveling wave along the basilar membrane, in which the magnitude of the displacement of each point along the membrane is frequency dependent. Hair cells in contact with the basilar membrane transform this mechanical motion into neural signals, which are relayed by the ganglion cells through the auditory nerve. The cochlea can be described as a bank of bandpass filters, which functions to a rough first approximation, as a mechanical spectrum analyzer, separating the signal into a spectrum of frequencies and their amplitudes. The first neural representation of an echo is (roughly) its spectrum as encoded by the ganglion cells.

Because of the brief duration of the echo and the relatively slow firing rates of neurons (up to about 1 kHz), it is unlikely (though not inconceivable) that the dolphin can detect the temporal properties of the echo directly; rather, it is more likely to derive the temporal properties of the echo from its spectrum. Recall that echoes typically last up to 200 μs, which would be one-fifth of the interspike interval at 1 kHz. Fortunately, according to Fourier theory, it is not difficult in principle to recover temporal information from the spectrum.

Another factor that helps dolphins to recognize objects from multiple angles is the use of two ears. Dolphins' ability to assess the angle to objects depends on using amplitude and time-of-arrival differences in signals arriving at the two ears. This same kind of information also provides cues to the three-dimensional structure of objects returning echoes in that each ear receives the echo at a small angle relative to the dolphin's echolocation beam. This difference between the two ears as well as its head movements help the dolphin to extract structural information. For example, more distant surfaces return echoes slightly later than closer surfaces do. Furthermore, as the dolphin approaches an object on which it is echolocating, the angles between the two ears and the object change, again providing important cues to the three-dimensional structure of the object. Because of differential attenuation of signals of different frequency over distance in the water, the frequency structure of the returning echo also changes as the dolphin approaches. Having two receivers thus not only increases the sensitivity of the system relative to one ear but also provides additional nonredundant information about the structure of the object.

We cannot know what the dolphin's subjective experience of perception is like, but we can know something about what it experiences by identifying the sensory dimensions that are available and assessing how those dimensions might be used by the animal in its day-to-day activities. It seems reasonable to suppose that echolocation, like vision, is used to perceive the properties of objects, but it is unlikely that dolphins use their sonar to "paint" pictures of the objects that they ensonify. There is no evidence to support the hypothesis that dolphins must scan across an object point for point (akin to laser tomography) in order to perceive its structure (cf. Herman et al. 1998). For example, at the rate at which they echolocate (up to about 66 clicks per second), such a scan would result in a very low-density collection of points. At the same time, there is no a priori reason to think that dolphins could not construct "images" of the objects that they echo-locate, as long as we recognize that images are not limited to just the visual modality (Roitblat et al. 1995).

People, for instance, can image the sound of a symphony, can image a scene from the sound of plates shattering on a floor and can even describe many of the properties of the objects that are breaking and the surface onto which they have fallen, all on the basis of sound. Identifying the properties of objects seems to be the essential characteristic of imagery, not whether it is visual. There is no reason to think that dolphins could not similarly identify the properties of the objects that they echolocate and construct images of the objects in their world. Studies of cross-modal processing in dolphins (Harley et al. 1996), including some ongoing studies, support the notion that dolphins derive corresponding information about the structure of objects from both vision and echolocation.

References

Au, W. W. L. (1993). *The Sonar of Dolphins.* New York: Springer-Verlag.

Au, W. W. L., Moore, P. W. B., and Pawloski, D. (1986). Echolocating transmitting beam of the Atlantic bottlenose dolphin. *Journal of the Acoustical Society of America* 80: 688–691.

Bellwood, J. (1988). Foraging behavior, prey selection, and echolocation in phyllostomine bats (Phyllostomidae). In *Animal Sonar: Processes and Performance*, P. E. Nachtigall and P. W. B. Moore, eds., pp. 601–605. New York: Plenum.

Fenton, M. B. (1988) Variations in foraging strategies in five species of insectivorous bats—implications for echolocation call design. In *Animal Sonar: Processes and Performance*, P. E. Nachtigall and P. W. B. Moore, eds., pp. 607–611. New York: Plenum.

Harley, H. E., Roitblat, H. L., and Nachtigall, P. E. (1996). Object representation in the bottlenose dolphin (*Tursiops truncatus*): Integration of visual and echoic information. *Journal of Experimental Psychology. Animal Behavior Processes* 22: 164–174.

Herman, L. M., Pack, A. A., and Hoffmann-Kuhnt, M. (1998). Seeing through sound: Dolphins perceive the spatial structure of objects through echolocation. *Journal of Comparative Psychology* 112: 292–305.

Möller, R. and Schnitzler, H.-U. (1999). Acoustic flow perception in cf bats: Properties of the available cues. *Journal of the Acoustical Society of America* 105: 2958–2966.

Morozov, V. P., Akopian, A. I., Zaytseva, K. A., and Sokovykh, Y. A. (1972). Tracking frequency of the location signals of dolphins as a function of the distance to the target. *Biofizika* 17: 139–143.

Penner, R. H. (1988). Attention and detection in dolphin echolocation. In *Animal Sonar: Processes and Performance*, P. E. Nachtigall and P. W. B. Moore, eds., pp. 707–714. New York: Plenum.

Roitblat, H. L., Helweg, D. A., and Harley, H. E. (1995). Echolocation and imagery. In *Sensory Systems of Aquatic Mammals*, R. Kastelein, J. Thomas, and P. Nachtigall, eds., pp. 171–181. Woerden, The Netherlands: De Spil.

Roitblat, H. L., Moore, P. W. B., Helweg, D. A., and Nachtigall, P. E. (1993). Representation and processing of acoustic information in a biomimetic neural network. In *From Animals to Animats 2: Simulation of Adaptive Behavior*, J.-A. Meyer, S. W. Wilson, and H. L. Roitblat, eds., pp. 90–99. Cambridge, Mass.: MIT Press.

Roitblat, H. L., Moore, P. W. B., Nachtigall, P. E., and Penner, R. H. (1991). Natural dolphin echo recognition using an integrator gateway network. In *Advances in Neural Information Processing Systems 3*, D. S. Touretsky and R. Lippman, eds., pp. 273–281. San Mateo, Calif.: Morgan Kaufmann.

Roitblat, H. L., Penner, R. H., and Nachtigall, P. E. (1990). Matching-to-sample by an echolocating dolphin. *Journal of Experimental Psychology: Animal Behavior Processes* 16: 85–95.

Rossbach, K. A. and Herzing, D. L. (1997). Underwater observations of benthic-feeding bottlenose dolphins (*Tursiops truncatus*), near Grand Bahama Island, Bahamas. *Marine Mammal Science* 13: 498–504.

Suthers, R. (1988). The production of echolocation signals by bats and birds. In *Animal Sonar: Processes and Performance*, P. E. Nachtigall and P. W. B. Moore, eds., pp. 23–45. New York: Plenum.

Ullman, S. (1979). *The Interpretation of Visual Motion.* Cambridge, Mass.: MIT Press.

Chimpanzee Ai and Her Son Ayumu: An Episode of Education by Master-Apprenticeship

Tetsuro Matsuzawa

I have been studying chimpanzee (*Pan troglodytes*) intelligence both in the laboratory and in the wild (Matsuzawa 2001). Chimpanzees in the wild use and manufacture a wide variety of tools, such as twigs to fish for termites or a pair of stones to crack open hard-shelled nuts. Recent studies comparing different communities of chimpanzees have shown that each community develops its own unique set of cultural traditions.

Chimpanzees in the laboratory can also master various kinds of skills, including, to some extent, linguistic and numerical abilities. One question arising from these studies concerns the social transmission of knowledge and skills across generations. How and when does such learning occur and who passes it to whom? To address these questions, this essay briefly summarizes our attempts at synthesizing two distinct approaches to understanding the nature of chimpanzee intelligence: ethological observation in the wild and psychological experiments in the laboratory. In addition, it also provides an account of one of the most impressive episodes of learning by an infant chimpanzee from a skillful mother.

Wild Chimpanzees at Bossou

The forests of Bossou, Guinea-Conakry in West Africa, are home to a group of about 20 chimpanzees. They are known to use a pair of stones as hammer and anvil to crack open oil palm nuts in order to gain access to the kernel inside the hard shell (figure 25.1). Through long-term observation of the nut cracking, my colleagues and I have identified various interesting aspects of this tool-using behavior (Matsuzawa 1994).

For example, each chimpanzee shows perfect laterality in using hammer stones. The "right-handers" always use the right hand for hammering, while the "left-handers" use their left hand exclusively. Such perfect laterality in tool use has never before been found in nonhuman animals. Humans show strong hand preference on the individual level, and there is also a strong right bias at the population level. The chimpanzees of Bossou show a slight bias toward the right for hammering at the population level, with about 67 percent of group members being right-handers. However, there is perfect correspondence in siblings' hand preference. Members of every sibling pair we have come across prefer to use the same hand for hammering. Hand preference is thought to be related to functional lateral asymmetry of brain function, but, with many unanswered questions, this remains a controversial issue in nonhuman animals.

Young chimpanzees require at least 3.5 years to master nut cracking. Furthermore, there is a critical period for learning between the ages of 3.5 and 5 years. Chimpanzees who fail to learn to crack nuts by the end of this period will not acquire the skill in later life. Learning is aided by a form of education by master-apprenticeship. Young chimpanzees learn the skill by carefully observing the behavior of adults for a long time after birth. This observation tends to be a one-way process. Adult chimpanzees seldom observe the behavior of the younger members of the community.

In addition to nut cracking, Bossou chimpanzees possess a unique repertoire of tool-manufacturing and tool-using skills. These include the use of leaves for drinking water, pestle-pounding of oil palms, fishing for safari ants with a wand, scooping algae floating on a pond with a stick, and so forth.

Chimpanzees in the wild have to learn many things besides tool use. For example, there are about 600 different species of plants in the forests of Bossou; of these, the chimpanzee food repertoire includes about 200 species. Various parts of the plants are eaten: the fruit, leaves, young stem, and the bark. Fruits such as figs are highly prized, but efficient foraging has several pre-

Figure 25.1
Chimpanzees at Bossou cracking oil palm nuts using a hammer stone and an anvil stone. Infant chimpanzees, which usually accompany their mothers everywhere, observe the mothers' behavior. There is no active teaching; however, mothers are very tolerant of the infants' observation.

requisites. Chimpanzees must remember where the fig trees are located in the forest. In addition, they have to know what time of year the fruits are ripe. They must recognize that, for instance, the red, mature fruit is tasty, while the green, unripe one is not. They have to learn how to reach the fruit in a large tree. Fig trees can grow to be enormous, too large for the chimpanzees to directly climb the trunks. Instead, they have to remember routes along the branches of nearby trees to reach their destination.

Observational studies in the natural habitat are a fountain of information about chimpanzee intelligence, as well as their society and ecology. However, the constraints associated with observation in the wild preclude us from seeing many of the details of chimpanzee behavior.

Ai Project

Chimpanzees in captivity apply the intelligence evident in the wild to surviving in the human environment. They must learn various skills to communicate with their human cohabitants. In many cases in captivity, chimpanzees have a very limited range or freedom to move from one place to another, to get food, to meet conspecific friends and then to leave them (referred to as the fission-fusion of parties), and so forth. They are in a sense forced to utilize their intelligence to adapt themselves to the human way of communication and lifestyle. Such is the general and common background to the studies of ape intelligence and "ape language" projects so far.

Wolfgang Koehler pioneered the study of chimpanzees. During the early 1910s, he maintained a group of young chimpanzees in a facility located on an island off the west coast of the African continent. His research methods involved providing test situations for the chimpanzees in which they were required to solve a problem. For example, he suspended a piece of banana high up in the air and laid out a selection of sticks and boxes. Chimpanzees were found to move boxes to the spot right underneath the banana, to stack the boxes, to use a stick to prod the banana while standing on the boxes, and even to join two sticks together to extend their instrument if the fruit was still out of reach. In sum, Koehler demonstrated that chimpanzees have the intelligence to make and to use tools for solving given tasks. Today, we can see many parallels between the observations in Koehler's classic work in captivity and the tool manufacturing and use seen in the wild.

Besides tool use, apes raised in a human environment can, to some extent, learn how to use human signs. Through such long-term rearing and training projects, the Gardners and Premacks, as well as other researchers, have contributed a great deal to our understanding of the nature of chimpanzee intelligence.

Ai is one such ape who has learned a variety of skills in captivity. Ai, pronounced "eye" and meaning "love" in Japanese, is a 24-year-old female chimpanzee. In a project that has been running continuously for more than two de-

cades, she has from the age of 1 year learned to communicate through letters and numerals using a computer-controlled device. For example, she learned to touch letters and numerals on a computer terminal to represent the color, identity, and number of items shown to her (figure 25.2) (Biro and Matsuzawa 1999; Kawai and Matsuzawa 2000; Matsuzawa 1985).

A New Project: Infants Reared by Their Mothers

A common disadvantage of captive research is the lack of "community." Most of the "ape-language" studies have concentrated on a single subject or a simple aggregation of multiple subjects. Chimpanzees in the wild live together in a community, which is itself often divided into small parties. Infants less than 3 years old always accompany their mothers. They cling to the mother and the mother in turn embraces the infant. This is the natural way of life and the natural context for learning in the wild.

Ai gave birth on April 24, 2000. In addition to Ai, two other female chimpanzees, Chloé and Pan, also gave birth soon thereafter. Together with these babies, we now have a group of 14 chimpanzees at the Primate Research Institute in Inuyama. The members' ages range from the newborns to a 36-year-old, encompassing three generations of both sexes. This is the Inuyama community of chimpanzees, which simulates the natural way of life of chimpanzees in the African forest (figure 25.3).

In our outdoor compound, we have planted more than 500 trees from 60 different species and built climbing frames more than 50 feet high. The chimpanzees are free to stay outside all day for as long as they wish. However, of their own free will, they prefer to come to experimental booths to interact with human partners.

My colleagues and I are now concentrating on a new project aimed at examining the processes underlying social transmission of knowledge and skills across generations. The three mothers, Ai,

Figure 25.2
The chimpanzee Ai is selecting numerals in an ascending order.

Chloé, and Pan, have learned a variety of computer skills, in addition to many different kinds of tool use much like those of wild chimpanzees. How can such knowledge and skills be transmitted from one generation to the next? When are they transmitted? And from who to whom?

Such questions are not easy to answer. Suppose that one day in the forest you observe a chimpanzee mother and her infant. You can never be sure whether you will have the opportunity to see them again the next day. It may be a week before you see them again, in some cases a month or more. This is the fundamental constraint of behavioral observation in the wild.

However, in our new project, we can observe and also videorecord the chimpanzees (1) 24 hours a day, (2) from a close distance, and (3) with the assistance of the mothers. Without explicit training, the chimpanzee Ai can discriminate about 30 words of human speech: head, mouth, hand, foot, come, go, wait, climb, and so forth. When I first began cognitive tests of the newborn, I announced "Lay down!" and pointed the floor. Although this was the very first instance in my interactive history with Ai of saying such a thing, she immediately realized what was required and lay down, holding her infant.

With such assistance from the mothers, we have been carrying out experiments to observe the cognitive development of chimpanzees reared by their mothers living in a community (figure 25.4). We have been attempting to simulate the natural mother–infant interaction in the context of sophisticated manipulation skills in the wild. The following is the most impressive event of the

Figure 25.3
Outdoor compound for the chimpanzee community of the Primate Research Institute of Kyoto University in Inuyama. Here 14 chimpanzees from 0 to 36 years old live together as a group in a setting that simulates the community-based life of wild chimpanzees in terms of their social and physical environments.

first 10 months in the life of an infant chimpanzee, Ai's son, Ayumu.

Ayumu's First Attempt on the Computer

The infant chimpanzee Ayumu surprised researchers. He attempted a computer task designed for his mother and selected the color brown immediately after he touched the Japanese kanji character meaning "brown." He correctly performed this complex skill on his first attempt.

Ayumu was 9 and a half months old at the time. Sixteen of his deciduous teeth had already erupted; this means that his physical development roughly corresponded to a little less than that of a 2-year-old human infant. Claudia Sousa, a graduate student, was in charge of carrying out the experiment, while Sanae Okamoto, another graduate student, was videorecording the entire process.

The task that was given to Ai consisted of the following two phases. The first was a discrimination task. The chimpanzee had to perform matching-to-sample of colors and the corresponding visual symbols on a touch-sensitive monitor (figure 25.5). The correct answer was rewarded by a 100-yen coin, equivalent to a dollar. The second phase was a coin-use task. The

Figure 25.4
The chimpanzee Ai and her infant son, Ayumu

chimpanzee had to insert the coin into a vending machine to obtain her favored food when pictures of different food items were presented to her on another touch-sensitive monitor (Sousa and Matsuzawa, 2001).

The details of the first task, which happened to be kanji-to-color matching that day, were as follows: First step: Ai touched a white circle on the monitor to start a trial. Second step: immediately after the touch, the white circle was erased and a kanji character, for example "red," appeared in the center of the bottom row of the screen. A touch to the kanji character resulted in the appearance of two alternatives, for example, red- and blue-colored rectangular patches on the screen. Third step: a coin was delivered if the chimpanzee touched the color corresponding to the kanji character, red in this case.

Ai spontaneously saved the coins. After saving three to four coins, she moved to the vending machine located at a distance of about 2 m, and then used the coins for choosing her favorite food. Ayumu had been watching his mother's behavior every day. For a long time since his birth, it had been a daily routine, Monday to Saturday, six times a week. However, he had never touched the screen before the incident described here.

It was 14:31 in the afternoon of February 16, 2001. Immediately after the mother, Ai, moved to the vending machine, Ayumu approached the computer used for the discrimination task. He stood on his feet, holding on to the edge of the computer terminal with his hands. First he touched the white circle on the screen. Once it had vanished, the sample character appeared. It happened to be "brown" in this trial. After staring at the character for about 3 seconds, he touched it. Then two colored patches appeared in a column in the upper right corner of the touch monitor. The color brown was located above the color pink, farthest from Ayumu. The height from the floor to the color brown was about 70 cm.

Ayumu's height was about 60 cm. He stretched his left arm, but failed to reach the color. In the second attempt, he stretched his body while standing on his feet, but again failed to reach it. Then, in his third attempt, he climbed one step up on the wall, resting his feet on the tray located under the monitor. He stood up on the tray and kept his body upright and finally reached the brown color. The videorecord clearly shows that he was definitely aiming to touch the brown-colored patch in the far position.

Since the answer was correct, a 100-yen coin was automatically delivered. Ayumu picked it up, held it in his hand, and continued to mouth and manipulate it throughout the rest of the session, until the very end.

This was Ayumu's first attempt at touching the monitor and he was successful. It is still unclear whether he recognized the relationship between the kanji character and the color, as so far this is the only time this behavior has been seen. However, it is clear that Ayumu knows the flow of a trial.

The infant had witnessed his mother's behavior every day. It reminds me of education by master-apprenticeship (Matsuzawa, 2001), the relationship that exists between, for example, a Sushi master and his apprentice. Just looking,

Figure 25.5
The chimpanzee Ai is performing a computer task (matching a lexigram, or visual symbol meaning "blue," to the kanji character "blue" rather than "orange" in this trial). Her son Ayumu, 9 and a half months old, is observing his mother's behavior.

carefully watching—it is the way of learning in chimpanzees.

You can access the following web site to see a movie clip of this fascinating episode in the life of an infant chimpanzee: click on "Chimpanzee Ai" at http://www.pri.kyoto-u.ac.jp.

References

Biro, D. and Matsuzawa, T. (1999). Numerical ordering in a chimpanzee (*Pan troglodytes*): Planning, executing, and monitoring. *Journal of Comparative Psychology* 113: 178–185.

Kawai, N. and Matsuzawa, T. (2000). Numerical memory span in a chimpanzee. *Nature* 403: 39–40.

Matsuzawa, T. (1985). Use of numbers by a chimpanzee. *Nature* 315: 57–59.

Matsuzawa, T. (1994). Field experiments on use of stone tools by chimpanzees in the wild. In *Chimpanzee Cultures*, R. Wrangham, W. McGrew, F. de Waal, and P. Heltne, eds., pp. 351–370. Cambridge, Mass.: Harvard University Press.

Matsuzawa, T. (2001). *Primate Origins of Human Cognition and Behavior*. Berlin: Springer-Verlag.

Sousa, C. and Matsuzawa, T. (2001). The use of tokens as rewards and tools by chimpanzees (*Pan troglodytes*). *Animal Cognition* 4: 213–221.

Elizabeth M. Brannon and Herbert S. Terrace

It is generally assumed that the development of human mathematical reasoning requires years of schooling. That being the case, mathematical reasoning would seem beyond the reach of the rest of the animal kingdom. This commonsensical conclusion poses an issue that is the focus of this chapter. What, if any, evolutionary precursors of human mathematical reasoning can be observed in animals?

To answer that question, we must first recognize that human mathematical ability is composed of many heterogeneous skills. Humans use symbols to represent numerosities and to represent operations such as addition and division; they are capable of manipulating numerical symbols in complicated ways (e.g., algebra and the calculus). It is even more important to recognize that the most basic numerical skills do not require *any* numerical symbols. It is, for example, possible to discriminate the relative numerosity of two sets of objects without the help of numerals (e.g., that a collection of 4 peanuts is numerically larger than a collection of 2 apples).

During the past 30 years, investigators of animal behavior have shown that many species possess some numerical ability (for reviews see Davis and Perusse 1988; Roberts 1997). Those observations have led some psychologists to hypothesize that human mathematical ability evolved from numerical abilities that can be observed in animals (Dehaene 1997; Gallistel and Gelman 1992, 2000). Our research program on the ordinal numerical abilities of rhesus monkeys has provided considerable evidence in support of that hypothesis (Brannon and Terrace 1998, 1999, 2000). As background, we first describe other experiments that have addressed this topic and show how our approach differs. We then discuss aspects of a monkey's numerical behavior that appear to be analogs of mathematical thinking in adult and developing humans. Fi-

nally, we define some promising future directions for research.

If monkeys use number to organize events in their natural environment, we should expect them to represent number on at least an ordinal scale. They should not only be able to differentiate n versus m objects, but they should also appreciate that a collection of $n + m$ objects is numerically greater than n objects. Thomas and colleagues (1980) tested this idea in an experiment in which squirrel monkeys were trained to respond to the lesser of two numerosities. The values of the numerosities were increased progressively as the monkeys learned each pair. Although Thomas et al. provided impressive evidence that squirrel monkeys could discriminate sets containing as many as 10 and 11 elements, it was unclear whether their subjects used an ordinal rule to solve each pair, or whether they had learned a series of pairwise discriminations, for example, that 4 is rewarded when it is paired with 5, but not when it is paired with 3. The latter interpretation cannot be ruled out because the pairs of numerosities were trained successively, one pair at a time.

Washburn and Rumbaugh (1991) used a different paradigm to investigate the numerical abilities of rhesus monkeys. In each trial they presented a pair of Arabic numerals whose values ranged from 1 to 9. The monkeys learned to choose the larger numeral and even responded correctly when novel combinations of Arabic numerals were tested. Although the monkeys learned to choose the larger numeral when it was presented in a novel pair, it does not follow that they learned a symbolic numerical rule. The monkeys' choices could have been based on the hedonic value associated with each of the numerals (yummie versus very yummie; also see Olthof et al. 1997). To show that the monkeys associated a discrete number of pellets with each Ar-

Rosencrantz and Macduff

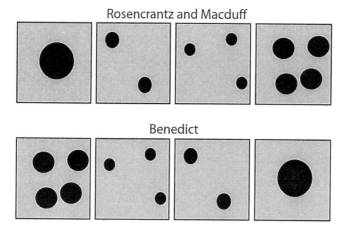

Benedict

Figure 26.1
Example of one stimulus set. Rosencrantz and Macduff responded in ascending order (1-2-3-4) and Benedict in descending order (4-3-2-1). Stimuli were presented in a random configuration on a touch-sensitive computer monitor. Many different stimulus sets were used, only one of which is shown here.

abic numeral, it would be necessary to test them with a paradigm that provided the same amount of food for each correct choice.

Other investigators have used a forced-choice discrimination procedure to study ordinal numerical knowledge (Meck and Church 1983; Roberts and Mitchell 1994; Emmerton et al. 1997). In these studies rats and pigeons were trained to make one response to a small number of stimuli (sounds and/or light flashes) and another response to a larger number. When subsequently tested with intermediate numerical values, the probability of making the large number response depended on the ratio of the intermediate value to the trained anchor (small and large) values. The confusions of magnitude produced by this paradigm provide an indirect measure of knowledge of numerical order.

However, for a more direct test of an animal's knowledge of numerical order, it is necessary to train subjects to order values in one numerical range and then test their ability to order values outside that range. Without such tests, one cannot distinguish between knowledge of an ordinal

rule and memory of a set of pairwise categorical discriminations as a basis for accurate responding to novel pairs of numerosities. To distinguish between those possibilities, we performed an experiment (Brannon and Terrace 1998) in which we trained three monkeys to order the numerosities 1, 2, 3, and 4 in an ascending or descending order (see figure 26.1). Exemplars of the numerosities 1, 2, 3, and 4 were selected from a large library of stimuli in which surface area was varied systematically to ensure that it could not serve as a cue for number. The monkeys were first trained on 35 different sets of the numerosities 1–4. The same numerical stimuli were used in each trial, albeit in randomly selected physical configurations. The monkeys' performance improved dramatically over the 35 training stimulus sets. On average, the monkeys responded in the correct order on 45 percent of the trials on the final 10 training sets (chance level of accuracy ~4 percent).

To rule out the possibility that the monkeys had memorized each of the 35 stimulus sets and the order in which to respond to each set, we

Smaller Numerosity Has:

Smaller Area   Larger Area

Figure 26.2
Example stimulus sets used in numerical comparison task with monkeys and humans. On half of the trials, the smaller numerosity had a larger cumulative surface area.

tested their ability to order 150 novel sets of the numerosities 1–4. Each novel set was presented only once. The subjects continued to respond at approximately the same level of accuracy (40 percent) even though they had no opportunity to memorize any of the novel stimulus sets. The absence of a decrease in performance with the novel stimulus sets provided clear evidence that the monkeys used the numerosity of each stimulus (as opposed to some non-numerical memory strategy) to determine the order in which to respond to the stimuli from each set.

Less clear is *how* the monkeys represented the order in which to respond to the numerosities 1–4. One possibility is that the monkeys assigned each numerosity to one of four distinct nominal categories. In this scenario, the monkeys would have learned an arbitrary ordering of the four categories, just as if we had taught them to respond to different exemplars of, say, birds, flowers, trees, and rocks. In this view, it should be just as easy for monkeys to respond in an arbitrary numerical order such as 3-1-4-2, compared with 1-2-3-4 or 4-3-2-1. Contrary to that hypothesis, one of our subjects (Macduff) showed no signs of improvement on a 3-1-4-2 sequence after training on 13 stimulus sets (see figure 26.2 for example stimulus sets) (Brannon and Terrace 2000). However, Macduff's performance rapidly improved once he was required to respond in an ascending numerical order. The ease of learning monotonic rules, compared with

nonmonotonic rules, strongly suggests that monkeys perceive the ordinal relations between the numerosities on which they were trained.

In our next experiment we used the same subjects to evaluate more directly a monkey's ability to perceive ordinal relations between novel numerosities. The subjects were tested on their ability to order pairs of novel numerosities according to the ascending or descending rule they had learned previously with respect to the numerosities 1–4. The test consisted of exemplars of all possible pairs of the numerosities 1–9 (see figure 26.2). The monkeys trained to respond 1-2-3-4 were expected to respond in an ascending order to the new pairs (e.g., 4 then 7 or 5 then 9) and the monkey trained to respond 4-3-2-1 was expected to respond in the reverse order. Reinforcement for correct responding was available only on trials in which the pairs were composed of familiar numerosities (1-2, 1-3, 1-4, 2-3, 2-4, 3-4). In trials on which a novel numerosity was presented, no positive or negative reinforcement was provided, and the monkeys were permitted to respond in either an ascending or a descending order. This prevented subjects from learning the ordinal relationships between novel numerosities.

Rosencrantz and Macduff, the monkeys who had been trained to respond to the numerosities 1–4 in an ascending numerical order, responded correctly on approximately 75 percent of the trials composed of two novel numerosities. Both

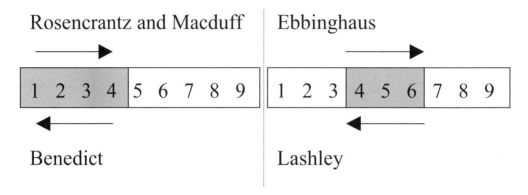

Figure 26.3
Schematic diagram of the experimental design. The shaded values were used in training. One monkey was trained to respond 4 then 5 then 6 and the other monkey was trained to respond in the reverse order. Both monkeys were then tested on all possible pairs of the numerosities 1–9.

subjects were just as accurate on pairs for which the larger numerosity covered a smaller area as on pairs for which the larger numerosity covered a larger area. Since neither subject had any previous training with the numerosities 5–9, their ability to respond to those numerosities in an ascending order provides clear evidence that a monkey can extrapolate an ascending rule to novel numerosities.

Curiously, Benedict, the single monkey who learned to respond in a descending order (4-3-2-1), did not exceed a chance level of accuracy on pairs of novel numerosities (Brannon and Terrace 2000). Since Benedict was the only subject who learned a descending sequence, it was unclear whether his inability to order novel numerosities was idiosyncratic or whether it was a consequence of learning a descending rule. Even if individual differences could be ruled out, it remained unclear why knowledge of an ascending rule should enable a monkey to extrapolate that rule to novel numerosities and why knowledge of a descending rule should not.

To address that question, we recently trained two rhesus monkeys to respond to the numerosities 4, 5, and 6, one in an ascending order (Ebbinghaus); the other in a descending order (Lashley), and then tested them on all possible pairs of the numerosities 1–9 (Brannon et al. in preparation; Kovary et al. 2000). This design (which is shown schematically in figure 26.3) creates novel numerical values that are both smaller and larger than the training values.

Both monkeys learned to respond to the stimulus sets composed of the numerosities 4, 5, and 6 in the required order. In each instance, however, accuracy with pairs of novel numerosities varied systematically with the relationship between the values of the novel numerosities and the initial value of the sets used to train the ascending and descending rules (see figure 26.4). For example, Ebbinghaus, who was trained on the ascending rule 4-5-6, responded at a high level of accuracy with the pairs 7-8, 7-9 and 8-9, but performed at chance levels of accuracy with the pairs 1-2, 1-3 and 2-3. Conversely, Lashley, who was trained on the descending rule 6-5-4, responded at a greater than chance level of accuracy with the pairs 3-1, 3-2, and 2-1, but at chance levels of accuracy with the pairs 9-7, 9-8, and 8-7. This pattern of results suggests that when ordering two novel numerosities, the sub-

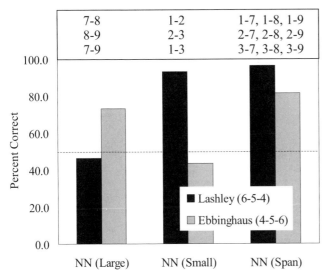

Figure 26.4
Performance on ordinal comparisons of two novel numerosities for monkeys trained on a 4-5-6 or 6-5-4 rule. Performance is shown separately for pairs that were composed of two values smaller than the training values (NN small), two values larger than the training values (NN large), and pairs that included one small and one large value (NN span). Note that performance on pairs composed of two familiar values or one familiar and one novel value is not shown here, but was high for both monkeys (range 75–85 percent).

jects compared each novel value with a representation of the initial value of the training set (e.g., 4 for Ebbinghaus and 6 for Lashley). That rule can also account for Benedict's failure to order the novel numerosities 5–9 after being trained on sequences of 4-3-2-1.

In an effort to further assess the possible link between nonhuman primate and human numerical representations, we investigated whether the well-established numerical distance effects found with adult humans are also found when nonhuman primates make numerical comparisons. Distance and size effects are found in a wide variety of circumstances when adult humans make numerical comparisons (e.g., Moyer and Landaeur 1967; Hinrichs et al. 1981; Tzeng and Wang 1983; Dehaene et al. 1990). Reaction time decreases with increases in the numerical distance between a pair of Arabic numerals and increases

as the absolute size of their value increases. Similar distance and size effects have been reported when adults compare the numerosity of collections of dots (Buckley and Gilman 1974). If monkeys and humans rely on a shared system for judging relative numerosity, we should expect to find similar effects in both species with respect to the accuracy and reaction times of numerical comparison judgments.

We tested this hypothesis in an experiment on college students using the same pairwise numerical comparison task and the same stimuli that we used with monkeys. The subjects were given verbal instructions to respond first to the stimulus with the fewer number of elements and to respond as quickly as possible while not making too many errors. As can be seen in figure 26.5, numerical distance and size had similar systematic effects on the reaction times and accuracy

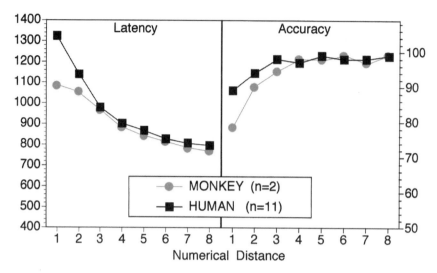

Figure 26.5
Latency to the first response (left) and accuracy (right) in a pairwise numerical comparison task as a function of numerical distance. Monkeys (Rosencrantz and Macduff: circles) and humans (squares) were required to respond first to the stimulus with the fewer number of elements.

levels of both college students and monkeys (E. M. Brannon and H. S. Terrace, unpublished). These data provide compelling evidence that animals and humans use the same numerical comparison process (see Whalen, Gallistel, and Gelman 1999).

What do the numerical distance and size effects tell us about the format of numerical representations? Just as Weber's law applies to perceptual discriminations of continuous dimensions such as line length and weight (Welford 1960), it also applies to numerical discriminations (Moyer and Landaeur 1967). This suggests that animals and humans rely on analog representations of number. The idea is that sets of discrete stimuli are converted into continuous-magnitude representations before the comparison process takes place (Gallistel and Gelman 2000).

If there is an analogical numerical processing system in animals and adult humans that is independent of language, at what point in development does it arise? E. M. Brannon recently investigated the existence of ordinal numerical knowledge in children 2 years of age who had not yet mastered the verbal counting system (Brannon and Van de Walle 2001). The children were tested with a simple game in which two trays were presented that contained one and two boxes, respectively. The tray with two boxes was designated as the winner and a sticker was hidden beneath each box. The children were asked to choose the winner tray to receive the stickers. After 5–10 trials of the 1 versus 2 comparison, the children were then tested with new numerosities. The results indicated that they were able to choose the larger numerosity on more trials than would be expected by chance, even when the size of the boxes was varied so that number was not confounded with surface area. This suggests that even before children learn how to count verbally, they have an appreciation of the ordinal relations between numerosities (see also Bullock and Gelman

1977). Research in many different laboratories is currently investigating how early in development this ordinal numerical knowledge is in place (see Brannon in press, Feigenson et al. in press).

In summary, the data reviewed reveal the similarities in the ways adult humans, animals, and children compare numerosities. Whether the similarities between humans and animals result from convergent evolution or reflect homologous processes is not yet known. Similarly, whether the numerical abilities demonstrated in infants and young precounting children are the foundation out of which the verbal counting system develops is not yet known (Carey 1998). We do not yet understand the processes that animals, infants, or adults use to enumerate nor do we understand the steps of the numerical comparison process. For example, do monkeys employ a serial countinglike process? Do infants serially scan the elements in a display when making discriminations of 2 versus 3 or 8 versus 16 dots (e.g., Strauss and Curtis 1981; Xu and Spelke 2000) or do they instead use a parallel enumeration process (e.g., Dehaene and Changeux 1993)?

Another important unanswered question is whether the same neural circuitry supports ordinal numerical comparisons in animals and humans. Neuropsychological and brain-imaging studies implicate the parietal lobe in nonverbal approximate number knowledge in adult humans (see Dehaene 2000 for a review). However, almost nothing is known about the neural underpinnings of numerical abilities in animals and human infants. Identification of the brain circuits involved in numerical behavior in animals and human infants promises to shed light on whether animals, human adults, and human infants share a numerical processing system.

References

Brannon, E. M. (in press). The development of ordinal numerical knowledge in infancy. *Cognition*.

Brannon, E. M. and Terrace, H. S. (1998). Ordering of the numerosities 1–9 by monkeys. *Science* 282: 746–749.

Brannon, E. M. and Terrace, H. S. (1999). Letter to the editor. *Science* 283: 1852.

Brannon, E. M. and Terrace, H. S. (2000). Representation of the numerosities 1–9 by rhesus macaques (*Macaca mulatta*). *Journal of Experimental Psychology: Animal Behavior Processes* 26: 31–49.

Brannon, E. M. and Van de Walle, G. (2001). Ordinal numerical knowledge in young children. *Cognitive Psychology* 43: 53–81.

Brannon, E. M., Kovary, I., Prasankumar, R., and Terrace, H. S. (in preparation). Asymmetrical extrapolation of ordinal numerical rules by rhesus macaques.

Buckley, P. B. and Gillman, C. B. (1974). Comparisons of digit and dot patterns. *Journal of Experimental Psychology* 103: 1131–1136.

Bullock, M. and Gelman, R. (1977). Numerical reasoning in young children: The ordering principle. *Child Development* 48: 427–434.

Carey, S. (1998). Knowledge of number: Its evolution and ontology. *Science* 282: 641–642.

Davis, H. and Perusse, R. (1988). Numerical competence: From backwater to mainstream of comparative psychology. *Behavioral Brain Sciences* 11: 602–615.

Dehaene, S. (1997). *The Number Sense: How the Mind Creates Mathematics*. New York: Oxford University Press.

Dehaene, S. (2000). Cerebral bases of number processing and calculation. In *The New Cognitive Neurosciences*, M. Gazzaniga, ed., pp. 987–998. Cambridge, Mass.: MIT Press.

Dehaene, S. and Changeux, J. (1993). Development of elementary numerical abilities: A neuronal model. *Journal of Cognitive Neuroscience* 5: 390–407.

Dehaene, S., Dupoux, E., and Mehler, J. (1990). Is numerical comparison digital: Analogical and symbolic effects in two-digit number comparison. *Journal of Experimental Psychology: Human Perceptual Performance* 16: 626–641.

Emmerton, J., Lohmann, A., and Niemann, J. (1997). Pigeons' serial ordering of numerosity with visual arrays. *Animal Learning & Behavior* 25: 234–244.

Feigenson, L., Carey, S., and Hauser, M. (in press). Spontaneous ordinal abilities by human infants. *Psychological Science*.

Gallistel, C. R. and Gelman, R. (1992). Preverbal and verbal counting and computation. *Cognition* 44: 43–74.

Gallistel, C. R. and Gelman, R. (2000). Non-verbal numerical cognition: From reals to integers. *Trends in Cognitive Sciences* 4: 59–65.

Hinrichs, J., Yurko, D. S., and Hu, J. M. (1981). Two-digit number comparison: Use of place information. *Journal of Experimental Psychology: Human Perceptual Performance* 7: 890–901.

Kovary, I., Brannon, E. M., and Terrace, H. S. (2000). The ability of a rhesus monkey to extrapolate a descending numerical rule to novel numerosities. Paper presented at Conference on Comparative Cognition.

Meck, W. H. and Church, R. M. (1983). A mode control model of counting and timing processes. *Journal of Experimental Psychology: Animal Behavior Processes* 9: 320–334.

Moyer, R. S. and Landaeur, T. K. (1967). Time required for judgments of numerical inequality. *Nature* 215: 1519–1520.

Olthof, A., Iden, C. M., and Roberts, W. A. (1997). Judgments of ordinality and summation of number symbols by squirrel monkeys (*Saimiri sciureus*). *Journal of Experimental Psychology: Animal Behavior Processes* 23: 325–339.

Roberts, W. A. (1997). Does a common mechanism account for timing and counting phenomena in the pigeon? In *Time and Behavior: Psychological and Neurobiological Analyses*, C. M. Bradshaw and E. Szabadi, eds., pp. 185–215. New York: Elsevier Science.

Roberts, W. A. and Mitchell, S. (1994). Can a pigeon simultaneously process temporal and numerical information? *Journal of Experimental Psychology: Animal Behavior Processes* 20: 66–78.

Strauss, M. S. and Curtis, L. E. (1981). Infant perception of numerosity. *Child Development* 52: 1146–1152.

Thomas, R. K., Fowlkes, D., and Vickery, J. D. (1980). Conceptual numerousness judgments by squirrel monkeys. *American Journal of Psychology* 93: 247–257.

Tzeng, O. J. L. and Wang, W. (1983). The first two R's. *American Scientist* 71: 238–243.

Washburn, D. and Rumbaugh, D. M. (1991). Ordinal judgements of numerical symbols by macaques *Macaca mulatta*. *Psychological Science* 2: 190–193.

Welford, A. T. (1960). The measurement of sensory-motor performance: survey and reappraisal of twelve years progress. *Ergonomics* 3: 189–230.

Whalen, J., Gallistel, C. R., and Gelman, R. (1999). Nonverbal counting in humans: The psychophysics of number representation. *Psychological Science* 10: 130–137.

Xu, F. and Spelke, E. S. (2000). Large number discrimination in 6-month-old infants. *Cognition* 74: B1–B11.

27 Domain-Specific Knowledge in Human Children and Nonhuman Primates: Artifacts and Foods

Laurie R. Santos, Marc D. Hauser, and Elizabeth S. Spelke

One of the most important things an organism needs to recognize is where to direct its attention. In order to act effectively, an organism needs to focus its attention on properties and events that are relevant to the problem at hand. The task of discovering what information to attend to and what to ignore presents a challenge because different types of information must be selected in different situations. For example, a monkey in the canopy searching for a branch to climb on must pay attention to the shape, size, strength, and position of potential branches, ignoring other information such as the color of the branches and smell of the fruit. Later, the same monkey looking for a ripe piece of fruit to eat must attend to color and smell, the very features he disregarded earlier. How do organisms decide which features to attend to in order to build effective strategies for classifying the complicated assortment of objects in their world?

Researchers in a number of fields, including cognitive development (Gelman 1990; Hirschfeld and Gelman 1994; Keil 1989), evolutionary psychology (Cosmides and Tooby 1994; Pinker 1997), animal cognition (Gallistel 1990; Hauser 2000; Shettleworth 1998), neuropsychology (Caramazza 1998; Santos and Caramazza, in press), anthropology (Sperber 1994), and archaeology (Mithen 1996) have answered this question by appealing to notions of domain-specific constraints on learning. From a domain-specific perspective, the mind consists of a collection of specialized learning systems designed for processing different types of input. Advocates of the domain-specificity view argue that organisms are endowed with domain-relevant content that both biases and guides their attention to conceptually relevant perceptual inputs. The domains that make up an animal's cognitive architecture are thought to have evolved in response to the computational problems that were most salient over the animal's phylogenetic history.

In the past decade, considerable research has investigated the ontogeny of human domains of knowledge (Gelman 1990; Hirschfeld and Gelman 1994; Keil 1989; Keil et al. 1998). Relatively little work, however, has explored whether the domains of knowledge that constitute the human mind are shared with our closest evolutionary relatives, the nonhuman primates. If accounts of domain specificity are correct, then the domains of understanding that comprise human cognition may be phylogenetically quite ancient and thus shared by other nonhuman animals, especially nonhuman primates. It is also possible, of course, that human evolution led to the emergence of new domain-specific systems (e.g., Mithen 1996).

We have attempted to address this problem by examining how human children and two nonhuman primate species—captive cotton-top tamarins (*Saguinus oedipus*) and free-ranging rhesus monkeys (*Macaca mulatta*)—reason about problems in two different domains. Specifically, we have focused on the features that primates use when categorizing foods and artifacts. Here we systematically contrast the knowledge about food and artifacts shown by mature tamarins and rhesus monkeys with the human child's developing knowledge of these domains. We argue that there are important similarities in the ways that these three species reason about objects in these domains.

Children's Understanding of the Relevant Features of Artifacts

Human children are surrounded by artifacts from birth. As one might predict from this rich early experience, humans develop some understanding of artifacts at a rather young age. Five-year-old children understand which properties are important for classifying artifacts (e.g., shape, rigidity, size) and perceive these as distinct from the set of features that are important for categorizing other kinds of things, such as animals (e.g., color, material composition, surface markings; see Carey 1985; Keil 1989; Keil et al. 1998).

There is also evidence that toddlers possess some understanding of the causally relevant properties of an artifact, and specifically its functional capacity. Brown (1990) designed a tool task in which 1–3-year-old children were trained to use a cane-shaped tool to obtain an out-of-reach toy. Once children successfully obtained the toy with a particular tool, they were tested with new tools that were designed to assess the salience of particular featural transformations. Children readily used tools of a different color to perform the same function, which suggests that color plays a relatively insignificant role in the child's understanding of a functional tool. However, children rejected tools that were too flimsy to pull the toy or whose tops were shaped inappropriately for the pulling task. This suggests that shape and rigidity play significant roles in the child's understanding of a functional tool. More important, new evidence suggests that children as young as 2 years of age generalize new labels for objects based on information about an object's function, not merely its shape and overall physical appearance (Kemler Nelson 1999; Kemler Nelson et al. 2000).

Nonhuman Primates' Understanding of the Relevant Features of Artifacts

In contrast to the wealth of studies examining what children understand about the functional properties of objects, relatively few studies have critically evaluated nonhuman primates' understanding of artifacts. Tool use is present throughout a number of primate species, both in the laboratory (Povinelli 2000; Tomasello and Call 1997; Visalberghi and Tomasello 1998) and in the wild (Goodall 1986; Matsuzawa 1994; McGrew 1992). Although this research has conclusively demonstrated that several primate species use tools, only a few studies (Hauser 2000; Povinelli 2000; Visalberghi and Tomasello 1998) have explored nonhuman primates' understanding of tools and in particular the features that give tools their particular functions.

To examine these issues, Hauser (1997) initiated a research program with cotton-top tamarins. In the first task, modeled after Brown's (1990) studies of children, the subjects were required to pull one of two blue canes to gain access to an out-of-reach piece of food (figure 27.1). Once the subjects learned to selectively pull the blue cane with food inside the hook, in preference to a cane with food outside the hook, they were tested with a variety of new tools of varying sizes, shapes, colors, and textures. In

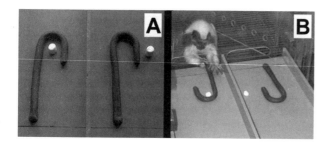

Figure 27.1
An example of an original tool training condition (A). Subjects learned to pull one of these two canes to obtain the food reward (B). (From Hauser 1997.)

critical test trials, the subjects were presented with a choice of two canes that differed from the original blue cane in only one feature (figure 27.2). They had to choose, for example, between a pink cane (new color) and a fat blue cane (new size) or between a blue cane with holes (new texture) and an oddly shaped cane (new shape). Based on both first trial and session performance, the results showed that tamarins chose tools with novel colors over those with novel sizes, and chose tools with novel textures over those with novel shapes. In the absence of explicit training, tamarins evidently understand that size and shape are more causally relevant to a tool's functionality than are color and texture.

After presenting his subjects with single-feature changes, Hauser presented the tamarins with new test trials in which additional features of the objects were altered, some of which changed their functional properties. The tamarins reliably preferred the more functional of the two tools, even when the more functional object was perceptually quite different from the original blue cane, and even when a novel but functionally appropriate tool was pitted against a familiar tool placed in an orientation that blocked the exercise of its function.

Hauser and colleagues (1999) turned next to a modified version of this task in which tamarins were trained to choose one of two pieces of cloth to obtain a food reward. As in the previous experiments, the tamarins focused on changes to the cloth that affected its functionality (see Gibson 1979). For example, subjects rejected cloths that did not allow pulling (e.g., pieces of cloth connected with chipped wood, sand, or a broken rope) and they chose cloths of radically different shapes (e.g., triangles, circles, teeth-shaped) that functionally supported the food reward. Furthermore, they distinguished between cloths that supported the food reward and those that were merely in contact with the food and thus functionally inappropriate.[1] Once again, the subjects distinguished the features that were relevant for the cloth tool's function from those that were not.

Santos et al. (under review) then set out to examine whether primates understand which features are most causally relevant to an artifact's function in the absence of direct physical experience with that type of artifact. To this end, Santos and her colleagues used an expectancy violation paradigm. The logic behind this paradigm is that subjects will look longer at events that violate their expectations about the physical world than at events that are consistent with those expectations (see Hauser and Carey 1998; Spelke 1985, 1991). Santos et al. habituated tamarins to an event in which a novel object—a purple L-shaped tool made of Play-Doh—pushed a grape down a ramp and onto a lower platform (figure 27.3A), they then presented subjects with one of two test trials.

In one test trial, the subjects saw a tool of a different color but a similar shape push the grape down the ramp (figure 27.3B). In the other test trial, they saw a tool of the same color but a different shape (an I-shaped tool) appear to push the grape down the ramp. This new shape test trial was considered unexpected from the perspective of a human observer because the base of the tool was too small to effectively push the grape (figure 27.3C). The results showed that the subjects looked longer during the new shape test trial than during the new color test trial, which suggests that a change in the tool's shape was more important to its functioning than a change in its color.

Santos and colleagues then extended their work to free-ranging rhesus macaques living on the island of Cayo Santiago, Puerto Rico, a population with far less experience with artifacts than the captive tamarins.[2] They conducted the same expectancy violation experiments involving the same Play-Doh objects and obtained similar results. Rhesus monkeys, who lack experience with artifacts of any kind, looked longer when the tool was used after a change in its shape—a change that should have impaired the tool's function—than after a change in its color. Even when free-ranging rhesus are presented with

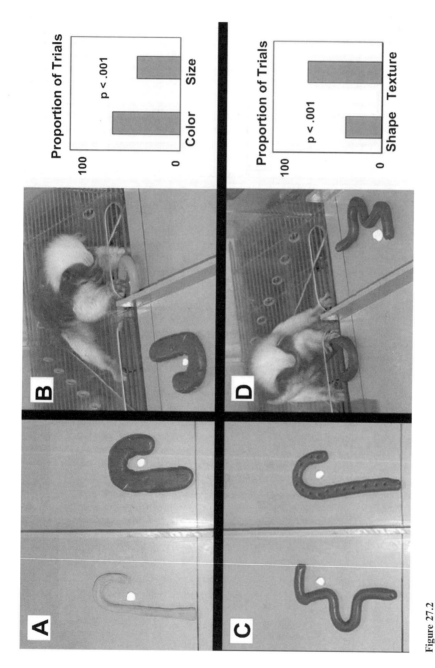

Figure 27.2
Feature change conditions. The subjects were presented with a cane of a new color and a cane of a new size (A) or a cane of a new shape and a cane of a new texture (B). The subjects preferred canes with new colors (C) and textures, respectively (D).

Habituation Trial

Color Change Test Trial

Shape Change Test Trial

Figure 27.3
Santos et al. (under review) expectancy violation experiment. In the habituation trial, subjects were habituated to an event in which a purple L-shaped tool (A) was used to push a grape across a stage (B) and down a ramp (C). They were then given two test trials, one in which the color of the tool changed (color change test trial) and one in which the shape of the tool changed (shape change test trial). The subjects looked reliably longer during the shape change test trial.

tools with which they have no direct physical experience in tasks that involve no training, they appear to understand at some level which features are relevant to an artifact's function.

Children's Understanding of the Relevant Features of Foods

In light of evidence that both human and non-human primates attend to the features of shape and orientation when reasoning about artifacts, we now turn to a different domain—food—in which these features play little or no causal role. Despite the fact that food is critical to the survival of all animals, relatively few studies have been devoted to examining children or nonhuman animals' understanding of this domain (see Macario 1991; Rozin 1990 for exceptions). The little work that has been done suggests that children possess some understanding of food objects from a rather early age.

Children as young as 2 and a half years of age predict that objects of the same color will have

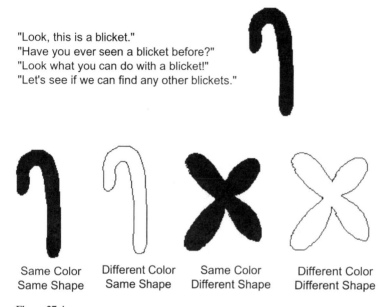

"Look, this is a blicket."
"Have you ever seen a blicket before?"
"Look what you can do with a blicket!"
"Let's see if we can find any other blickets."

| Same Color | Different Color | Same Color | Different Color |
| Same Shape | Same Shape | Different Shape | Different Shape |

Figure 27.4
Santos et al. (1999) word learning experiment. The subjects were taught a label for a novel food or artifact and then were asked to transfer this label to objects of the same and different colors and shapes.

the same smell and taste (see Macario 1991). Santos et al. (1999) (see figure 27.4) examined the features that 4-year-old children attend to when learning the words for food objects. They first taught children the labels for novel objects made of pretzel material of a particular shape and color (e.g., blue cane shape) and then asked whether other objects of similar shapes and colors shared the same label. When children were told that the novel object was a tool, they transferred the label to objects of the same shape (see also Landau et al. 1998). When children were told that the novel object was a kind of food, in contrast, they transferred the new label to objects of the same color as the originally labeled object. In other words, children used the feature of color, not shape, when generalizing labels to new food objects. Similarly, Lavin and Hall (1999) found that 3-year-old children use color and texture information when learning the labels for novel food objects, disregarding information

about the object's shape. Taken together, these results suggest that young preschoolers have some understanding that substance properties such as color and texture, are more relevant to categorizing food objects than form properties such as shape. Children's substance bias for food objects stands in contrast to their selective attention to form when categorizing and reasoning about artifacts.

Nonhuman Primates' Understanding of the Relevant Features of Foods

Despite the enormous attention that behavioral ecologists in the field and laboratory have devoted to studies of foraging (Stephens and Krebs 1986; Ydenberg 1998), relatively little research has explored what nonhuman animals understand about food objects. Garcia's groundbreaking work on avoidance learning established

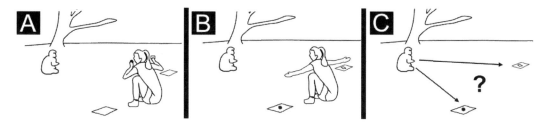

Figure 27.5
Santos et al. (2001) social facilitation test. Rhesus macaques watched a human experimenter eat one of two novel
objects (A). The objects were then put down (B) and the subjects were allowed to choose one of the two items (C).

that rats are more likely to associate nausea with
the ingestion of a novel food than with a bright
light or other stimulus (Garcia and Koelling
1966). Although this result suggests that organisms take into account different information
when learning about food than when learning
about other kinds of stimuli, it does not speak
to the question of whether organisms spontaneously divide objects into different categories
(e.g., foods and nonfoods) and attend to different
properties of each.

To better examine these questions, Bovet
and Vauclair (1998) examined whether captive
baboons naturally categorize objects as edible or
inedible. They trained baboons to pull one rope
when presented with an apple and to pull a different rope when presented with a padlock. After
this initial training, they presented their subjects
with 80 novel objects, half of which were food,
half of which were not food. The baboons
spontaneously generalized their responses to the
appropriate objects, pulling the apple rope for
food objects and the padlock rope for nonfood
objects. These results suggest that baboons
classify objects as foods or nonfoods in the absence of training, although they do not reveal
the features the baboons use when making these
categorizations.

We have initiated a research strategy that explores how nonhuman primates spontaneously
categorize novel food objects using a somewhat

different technique and have focused especially
on the features that guide categorization in the
absence of training (Santos et al. 2001) (figure
27.5). We tested adult rhesus macaques from the
Cayo Santiago population with natural but unfamiliar food objects. In the first condition, an
experimenter presented a subject with two novel
foods (e.g., a baby carrot and a lemon slice); the
experimenter ate one while holding the other
near her face, placed one of each of these foods
on spatially separated platforms, stepped away,
and allowed the monkey to approach and eat
one of the food objects. We found that subjects
selectively approached the platform containing
the object that the experimenter had previously
eaten. These findings provide evidence that
monkeys show an effect of past experience on
their food choice, and they set the stage for our
critical studies.

In these experiments, rhesus monkeys were
initially presented with two objects of different
colors and shapes, each of which was made of
the same material (Play-Doh) as the objects in
our previous studies of tool use in this population (Santos et al. under review). As the experimenter presented the two objects to a subject,
she pretended to eat one of the objects and acted
on the other object in a different, attention-
getting way (e.g., rolling it on the ground or
sticking it in her ear). Then the experimenter
placed copies of the two objects on two plat-

forms, as in our previous studies, and watched the subject's patterns of approach to the platforms. The monkeys selectively approached the object with the same shape and color as the object that the experimenter had pretended to eat. These findings provide evidence that monkeys show a social facilitation effect across species (i.e., from a human facilitator) and across novel kinds of objects (i.e., objects that fail to resemble any foods the monkey might have eaten). They set the stage for our critical tests of domain-specific learning and generalization.

In these tests, the monkeys again observed an experimenter pretend to eat one of two Play-Doh objects of different colors and shapes, but then they were given an approach task in which objects were presented that differed in shape or color, or both. When both test objects had the same shape as the originally eaten object, but only one had the same color, the monkeys selectively approached the object with the same color. When both test objects had the same color as the originally eaten object, but only one presented the same shape, the monkeys selected at random between the two objects. Finally, when one object differed from the originally eaten object in color and the other object differed in shape, the monkeys selectively approached the object with the same color. These findings provide clear evidence that monkeys generalize their learning about edible objects along the dimension of color, not along the dimension of form.

Conclusions

Our social facilitation studies with monkeys support three conclusions. First, just as monkeys can learn about the functional properties of tools by observation, without direct physical experience, they can also learn about the functional properties of food by observation. Monkeys who observed a person raking a grape with a stick learned to represent the stick as a tool, and those who observed a person eating a pink Play-Doh

ball learned to represent this object as food. Second, observational learning is a robust process, which can take place even when a monkey observes a demonstrator of a different species (a human) acting on an object that is entirely artificial (Play-Doh). Third and most important, monkeys show different patterns of attention, and therefore generalization, to the features of objects in different domains. Presented with Play-Doh objects that are used as food, monkeys generalize their learning to new objects of the same color and different shapes. Presented with Play-Doh objects that are used as tools, monkeys generalize their learning to new objects of the same shape and different colors. These findings closely resemble those obtained in our studies of human children (Santos et al. 1999), who generalized from one object to new objects by color when the object was presented as food and by shape when the object was presented as a tool. Like humans, monkeys represent and reason about objects differently in different domains.

The data presented here suggest that human and nonhuman primates share important similarities in the way they categorize objects in two different domains.[3] Adult monkeys and human children recognize that the properties that are important for categorizing food objects are different from those that are important for categorizing artifacts. These findings suggest that at least two domains of human knowledge are shared with other primate species.

While these results provide an important first step toward understanding what other species know about different kinds of objects, more work is needed to assess the deeper similarities between human and nonhuman reasoning in different domains. For example, when human children reason about an artifact, they often take into account its intended history, the function for which it was originally designed (see Bloom 1996). Given that no nonhuman animals create tools as extensively and flexibly as humans do, it is important to ask whether any nonhuman animals share our intuition that an artifact's original purpose is important for its current use.

Further questions about human categorization of objects are suggested by our studies of nonhuman primates. For example, when learning whether an object is edible, nonhuman primates pay attention to the behavior of other individuals and particularly, whether another individual eats a novel food (see Santos et al., 2001). Do human children similarly use the eating behavior of other animals, human and nonhuman, to help them categorize food objects? An examination of questions like these will provide a richer understanding of the deeper similarities and differences between human and nonhuman domains of knowledge, and of the contributions of our phylogenetic history and ontogenetic experience in the development of these knowledge systems.

Acknowledgments

The authors wish to thank Dr. Alfonso Caramazza and Dr. Asif Ghazanfar for their insightful comments on the paper. All of this research conforms to federal guidelines for use of animals and humans in research. L. R. Santos was supported by the Mind, Brain, and Behavior Program, the Harvard University McMasters Fund, the Mellon Scholars Program, and a National Science Foundation predoctoral fellowship. M. D. Hauser was supported by the National Science Foundation (grant SBR-9357976) and by Harvard University. E. S. Spelke was supported by the National Institutes of Health (grant 237-HD23103) and by Massachusetts Institute of Technology.

Notes

1. Surprisingly, chimpanzees tested on a comparable task fail to appreciate this distinction (Povinelli 2000).

2. The only artifact the monkeys on Cayo Santiago have regular contact with is a food trough that holds the chow that they eat.

3. Human children and nonhuman primates also share an understanding of other domains of knowledge. These at least include the domain of animate objects (see Gelman 1990; Hauser 1998; Mandler and McDonough 1993; Santos and Caramazza, in press), spatial navigation (see Cheng 1986; Cheng and Gallistel 1984; de Ipolyi et al. 2001; Hermer and Spelke 1996; Wang et al. 1999), number (Hauser and Carey 1998; Hauser et al. 1996; Wynn 1998), and some of the building blocks of a theory of mind (see Hare and Wrangham, chapter 44 in this volume).

References

Bloom, P. (1996). Intention, history, and artifact concepts. *Cognition* 60: 1–29.

Bovet, D. and Vauclair, J. (1998). Functional categorization of objects and of their pictures in baboons (*Papio anubis*). *Learning and Motivation* 29: 309–322.

Brown, A. (1990). Domain-specific principles affect learning and transfer in children. *Cognitive Science* 14: 107–133.

Caramazza, A. (1998). The interpretation of semantic category-specific deficits: What do they reveal about the organization of conceptual knowledge in the brain? *Neurocase* 4: 265–272.

Carey, S. (1985). *Conceptual Change in Childhood.* Cambridge, Mass.: MIT Press.

Cheng, K. (1986). A purely geometric module in the rat's spatial representation. *Cognition* 23: 149–178.

Cheng, K. and Gallistel, C. R. (1984). Testing the geometric power of an animal's spatial representation. In *Animal Cognition*, H. L. Roitblat, T. G. Bever, and H. S. Terrace, eds., pp. 409–423. Hillsdale, N.J.: Lawrence Erlbaum Associates.

Cosmides, L. and Tooby, J. (1994). Origins of domain specificity: The evolution of functional organization. In *Mapping the Mind: Domain Specificity in Cognition and Culture*, L. A. Hirschfeld and S. A. Gelman, eds., pp. 85–116. Cambridge, Mass.: MIT Press.

de Ipolyi, A., Santos, L., and Hauser, M. D. (2001). The role of landmarks in cotton-top tamarin spatial foraging: Evidence for geometric and non-geometric features. *Animal Cognition* 4: 99–108.

Gallistel, C. R. (1990). *The Organization of Learning.* Cambridge, Mass.: MIT Press.

Garcia, J. and Koelling, R. (1966). Relation of cue to consequence in avoidance learning. *Psychonomic Science* 4: 123–124.

Gelman, R. (1990). First principles organize attention to and learning about relevant data: Number and the animate–inanimate distinction as examples. *Cognitive Science* 14: 79–106.

Gibson, J. J. (1979). *An Ecological Approach to Visual Perception*. Boston, Mass.: Houghton-Mifflin.

Goodall, J. (1986). *The Chimpanzees of Gombe*. Cambridge, Mass.: Harvard University Press.

Hauser, M. D. (1997). Artifactual kinds and functional design features: What a primate understands without language. *Cognition* 64: 285–308.

Hauser, M. D. (1998). A non-human primate's expectations about object motion and destination: The importance of self-propelled movement and animacy. *Developmental Science* 1: 31–38.

Hauser, M. D. (2000). *Wild Minds: What Animals Really Think*. New York: Henry Holt.

Hauser, M. D. and Carey, S. (1998). Building a cognitive creature from a set of primitives: Evolutionary and developmental insights. In *The Evolution of Mind*, D. Cummins and C. Allen, eds., pp. 51–106. New York: Oxford University Press.

Hauser, M. D., MacNeilage, P., and Ware, M. (1996). Numerical representations in primates. *Proceedings of the National Academy of Sciences* U.S.A. 93: 1514–1517.

Hauser, M. D., Kralik, J., and Botto-Mahan, C. (1999). Problem solving and functional design features: Experiments on cotton-top tamarins (*Saguinus oedipus*). *Animal Behaviour* 57: 565–582.

Hermer, L. and Spelke, E. (1996). Modularity and development: The case of spatial reorientation. *Cognition* 61: 195–232.

Hirschfeld, L. A. and Gelman, S. A. (1994). *Mapping the Mind: Domain Specificity in Cognition and Culture*. Cambridge: Cambridge University Press.

Keil, F. C. (1989). *Concepts, Kinds, and Cognitive Development*. Cambridge, Mass.: MIT Press.

Keil, F. C., Smith, W. C., Simons, D. J., and Levin, D. T. (1998). Two dogmas of conceptual empiricism: Implications for hybrid models of the structure of knowledge. *Cognition* 60: 143–171.

Kemler Nelson, D. G. (1999). Attention to functional properties in toddlers' naming and problem solving. *Cognitive Development* 14: 77–100.

Kemler Nelson, D. G., Frankenfield, A., Morris, C., and Blair, E. (2000). Young children's use of functional information to categorize artifacts: Three factors that matter. *Cognition* 77: 133–168.

Landau, B., Smith, L., and Jones, S. (1998). Object perception and object naming in early development. *Trends in Cognitive Science* 2: 19–24.

Lavin, T. and Hall, G. (1999). Perceptual properties and children's acquisition of words for solids and non-solids. Poster presented at the biennial meeting for the Society for Research in Child Development.

Macario, J. F. (1991). Young children's use of color and classification: Foods and canonically colored objects. *Cognitive Development* 6: 17–46.

Mandler, J. M. and McDonough, L. (1993). Concept formation in infancy. *Cognitive Development* 8: 291–318.

Matsuzawa, T. (1994). Field experiments on use of stone tools by chimpanzees in the wild. In *Chimpanzee Cultures*, R. W. Wrangham, W. C. McGrew, F. B. M. de Waal, and P. G. Heltne, eds., pp. 351–370. Cambridge, Mass.: Harvard University Press.

McGrew, W. C. (1992). *Chimpanzee Material Culture: Implications for Human Evolution*. Cambridge: Cambridge University Press.

Mithen, S. (1996). *The Prehistory of the Mind: A Search for the Origins of Art, Religion, and Science*. London: Thames and Hudson.

Pinker, S. (1997). *How the Mind Works*. New York: W.W. Norton.

Povinelli, D. J. (2000). *Folk Physics for Apes: Chimpanzees, Tool-Use, and Causal Understanding*. Oxford: Oxford University Press.

Rozin, P. (1990). Development in the food domain. *Developmental Psychology* 26: 555–562.

Santos, L. R. and Caramazza, A. (in press). The domain-specific hypothesis: A developmental and comparative perspective on category-specific deficits. In *Category-Specificity in Brain and Mind*, G. Humphreys and E. Forde, eds., New York: Psychology Press.

Santos, L. R., Miller, C. T., and Hauser, M. D. (1999). Knowledge of functionally relevant features for different object kinds. Poster presented at the Biennial Meeting for the Society for Research in Child Development.

Santos, L. R., Hauser, M. D., and Spelke, E. S. (2001). Recognition and categorization of biologically significant objects by rhesus monkeys (*Macaca mulatta*): The domain of food. *Cognition* 82: 127–155.

Santos, L. R., Miller, C. T., and Hauser, M. D. (under review). The features that guide them: Distinguishing between functionally relevant and irrelevant features of artifacts in cotton-top tamarins (*Saguinus oedipus oedipus*) and rhesus macaques (*Macaca mulatta*). *Journal of Comparative Psychology*.

Shettleworth, S. J. (1998). *Cognition, Evolution, and Behavior*. New York: Oxford University Press.

Spelke, E. S. (1985). Preferential looking methods as tools for the study of cognition in infancy. In *Measurement of Audition and Vision in the First Year of Post-Natal Life*, G. Gottlieb and N. Krasnegor, eds., pp. 37–61. Norwood, N.J.: Ablex Publishing.

Spelke, E. S. (1991). Physical knowledge in infancy: Reflections on Piaget's theory. In *The Epigenesis of Mind: Essays on Biology and Cognition*, S. Carey and R. Gelman, eds., pp. 37–61. Hillsdale, N.J.: Lawrence Erlbaum Associates.

Sperber, D. (1994). The modularity of thought and the epidemiology of representations. *Mapping the Mind: Domain Specificity in Cognition and Culture*, L. A. Hirschfeld and S. A. Gelman, eds., pp. 39–67. New York: Cambridge University Press.

Stephens, D. W. and Krebs, J. R. (1986). *Foraging Theory*. Princeton, N.J.: Princeton University Press.

Tomasello, M. and Call, J. (1997). *Primate Cognition*. New York: Oxford University Press.

Visalberghi, E. and Tomasello, M. (1998). Primate causal understanding in the physical and psychological domains. *Behavioural Processes* 42: 189–203.

Wang, R. F., Hermer, L., and Spelke, E. S. (1999). Mechanisms of reorientation and object localization by children: A comparison with rats. *Behavioral Neuroscience* 113: 475–485.

Wynn, K. (1998). Psychological foundations of number: Numerical competence in human infants. *Trends in Cognitive Sciences* 2: 296–303.

Ydenberg, R. C. (1998). Behavioral decisions about foraging and predator avoidance. In *Cognitive Ecology: The Evolutionary Ecology of Information Processing and Decision Making*, R. Dukas, ed., pp. 343–378. Chicago: University of Chicago Press.

The Cognitive Sea Lion: Meaning and Memory in the Laboratory and in Nature

Ronald J. Schusterman, Colleen Reichmuth Kastak, and David Kastak

The pinnipeds, or seals, sea lions, and walruses, are descendents of terrestrial carnivores. In contrast to their fully aquatic counterparts, the dolphins, they are shorter lived, have less complex social organization, and have a less encephalized brain. Despite these facts, some pinniped species display obviously intelligent behavior, which can be seen in the ease of their trainability in oceanaria settings and in their ability to trick unlucky fishermen out of their catch. The quick wits and adaptability of one species in particular, the California sea lion (*Zalophus californianus*), make it an ideal subject for our laboratory studies of problem solving and memory.

Language Learning in the Sea Lion

Historically, a top-down approach has been used to study cognition in a select group of mammals, including the great apes and bottlenosed dolphins. This approach emphasized the search for rudiments of language in nonhuman animals. Our own early work on sea lion cognition, namely, teaching them to comprehend an artificial gestural sign language (Schusterman and Gisiner 1988, 1989, 1997; Schusterman and Krieger 1984, 1986), was inspired by the apparent success of similar research with bottlenosed dolphins (see Herman et al. 1984). Throughout much of the 1980s, we focused our efforts on teaching three sea lions to relate particular gestural signals to objects (such as bats, balls, and rings), modifiers (large, small, black, and white), and actions (such as fetch, tail touch, and flipper touch). These signals could be combined in over 7000 different combinations, each instructing the animal to carry out a specific behavioral sequence. For example, in what was termed a "single object" instruction, the presentation of four signs such as *small white ball flipper-touch* would usually result in the sea lion touching the small white ball with its flipper, while ignoring

the irrelevant objects in the pool (see figure 28.1). More complicated instructional sequences required the sea lion to press one of two paddles to indicate whether an object was present or absent. The most complicated instructions required the sea lion to select one object in the tank and bring it to another object. These "relational" sequences could include up to seven signs; for example, the gestural sequence *large white cone, black small ball fetch* instructed the sea lion to bring the black small ball to the large white cone. Our sea lions were eventually able to respond appropriately to familiar as well as novel combinations of signs with a great deal of accuracy, as shown in figure 28.2, which describes the performance of our most experienced sea lion, Rocky.

Our results with the sea lions in the language-learning task gave us insight into several aspects of their cognitive abilities. Positive results with animals that were smaller brained, shorter lived, and less social than apes and dolphins led us to speculate that we were dealing with general learning processes rather than specialized cognitive abilities such as language. Instead of comprehending the instructional sequences within a linguistic framework, we believed that our sea lions, and perhaps other animals trained on similar tasks, were learning specific problem-solving rules via the associative mechanisms we describe later.

The artificial language tasks also led us to make predictions about how sea lions represented and remembered critical information. For example, our sea lion Rocky exhibited a stereotyped search response following her observation of the gestural cue denoting an object. As soon as she was given the cue, Rocky would turn her head to the left and slowly scan the pool until she located the correct object, at which point she would return to station to wait for the action signal to be given (see figure 28.1). Her purposeful search for the specified object suggested that

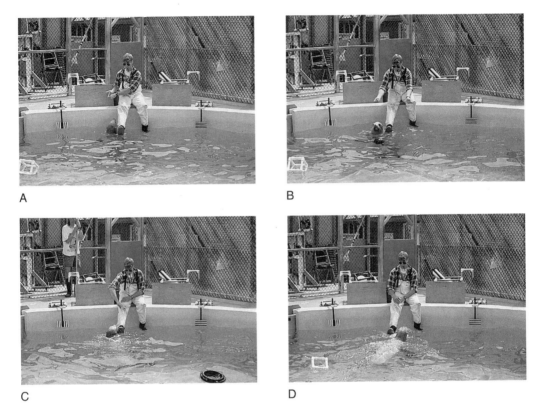

Figure 28.1
An instructional sequence given to sea lion Rocky by a trainer in the context of the language experiment: the gestural cue for the modifier WHITE (A), followed by the cue for the object DISK (B), and then the cue for the action FETCH (C). Rocky responds to the modifier cue with a slight head turn (A), but responds to the object cue by turning her head until she finds and points at the white disk (B). After finding the disk, she returns to her stationing position at the trainer's foot and receives the fetch sign (C). When the trainer's foot drops, Rocky carries out the instructional sequence WHITE DISK FETCH by bringing the white disk to the trainer (D).

she was coding the object signal prospectively; that is, she was translating the gestural cue into a representation, or search image, of the target object. This search image hypothesis is further supported by Rocky's performance on trials based on the presence or absence of an object. On such trials, her search time was significantly longer when the signed object was absent than when it was present (2.4 ± 0.8 seconds versus 1.6 ± 0.6 seconds; $t = 5.6$, $df = 97$, $p < .01$).

Rocky's search behavior on different types of instructional sequences also led us to other observations about her memory. For example, while scanning the pool, Rocky seemed to hold onto the search image of the object for a maximum of about 10 seconds. If she failed to locate the target object within that time period, she often refused to respond or simply "defaulted" to responding to any nearby object. This observation led us to hypothesize that California sea

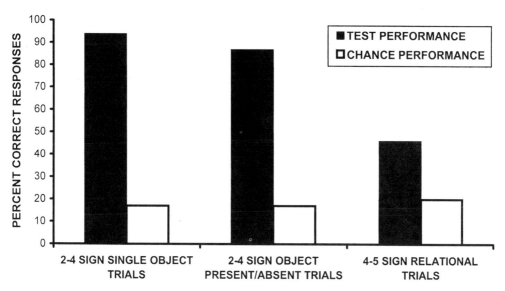

Figure 28.2
Sea lion Rocky's performance accuracy on three instructional types: single object ($n = 1087$ trials), object present or absent ($n = 99$ trials), and relational ($n = 35$ trials). The present or absent results are a random sample drawn from Rocky's data obtained in March 1988 and presented here for the first time. The single object and relational trials are taken from Schusterman and Gisiner (1988). The black bars show the percentage of correct responses to each instructional type and the white bars present performance expected by chance (all χ^2's for these data indicated that performance was significantly better than chance, $p < 0.01$. [See Schusterman and Gisiner (1988) for the calculation of chance probabilities.]

lions had a limited short-term memory. In addition, on trials involving two different objects, Rocky would usually remember the most recently signed object while sometimes forgetting the earlier signed object (Schusterman et al. 1993a). This type of performance reflects attempts at remembering earlier material while being distracted by more recent material. It is typical of what is termed retroactive interference, which occurs when the encoding of new information disrupts the retention of previously coded information.

Collectively, these findings generated more questions than answers. How did the sea lions relate gestural signals to the objects they represented? Did the sea lions learn new signals in a manner analogous to the way that children learn

new words? What was the relationship between the type of language learning demonstrated by the sea lions and their more general abilities to form abstract concepts? What were the limits of their short- and long-term memory? Finally, how did performance on psychological tests given in a laboratory setting relate to natural social and ecological problem solving?

From Language to Logic

To answer these questions regarding basic learning and memory processes, we changed our approach from top-down, beginning with complex languagelike performances and attempting to identify the components and variables influencing behavioral responses, to bottom-up, in which we could look at the formulation of problem-solving rules under more simplified and controlled conditions. This change was motivated by the idea that the gestures used in the language experiment to designate particular objects served as conditional cues that were arbitrarily related to each object by positive reinforcement training. We believed that the performance of our sea lions in the language experiment depended on their acquisition of straightforward, albeit complex, conditional discriminations or *if . . . then* rules, rather than on linguistic processes. For example, a sea lion could be trained to follow the rule *"If the gesture for disk is presented . . . then respond to the object disk."* Although the behavior of the sea lion seems to indicate comprehension of language, this type of rule learning is different from the symbolic nature of human language, in which a symbol becomes equivalent to its referent.

Conditional Discrimination Learning

In order to analyze rule learning in our sea lions, we abandoned our artificial language project and began training two sea lions in a procedure known as arbitrary matching-to-sample (MTS). This procedure allowed us to examine the relationship between potential symbols and their referents. The MTS task requires a subject to respond to one of two or more choice, or comparative, stimuli in the presence of a particular conditional cue, or sample stimulus (Carter and Werner 1978). A sample stimulus and its paired associate are arbitrarily related in much the same way that a gestural cue in the language project is related to an object. For example, through positive reinforcement training, a sea lion can learn to respond to a pattern shaped in the form of a crown, given the presence of a sample stimulus shaped in the form of a propeller, i.e., *"if propeller . . . then crown"* (see figure 28.3).

The sea lions' performance in the MTS task was similar to their performance in the language project. Despite the differences in configuration between the two tasks, the observed patterns of errors (including object and position biases) were quite similar. This was a strong indication that the sea lions were learning comparable *if . . . then* rules to solve the two types of problems.

Learning by Exclusion

The conditional discrimination problems just described were established in two different procedures, but shared an important training characteristic. In MTS and in the artificial language task, the first two conditional discriminations were trained simultaneously by trial and error; as expected, each subject made a significant number of mistakes during the acquisition of the first paired associates. The development of such specific rules through training generally requires a subject to make some errors in order to sort out the contingencies that predict reinforcement. However, following initial training in the language task, we found that our sea lions could learn associations between new gestures and new objects virtually without error.

When a novel gestural cue was produced in the presence of both a novel object and a familiar object (that is, one already associated with a different gestural cue), the sea lion immediately related the novel signal to the novel object and excluded the familiar object as a probable choice (Schusterman and Gisiner 1997). Thus, when a novel gestural cue was incorporated into the sequence, Rocky responded immediately to the only novel object in the pool. We obtained the same results in MTS—the sea lions excluded the familiar comparison, which had already been associated with a familiar sample, and immediately related the novel sample to the novel comparison (Kastak and Schusterman 1992; Schusterman et al. 1993b).

The same phenomenon has been reported in language learning experiments with both dolphins and great apes. Indeed, developmental psycholinguists have documented a similar phenomenon in word learning by children who

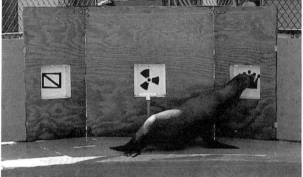

Figure 28.3
Examples of Rocky solving conditional discrimination, or *if ... then*, problems in the matching-to-sample procedure. The sample stimulus, or conditional cue, appears in the center of the apparatus, and comparison, or choice, stimuli appear on either side of the sample. Following an observation interval during which time all three stimuli are visible, Rocky is released from her stationing position in front of the sample stimulus to touch one of the choice stimuli. Correct stimulus matches are rewarded with fish reinforcement. In the upper panel, selection of the rectangle is contingent upon presentation of the spiral, i.e, *"if spiral ... then rectangle";* in the lower panel, selection of the crown is contingent upon presentation of the propeller, i.e., *"if propeller ... then crown."*

relate new words to new objects in their environment without error; this process has been described by various terms, such as *psycholinguistic inference, emergent matching, fast mapping,* and *mutual exclusivity* (Wilkinson et al. 1998). Any individual who uses an exclusion rule to deduce that two items must be related can eventually learn the more direct association between the two items. Once the direct association has been firmly established, the strategy of excluding the familiar item will no longer be necessary. Thus, the use of an exclusion rule minimizes frustration during the learning process by eliminating errors, and this is true whether the subjects are sea lions, dolphins, chimpanzees, or children. In summary then, we can see that part of the performance of sea lions in a languagelike task can be attributed to relatively simple forms of learning, such as specific *if ... then* rules and a more general exclusion rule.

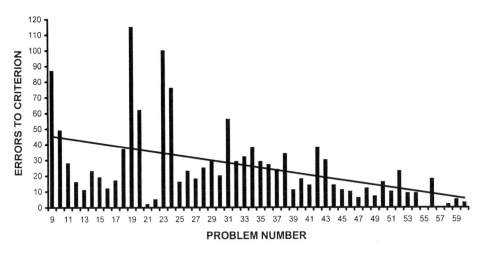

Figure 28.4
Performance for sea lion Rio on 52 successive conditional discrimination problems. The number of errors Rio made prior to reaching criterial levels of performance (90 + percent correct responses) declined significantly as she learned more and more problems of the same type ($R^2 = 0.23$, $p < 0.01$). These data for the learning set in MTS are presented here for the first time.

Learning to Learn

With repeated practice with a particular type of problem, an individual can often solve subsequent problems of the same type more rapidly (Harlow 1949). Thus, animals not only learn how to respond appropriately to individual problems but can also learn about the more general nature of the problems and how to solve them efficiently. We have observed this "learning set" phenomenon in our sea lions in a wide variety of situations, including language learning and MTS (see also Schusterman 1968). For example, using MTS, we trained our sea lion Rio to solve 60 different conditional discriminations. The problems were set up in such a way that Rio could solve the first 8 of these problems by using an exclusion rule, but thereafter had to solve the remaining 52 problems by learning specific *if ...then* rules for each of two paired associates. Rio's performance on these problems can be seen in figure 28.4, which shows that the number of

errors required for her to learn each successive problem gradually decreased. This pattern indicates that Rio formed a learning set, a general strategy for solving similar types of problems.

Equivalence Relations

Our language studies with the sea lions brought into question the symbolic nature of the gestural cues used to denote the modifiers, objects, and actions in an instructional sequence. We found that the responses of our sea lions to novel instructional sequences depended on the order in which the gestural cues were presented. The sea lions were trained to expect cues in the following order: modifier(s)-object(s)-action. A novel gestural cue placed in the modifier, object, or action position of a sequence would invariably result in the appropriate type of response. For example, after Rocky was trained by shaping procedures to porpoise over an object floating in her pool, and the behavior was placed under the control of

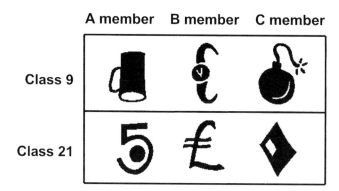

Figure 28.5

The A, B, and C members of two classes (9 and 21) that were part of the 30 potential 3-member classes used with sea lion Rio. Following MTS training of the relations *if A ... then B* and *if B ... then C* for each class, Rio was tested for the ability to form symmetrical (*if B ... then A; if C ... then B*), transitive (*if A ... then C*), and equivalence (*if C ... then A*) relations (see Schusterman and Kastak 1993, 1998).

a gestural cue, she was able to perform perfectly the first time she was given the sign sequences *ball over* or *white cube over*. However, if the object and action cues in a sequence were transposed, Rocky would frequently balk, indicating that she did not comprehend anomalous sequences. Any modifier, object, or action could be substituted for another in an instructional sequence, but they could not be substituted between signal types. These observations suggested that Rocky was organizing the gestural cues into classes based on the type, or function, of the signal. Because the type of signal is specifically related to its position in the sequence, such categorization is evidence of syntactic comprehension (Gisiner and Schusterman 1992).

In addition to such syntactic meaning, conceptual behavior can also be demonstrated in the context of symbolic meaning. The most commonly used paradigm in studying the symbolic nature of human words is called "stimulus equivalence," which is studied in the context of the MTS procedure (see Sidman 1994, 2000). Following the training of the conditional discriminations *if A ... then B* and *if B ... then C*,

stimulus equivalence is shown by the emergence of novel reflexive, or identity, relationships (*if A ... then A, if B ... then B, if C ... then C*), symmetrical relationships (*if B ... then A* and *if C ... then B*), and transitive relationships (*if A ... then C*). A complete equivalence relation combines symmetrical and transitive properties, and would be demonstrated if an individual could spontaneously relate *C* to *A* following training of the original rules *if A ... then B* and *if B ... then C*. In the context of word learning, the stimuli *A*, *B*, and *C* might refer to an object, its verbal label, and its written label. To examine the potential for symbolic meaning in our sea lions, we tested one of our animals for the ability to form equivalence relations using MTS and stimuli such as those shown in figure 28.5.

Our general strategy for documenting equivalence classification was to teach Rio to solve problems of a given type by direct reinforcement training and then test her to see if she could solve novel problems of the same sort. For example, after being trained with 60 identity problems in the form of conditional discriminations (i.e., *if A ... then A*), Rio was tested with 30 unfamiliar

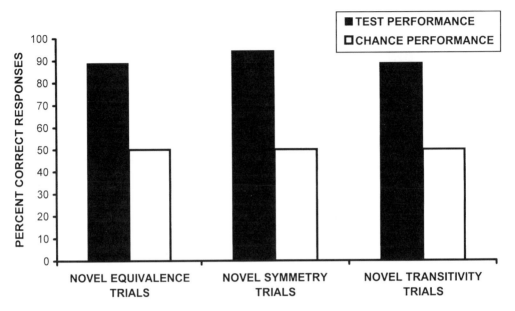

Figure 28.6
Sea lion Rio's trial one performance on tests of equivalence (*if C ... then A*), symmetry (*if B ... then A*, and *if C then A*), and transitivity (*if A ... then C*) for 18 potential classes following exemplar training with 12 other potential classes. Performance on all conditions is significantly better than expected by chance (Fisher's exact tests, $p < 0.01$; $n = 18$ for equivalence trials, $n = 36$ for symmetry trials, and $n = 18$ for transitivity trials).

stimuli; she correctly related identical pairs of objects (i.e., *if novel stimulus X ... then novel stimulus X, if Y ... then Y, if Z ... then Z*, etc.) 80 percent of the time even though she had never seen the objects before (Kastak and Schusterman 1994). Following these tests, Rio was taught conditional discriminations with 30 potential equivalence classes, each consisting of three different shapes (see figure 28.5). For each of the 30 classes, the conditional relations *if A ... then B* and *if B ... then C* were directly trained. Subsequently, 12 of the 30 classes were used to train the relational rules of symmetry and transitivity in order to give Rio experience solving these types of problems. Finally, the remaining 18 classes were presented to Rio in new configurations to determine if equivalence relations would emerge without training. The results shown in figure 28.6 demonstrate Rio's remarkable ability to reorganize trained stimulus relations into emergent combinations based on the properties of symmetry, transitivity, and equivalence (Schusterman and Kastak 1993).

Additional experiments have extended these results on equivalence relations to show such classification occurring in other discrimination learning contexts (Schusterman and Kastak 1998), and class membership expanding to include specific reinforcers such as different types of fish (Reichmuth Kastak et al. 2001). This conceptual reorganization of information allows an animal to connect different objects, events, and individuals in structured categories on the basis of common elements or shared functions. While humans use linguistic codes to facilitate such abstract problem solving, these studies show us that some nonhuman animals that lack language are nonetheless capable of logically organizing their perceptual worlds.

Remembrance of Things Past

Information, once organized into relationships, rules, or categories, must be remembered over time in order to be useful to an animal. Thus, processes related to the coding, storage, retention, and retrieval of information are fundamental to studies of animal cognition. In the language studies, the sea lions seemed to retain representations for very short time intervals, and distracting events appeared to disrupt these representations. To more systematically investigate aspects of short-term, or working, memory in our sea lions, we trained two of them to perform a variant of the MTS task, called "delayed matching-to-sample."

After the sea lions learned to relate sample stimuli to correct comparison stimuli in the standard procedure (see figure 28.3), memory trials were conducted in which the sample stimulus was removed prior to the presentation of the comparison stimuli. The delay between the removal of the sample and the presentation of the comparisons was varied in order to determine the time interval over which the encoded sample could be retained. We found that our sea lions continued to accurately solve the problems for as long as we could reasonably ask them to wait for the comparison stimuli to be shown. Delay periods were extended to 2 minutes for our older and more patient sea lion Rocky, and 45 seconds for our younger and more active sea lion, Rio. The performance of both animals across all delay intervals was consistently at about 90 percent correct responses (Schusterman et al. 1991).

We did not observe a gradual decline in performance with increasing delay intervals, as might be expected if the information encoded from the sample gradually faded from working memory. However, we did note that if irrelevant

stimulus material was presented during the delay interval, performance on the memory trials was completely disrupted. This finding is consistent with our observations in the language experiment, showing that retroactive interference leads to forgetting by disrupting the retention of encoded information in short-term memory.

In contrast to short-term memory, which is easily disrupted, long-term memory for individuals, locations, and tasks appears to be relatively stable for periods ranging up to several years. Recently, we attempted to quantify a sea lion's ability to remember relationships that it has not experienced for a long period of time. We tested our sea lion Rio to determine if she could remember two categories containing ten stimulus members each. Although Rio had not been exposed to the stimuli or the procedure for 12 months, she correctly related stimuli belonging to the same category without a single error on her memory test. Rio's excellent memory for these and other previously learned categories suggests that meaningful or conceptual material may persist in memory for very long periods of time.

Solving Problems in Nature

Our laboratory experiments with California sea lions show how their brain is used to transform perception into meaning and action. We hypothesize that sea lions use the same processes of classification and memory in attempting to solve problems in nature (Schusterman et al. 2000).

For instance, sea lions are capable of recognizing each other as individuals. If a sea lion can form equivalence classes corresponding to labels such as conspecifics, neighbors, allies, and kin, it can more efficiently solve problems by dealing with categories of individuals that may be treated in a similar manner; responses appropriate to one member of a class may, depending on the context, be appropriate for other members of a class.

Figure 28.7
A California sea lion mother–pup reunion. (A) Upon returning to the rookery, the mother emits her signature pup attraction call. (B) The pup responds to the call of its mother, and the pair exchange signature vocalizations and move closer to one another. (C) While continuing to call to her pup, the mother leads the pup around a physical barrier. (D) Mother and pup touch noses and confirm auditory recognition by olfactory, tactile, and visual cues.

The complex behaviors leading to individual and kin recognition appear to be fairly well developed in California sea lions. Immediately following birth, both mother and pup acquire sensory information and form meaningful representations of one another based on hearing, vision, olfaction, and touch. This imprinting process results in mutual recognition, which is critical to maintaining contact over and beyond the lactation period. As a result, when females return to the rookery after a foraging bout that may last several days, they are able to locate and reunite with their dependent pups. These reunions depend on the exchange of signature vocalizations at long range, followed by the confirmation of identity at short range via smell, sight, and touch, as shown in figure 28.7. Furthermore, sea lions born to the same mother in different years may learn to recognize one another through their mutual connection with their mother.

We believe that individual and kin recognition are closely tied to the process of equivalence class formation. For example, a pup's recognition of its mother may depend on the association of many sensory cues with the common reinforcing elements of warmth, contact, and nourishment, while a female's recognition of her sisters may depend on their mutual association with their mother (Schusterman et al. 1992).

There is accumulating evidence that a territorial male sea lion or fur seal recognizes the vocal displays of neighboring rivals. Over several breeding seasons, a male behaves as if it has categorized its competitors into familiar and novel groups. In a seminal investigation on the male territorial and reproductive behavior of Steller sea lions (*Eumetopias jubatus*), Gisiner (1985) found that the most reproductively successful males were those that had at least three seasons of experience maintaining nearly the same territorial site. These experienced males probably grouped neighboring males into classes of "familiar" and "novel" and became successful by not expending time and energy on aggression against familiar rivals, and instead fighting newcomers vigorously and successfully.

Furthermore, female Steller sea lions may also group males into two classes. Gisiner observed that a female was most likely to approach sites occupied by males with long territorial histories, while avoiding sites occupied by males in their first or second year on a territory. The behavior of the male Steller sea lions observed by Gisiner most likely depends on the classification of individuals into categories and a long-term memory for territorial sites, individuals and their signature vocalizations, and outcomes of previous aggressive contests. Female Steller sea lions probably depend on similar processes in choosing their mates.

There are many other examples of cognition gleaned from observations of wild sea lions that may apply to ecological as well as social problem solving. Our research in the laboratory has given us insight into the cognitive processes used by wild animals in such problem solving, and our results continue to raise more questions about their cognitive characteristics. As we get closer and closer to describing "the cognitive sea lion," we have been forced by our data to take a general process approach. We believe that many of the seemingly complex behavioral phenomena that we have elicited in the laboratory can be considered in the context of straightforward associative principles, which nevertheless give rise to a great deal of behavioral complexity. These same associative processes seem to be at work in a variety of natural contexts and are most likely common to many different animal species.

Acknowledgments

The preparation of this manuscript was supported by the U.S. Office of Naval Research grants N00014-99-1-0164 and N00014-00-1-0836 to R. J. Schusterman and a U.S. Department of Defense Augmentation Award for Science, Engineering and Research Training fellowship to C. Reichmuth Kastak.

References

Carter, D. E. and Werner, T. J. (1978). Complex learning and information processing by pigeons: A critical analysis. *Journal of the Experimental Analysis of Behavior* 29: 565–601.

Gisiner, R. (1985). Male Territoriality and Reproductive Behavior in the Steller sea lion, *Eumetopias jubatus*. Unpublished Ph.D. thesis, University of California, Santa Cruz.

Gisiner, R. C. and Schusterman, R. J. (1992). Sequence, syntax and semantics: Responses of a language trained sea lion (*Zalophus californianus*) to novel sign combinations. *Journal of Comparative Psychology* 104: 368–372.

Harlow, H. F. (1949). The formation of learning sets. *Psychological Review* 56: 51–65.

Herman, L. M., Richards, D. G., and Wolz, J. P. (1984). Comprehension of sentences by bottlenosed dolphins. *Cognition* 16: 129–219.

Kastak, D. and Schusterman, R. J. (1992). Comparative cognition in marine mammals: A clarification on match-to-sample tests. *Marine Mammal Science* 8: 414–417.

Kastak, D. and Schusterman, R. J. (1994). Transfer of visual identity matching-to-sample in two California sea lions (*Zalophus californianus*). *Animal Learning & Behavior* 22: 427–435.

Reichmuth Kastak, C., Schusterman, R. J., and Kastak, D. (2001). Equivalence classification by California sea lions using class-specific reinforcers. *Journal of the Experimental Analysis of Behavior* 76: 131–158.

Schusterman, R. J. (1968). Experimental laboratory studies of pinniped behavior. In *The Behavior and Physiology of Pinnipeds*, R. J. Harrison, R. C. Hubbard, R. S. Peterson, C. E. Rice, and R. J. Schusterman, eds., pp. 87–171. New York: Appleton-Century-Crofts.

Schusterman, R. J. and Gisiner, R. (1988). Artificial language comprehension in dolphins and sea lions: The essential cognitive skills. *Psychological Record* 39: 311–348.

Schusterman, R. J. and Gisiner, R. C. (1989). Please parse the sentence: Animal cognition in the Procrustean bed of linguistics. *Psychological Record* 39: 3–18.

Schusterman, R. J. and Gisiner, R. C. (1997). Pinnipeds, porpoises and parsimony: Animal language research viewed from a bottom-up perspective. In *Anthropomorphism, Anecdotes and Animals*, R. W. Mitchell, N. S. Thompson, and H. L. Miles, eds., pp. 370–382. Albany: State University of New York Press.

Schusterman, R. J. and Kastak, D. (1993). A California sea lion (*Zalophus californianus*) is capable of forming equivalence relations. *Psychological Record* 43: 823–839.

Schusterman, R. J. and Kastak, D. (1998). Functional equivalence in a California sea lion: Relevance to social and communicative interactions. *Animal Behaviour* 55: 1087–1095.

Schusterman, R. J. and Krieger, K. (1984). California sea lions are capable of semantic comprehension. *Psychological Record* 34: 2–23.

Schusterman, R. J. and Krieger, K. (1986). Artificial language comprehension and size transposition by a California sea lion (*Zalophus californianus*). *Journal of Comparative Psychology* 100: 348–355.

Schusterman, R. J., Hanggi, E. B., and Gisiner, R. (1992). Acoustic signaling in mother–pup reunions, interspecies bonding, and affiliation by kinship in California sea lions (*Zalophus californianus*). In *Marine Mammal Sensory Systems*, J. A. Thomas, R. A. Kastelein, and Y. A. Supin, eds., pp. 533–551. New York: Plenum.

Schusterman, R. J., Gisiner, R., Grimm, B. K., and Hanggi, E. B. (1991). Retroactive interference of delayed matching-to-sample in California sea lions. Paper presented at annual meeting of the Psychonomic Society, San Francisco.

Schusterman, R. J., Hanggi, E. B., and Gisiner, R. C. (1993a). Remembering in California sea lions: Using priming cues to facilitate language-like performance. *Animal Learning & Behavior* 21: 377–383.

Schusterman, R. J., Gisiner, R., Grimm, B. K., and Hanggi, E. B. (1993b). Behavior control by exclusion and attempts at establishing semanticity in marine mammals using match-to-sample paradigms. In *Language and Communication: Comparative Perspectives*, H. Roitblat, L. Herman, and P. Nachtigall, eds., pp. 249–274. Hillsdale, N.J.: Lawrence Erlbaum Associates.

Schusterman, R. J., Reichmuth, C. J., and Kastak, D. (2000). How animals classify friends and foes. *Current Directions in Psychological Science* 9: 1–6.

Sidman, M. (1994). *Equivalence Relations and Behavior: A Research Story*. Boston: Author's Cooperative.

Sidman, M. (2000). Equivalence relations and the reinforcement contingency. *Journal of the Experimental Analysis of Behavior* 74: 127–146.

Wilkinson, K. M., Dube, W. V., and McIlvane, W. J. (1998). Fast mapping and exclusion (emergent matching) in developmental language, behavior analysis and animal cognition research. *Psychological Record* 48: 407–422.

Robert G. Cook

One of the most interesting and difficult of scientific questions is how other kinds of animals think about the world we share. One long-standing approach to this question has focused on measuring and comparing the cognitive capacities of different animals (Darwin 1872; Morgan 1894; Romanes 1883; Thorndike 1911).

Birds play an important role in such comparative cognitive studies because they offer a unique, nonmammalian perspective on our understanding of these issues. Like mammals, over the past 200 million years these endothermic animals have evolved separately to interface with the events and objects of the world by employing a highly dynamic and interactive mode of living. This has placed similar demands on the sensory and cognitive processes of both of these classes of vertebrates. For instance, it is no accident that these two groups are the most visually sophisticated animals on the planet. Unlike mammals, however, the demands of flight have required birds to keep their body weight to a minimum, limiting them to small but apparently powerful central nervous systems for processing information.

One of the objectives of my research is to understand this paradox of how birds meet the perceptual and cognitive demands of their interactive mode of living with such small and limited neural equipment. My research is directed at understanding visual cognition in one kind of bird, the pigeon (*Columba livia*). During the past 50 years, pigeons have become an important species in the comparative study of perception and learning. This is because a great deal has been established about their basic behavioral processes and nervous system (Zeigler and Bischof 1993), and powerful and precise laboratory methods have been developed for experimentally investigating these processes.

Possessing a sophisticated visual system with established capabilities for color vision, form perception, and pattern recognition, pigeons are capable of learning a wide variety of simple and complex visual discriminations (Cook 2000). This can be seen in our research over the past few years, which has explored perceptual segregation and the mechanisms of visual search (e.g., Cook 1992a, b), the discrimination and perception of objects and the contribution of motion to these processes (Cook and Katz 1999; Cook et al. in press), and the learning and use of abstract concepts (Cook et al. 1995, 1997a, 1999; Cook and Wixted 1997). Because of its direct implications for the overarching themes developed in this volume, the remainder of this essay focuses only on the latter line of research.

The Comparative Psychology of Same-Different Concept Learning

Human behavior is often rule based. We can easily answer questions about things with which we have no direct experience, often using simplifying rules or general principles abstracted from the relations among a set of elements. The benefits of this cognitive ability are that it releases behavior from the direct control of the stimulus and its history of reinforcement, allows us to engage in behaviors unbounded by our experience with specific stimuli, and permits highly flexible and adaptive solutions to novel problems. Such relational rule-based concepts allow us to make accurate inductions about new events and their relations; they also form the basis for our use and appreciation of language, mathematics, analogical reasoning, social relations, and even fine arts such as music. As a species, we are expert at detecting and abstracting the general patterns present in the world's particulars.

While many animals often respond to specific stimulus situations with a fixed or limited repertoire of innate or learned behaviors, it has also become clear that some animals can detect and abstract the patterns present in the world. An understanding of the distribution, mechanisms, and conditions of this conceptual behavior in animals is essential to unraveling its evolution and function. The most widely known example of animal conceptual behavior has involved the categorization of objects, such as chairs, cars, flowers, fish, birds, mammals, trees and so on, from sets of pictures (Bhatt et al. 1988; Cook et al. 1990; Herrnstein and Loveland 1964; Herrnstein and De Villiers 1980). Far less well studied has been whether animals can form abstract rules regarding the relation of one event or stimulus to another. Research of the latter type has focused on investigations of such topics as serial pattern learning (e.g., Fountain and Rowan 1995), transitive inference (e.g., von Fersen et al. 1991), the development and use of syntactic rules (e.g., Kako 1999), and the learning and formation of relational concepts such as same-different (S-D) (e.g., Cook et al. 1997a) and identity (e.g., Wright et al. 1988).

Because of this gap, my recent work has focused on how pigeons perceive and potentially conceptualize same-different relations among visual elements. The detection and recognition of difference and identity are among the oldest and most fundamental of psychological discriminations. They are central to many types of advanced intellectual functions and behaviors, and have important roles in the processes of perception, discrimination, choice, sequential behavior, intelligence-related behavior, and its symbolic mediation by language. James even suggested that the recognition and integration of the "sense of sameness is the very keel and backbone of consciousness" (James 1910, p. 240).

The S-D task is one of the most powerful means of studying the discrimination of such stimulus relations. In this task, the subject is asked to respond "same" when two or more stimuli are identical and "different" if one or more of the stimuli is different from the others. After learning this discrimination, the degree to which this behavior transfers to novel situations

is taken as evidence of concept formation. Early attempts to use S-D procedures with pigeons met with limited success (Edwards et al. 1983; Fetterman 1991; Santiago and Wright 1984). Such results led some to suggest that this relational concept might be beyond the intellectual faculties of this particular animal (Pearce 1991; Mackintosh et al. 1985; Premack 1978, 1983; Wright et al. 1983).

In contrast, we have recently met with more robust success in producing S-D discrimination and concept formation in pigeons across a wide variety of stimuli and procedures. This variety may be in part responsible for our success, but it is also essential to building a convincing argument for any conceptual explanation of this behavior. This is because five operational criteria should be met in order to argue that an animal has formed a conceptual representation. These include (1) evidence of the successful discrimination of the targeted categories or rule during training, (2) evidence of discrimination transfer to novel exemplars of the target concept as recorded on the first trial or prior to any differential reinforcement, (3) evidence that the individual items within the stimulus classes used during training can be discriminated from each other, (4) evidence that the transfer items can be discriminated from the training items, and finally (5) evidence ruling out stimulus control by alternative features that are irrelevant to the concept under study.

Criteria 2 and 5 are typically judged the most important, with 3 and 4 generally overlooked in most studies. In the work described here, the evidence of transfer to novel stimuli is rather plentiful (criterion 2), but because of the perceptual origins of the concepts of same and different, ruling out alternative accounts of the features controlling this transfer (criterion 5; Mackintosh 2000) has turned out to be the more critical part of our studies.

The next two sections describe complementary approaches to obtaining evidence of conceptual behavior in animals. They illustrate empirical strategies that use behavioral techniques to investigate mental notions such as concepts and rules. The first section describes some of our published data on simultaneous S-D conceptual behavior, while the second reports some new data looking at successive S-D behaviors. In both, testing is done with computer-driven, touch-screen-equipped operant chambers in which the stimuli are presented on computer monitors, allowing both maximum control and flexibility in our stimulus presentations.

Simultaneous S-D Discrimination and Transfer

Color and shape-textured S-D displays, like those depicted in the top row of figure 29.1, were the first stimuli with which we established that pigeons could discriminate very large numbers of S-D displays and readily transfer this behavior to novel displays (Cook et al. 1995; see also Wasserman et al. 1995). A typical S-D trial in this procedure starts with a peck at a white "ready" signal, which is then followed by the presentation of a same (all elements identical) or different (containing a randomly located block of contrasting color or shape elements) display in which all the elements are simultaneously present on the display. The pigeons then indicate their reaction to the display by choosing between two "choice" hoppers located on opposite sides of the chamber (e.g., left, different; right, same). A correct choice is then reinforced with food. Based on a variety of evidence at that time, we argued that the pigeons might have used an abstract concept to solve this textured S-D discrimination (Cook et al. 1995). A quick examination of these texture displays revealed, however, that several alternative sources of control still demanded investigation. For example, the simple presence or absence of a perceptually contrasting square "box" could have been the source of control, and not a more cognitive concept of sameness and difference.

To specifically examine such perceptual alternatives, we (Cook et al. 1997a) conducted an experiment testing pigeons with four types or classes of highly variable S-D stimulus displays, examples of which are shown in figure 29.1. The top row shows the texture display type tested in the first study (Cook et al. 1995) and also used in this second study. The second row shows the feature display type, which because of its design involving randomized local elements required the animal to detect the global S-D relations of the display (Cook 1992b). The third row shows the geometric display type, which defined display difference with only a single odd element. The last row shows the object display type, consisting of digitized natural objects (e.g., flowers and birds; the details and rationale for these four display types are contained in Cook et al. 1997a). Taken together, these different classes created an extreme variety and number of ill-defined, polymorphic, global, S-D displays that break any direct correlation between simple perceptual features and the conceptual status of the displays.

We found that pigeons could still easily learn an S-D classification of these multidimensional classes. Furthermore, learning proceeded at the same rate for all four types, suggesting that only a common discrimination rule was being applied to each distinct type. We also found that these birds transferred their discrimination behavior to novel examples of each class (see figure 29.2).

In subsequent experiments (Cook et al. 1999), we were further able to establish that the majority of these pigeons could transfer this S-D discrimination behavior to a fifth novel stimulus class (color and gray-scale photographs) they had never seen before. This latter transfer is particularly important in showing the relative degree of abstractness of their discrimination. The more abstract a conceptual representation, the greater the range of novel conditions to which it should apply. Our results are among the first, to our knowledge, that establish transfer to stimuli far outside of the range of values experienced during training. As such, the pattern of results from these three studies is consistent with the idea that pigeons can detect, recognize, and abstract simultaneously presented S-D relations. While appeals to simple features seem no longer tenable, a skeptic could, nevertheless, still argue

SAME   DIFFERENT

TEXTURE

FEATURE

GEOMETRIC

OBJECT

that perhaps they only learned to detect the generalized presence or absence of large-scale spatial discontinuities among the repeated elements of these displays and did not acquire a true concept.

Successive S-D Discrimination and Transfer

To begin dealing with such possibilities, we have recently developed a new S-D procedure that eliminates such spatial discontinuities. Using a go–no-go discrimination, new pigeons were shown an alternating sequence of either identical (AAAA ... or BBBB ...) or different (e.g., ABAB ...) photographic stimuli over time (see figure 29.3). Pecks to same sequences (S+) were reinforced on a VI-10 variable interval schedule, while pecks to different sequences (S−) eventually produced a brief time-out following their presentation. During each 20-second trial, photographs were successively presented for 2 seconds each, with a 0.5-second blank interstimulus interval separating each one. We have now successfully trained four pigeons to discriminate the successive pairwise S-D arrangement of 60 photographic stimuli. Most important, evidence of concept formation was confirmed by finding significant transfer to novel photographs.

These data can be seen in figure 29.4, which shows peck rates to baseline same and different trials for both nonreinforced probe training stimuli and novel transfer stimuli. Although discrimination behavior was reduced with the novel stimuli, the peck rates for the same sequences were significantly higher than for the different sequences for all birds. These data show that

pigeons can learn a two-item S-D discrimination and form a concept, even when these relations are presented successively over time. These results help to argue against concerns that our previous S-D results may have been due only to detecting generalized spatial or perceptual patterns within the displays.

Conclusion

Our results suggest that pigeons may have a previously unappreciated capacity for learning and using abstract S-D relations among individual elements across a wide variety of stimuli (texture, feature, geometric, object, photographs) and that they can do so whether these relations are presented simultaneously or successively. Returning to our criterion for concept formation outlined earlier, our pigeons easily learn these types of discrimination (criterion 1), even with markedly different stimuli (criterion 3). Most important, they readily show transfer of discrimination to novel items (criterion 2). This transfer is typically reduced in comparison with that for the familiar training items, indicating that the birds recognize and discriminate between the training and transfer items (criterion 4). Furthermore, multiple tests for simpler alternative accounts have not been successful (criterion 5).

Taken together, these different lines of evidence support the hypothesis that our birds are learning a single discriminative rule that is broadly applied to both familiar and novel stim-

Figure 29.1
Representative examples of the original four display types used by Cook et al. (1997a) in training the pigeons tested in these experiments. The left column shows examples of same displays for each display type (the example for the feature display type depicts a shape-same display). The right column shows examples of different displays differing in shape for each display type. For the texture, feature, and geometric display types, there were also corresponding displays in which the elements differed in color. Multiple different colors and shapes and pictures were used to create the large number of displays. The different colors are represented by different levels of grey scale. (Adapted from the *Journal of Experimental Psychology: Animal Behavior Processes* 23: 417–433. Copyright 1997 by the American Psychological Association.)

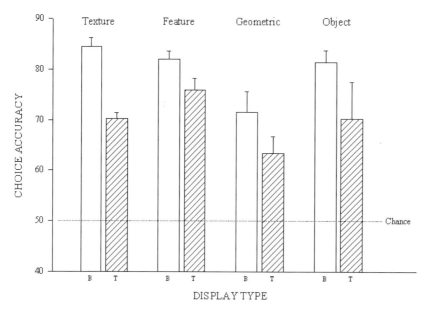

Figure 29.2
Mean choice accuracy on baseline (B) and nonreinforced novel transfer (T) trials for each of the four display types in Cook et al. (1997a). The dotted reference line represents chance performance in the task. (From the *Journal of Experimental Psychology: Animal Behavior Processes* 23: 417–433. Copyright 1997 by the American Psychological Association.)

uli from within and outside the range of their past training experiences. Our newer results indicate that this decision can be reached with as few as two stimuli. Such evidence strengthens the claim that pigeons may be capable of formulating rule-based concepts that give them previously undocumented behavioral flexibility in regard to identity and difference judgments, much like higher primates. Thus this integral component of intelligent behavior may be more widespread in the animal kingdom than previously supposed. Furthermore, it challenges prior claims that S-D behavior (Premack 1978, 1983), and rule-based behavior more generally (Ashby et al. 1998), are critically tied to language.

Looking to the future, our results create something of a paradox that needs further exploration. Despite our results, there is good evidence

that pigeons are often extremely stimulus specific and capable of memorizing large numbers of exemplars and their relations (Carter and Werner 1978; Edwards and Honig 1987; Vaughn and Green 1984). This type of stimulus-specific learning restricts transfer to novel situations, which is exactly the opposite of what we have found. Thus there is evidence that both exemplar *and* conceptual-driven behaviors seem to coexist in the pigeon.

How then are these two distinct forms of learning to be reconciled? Do they reflect different aspects of the same process or separate and distinct processes, as proposed for humans (Ashby et al. 1998)? One popular solution in the human categorization literature is that they reflect the same exemplar learning process. The notion here is that categorizing behavior is

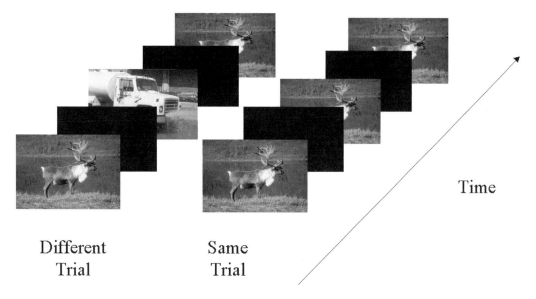

Different
Trial

Same
Trial

Time

Figure 29.3
A schematic representation of same and different trials in our successive S-D procedure. Each item appeared for 2 seconds followed by a 0.5-second blank period that separated it from the presentation of the next item. Using a same+-different− go-no-go procedure, the stimuli were alternated in this manner for 20 seconds for each type of trial. The actual pictures were presented in color.

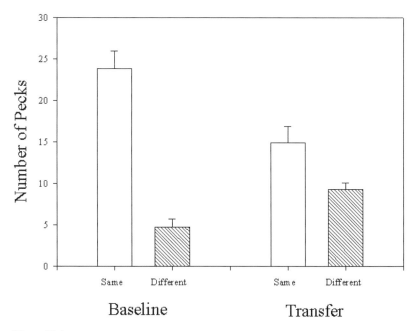

Figure 29.4
Mean peck rates for same (S+) and different (S−) trials for four birds from nonreinforced probe baseline (left side) and novel transfer (right side) trials collected using the successive S-D procedure.

derived from judging the similarity of any new stimulus to a set of stored exemplars. Several exemplar-based theories of pigeon discrimination learning have been proposed (Astley and Wasserman 1992; Chase and Heinemann 2001; Pearce 1991), and it will be interesting to see how they fare in accounting for concept results like those briefly described here. Future work clearly needs to focus on a resolution of such issues. Another important issue for us to address is the degree of transfer between simultaneous and successive procedures. If the pigeons are truly abstracting general S-D relations, it should not matter which procedure is used for training and which is used for testing; the behavior should be readily transferred back and forth.

Nevertheless, in conjunction with research investigating insects, amphibians, mammals, and computers, the promise of the current comparative approach is that we can differentiate those general information-processing principles and mechanisms that are shared by many species from those that are unique or specific to individual species or groups, and determine their functions and the conditions for their development. Such information is critical to understanding the evolution of cognition in both human and nonhuman animals. For the moment, however, our goals are far more limited as we continue to try and understand the challenge of just how one small, complex, autonomous biological system, the pigeon, acquires information about object and event relations in the world and the basis by which it flexibly extrapolates this knowledge to new situations.

References

Ashby, F. G., Alfonso-Reese, L. A., Turken, U., and Waldron, E. M. (1998). A neuropsychological theory of multiple systems in category learning. *Psychological Review* 105: 442–481.

Astley, S. L. and Wasserman, E. A. (1992). Categorical discrimination and generalization in pigeons: All negative stimuli are not created equal. *Journal of Experimental Psychology: Animal Behavior Processes* 18: 193–207.

Bhatt, R. S., Wasserman, E. A., Reynolds, W. F., Jr., and Knauss, K. S. (1988). Conceptual behavior in pigeons: Categorization of both familiar and novel examples from four classes of natural and artificial stimuli. *Journal of Experimental Psychology: Animal Behavior Processes* 14: 219–234.

Carter, D. E. and Werner, J. T. (1978). Complex learning and information processing in pigeons: A critical analysis. *Journal of the Experimental Analysis of Behavior* 29: 565–601.

Chase, S. and Heinemann, E. G. (2001). Exemplar memory and discrimination. In *Avian Visual Cognition*, R. G. Cook, ed. [On-line]. Available at www.pigeon. psy.tufts.edu/avc/chase/.

Cook, R. G. (1992a). Acquisition and transfer of visual texture discriminations by pigeons. *Journal of Experimental Psychology: Animal Behavior Processes* 18: 341–353.

Cook, R. G. (1992b). Dimensional organization and texture discrimination in pigeons. *Journal of Experimental Psychology: Animal Behavior Processes* 18: 354–363.

Cook, R. G. (2000). The comparative psychology of visual cognition. *Current Directions in Psychological Science* 9: 83–88.

Cook, R. G. and Katz, J. S. (1999). Dynamic object perception in pigeons. *Journal of Experimental Psychology: Animal Behavior Processes*. 25: 194–210.

Cook, R. G. and Wixted, J. T. (1997). Same-different texture discrimination in pigeons: Testing competing models of discrimination and stimulus integration. *Journal of Experimental Psychology: Animal Behavior Processes* 23: 401–416.

Cook, R. G., Wright, A. A., and Kendrick, D. F. (1990). Visual categorization in pigeons. In *Quantitative Analyses of Behavior: Behavioral Approaches to Pattern Recognition and Concept Formation*, M. L. Commons, R. Herrnstein, S. M. Kosslyn and D. B. Mumford, eds., pp. 187–214. Hillsdale, N.J.: Lawrence Erlbaum Associates.

Cook, R. G., Cavoto, K. K., and Cavoto, B. R. (1995). Same/Different texture discrimination and concept learning in pigeons. *Journal of Experimental Psychology: Animal Behavior Processes* 21: 253–260.

Cook, R. G., Katz, J. S., and Cavoto, B. R. (1997a). Pigeon same-different concept learning with multiple stimulus classes. *Journal of Experimental Psychology: Animal Behavior Processes* 23: 417–433.

Cook, R. G., Cavoto, B. R., Katz, J. S., and Cavoto, K. K. (1997b). Pigeon perception and discrimination of rapidly changing texture stimuli. *Journal of Experimental Psychology: Animal Behavior Processes* 23: 390–400.

Cook, R. G., Katz, J. S., and Kelly, D. M. (1999). Pictorial same-different concept learning and discrimination in pigeons. *Cahiers de Psychologie Cognitive* 18: 805–844.

Cook, R. G., Shaw, R., and Blaidell, A. P. (in press). Dynamic object perception by pigeons: Discrimination of action in video presentations. *Animal Cognition.*

Darwin, C. (1872). *The Expression of the Emotions in Man and Animals.* London: Murray.

Edwards, C. A. and Honig, W. K. (1987). Memorization and "feature selection" in the acquisition of natural concepts in pigeons. *Learning and Motivation* 18: 235–260.

Edwards, C. A., Jagielo, J. A., and Zentall, T. R. (1983). "Same/different" symbol use by pigeons. *Animal Learning & Behavior* 11: 349–355.

Fetterman, J. G. (1991). Discrimination of temporal same-different relations by pigeons. In *Signal Detection: Mechanisms, Models, and Applications. Quantitative Analyses of Behavior Series*, M. L. Commons and J. A. Nevin, eds., pp. 79–101. Hillsdale, N.J.: Lawrence Erlbaum Associates.

Fountain, S. B. and Rowan, J. D. (1995). Coding of hierarchical versus linear pattern structure in rats and humans. *Journal of Experimental Psychology: Animal Behavior Processes* 21: 187–202.

Herrnstein, R. J. and De Villiers, P. A. (1980). Fish as a natural category for people and pigeons. In *The Psychology of Learning and Motivation*, G. H. Bower, ed., pp. 59–95. New York: Academic Press.

Herrnstein, R. J. and Loveland, D. H. (1964). Complex visual concept in the pigeon. *Science* 146: 549–551.

James, W. (1910). *Psychology.* New York: Henry Holt.

Kako, E. (1999). Elements of syntax in the systems of three language-trained animals. *Animal Learning & Behavior* 27: 1–14.

Mackintosh, N. J. (2000), Abstraction and discrimination. In *The Evolution of Cognition*, C. Heyes and L. Huber, eds., pp. 123–142. Cambridge, Mass.: MIT Press.

Mackintosh, N. J., Wilson, B., and Boakes, R. A. (1985). Differences in mechanisms of intelligence among vertebrates. In *Animal Intelligence*, L. Weiskrantz, ed., pp. 53–65. Oxford: Clarendon Press.

Morgan, C. L. (1894). *An Introduction to Comparative Psychology.* London: Walter Scott.

Pearce, J. M. (1991). The acquisition of concrete and abstract categories in pigeons. In *Current Topics in Animal Learning: Brain, Emotion, and Cognition*, L. Dachowski and C. F. Flaherty, eds., pp. 141–164. Hillsdale, N.J.: Lawrence Erlbaum Associates.

Premack, D. (1978). On the abstractness of human concepts: Why it would be difficult to talk to a pigeon. In *Cognitive Processes in Animal Behavior*, S. H. Hulse, H. Fowler, and W. K. Honig, eds., pp. 423–451. Hillsdale, N.J.: Lawrence Erlbaum Associates.

Premack, D. (1983). The codes of beast and man. *Behavioral and Brain Sciences* 6: 125–167.

Romanes, G. J. (1883). *Animal Intelligence.* New York: Appleton.

Santiago, H. C. and Wright, A. A. (1984). Pigeon memory: Same/different concept learning, serial probe recognition acquisition, and probe delay effects on the serial-position function. *Journal of Experimental Psychology: Animal Behavior Processes* 10: 498–512.

Thorndike, E. L. (1911). *Animal Intelligence: Experimental Studies.* New York: Macmillan.

Vaughan, W., Jr. and Green, S. L. (1984). Pigeon visual memory capacity. *Journal of Experimental Psychology: Animal Behavior Processes* 10: 256–271.

von Fersen, L., Wynne, C. D., and Staddon, J. E. (1991). Transitive inference formation in pigeons. *Journal of Experimental Psychology: Animal Behavior Processes* 17: 334–341.

Wasserman, E. A., Hugart, J. A., and Kirkpatrick-Steger, K. (1995). Pigeons show same-different conceptualization after training with complex visual stimuli. *Journal of Experimental Psychology: Animal Behavior Processes* 21: 248–252.

Wright, A. A., Santiago, H. C., Urcuioli, P. J., and Sands, S. F. (1983). Monkey and pigeon acquisition of same/different concept using pictorial stimuli. In *Quantitative Analyses of Behavior*. Vol. 4, M. L. Commons, R. J. Herrnstein, and A. R. Wagner, eds., pp. 295–317. Cambridge, Mass.: Ballinger.

Wright, A. A., Cook, R. G., Rivera, J., Sands, S. F., and Delius, J. (1988). Concept learning by pigeons: Matching to sample with trial-unique video picture stimuli. *Animal Learning & Behavior* 16: 436–444.

Zeigler, H. P. and Bischof, H. J. (1993). *Vision, Brain, and Behavior in Birds.* Cambridge, Mass.: MIT Press.

Categorization and Conceptual Behavior in Nonhuman Primates

Jacques Vauclair

Animals often behave adaptively in response to a novel stimulus because the stimulus resembles others for which the appropriate response is already known. Such an adaptation expresses an ability to categorize. In effect, in the absence of categorization, each object or event would be perceived as unique, and generalizations would be impossible. Therefore it is not surprising to find categorizing abilities in various animal species, although most of the empirical evidence concerns birds (mainly pigeons) and primates. Since categorization is a fundamental aspect of information processing, its study is crucial for increasing our understanding of animals' cognitive abilities.

This essay describes some of the studies that have been carried out in the past 5 years with two species of baboons, both in laboratory-controlled conditions (Guinea baboons, *Papio papio*) and in outdoor settings (olive baboons, *Papio anubis*). These studies aimed to explore different levels of categorizing behaviors (and their underlying processes) in monkeys confronted with various tasks. Furthermore, we were interested in comparing monkeys and humans tested with similar stimuli and procedures.

A useful general framework for the investigation of these behaviors was provided by Herrnstein (1990), who described categorization abilities in animals in five levels of increasing abstractness, including (1) discrimination, (2) categorization by rote, (3) open-ended categorization (namely, category formation resting on a perceptual similarity between individuals that belong to a given class), (4) concepts, and (5) abstract relations. Herrnstein (1990) uses two criteria to define conceptual categorization (level 4). The first criterion is met when a rapid generalization about members of a class of items is observed. The second criterion, which is related to conceptual processing, implies categorization abilities that go beyond perceiving a similarity between exemplars of a class. Thus, level 4 is more complex than open-ended classification, the latter being related to the use of perceptual dimensions of stimuli. [See Schrier et al. (1984) for an example in macaques, D'Amato and Van Sant (1988) for an example in cebus monkeys, and Vauclair and Fagot (1996) for an example in Guinea baboons.] Level 5 of Herrnstein's categorization is attained when a subject is able to use abstract relations not only between objects but also between concepts, such as in conceptual matching or in conceptual identity (for example, the mastery of a "sameness" relation). The evidence for capacities to perform the first three levels of categorization is abundant for several animal species (see Zayan and Vauclair 1998 and Thompson and Oden 2000 for reviews). It is, however, much less clear concerning levels 4 and 5.

Laboratory Studies with Guinea Baboons

Several experiments were conducted with baboons in order to assess the abilities of these monkeys to discriminate objects on the basis of their membership in a category and to study the nature of the representations of categories the baboons formed. In all the experiments reported in this section, we used a video task requiring the baboons to manipulate a joystick that controlled the movements of a cursor on a screen (Vauclair and Fagot 1994). Briefly, with this technique, the subject was required to manipulate the joystick so as to "touch" with the cursor a response stimulus that matched the sample stimulus on an arbitrary (experimenter-defined) basis.

In one of our studies, we examined our monkeys' abilities to categorize artificial stimuli (Vauclair and Fagot 1996). More specifically, we explored how baboons categorize alphanumeric characters displayed in various typefaces. For this purpose, the baboons were first trained in a symbolic matching-to-sample task with 21 different fonts of the characters "B" and "3" as sample forms, and color squares as comparison forms. After training, novel fonts were displayed. The monkeys showed positive transfer of categorizing performance to the novel stimuli of the characters used in the original training. Such results demonstrate that the original learning was not achieved by rote, because in that case the animals would have demonstrated no transfer to the novel typefaces. Thus the baboons' performance indicates that these monkeys were able to exhibit level 3 behavior (open-ended categories in Herrnstein's sense).

A proper assessment of categorization requires not only that subjects conceive that different objects have common class attributes but also that the subjects can discriminate among individual members within a category (Thompson 1995). Thus, in order to dismiss simple stimulus generalization, it must be demonstrated that stimuli to be classified in the same category are discriminably different from one another. This control was used in our study (Vauclair and Fagot 1996), but it must be pointed that it is rarely used in investigations of categorizing abilities of animals. We also demonstrated in further studies similar abilities of our monkeys to categorically process spatial relations such as "above" and "below" categories (Dépy et al. 1999) as well as "long" and "short" distances (Dépy et al. 1998).

A hallmark of categorical processing in humans is the ability to extract a prototype of a given category. A possible way to investigate the issue of categorical representations is therefore to search for a prototype effect. This effect, which was initially reported in the human literature (e.g., Rosch and Mervis 1975), is expressed by a better categorizing performance with prototypical stimuli representing the central tendency of the category than with other, less typical exemplars. For example, humans think that the sparrow is a better exemplar for the "bird" category than the ostrich.

Demonstration of prototype effects in animals is controversial. Some authors (von Fersen and Lea 1990) have described it in pigeons, while others (e.g., Huber and Lenz 1993) did not find it. Only one study (Jitsumori 1994) has assessed prototype effects in nonhuman primates. In this research, artificial stimuli defined by three two-valued positive or negative dimensions (color, shape, and background color) were presented to rhesus monkeys. The stimuli used in training included only two of the three positive or negative dimension values. In transfer, both the prototypes and novel exemplars of the two categories were shown to the animals. Three of the five monkeys demonstrated a high level of transfer with the prototypes containing all three positive or negative features. However, for those monkeys, there was no statistical difference between the performance achieved with prototypes and that obtained with the nonprototypical stimuli.

Given the contrasting evidence for the use of prototypical representations in animals, we further investigated prototype effects by testing two different species of primates. Identical polymorphous artificial stimuli were presented to both humans and baboons in a symbolic matching-to-sample task. In line with Jitsumori (1994), the rationale of our study with baboons (Dépy et al. 1997) was first, to train subjects to classify two out of three feature stimuli (color, shape, position) and second, to assess transfer of performance with the prototypes of each category. Analyses of data searched for possible prototype effects in both species and focused on species differences and similarities in the procedures creating the categories.

Whereas our human participants solved the task in a propositional way, the results showed that baboons were faster in categorizing prototypical forms than nonprototypical forms. However, an analysis of the training data indicated that the baboons did not extract the prototypes, but responded according to a peak shift phenomenon. This phenomenon (e.g., Mackintosh 1995) could explain the choice of prototypical stimuli as being made, not because these stimuli represented the central tendency of the class, but because they corresponded to the exemplars sharing the least common features with the alternative category. In fact our monkeys used a mixed procedure that consisted in memorizing salient cues between stimuli (e.g., the background color) or specific associations between exemplars and response associations.

Studies with Olive Baboons in a Natural Setting

Tasks requiring the categorization of items on the basis of their functional similarity, such as the distinction between food and nonfood items, are good candidates for studying level 4 behavior. This kind of classification indicates the presence of conceptual categorization given that the items to be classified do not necessarily resemble each other. The animal literature provides only few cases of conceptually based, functional categorizations. In one study by Watanabe (1993), pigeons were trained with real objects (4 edible and 4 inedible stimuli) and then tested with printed color photographs (6 novel stimuli, each belonging either to the food or nonfood category). Generalization to the novel stimuli regardless of the type of stimulus presented (picture or real object) was taken as evidence that pigeons displayed object–picture equivalence based on functional classification. The evidence for similar abilities in nonhuman primates comes from the work of Savage-Rumbaugh et al. (1980). Although the aim of these authors was to demonstrate the mastery of reference in linguistically trained chimpanzees, their study offers clear-cut evidence of categorizing abilities in apes. In this experiment, the chimpanzees were first trained to classify real items in two categories (food and tools). Then the subjects easily transferred their categorization to novel objects and later to pictures and arbitrary symbols (lexigrams) of the respective categories.

In one of our studies (Bovet and Vauclair 1998), olive baboons living in small social groups in an outdoor enclosure were individually trained and tested on the natural category of food versus nonfood with real objects using an adapted version of a Wisconsin General Test Apparatus. The monkeys were first trained to categorize two objects, one food and one nonfood; then 80 other objects (40 food and 40 nonfood) were presented and the categorizing response to each object was recorded. The baboons showed a high and rapid transfer of their categorizing abilities to the novel items. A similar performance for vervet monkeys was described by Zuberbühler et al. (1999). These abilities also persisted in subsequent experiments in which we used cutout photos and various modes of picture presentations. This set of data further demonstrates the abilities of the baboons to relate real objects to their pictorial representations (Bovet and Vauclair 2000).

The same procedure of successive simple discriminations in a two-alternative forced-choice procedure was used in follow-up studies (Bovet and Vauclair 2001). In a first experiment, the monkeys had to judge two physical objects as "same" or "different" (perceptual identity). For example, they were required to judge two apples as being the same, or an apple and a padlock as being different. In a crucial test (second experiment) of conceptual identity (corresponding to Herrnstein's level 5), the baboons had to combine their previously acquired skills in order to classify as "same" two (different) objects that belonged to the same functional category (food or nonfood) and apply that learning to new exemplars. For example, they had to classify as "same" an apple and a banana, or a padlock and a cup, and as "different" an apple and a padlock. This ability corresponds to level 5 of Herrnstein's classification scheme. The monkeys attained a high level of performance at the end of the experiment with totally novel objects (i.e., objects novel in the task but left in the monkeys' enclosure before the experiment). Such results demonstrate the mastery of the "same-different" relation and the ability to conceptually judge as same or different objects in the previously learned categories.

Discussion and Prospects for Future Work

A number of issues can be raised about categorization processes from the studies reported in this essay. They concern in turn (1) the presence of categorization procedures in animals outside of visual perception, (2) the adaptive value of categorization treatments, (3) the interest and importance of studying categorization in the social domain, and (4) the limits of categorization procedures in animals, and possible differences with human categorization.

First, it is important to stress that the examples provided here dealt with the processing of visual objects. However, acoustical stimuli are also categorically organized, not only in nonhuman primates (e.g., May et al. 1989; Ramus et al. 2000), but also in rodents (e.g., Ehret 1992) and in birds (Weary 1989). Note also that the involvement of categorization procedures for organizing information in the environment goes well beyond perception and concerns the coding of objects and space. [An example of the categorization of geometric versus nongeometric information in a reorientation task by rhesus macaques can be found in Gouteux et al. (2001).]

Second, our experiments with two baboon species show that they are able to categorize stimuli by using varied kinds of procedures, the extent of which may depend on the species and individual experiences with the objects and procedures. In all cases, categorizing strategies appear to be adaptive given that they allow the animals to respond efficiently to sets of different stimuli with limited cognitive effort and memory. This latter feature can be exemplified in the experiment with dimensional stimuli (Dépy et al. 1997). Baboons adopted a strategy for processing these stimuli based on the use of background color as a dominant cue for discrimination. It could be stated that the multiplicity of these strategies, as well as their opportunistic use, ensures the production of responses that are adapted to changing and varied environmental conditions.

Third, category formation and some of its underlying cognitive processes have been discussed so far mainly with respect to physical objects or features. There is, however, another context in which animals elaborate natural categories, that of their relations with conspecifics, a domain considered to allow the most complex expression of cognitive processes (e.g., Humphrey 1976). Animals may represent the classes of their conspecifics as well as the classes of relationships established between conspecifics or between the latter and themselves. As claimed elsewhere (Vauclair 1996; Zayan and Vauclair 1998), two types of social categories could be fruitfully studied: intraspecies recognition, including recognition of individual conspecifics, and representation of dominance hierarchies and their transitivity in linear orders. Moreover, it is particularly important to study categorizing processes in monkeys and apes because of their phylogenetic proximity to humans and because of their complex cognitive abilities.

Rare experimental studies are available on social categorization in nonhuman primates; they concern the mother–infant bond and dominance–subordination relation in vervets (Cheney and Seyfarth 1980) and macaques (Dasser 1988). However, because of the controversies surrounding the results and interpretations of these experiments (e.g., Thompson 1995) and the limited number of subjects tested, these experiments need to be both replicated and extended. In particular, it would be useful to test and compare monkeys and apes confronted with similar problems for which they would have to infer a kin or a hierarchy relation in their conspecifics as well as their ability to use transitive inference within these relations. For example, movies of conspecifics could be shown to these animals in which animal *A* is dominant over *B* and *B* is dominant over *C*. Several indexes could be used to assess dominance relationships, such as avoidance, access to food, and facial expressions or submissive gestures (de Waal 1982). The subjects could be trained with a conditional procedure to choose the dominant individual in a pair. Then they could be tested for their abilities to infer (transitive inference), for example, that *A* is dominant to *C*.

Fourth, the experiments that we carried out on monkeys tested in two different setups (laboratory and more natural conditions) permit one to evaluate in some way the respective role of these different environments in the expressions of categorical competence. Thus the studies on functional categorization with untrained animals (Bovet and Vauclair 1998, see earlier discussion) suggest that when they were tested with biologically relevant stimuli (grouping objects in food and nonfood categories has obvious ecological significance for these animals), monkeys were able not only to judge the sameness between physical objects but also the sameness of functional concepts. Such a high degree of abstraction and conceptualization by monkeys has, to my knowledge, not yet been reported in the literature. Note, however, that chimpanzees were successful in a conceptual matching task (Thompson et al. 1997), whereas sea lions approached this level of abstraction when tested in an equivalence class membership task (Schusterman and Kastak 1998).

Moreover, and contrary to Premack's (1983) contention, we demonstrated that cognitive competence comparable to relational matching does not necessarily require previous training with explicit tokens and symbols. In effect, our monkeys' previous training only involved categorizing objects that belonged to one of two categories, and using the same-different relation between objects (within and between the two categories).

A final issue that needs to be briefly considered is related to the existence of likely differences between monkeys' and humans' strategies in solving categorization problems. Our experiments most of the time involved both monkeys and human participants who were tested with similar procedures. This practice, which is seldom followed in research, allows direct comparisons of the procedures used by each species. A comparison of the results calls for qualitative differences between strategies expressed by humans, who used propositional reasoning for categorization of polymorphic stimuli (see also Fagot et al. 1998) and those expressed by baboons, who relied on perceptual discriminations. Such discrepancies, which were also observed when we compared baboons and 3-year-old children (Bovet et al. 2001), indicate the limits of a comparative approach to categorization processes.

Categorizing procedures used by animals are simultaneously dependent on environmental constraints on their cognitive competence (for example, the nature and the salience of available stimuli) and their past experience. Attempts at homogenizing stimuli and methods seem to be insufficient to elicit the use of homologous procedures in the species under comparison. Such effort is, however, needed, along with a consideration of the cerebral structures involved in solving the task (Roberts 1996) in order to determine if the processing is homologous or if it is an effect that directly derives from interspecific cognitive differences.

In any case, it can be expected, through such investigations, to obtain novel data and interpretations about categorization and to increase our knowledge concerning general cognitive abilities and the levels of representation of the social relationships of monkeys and apes. More generally, this enterprise of comparative cognition should lead to a better comprehension of human cognition, given that the latter is the product of both our ontogenetic history and our phylogenetic past.

References

Bovet, D. and Vauclair, J. (1998). Functional categorization of objects and of their pictures in baboons (*Papio anubis*). *Learning and Motivation* 29: 309–322.

Bovet, D. and Vauclair, J. (2000). Picture recognition in animals and in humans: A review. *Behavioral Brain Research* 109: 143–165.

Bovet, D. and Vauclair, J. (2001). Judgement of conceptual identity in monkeys. *Psychonomic Bulletin and Review* 8: 470–475.

Bovet, D., Vauclair, J., and Blaye, A. (2001). Categorization and abstraction abilities in three-year-old children and baboons. Submitted.

Cheney, D. L. and Seyfarth, R. M. (1980). Vocal recognition in free-ranging vervet monkeys. *Animal Behaviour* 28: 362–367.

D'Amato, M. R. and van Sant, P. (1988). The person concept in monkeys (*Cebus apella*). *Journal of Experimental Psychology: Animal Behavior Processes* 14: 43–55.

Dasser, V. (1988). A social concept in Java monkeys. *Animal Behaviour* 36: 225–230.

Dépy, D., Fagot, J., and Vauclair, J. (1997). Categorisation of three-dimensional stimuli by humans and baboons: Search for prototype effects. *Behavioural Processes*, 39: 299–306.

Dépy, D., Fagot, J., and Vauclair, J. (1998). Comparative assessment of distance processing and hemispheric specialization in humans (*Homo sapiens*) and baboons (*Papio papio*). *Brain and Cognition* 38: 165–182.

Dépy, D., Fagot, J., and Vauclair, J. (1999). Processing of above/below categorical spatial relations by baboons (*Papio papio*). *Behavioural Processes* 48: 1–9.

de Waal, F. B. M. (1982). *Chimpanzee Politics*. London: Jonathan Cape.

Ehret, G. (1992). Categorical perception of mouse-pup ultrasounds in the temporal domain. *Animal Behaviour* 43: 409–416.

Fagot, J., Dépy, D., Vauclair, J., and Kruschke, J. K. (1998). Associative learning in baboons and humans: Species differences in learned attention to visual features. *Animal Cognition* 1: 123–133.

Gouteux, S., Thinus-Blanc, C., and Vauclair, J. (2001). Rhesus monkeys use geometric and nongeometric information during a reorientation task. *Journal of Experimental Psychology: General* 130: 505–519.

Herrnstein, R. J. (1990). Levels of stimulus control: A functional approach. *Cognition* 37: 133–166.

Huber, L. and Lenz, R. (1993). A test of the linear feature model of polymorphous concept discrimination with pigeons. *Quarterly Journal of Experimental Psychololgy* 46B: 1–18.

Humphrey, N. (1976). The social function of intellect. In *Growing Points in Ethology*, P. P. G. Bateson and R. A. Hinde, eds., pp. 303–317. New York: Cambridge University Press.

Jitsumori, M. (1994). Category discrimination of artificial polymorphous stimuli by rhesus monkeys (*Macaca mulatta*). *Quarterly Journal of Experimental Psychololgy* 47B: 371–386.

Mackintosh, N. J. (1995). Categorization by people and pigeons: The twenty-second Bartlett memorial lecture. *Quarterly Journal of Experimental Psychololgy* 48B: 193–214.

May, B., Moody, D. B., and Stebbins, W. C. (1989). Categorical perception of conspecific communication sounds by Japanese macaques, *Macaca fuscata*. *Journal of the Acoustical Society of America* 85: 837–847.

Premack, D. (1983). The codes of man and beasts. *Behavioral and Brain Sciences* 6: 125–167.

Ramus, F., Hauser, M. D., Miller, C., Morris, D., and Mehler, J. (2000). Language discrimination by human newborns and by cotton-top tamarin monkeys. *Science* 288: 349–351.

Roberts, A. (1996). Comparison of cognitive functions in human and non-human primates. *Cognitive Brain Research* 3: 319–327.

Rosch, E. and Mervis, C. B. (1975). Family resemblances: Studies in the internal structure of categories. *Cognitive Psychology* 7: 573–605.

Savage-Rumbaugh, E. S., Rumbaugh, D. M., Smith, S. T., and Lawson, J. (1980). Reference: The linguistic essential. *Science* 210: 92–925.

Schrier, A. M., Angarella, R., and Povar, M. (1984). Studies of concept formation by stumptail monkeys: Concepts monkeys, humans and letter A. *Journal of Experimental Psychology: Animal Behavior Processes* 10: 564–584.

Schusterman, R. J. and Kastak, D. (1998). Functional equivalence in a California sea lion: Relevance to animal social and communicative interactions. *Animal Behaviour* 55: 1087–1095.

Thompson, R. K. R. (1995). Natural and relational concepts in animals. In *Comparative Approaches to Cognitive Science*, H. Roitblat and J. A. Meyer, eds., pp. 175–224. Cambridge, Mass.: MIT Press.

Thompson, R. K. R., Oden, D. L., and Boysen, S. T. (1997). Language-naive chimpanzees (*Pan troglodytes*) judge relations between relations in a conceptual matching-to-sample task. *Journal of Experimental Psychology: Animal Behavior Processes* 23: 31–43.

Thompson R. K. R. and Oden, D. L. (2000). Categorical perception and conceptual judgments by nonhuman primates: The paleological monkey and the analogical ape. *Cognitive Science* 24: 363–396.

Vauclair, J. (1996). *Animal Cognition: Recent Developments in Modern Comparative Psychology*. Cambridge, Mass.: Harvard University Press.

Vauclair, J. and Fagot, J. (1994). A joystick system for the study of hemispheric asymmetries in nonhuman primates. In *Current Primatology: Behavioral Neuroscience, Physiology and Reproduction*. Vol. 3, J. R. Anderson, J.-J. Roeder, B. Thierry, and N. Herrenschmidt, eds., pp. 69–75. Strasbourg: University Louis Pasteur.

Vauclair, J. and Fagot, J. (1996). Categorization of alphanumeric characters by baboons (*Papio papio*): Within and between class stimulus discrimination. *Current Psychology of Cognition* 15: 449–462.

von Fersen, L. and Lea, S. E. G. (1990). Category discrimination by pigeons using five polymorphous features. *Journal of the Experimental Analysis of Behavior* 54: 69–89.

Watanabe, S. (1993). Object-picture equivalence in the pigeon: An analysis with natural concept and pseudo concept discriminations. *Behavioural Processes* 30: 225–232.

Weary, D. M. (1989). Categorical perception of bird song: How do great tits (*Parus major*) perceive temporal variation in their song? *Journal of Comparative Psychology* 103: 320–325.

Zayan, R. and Vauclair, J. (1998). Categories as paradigms for comparative cognition. *Behavioural Processes* 42: 87–99.

Zuberbühler, K., Cheney, D. L., and Seyfarth, R. M. (1999). Conceptual semantics in a nonhuman primate. *Journal of Comparative Psychology* 113: 33–42.

31 Cognitive and Communicative Abilities of Grey Parrots

Irene Maxine Pepperberg

How do we measure avian "intelligence" and communicative capacities? Two decades of study on Grey parrots (*Psittacus erithacus*) provide more questions than answers. What *is* intelligence? Can we evaluate nonhumans using human tasks and definitions? Or fairly test nonhuman sensory systems? How do nonmammalian brains process information? Do avian and mammalian cognitive capacities overlap significantly? Preliminary answers exist for the first four questions and considerable data are related to the fifth. To summarize current knowledge, here I examine concepts of intelligence, review techniques for evaluating parrot cognition, and discuss these results and implications.

How Can We Study Avian Intelligence?

"Intelligence" has as many definitions as researchers in the field (Kamil 1988; Sternberg and Kaufman 1998) because it is not a unitary entity, but rather many abilities that interact with stored information to "produce behaviour we see as 'intelligent'" (Byrne 1995, p. 38). For me, intelligence involves not only using experience to solve current problems, but also knowing how to choose, from many sets of information acquired in many domains, the appropriate set for solving the current problem (Pepperberg 1990). Organisms limited to the first ability have learned important associations, but cannot transfer and adapt information—Rozin's (1976) definition of intelligence.

But how should one study "intelligence"? My choices of subject and technique exploited Griffin's (1976) suggestion to use communication as the primary tool. When I began my work, however, both choices were nontraditional.

Studying Avian Cognition

Prior to the mid-1970s, researchers studied mostly pigeons in Skinner boxes; these birds demonstrated capacities far inferior to those of mammals (Premack 1978). Such results were thought to represent the abilities of all birds, despite evidence suggesting that some avian species might exhibit more impressive cognitive and communicative feats (Koehler 1953). For example, given how parrots' large brains, long lives, and highly social natures resemble those of primates and cetaceans, shouldn't parrots also have evolved complex cognitive capacities? Might proper training enable them to demonstrate language-like abilities comparable to those of nonhuman primates and cetaceans?

Specifically, parrots' vocal plasticity would make them candidates for evaluating intelligence via interspecies communication (Pepperberg 1981). Interspecies communication (1) directly states the precise content of the questions being asked; thus an animal need not determine the nature of the question through trial and error; (2) incorporates research showing that social animals respond more readily and often more accurately within an ecologically valid social context (Menzel and Juno 1985); and (3) allows data comparisons among species, including animals and humans. Interspecies communication is also an open, arbitrary, creative code—with an enormous variety of signals that allow researchers to examine the nature and extent of information an animal perceives. And two-way communication allows rigorous testing: Subjects can be required to choose responses from their entire repertoire rather than from a subset relevant only to a particular question's topic. Moreover, an animal that learns such a code may respond in novel, possibly innovative ways that demonstrate greater competence than the responses required in operant paradigms. Interspecies communication may thus more facilely demonstrate nonhumans' inherent capacities or even enable them to learn more complex tasks.

Of course, nonhumans must be taught interspecies communication; here, too, techniques varied. I was among the few to emphasize socially interactive training (Pepperberg 1999). My model/rival (M/R) procedure, adapted from Todt (1975), uses social interaction to demonstrate targeted vocal behaviors: labeling, concept formation, clear pronunciation (Pepperberg 1981, 1991). In this procedure, a bird watches one human train another (the model/rival), i.e., ask questions (e.g., "What color?") about an item of interest to the bird. The trainer rewards correct responses with the item, demonstrating referential, functional use of labels, respectively, by providing a 1 to 1 correspondence between a label and an object, and demonstrating the use of the label as a means to obtain the item. The second human is a model for a bird's responses and its rival for the trainer's attention, and illustrates the adverse consequences of errors: Trainers respond to errors with scolding and temporarily remove the object. The model/rival is told "Try again" after garbled or incorrect responses, thus demonstrating corrective feedback. A bird is included in the interactions and is initially rewarded for an approximate response; the training is thus adjusted to its level. Then the model/rival and trainer reverse roles, showing how both use the communication process to request information or effect environmental change. Without this role reversal, the birds neither transfer responses beyond the human who posed the questions nor learn both parts of the interaction (Todt 1975)—behavior that is inconsistent with interactive, referential communication. Using this technique, I trained a Grey parrot, Alex, to identify objects,

materials, colors, and shapes, and used these abilities to examine his conceptual capacities. Such studies showed how well he transferred information across domains and facilitated comparisons with mammals, primates, and sometimes humans.

Specific Avian Abilities

Categorization—Sorting the World into Definable Bins

Birds sort items into shelter or not-shelter, food or not-food, predator or not-predator, mate or not-mate, conspecific or allospecific. But can birds respond to more than specific properties or stimuli patterns? Can they respond to classes or categories to which these properties or patterns belong (Premack 1978; Thomas 1980)? Can birds go beyond, for example, sorting green or not-green to recognizing the relationship between green pens and lettuce? Noting "greenness" is stimulus generalization; recognizing a category "color" involves formation of a categorical class (Pepperberg 1996). The former is relatively simple, whereas the latter is complex.

One approach to separating these abilities uses symbolic labels (Pepperberg 1983, 1996, 1999). Arbitrary, abstract sound patterns ("red", "4-corner"), hand, or pictorial signals representing concrete physical attributes (e.g., redness, squareness) are grouped into multiple higher-order abstract classes also labeled by arbitrary patterns (e.g., "color", "shape"). The ability to form these classes is not elementary, even for humans. To acquire the category "color" *and* color labels, for example, requires (1) distinguishing color from other categories; (2) isolating certain colors as focal and others as variants; (3) understanding that each color label is part of a class of labels linked under the category label "color"; and (4) producing each label appropriately (de Villiers and de Villiers 1978). Can a bird respond this way?

Alex does (Pepperberg 1983). He not only labels objects, hues, materials, and shapes, but, for example, also understands that "blue" is one instance of the category "color," and that for any colored and shaped item, specific attributes (e.g., blue, 3-corner) represent different categories. If asked "What color?" or "What shape?", he vocally classifies items having one of seven colors and five shapes with respect to either category. This task requires comprehending categorical concepts, not just sorting items into categories, and flexibility in changing the classification basis because he must categorize the same item by shape at one time and color at another. Such *re*classification indicates abstract aptitude comparable to that found in chimpanzees (Hayes and Nissen 1956/1971).

In two more complicated tasks, Alex sees unique combinations of seven items. In the first, he is asked "What color is object X?", "What shape is object Y?", "What object is color A?", or "What object is shape B?" (Pepperberg 1990). In the second, he must label the specific instance of one category of an item defined by two other categories; e.g., "What object is color A and shape B?" Other items exemplify one, not both defining categories. To succeed, he must understand all elements in the question and categorize conjunctively (Pepperberg 1992). His accuracy on both tasks matches that of marine mammals (Schusterman and Gisiner 1988).

Same-Different

Researchers once thought comprehension of the concept of same-different required relational abilities absent in any nonprimates (Premack 1978; Mackintosh et al. 1985). Testing for recognition of same-different is more complex than testing knowledge of match-to- or oddity-from-sample. The former requires arbitrary symbols to represent same-different relationships between sets of items and the ability to denote which attribute is the same or different (Premack 1983). The latter requires only that a subject needs

fewer trials for the subject to respond to *B* and *B* as a match after it has learned to respond to *A* and *A* as a match; likewise for *C* and *D* after it has learned to respond to *A* and *B* as nonmatching. Match-to- and oddity-from-sample responses might even be based on old-new or familiar-unfamiliar contrasts (Premack 1983)— the relative number of times *A* versus different *B*s are seen. A subject that understands same-different, however, knows not only that two nonidentical blue items are related just as are two nonidentical green objects—by color—but also that the blue items are the same in a separate way than two nonidentical square items; moreover, it can transfer this process to any attribute and to differences (Premack 1978; Pepperberg 1999).

Natural avian behavior patterns of recognition of individuals, vocal dueling, and song matching (see the review in Stoddard 1996) require same-difference-based discrimination, implying that the ability is adaptive. In the laboratory, however, apparent same-different discriminations of vocalizations may be based instead on learning unique call characteristics (Park and Dooling 1985) or a differential weighing of information in various song features; the results could depend on the experimenter's choice of features (Nelson 1988). Also, experimental design may not reveal the subjects' same-different perception. Starlings (*Sturnus vulgarus*), for example, classify novel series of tones as the same or different from an ascending or descending reference series, but unless they are pressed, do so only for sequences within their training frequency range (Hulse el al. 1990). And none of these studies required labeling the same-different relation or indicating which attributes were the same or different, or transfer to novel situations (e.g., different species' calls or songs; Pepperberg 1987).

Alex, however, has learned abstract concepts of same-different and has learned to respond to an absence of information about these concepts. Given two identical items or ones that vary with respect to some or all attributes of color, shape, and material, he utters the appropriate category label for what is the same or different (Pepperberg 1987); if nothing is the same or different, he replies "none" (Pepperberg 1988). He responds accurately to novel items, colors, shapes, and materials, including those he cannot label, and to specific queries. This is not done from rote training or the objects' physical attributes. If he were ignoring our queries—for example, "What's same?" for a red and a blue wooden square—and responding based on prior training, he would ascertain and label the one anomalous attribute (here color). Instead, in such cases he gives one of two appropriate answers ("shape," "mah-mah" [matter]). The test conditions match Premack's (1983) chimpanzee study in rigor. Alex also transfers concepts of same-different and absence to untrained situations. When he was first shown two same-sized objects after learning to answer the question "What color-matter is bigger or smaller?" for any two items, he asked "What's same?", then said "None" (Pepperberg and Brezinsky 1991).

Numerical Capacities

Numerical studies in animals are difficult (Pepperberg 1999). Even for humans, researchers disagree on the stages (content, ordering) of numerical ability and whether language affects numerical tasks (e.g., Davis and Pérusse 1988; Fuson 1988, 1995; Gelman and Gallistel 1986; Siegler 1991; Starkey and Cooper 1995). No avian subjects count in the sense described for humans (Fuson 1988), but they nevertheless demonstrate numerical abilities. To distinguish counting from subitizing (a simpler perceptual mechanism involving pattern recognition), I constructed collections of four groups of items varying in two color and two object categories (e.g., blue and red keys and cars), and asked Alex to quantify items defined by one color and one object category (e.g., "How many blue key?"; Pepperberg 1994). His accuracy (83.3 percent)

replicates that of humans in a comparable study by Trick and Pylyshyn (1993), who argue that humans cannot subitize when quantifying a subset of items distinguished from other subsets by a conjunction of qualifiers. Although the same behavior may be mediated by different mechanisms in different species, the data for Alex suggest that a nonhuman, nonprimate, nonmammal has abilities that in an ape would be taken to indicate human competence.

Communication

Communication elucidates many avian capacities. Some psittacids (e.g., *Amazona vittata*, Snyder et al. 1987), corvids (crows, Maccarone 1987; Florida scrub jays, *Aphelocoma coerulescens coerulescens*, McGowan and Woolfenden 1989) and chickens (*Gallus gallus*, Evans et al. 1993) may, like vervet monkeys (*Cercopithecus aethiops*), vocally categorize different predators (Seyfarth et al. 1980). Grey parrots use English speech to label and categorize items, quantify arrays, and respond to queries concerning same-different and relative and conjunctive concepts (Pepperberg 1990, 1992, 1994; Pepperberg and Brezinsky 1991). Thus, for any two items, Alex can be asked "What's same-different?", "How many?", *or* "What color-matter is bigger-smaller?" To respond appropriately, he must not only understand each concept, but also must determine which is targeted and from what domain an answer must originate, which fits my original definition of intelligence. His abilities suggest striking parallels between birds and primates.

Avian Intelligence from a Human Perspective

Despite the above data, two problems remain when evaluating nonhuman intelligence, particularly in creatures so different from humans as parrots. First, tasks used to evaluate nonhumans are interpreted with respect to human sensory systems and perceptions of intelligence. Second, specific abilities vary across avian species.

Human biases, which underlie all evaluations of nonhumans, can be addressed by designing—with extreme care—tasks relevant to nonhuman ecology and physiology. We study songbird cognition, for example, not by how birds resolve match-to-sample problems on colored lights, but by how they categorize, repeat, discriminate, and order songs in territorial encounters (e.g., Todt and Hultsch 1996; Kroodsma and Byers 1998). For the importance of such design, consider a tongue-in-cheek analogy: A test of how a human male's choice of song attracts mates and repels intruders would reveal significant incompetence (Pepperberg 1999).

Avian abilities also differ across species. No one species illustrates the range of avian intelligence. Birds with large vocal repertoires learn auditory discriminations faster than birds with small repertoires (Cynx 1995); caching birds outperform noncachers in spatial but not in nonspatial tasks (Olson et al. 1995). The knowledge that cachers rarely have large vocal repertoires and that versatile songsters generally do not cache precludes cross-species comparisons of similar intelligence types, intraspecies comparisons of different intelligence types, and limits research on information transfer across domains. These problems might be resolved by positing that such differences correspond to various specialized human intelligences (e.g., Gardner 1983; Kamil 1988), and, as for humans (Sternberg 1997), that the same basic, underlying processing capacities mediate different abilities. Caching and song storage involve different brain areas, but we do not know if different mechanisms encode changes representing learning and memory within these different structures. Even so, whether specialized avian abilities reflect specific or general mechanisms, data demonstrating the range of avian capacities suggest the need for further study.

Summary

Despite the concerns expressed here, judgments about human intelligence most likely apply to nonhumans. As an example, substitute "species" (my italics) for "culture" in the following quote from Sternberg and Kaufman:

[Species] designate as "intelligent" the cognitive, social, and behavioral attributes that they value as adaptive to the requirements of living.... To the extent that there is overlap in these attributes across [species], there will be overlap in the [species'] conceptions of intelligence. Although conceptions of intelligence may vary across [species], the underlying cognitive attributes probably do not.... As a result there is probably a common core of cognitive skills that underlies intelligence in all [species], with the cognitive skills having different manifestations across the [species]. (Sternberg and Kaufman 1998, p. 497)

References

Byrne, R. (1995). *The Thinking Ape: Evolutionary Origins of Intelligence*. Oxford: Oxford University Press.

Cynx, J. (1995). Similarities in absolute and relative pitch perception in songbirds (starling and zebra finch) and a nonsongbird (pigeon). *Journal of Comparative Psychology* 109: 261–267.

Davis, H. and Pérusse, R. (1988). Numerical competence in animals: Definitional issues, current evidence, and a new research agenda. *Behavioral and Brain Sciences* 11: 561–579.

de Villiers, J. G., and de Villiers, P. A. (1978). *Language Acquisition*. Cambridge, Mass.: Harvard University Press.

Evans, C. S., Evans, L., and Marler, P. (1993). On the meaning of alarm calls: Functional reference in an avian vocal system. *Animal Behaviour* 46: 23–38.

Fuson, K. C. (1988). *Children's Counting and Concepts of Numbers*. New York: Springer-Verlag.

Fuson, K. C. (1995). Aspects and uses of counting: An AUC framework for considering research on counting to update the Gelman/Gallistel counting principles. *Cahiers de Psychologie Cognitive* 14: 724–731.

Gardner, H. (1983). *Frames of Mind: The Theory of Multiple Intelligences.* New York: Basic Books.

Gelman, R. and Gallistel, C. R. (1986). *The Child's Understanding of Number*, 2nd ed. Cambridge, Mass.: Harvard University Press.

Griffin, D. R. (1976). *The Question of Animal Awareness: Evolutionary Continuity of Mental Experience.* New York: Rockefeller University Press.

Hayes, K. J. and Nissen, C. H. (1956/1971). Higher mental functions of a home-raised chimpanzee. In *Behavior of Nonhuman Primates.* Vol. 4, A. Schrier and F. Stollnitz, eds., pp. 57–115. New York: Academic Press.

Hulse, S. H., Page, S. C., and Braaten, R. F. (1990). Frequency range size and the frequency range constraint in auditory perception by European starlings (*Sturnus vulgaris*). *Animal Learning & Behavior* 18: 238–245.

Kamil, A. C. (1988). A synthetic approach to the study of animal intelligence. In *Comparative Perspectives in Modern Psychology. Nebraska Symposium on Motivation.* Vol. 7, D. Leger, eds., pp. 257–308. Lincoln: University of Nebraska Press.

Koehler, O. (1953). Thinking without words. *Proceedings of the XIVth International Congress of Zoology.* pp. 75–88.

Kroodsma, D. E. and Byers, B. E. (1998). Songbird song repertoires: An ethological approach to studying cognition. In *Animal Cognition in Nature*, R. P. Balda, I. M. Pepperberg, and A. C. Kamil, eds., pp. 305–336. London: Academic Press.

Maccarone, A. D. (1987). Sentinel behaviour in American crows. *Bird Behaviour* 7: 93–95.

Mackintosh, N. J., Wilson, B., and Boakes, R. A. (1985). Differences in mechanism of intelligence among vertebrates. *Philosophical Transactions of the Royal Society, London B*308: 53–65.

McGowan, K. J. and Woolfenden, G. E. (1989). A sentinel system in the Florida scrub jay. *Animal Behaviour* 37: 1000–1006.

Menzel, E. W., Jr., and Juno, C. (1985). Social foraging in marmoset monkeys and the question of intelligence. *Philosophical Transactions of the Royal Society, London B*308: 145–158.

Nelson, D. A. (1988). Feature weighting in species song recognition by the field sparrow (*Spizella pusilla*). *Behaviour* 106: 158–182.

Olson, D. J., Kamil, A. C., Balda, R. P., and Nims, P. J. (1995). Performance of four seed caching corvid species in operant tests of nonspatial and spatial memory. *Journal of Comparative Psychology* 109: 173–181.

Park, T. J. and Dooling, R. J. (1985). Perception of species-specific contact calls by budgerigars (*Melopsittacus undulatus*). *Journal of Comparative Psychology* 99: 391–402.

Pepperberg, I. M. (1981). Functional vocalizations by an African Grey parrot (*Psittacus erithacus*). *Zeitschrift für Tierpsychologie* 55: 139–160.

Pepperberg, I. M. (1983). Cognition in the African Grey parrot: Preliminary evidence for auditory/vocal comprehension of the class concept. *Animal Learning & Behavior* 11: 179–185.

Pepperberg, I. M. (1987). Acquisition of the same/different concept by an African Grey parrot (*Psittacus erithacus*): Learning with respect to color, shape, and material. *Animal Learning & Behavior* 15: 423–432.

Pepperberg, I. M. (1988). Comprehension of "absence" by an African Grey parrot: Learning with respect to questions of same/different. *Journal of the Experimental Analysis of Behavior* 50: 553–564.

Pepperberg, I. M. (1990). Cognition in an African Grey parrot (*Psittacus erithacus*): Further evidence for comprehension of categories and labels. *Journal of Comparative Psychology* 104: 41–52.

Pepperberg, I. M. (1991). Learning to communicate: The effects of social interaction. In *Perspectives in Ethology.* Vol. 9, P. P. G. Bateson and P. H. Klopfer, eds., pp. 119–162. New York: Plenum.

Pepperberg, I. M. (1992). Proficient performance of a conjunctive, recursive task by an African Grey parrot (*Psittacus erithacus*). *Journal of Comparative Psychology* 106: 295–305.

Pepperberg, I. M. (1994). Evidence for numerical competence in a Grey parrot (*Psittacus erithacus*). *Journal of Comparative Psychology* 108: 36–44.

Pepperberg, I. M. (1996). Categorical class formation by an African Grey parrot (*Psittacus erithacus*). In *Stimulus Class Formation in Humans and Animals,* T. R. Zentall and P. R. Smeets, eds., pp. 71–90. Amsterdam: Elsevier.

Pepperberg, I. M. (1999). *The Alex Studies: Cognitive and Communicative Abilities of Grey Parrots.* Cambridge, Mass.: Harvard University Press.

Pepperberg, I. M. and Brezinsky, M. V. (1991). Acquisition of a relative class concept by an African Grey parrot (*Psittacus erithacus*): Discriminations based on relative size. *Journal of Comparative Psychology* 105: 286–294.

Premack, D. (1978). On the abstractness of human concepts: Why it would be difficult to talk to a pigeon. In *Cognitive Processes in Animal Behavior*, S. H. Hulse, H. Fowler, and W. K. Honig, eds., pp. 421–451. Hillsdale, N.J.: Lawrence Erlbaum Associates.

Premack, D. (1983). The codes of man and beasts. *Behavioral and Brain Sciences* 6: 125–176.

Rozin, P. (1976). The evolution of intelligence and access to the cognitive unconscious. In *Progress in Psychobiology and Physiological Psychology*. Vol. 6, J. M. Sprague and A. N. Epstein, eds., pp. 245–280. New York: Academic Press.

Schusterman, R. J. and Gisiner, R. (1988). Artificial language comprehension in dolphins and sea lions: The essential cognitive skills. *Psychological Record* 38: 311–348.

Seyfarth, R. M., Cheney, D. L., and Marler, P. (1980). Vervet monkey alarm calls: Semantic communication in a free-ranging primate. *Animal Behaviour* 28: 1070–1094.

Siegler, R. S. (1991). In young children's counting, procedures precede principles. *Educational Psychology Review* 3: 127–135.

Snyder, N. F., Wiley, J. W., and Kepler, C. B. (1987). *The Parrots of Luquillo: Natural History and Conservation of the Puerto Rican Parrot*. Los Angeles: Western Foundation for Vertebrate Zoology.

Starkey, P. and Cooper, R. G. (1995). The development of subitizing in young children. *British Journal of Developmental Psychology* 13: 399–420.

Sternberg, R. J. (1997). The concept of intelligence and its role in lifelong learning and success. *American Psychologist* 52: 1030–1037.

Sternberg, R. J. and Kaufman, J. C. (1998). Human abilities. *Annual Review of Psychology* 49: 479–502.

Stoddard, P. K. (1996). Vocal recognition of neighbors by territorial passerines. In *Ecology and Evolution of Acoustic Communication in Birds*, D. E. Kroodsma and E. H. Miller, eds., pp. 356–374. Ithaca, N.Y.: Cornell University Press.

Thomas, R. K. (1980). Evolution of intelligence: An approach to its assessment. *Brain, Behavior, and Evolution* 17: 454–472.

Todt, D. (1975). Social learning of vocal patterns and modes of their applications in Grey parrots. *Zeitschrift für Tierpsychologie* 39: 178–188.

Todt, D. and Hultsch, H. (1996). Acquisition and performance of repertoires: Ways of coping with diversity and versatility. In *Ecology and Evolution of Communication in Birds*, D. E. Kroodsma and E. H. Miller, eds., pp. 79–96. Ithaca, N.Y.: Cornell University Press.

Trick, L. and Pylyshyn, Z. (1993). What enumeration studies can show us about spatial attention: Evidence for limited capacity preattentive processing. *Journal of Experimental Psychology: Human Perception and Performance* 19: 331–351.

III COMMUNICATION, LANGUAGE, AND MEANING

32 Cognition and Communication in Prairie Dogs

C. N. Slobodchikoff

My central research question concerns the relationship between the complex communication system of Gunnison's prairie dogs and their cognitive abilities. Gunnison's prairie dogs (*Cynomys gunnisoni*) are social, colonial animals that are found in the American Southwest, within the states of Arizona, New Mexico, Utah, and Colorado. There are four other species of prairie dogs: the black-tailed prairie dog (*Cynomys ludovicianus*), found in the midwestern United States; the Utah prairie dog (*Cynomys parvidens*), found in the state of Utah; the white-tailed prairie dog (*Cynomys leucurus*), found in the states of Montana and Wyoming; and the Mexican prairie dog (*Cynomys mexicanus*), found in the state of Chihuahua in Mexico.

Gunnison's prairie dogs typically spend the winter in a state of torpor inside extensive underground burrow systems, then emerge in the spring to set up territories (Slobodchikoff 1984; Rayor 1988). Each territory is defended by the group living on it, and the social structure can vary considerably within the same colony. Some territories are occupied by a single male or female; others are occupied by a single male and a single female; still others are occupied by a single male and several females; and some are occupied by several adult males and several adult females (Slobodchikoff 1984; Travis and Slobodchikoff 1993). The structure of the social system within a territory seems to be correlated with the distribution of food resources: uniformly distributed food resources correlate with single male–single female territories, while patchily distributed food resources correlate with single male–multiple female and multiple male–multiple female territories (Slobodchikoff 1984; Travis and Slobodchikoff 1993). The colonies are spatially fixed, and the extensive burrow systems can persist for perhaps hundreds of years.

The spatial concentration of prairie dogs into colonies means that a number of predators can encounter a dependable food source throughout much of the year. Prairie dogs are preyed upon by coyotes, foxes, badgers, golden eagles, red-tailed hawks, ferruginous hawks, harriers, black-footed ferrets, domestic dogs, domestic cats, rattlesnakes, and gopher snakes. Also, prairie dogs are hunted extensively by humans for target practice and sport. Prior to the introduction of rifles, prairie dogs were hunted as a source of meat by Native American peoples for at least 800 years (Slobodchikoff et al. 1991).

Such predation pressure was probably important for the evolution of antipredator defenses. Prairie dogs have dichromatic color vision (Jacobs and Pulliam 1973) and can detect the presence of a predator from long distances. They also have an alarm call system that allows them to advertise the approach of a potential predator. The alarm calls are very loud and can carry for distances of more than a kilometer (Hoogland 1996). The burrows provide an escape route from most terrestrial and aerial predators. The burrow architecture within a territory has several openings to the surface which are connected to a series of underground tunnels that can run a horizontal distance of more than 10 m below the ground's surface (Fitzgerald and Lechleitner 1973).

The alarm call system has proven to be a Rosetta stone for deciphering the information encoded in the prairie dog vocalizations. When a prairie dog detects a predator, he or she emits a call that alerts other prairie dogs to the presence of danger. The call can be given as a single bark or as a series of barks that comprise a calling bout. The external referent, the predator, can be seen and videotaped by field observers, as can the escape behaviors of the prairie dogs in response to the predator. The alarm calls can be recorded on audiotape and brought back to the laboratory for analysis. The calls can be analyzed through fast Fourier transform to assess the acoustic structure of the vocalizations. Different parts of the waveforms of the calls can be measured, and statistical analyses or fuzzy logic neural net analyses can be performed to determine whether calls elicited by different predators are similar to one another or different from each other (Slobodchikoff et al. 1991; Placer and Slobodchikoff 2000). The calls recorded for each predator can be played back to the prairie dogs when no predator is present and when no prairie dog is calling. The escape behaviors of the prairie dogs can be recorded on videotape, allowing comparisons of escape behaviors elicited by a predator with escape behaviors elicited by the playback of the alarm calls (Kiriazis 1991). Field experiments can be designed that expose prairie dogs to different predators or different attributes of individual predators, and the calls elicited can be recorded and analyzed. All of the prairie dogs can be marked with black fur dye, using a code of letters and numbers that allows observers to identify the individual prairie dog who is calling. Also, several colonies are available for study, and repeating experiments at multiple colonies can increase the external validity of the experiments.

Using these techniques, we have shown that the calls contain a variety of information. A call can identify the category of predator, such as coyote, human, domestic dog, or red-tailed hawk (Slobodchikoff et al. 1986; Placer and Slobodchikoff 2000). Each category of predator-specific calls elicits different escape responses (Kiriazis 1991), just as the different alarm calls of vervet monkeys elicit different escape responses (Cheney and Seyfarth 1990). The escape responses are of two different types, unlike other ground squirrels that seem to have only one response, running to their burrows (Owings and Morton 1998). Among the prairie dogs, hawk and human alarm calls elicit running to the burrows and diving inside. For a human-elicited call, running to the burrows is a colony-wide response, and for a hawk-elicited call, only the animals in the immediate flight path of a diving hawk run to their burrows. Coyote and domestic dog alarm calls

elicit either a running to the lip of the burrow and standing at the burrow (coyote) or standing in place where the animal was feeding (domestic dog). In both of the latter cases, other animals emerge from their burrows and watch the progress of the predator through the colony (Kiriazis 1991).

The predator-specific calls appear to be referential, describing objects or events external to the animal. [See Evans (1997) for a discussion of referential signals.] Such categorizations functionally serve as nounlike elements in the alarm calls. In addition, within a predator category, the prairie dogs can incorporate information about the physical features of a predator, such as color, size, and shape (Kiriazis 1991; Slobodchikoff et al. 1991). For example, with humans, the calls contain information about the color of clothes that the humans are wearing, and the general size and shape of the humans (Slobodchikoff et al. 1991). These categorizations serve as adjectivelike elements in the alarm calls. Finally, the prairie dogs can incorporate information about the relative speed of travel of a predator, or the relative urgency of the response, by shortening the time interval between individual alarm barks in a calling bout, in direct proportion to the speed of travel of the predator (Kiriazis 1991). The time element between barks thus serves as a verblike element.

These sources of information in alarm calls appear to function as a primitive grammar, composed of nounlike, adjectivelike, and verblike elements (figures 32.1–32.4). For example, if a coyote appears and is moving slowly, one prairie dog will produce a single bout of calls that contains descriptive information about the size, shape, and color of the coyote. If the coyote starts to run, a number of prairie dogs start to call, each providing a description of the coyote. In addition, the interval between each alarm bark in a bout is shortened proportionally to the speed of travel of the coyote (Kiriazis 1991). The information that is encoded into the alarm calls could let other prairie dogs know who the individual

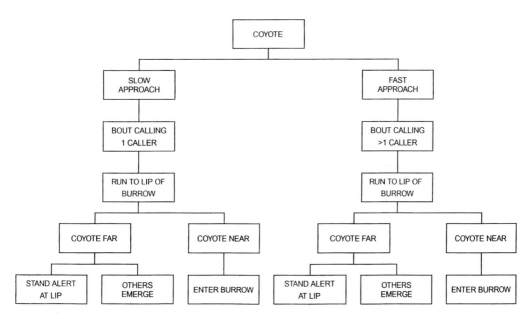

Figure 32.1
Responses of prairie dogs to an approaching coyote. If the coyote is approaching quickly, a single animal will produce a multiple-note bout of alarm calls. If the coyote is approaching slowly, several animals will simultaneously produce multiple-note bouts of alarm calls. All animals run to the lips of their burrows, and animals below ground emerge to watch the progress of the coyote through the colony.

coyote is, from the description of the predator and the speed of its travel through the colony.

The identity of the individual coyote may be important to the prairie dogs because different coyote individuals hunt prairie dogs in different ways (Leydet 1977). Some coyotes walk or run through the colony in a straight path, and then lunge at any prairie dog that appears to be away from the lip of its burrow. Other coyotes lie down next to a burrow that contains prairie dogs and wait for more than an hour by the side of the burrow. Knowing the identity of the individual coyote might provide some information about the type of hunting style that that individual typically adopts.

The categorization and description of predators implies a sophisticated cognitive function. This cognitive function can be achieved through genetic hardwiring into the brains of prairie dogs, through cultural transmission of learned information, or a combination of both. Some measure of cultural transmission of information might exist in the alarm calls because the calls vary somewhat from colony to colony in what has been described as dialects (Slobodchikoff and Coast 1980; Slobodchikoff et al. 1998). Within a local area, such as the vicinity of Flagstaff, Arizona, colonies separated by less than 2 km can have differences in frequency and time components within a category of call, such as that elicited by a human. All of the calls elicited by a human from these colonies recognizably fall into the structure of the human category, but significant differences exist. On a broader regional basis, such as over the entire range of the Gunnison's prairie dogs, dialect differences are more

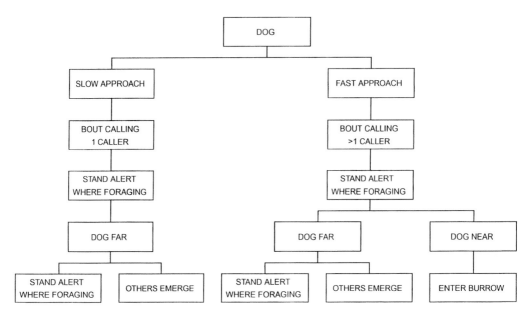

Figure 32.2

Responses of prairie dogs to an approaching domestic dog. As with a coyote, different rates of approach elicit calling by either one or several prairie dogs. However, unlike the coyote response, the prairie dogs do not run to their burrows, but stand upright in an alert posture wherever they were foraging. Other prairie dogs emerge from their burrows to watch the progress of the dog through the colony.

pronounced, although even there a human-elicited call still retains the basic structure of the human category. Although some genetic differences exist between colonies at the local level, these differences are relatively slight (Travis et al. 1997), suggesting that genetic differences alone might not explain the existence of dialects.

In order to describe the individual features of predators, the prairie dogs might have an innate rule-based cognitive system. This rule-based system might have a template of different time and frequency components that correspond to differences in color, shape, and size, stored in the brains of the prairie dogs. As an individual predator appears, the prairie dogs might pull out of that stored repertoire the components that correspond to the physical description of the individual predator.

A rule-based system is suggested by the experiments done by Ackers and Slobodchikoff (1999). Although earlier work showed that each prairie dog consistently incorporated into its call information about a predator in the same way as other prairie dogs calling to warn of the same predator, such incorporation could come about through cultural transmission. To address this, Ackers and Slobodchikoff (1999) used three kinds of plywood models as stimuli that elicited alarm calls. All of the models were painted black and were presented as silhouettes. One model was an oval, one was a realistic silhouette of a coyote, and one was a realistic silhouette of a skunk. The models were presented to a colony of prairie dogs by having the model come from concealment and travel across a part of the colony on a pulley system.

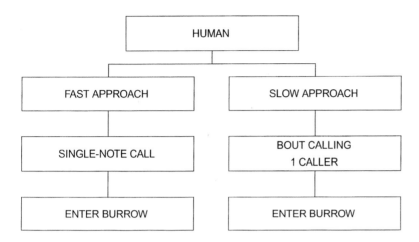

Figure 32.3
Responses of prairie dogs to an approaching human. Depending on the rate of approach of the human, a single animal will give either a one-note call if the human is approaching quickly, or a multiple-note bout of calls if the human is approaching slowly. In either case, all the prairie dogs run to their burrows and disappear inside.

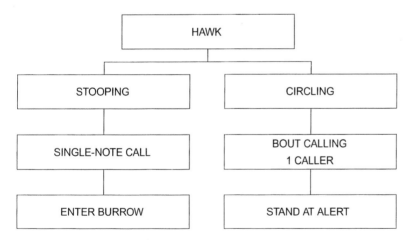

Figure 32.4
Responses of prairie dogs to an approaching hawk. If the hawk is stooping or diving, a single animal will give a single-note call, eliciting running to burrows and disappearing inside. If the hawk is circling overhead, a single animal will give a multiple-note bout of calls. In response to this signal, other prairie dogs do not run to their burrows, but stand upright in an alert posture wherever they were foraging.

All of the models elicited alarm calls. The calls were recorded and each prairie dog's alarm call in response to each model was noted. For each type of model, all of the prairie dogs consistently called the same for that model, using the same frequency and time components. Although none of the prairie dogs had ever seen a large black oval before, they all had a call that corresponded to the presentation of black oval, and this call was significantly different from their calls for either the coyote silhouette or the skunk silhouette. Similarly, the calls for the coyote silhouette were significantly different from those for the skunk silhouette. The coyote silhouette elicited calls that were mostly similar to those elicited by a live coyote, but differed in several components. Neither the oval silhouette nor the skunk silhouette represented potential predators, yet each elicited its own model-specific calls. If the calls elicited by the oval and the skunk were just expressing the novelty of the stimulus, the calls for the two models might have been expected to be either the same or at least very similar. Instead, they were very different in terms of their acoustic frequency and time components. This suggests that the prairie dogs were describing their perception of the attributes of each silhouette, rather than the novelty of the stimulus. These results suggest that prairie dogs may have an internal cognitive representational system that they can utilize to encode information about external events.

The alarm calls of prairie dogs can be viewed as an intentional system (Allen and Bekoff 1997). The elements of this representational system allow the coding of information about known objects in the environment, such as predators, as well as unknown objects that the prairie dogs have not previously experienced. We can postulate that prairie dogs might have an internal lexicon corresponding to a set of external conditions. This lexicon might be innate or it might be a combination of innate and learned elements. Upon the occurrence of an external event, such as the appearance of a predator, the prairie dogs might apply a series of transformational rules (as defined by Chomsky, 1965). These transformational rules convert a base structure (elements of the lexicon) into an output structure (the alarm call) that provides a message for other prairie dogs to decode with their own internal lexicon.

In terms of future work, two immediate questions arise. One is related to the lexicon. The Ackers and Slobodchikoff (1999) experiments suggest that prairie dogs can utilize novel combinations of descriptors. Of interest here is how many different external objects the sender can encode and the receiver decode. Are there instances when related objects are grouped together into a single type of call? How fine scaled is the discriminatory ability of both the sender and the receiver? These questions can be addressed with further experiments using the methodology described here.

The other question is that of states of mind (as defined by Cheney and Seyfarth 1990 and Dennett 1987, 1996). This one question leads to a whole host of other questions. Are prairie dogs aware of their own danger when they see a predator? Can they assess the risk and decide to either call or not call? Can they categorize to themselves the different predators that they encounter? Are they aware of communicating a message to another prairie dog, or is their message produced without any awareness?

In terms of Dennett's (1987) intentional systems, prairie dogs could have a zero-order intentional system, in which the calls are simple expressions of fear. This appears to be unlikely, given the complexity of the calling system. Alternatively, the prairie dogs could have a first-order intentional system, in which they have beliefs, and when giving a call they believe that a predator is nearby and want other prairie dogs to take evasive action. It is also possible that prairie dogs might have higher-order intentional systems, with beliefs about the beliefs of other prairie dogs. These are all questions that are relatively difficult to address. Cheney and Seyfarth (1988, 1990) have used with vervet monkeys a habituation-dishabituation experimental design that had been successfully used previously with preverbal children to assess the degree to which representational mental states might exist. Such experiments have not proven to be successful with prairie dogs because of the rapid habituation that the prairie dogs showed to playbacks of alarm calls (Ackers 1997). Another methodology has to be designed to address these questions.

From the standpoint of cognitive ethology (Allen and Bekoff 1997), we can ask the question: Why do prairie dogs have such a complex system of communication? Although the primates are our closest evolutionary relatives, so far no primate species is known to have such a complex referential communication system. One of the best examples of a referential system in primates is that of the vervet monkeys, and they have only three categories of calls: leopard, eagle, and snake. Within these categories, the calls can be given for a variety of different predators and nonpredators (Seyfarth et al. 1980). Perhaps the prairie dogs evolved such a complex system because of the ecological circumstances in which they live. They are a social species occupying spatially fixed colonies that attract the same predator individuals day after day. Since natural selection favors mechanisms that improve an animal's fitness, perhaps phylogenetic relatedness is less important than the ecological conditions in which an animal species lives.

References

Ackers, S. H. (1997). The Communicative Function of Variation in Gunnison's Prairie Dog (*Cynomys gunnisoni*) Alarm Calls. Ph. D. dissertation, Northern Arizona University, Flagstaff.

Ackers, S. H. and Slobodchikoff, C. N. (1999). Communication of stimulus size and shape in alarm calls of Gunnison's prairie dogs. *Ethology* 105: 149–162.

Allen, C. and Bekoff, M. (1997). *Species of Mind*. Cambridge, Mass.: MIT Press.

Cheney, D. L. and Seyfarth, R. M. (1988). Assessment of meaning and the detection of unreliable signals by vervet monkeys. *Animal Behaviour* 36: 477–486.

Cheney, D. L. and Seyfarth, R. M. (1990). *How Monkeys see the World*. Chicago: University of Chicago Press.

Chomsky, N. (1965). *Aspects of the Theory of Syntax*. Cambridge, Mass.: MIT Press.

Dennett, D. C. (1987). *The Intentional Stance*. Cambridge, Mass.: MIT Press.

Dennett, D. C. (1996). *Kinds of Minds*. New York: Basic Books.

Evans, C. S. (1997). Referential signals. In *Perspectives in Ethology*. Vol. 12, D. H. Owings, M. D. Beecher, and N. S. Thompson, eds., pp. 99–143.12. New York: Plenum.

Fitzgerald, J. P. and Lechleitner, R. R. (1973). Observations on the biology of Gunnison's prairie dog in central Colorado. *American Midland Naturalist* 92: 146–163.

Hoogland, J. L. (1996). Why do Gunnison's prairie dogs give anti-predator calls? *Animal Behaviour* 51: 871–880.

Jacobs, G. H. and Pulliam, K. A. (1973). Vision in the prairie dog: Spectral sensitivity and color vision. *Journal of Comparative and Physiological Psychology* 84: 240–245.

Kiriazis, J. (1991). Communication and Sociality in Gunnison's Prairie Dogs. Ph. D. dissertation, Northern Arizona University, Flagstaff.

Leydet, F. (1977). *The Coyote: Defiant Songdog of the West*. San Francisco: Chronicle Books.

Owings, D. H. and Morton, E. S. (1998). *Animal Vocal Communication: A New Approach*. Cambridge: Cambridge University Press.

Placer, J. and Slobodchikoff, C. N. (2000). A fuzzy-neural system for identification of species-specific alarm calls of Gunnison's prairie dogs. *Behavioural Processes* 52: 1–9.

Rayor, L. S. (1988). Social organization and space-use in Gunnison's prairie dog. *Behavioral Ecology and Sociobiology* 22: 69–78.

Seyfarth, R. M., Cheney, D. L., and Marler, P. (1980). Vervet monkey alarm calls: Semantic communication in a free-ranging primate. *Animal Behaviour* 28: 1070.

Slobodchikoff, C. N. (1984). Resources and the evolution of sociality. In *A New Ecology*, P. W. Price, C. N. Slobodchikoff, and W. S. Gaud, eds., pp. 227–251. New York: Wiley-Interscience.

Slobodchikoff, C. N. and Coast, R. (1980). Dialects in the alarm calls of prairie dogs. *Behavioral Ecology and Sociobiology* 7: 49–53.

Slobodchikoff, C. N., Fischer, C., and Shapiro, J. (1986). Predator-specific alarm calls of prairie dogs. *American Zoologist* 26: 557.

Slobodchikoff, C. N., Kiriazis, J., Fischer, C., and Creef, E. (1991). Semantic information distinguishing individual predators in the alarm calls of Gunnison's prairie dogs. *Animal Behaviour* 42: 713–719.

Slobodchikoff, C. N., Ackers, S. H., and Van Ert, M. (1998). Geographical variation in alarm calls of Gunnison's prairie dogs. *Journal of Mammalogy* 79: 1265–1272.

Travis, S. E. and Slobodchikoff, C. N. (1993). Effects of food resources on the social system of Gunnison's prairie dogs. *Canadian Journal of Zoology* 71: 1186–1192.

Travis, S. E., Slobodchikoff, C. N., and Keim, P. (1997). DNA fingerprinting reveals low genetic diversity in Gunnison's prairie dog. *Journal of Mammalogy* 78: 725–732.

33 Meaningful Acoustic Units in Nonhuman Primate Vocal Behavior

Cory T. Miller and Asif A. Ghazanfar

An experience common to all of us who travel to foreign countries is trying to make sense of the confusing sounds being uttered by native speakers. Unless one can find a translator or at least a good bilingual dictionary, one often will have an extraordinarily difficult time expressing one's basic needs for food and shelter. In such situations, we lack not only the knowledge of what the different words mean but also an understanding of where the boundaries are for the different acoustic units within a foreign stream of speech. The extreme version of this Quinean problem of translation (Quine 1973) applies not only to linguists and foreign travelers but also to those of us who wish to shed light on the vocal communication systems of other species (Hauser 1996).

Ethologists studying nonhuman animal communication systems are faced with the daunting task of dividing the vocal repertoire into different types of acoustic units (e.g., bouts, vocalizations, syllables). Specifically, how can one determine whether a sequence of temporally distinct units emitted by an animal represents a single functional unit (such as meaningless syllables put together to form a word in speech), a string of functionally independent units (such as words forming a sentence in speech), or something simpler, such as the repetition of one small unit? A true understanding of how vocal signals are parsed must be derived from the animal's perspective. Vocal signals must be parsed into the acoustic units that are meaningful in terms of eliciting specific behaviors from the intended receivers (Green and Marler 1979; Hauser 1996). Thus, an animal's behavior serves as a "translator" for ethologists entering a species' perceptual world. Using this approach, we have learned much about the meaningful acoustic units in many avian (e.g., Podos et al. 1992; Searcy et al. 1999) and anuran vocal repertoires (e.g., Narins and Capranica 1978; Ryan and Rand 1990).

Studies in these taxa have given us significant insights into how vocal behavior relates to brain design.

Like birds and anurans, many nonhuman primate (hereafter, primate) species produce bouts of vocalizations containing sequences of similar acoustic units and/or different-sounding acoustic units (figure 33.1), but we know very little about the meaningful units in primate vocal signals. An understanding of how primates perceive and produce such vocalizations is important for several reasons. First, the evolution of speech and language may have involved selection for capabilities that existed in extant primates (Lieberman 1984; Ghazanfar and Hauser 1999; Fitch 2000). One such capability may be to produce vocal signals that mean one thing when produced individually, but something different when recombined into sequences of sounds serve an entirely different function. Second, from a more general perspective, an understanding of the constraints on the perceptual and motor domains of primates' vocal behavior may provide us with insights into the species-specific perceptual world and thus their cognitive abilities and limitations. In this essay we review our understanding of the meaningful acoustic units of production and perception in primate vocal communication. Given the space constraints, a complete literature review cannot be provided here. For a list of studies relevant to this issue, see table 33.1.

To begin this discussion, we would like to clarify our use of the term *acoustic unit*. For the purposes of this review, we refer to all temporally distinct acoustic pulses as syllables. This criterion is arbitrary in the sense that it is based solely on an acoustic measure, not a behavioral one. Nevertheless, it serves as a good starting point. Our own research is aimed at refining this definition as we gain a better understanding of the functional and perceptual significance of all

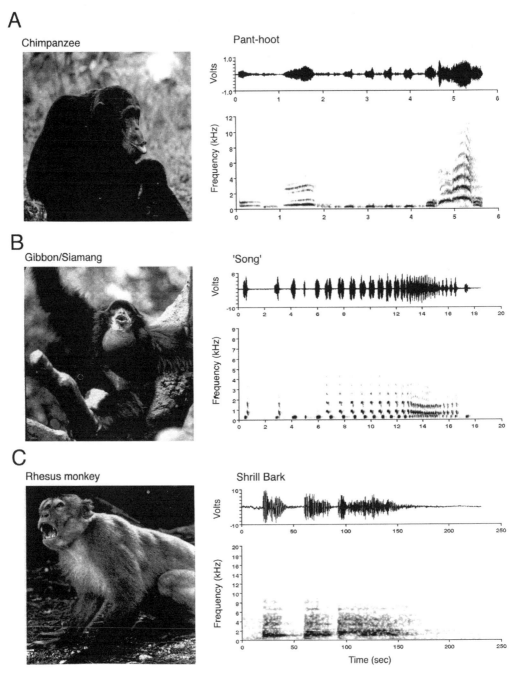

Figure 33.1
Examples of multisyllabic calls in primates. Time-amplitude waveforms (top right) and spectrograms (bottom right). (A) Chimpanzee pant-hoot. (Photograph courtesy of P. Marler; vocalization courtesy M. Wilson and M. Hauser.) (B) Siamang (gibbon) song. (Photograph courtesy of Roy Fontaine/Monkey Jungle; vocalization courtesy of Thomas Geissmann.) (C) Rhesus monkey shrill bark. (Photograph courtesy of Cayo Santiago photo archives; vocalization courtesy of M. Hauser.)

Table 33.1
Investigations of meaningful acoustic units in primate communication

Vocal Production	Vocal Perception
Observations	
Titi monkeys	Titi Monkeys
Robinson (1979)	Robinson (1979)
Capuchins	Capuchins
Robinson (1984)	Robinson (1984)
Gibbons	Gibbons
Mitani and Marler (1989)	Mitani and Marler (1989)
Goustard (1976)	
Chimpanzees	
Marshall et al. (1999)	
Experiments	
Tamarins	Titi monkeys
Miller and Hauser (in prep)	Robinson (1979)
	Gibbons
	Mitani and Marler (1989)
	Tamarins
	Ghazanfar et al. (2001)
	Rhesus monkeys
	Hauser et al. (1998)

acoustic units within primate vocal signals; i.e., determining how acoustic units are organized to form units that are "meaningful" from the specie's perspective.

Observations and Experiments in the Field

Insights into the units of production and perception are possible from studies of the orderly arrangement of syllables within primate long calls. Long calls serve as localization cues for conspecifics and are often produced in the context of territorial encounters, mate attraction, and isolation or group cohesion (Marler 1968; Waser 1982). These multisyllabic calls are produced by a range of different primate species, such as the "whoop-gobble" of mangabeys (*Cercocebus*

albigena; Waser 1977), the "chirrup-pant-bellow-pump-honk" call sequences of titi monkeys (*Callicebus moloch*; Robinson 1979), the "chirp-trills-squaws" of wedge-capped capuchins (*Cebus olivaceus*; Robinson 1984), the "twitters" of squirrel monkeys (*Saimiri sciureus*; Newman et al. 1978), the "whinny" of spider monkeys (*Ateles geoffroyi*; Chapman and Weary 1990; Teixidor and Byrne 1999), the "pant-hoot" of chimpanzees (figure 33.1A) (*Pan troglodytes*; Clark Arcadi 1996; Mitani and Nishida 1993), and the "songs" of gibbons (figure 33.1B) (*Hylobates* spp.; Mitani 1985; Mitani and Marler 1989). Given the similarities in the acoustic structure of long calls throughout the primate order, data on meaningful units in one species may be generalized to other species' communication systems.

In vocal signals consisting of multiple syllables, both the individual syllables and combinations of syllables may mediate specific aspects of behavior. As a result, for any vocalization, information may be encoded at multiple levels. For example, the chimpanzee pant-hoot consists of a series of "hoo" calls followed by a series of screams (figure 33.1A). Since both hoos and screams are produced individually in other contexts, the pant-hoot could be either a single vocalization or a bout of several vocalizations. To reframe the question from a production standpoint, is the vocal control during pant-hoot production akin to a tape recorder that once activated must run to completion, or can control be exerted for each syllable? Recently, Marshall and colleagues (1999) reported that two different, genetically unrelated groups of captive chimpanzees incorporated a novel vocal element (dubbed the "Bronx cheer") at either the beginning or end of pant-hoots, but never in the middle of the call. This suggests that although each syllable can be produced independently, perhaps the pant-hoot is encoded as a single vocal signal that can be added to, but cannot be internally modified.

Not only is the syntactic arrangement of syllables within long calls relevant to our understanding of vocal production, it can also provide insights into the units of perception. Robinson (1979) and Mitani and Marler (1989) observed that the long calls of titi monkeys and the songs of gibbons contain sequences of syllables that vary in their order. The males of these species recombine different syllable types to create unique phrases or songs, and certain orders of syllables have a higher probability of occurring than others. Based on these observations, a series of field playback experiments were conducted to determine whether the different syllable arrangements serve different functions in these primates. Playbacks in which the syllables were arranged in a low probability sequence revealed that both gibbons and titi monkeys recognize when the syllables in conspecific vocalizations occur in an

atypical order. Gibbons produced significantly more "squeak" calls (given during intergroup encounters) upon hearing a song with experimentally rearranged notes (Mitani and Marler 1989). Similarly, titi monkeys produced significantly more moaning responses (given in response to interspecies and intergroup encounters) following playbacks of conspecific vocalizations with abnormal arrangements of syllables (Robinson 1979).

The first step in assessing the meaningful units of a vocal signal is to determine the smallest invariable acoustic unit that can influence behavior. For example, Hauser et al. (1998) recently conducted a field playback study on rhesus macaques (*Macaca mulatta*) demonstrating that the intersyllable interval plays an important role in conspecific vocal recognition for some vocalization types. Using the orienting preference experimental assay developed by Hauser and Andersson (1994), they increased or decreased the intersyllable interval beyond the species' typical range for three call types: grunts, shrill barks (figure 33.1C), and copulation screams. Each of these three call types consists of repeated syllables with similar acoustic structures. Playbacks of these manipulated calls revealed that rhesus monkeys recognize grunts and shrill barks based on the overall, multisyllabic call structure, while copulation screams are recognized at the level of the individual syllable. In other words, for grunts and shrill barks, the smallest meaningful unit of perception is the whole, multisyllabic call, while for copulation screams, a single syllable is the smallest meaningful unit of perception.

These data provide an important foundation for primate communication studies, but there are still many unanswered questions concerning the meaningful units of primate vocal signals. Do each of the individual syllables represent distinct units that can be recombined under the caller's volition, or are calls with different syllable orders each representative of completely different vocalizations? Such questions are directly relevant to understanding the constraints on vocal control.

In terms of perception, no studies of primate calls have systematically varied syllable order or presented subjects with isolated syllables to determine the functionality of syllables within a vocal signal. Such an approach is necessary if we are to fully understand the organization of complex design features in primate vocal signals.

Unit of Production in Cotton-Top Tamarins

Our research focuses on a small New World primate, the cotton-top tamarin (*Saguinus oedipus*). This species has a complex vocal repertoire in which similar-sounding syllables can be used in several different vocalizations (Cleveland and Snowdon 1981). One vocal signal of particular interest is the combination long call, a contact call emitted when individuals are isolated from members of their colony (Cleveland and Snowdon 1981; Weiss et al. in press) (figure 33.2A). The vocalization is multisyllabic, composed of 1–3 chirps followed by 1–5 whistles, and thus is ideal for addressing questions about meaningful units of production and perception.

We first set out to examine the organization of syllables within the cotton-top tamarin long call from the perspective of vocal production. Specifically, is the entire long call produced as a single unit, or is each syllable its own encapsulated unit that can be produced in isolation? To address this question, we borrowed an elegant experimental technique originally used by Cynx (1990) to test a similar question in a songbird, the zebra finch (*Taeniopygia guttata*).

In this experiment, a bright light was flashed at subjects while they produced their multisyllabic song. Cynx predicted that if the song was its own unit of production, then subjects should be unaffected by the light flash and continue singing until the song was complete. If the smaller acoustic units within the song (i.e., notes, syllables, or motifs) represented the unit of production, then subjects should be able to stop singing in the middle of the song following the

completion of one of these units. He found that zebra finches interrupted their song approximately 60 percent of the time following a light flash. Interestingly, when the song was interrupted, subjects consistently completed the syllable already being produced before ending the vocalization. Based on these results, Cynx argued that the minimal unit of production in zebra finch song is the syllable.

We used the same logic to assess the units of production in tamarin long call production and attempted to interrupt vocal production using both light flashes and white noise bursts (Miller and Hauser, in preparation). Overall, only 14 percent of all calls were interrupted. Of those trials in which an interruption occurred, the auditory stimulus was more effective (22 percent of occurrences) than the light flash (5 percent) at causing interruptions. During instances when calls were interrupted, the subjects always completed the syllable already in production, suggesting that the minimal unit of production in the tamarin long call is the syllable. The low rate of interruptions in comparison with birds, however, suggests that although the syllable is the smallest unit of production, the organization of the syllables is dramatically different than in bird song. While zebra finches appear to have good control over the acoustic units within their song, tamarins exhibit a more limited degree of vocal control over syllables within the long call.

Unit of Perception in Cotton-Top Tamarins

In a variety of vertebrate species, individuals often respond to territorial and/or contact calls by producing an identical or similar-sounding call. Such "antiphonal calling" can be used as a robust behavioral assay for investigating which acoustic features are important in a given vocal signal. For example, in an elegant study of the two-note mating call of the Coqui frog (*Eleutherodacylus coqui*), Narins and Capranica (1978) used the antiphonal calling of male frogs to de-

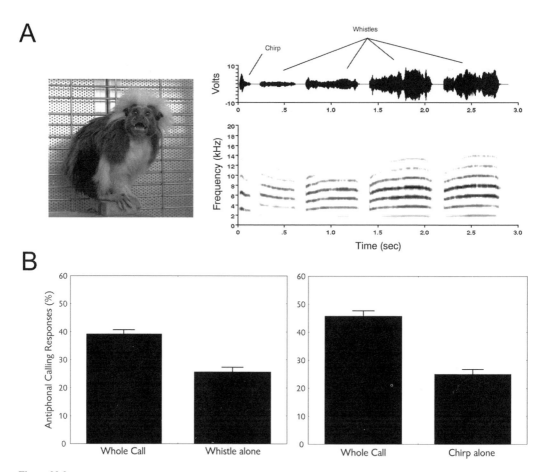

Figure 33.2
(A) Image of a cotton-top tamarin (left). Cotton-top tamarin combination long call: Time-amplitude waveforms (top right) and spectrogram (bottom right) of a representative combination long call recorded in captivity. Chirps and whistles are labeled. (B) Antiphonal calling responses to playbacks of whole long calls versus single syllables (whistles and chirps). Error bars show 1 standard error of mean.

termine which of the two notes in the call elicited the antiphonal response. Using field playbacks, they found that the first note elicits antiphonal calling responses from conspecific males, but the second note does not.

Recently, we used a similar antiphonal calling assay to examine the unit of perception for the cotton-top tamarin long call (Ghazanfar et al. 2001). We measured the antiphonal long calling responses of isolated tamarin subjects to playbacks of (1) whole long calls, (2) isolated whistles, (3) isolated chirps, (4) white noise bursts, and (5) artificial sequences of chirps. Our goal was to determine the acoustic cues necessary to elicit antiphonal long calls from isolated individuals.

In some situations, cotton-top tamarins produce chirps and whistles independently, but combination long calls consist of a concatenation of both syllable types occurring sequentially in a single vocalization (Cleveland and Snowdon 1981). We hypothesized that the combination of both syllable types in the form of a combination long call would be a more effective stimulus for eliciting antiphonal long calls from our subjects than either unit alone (see figure 33.2A). This was indeed the case. Whole calls were much more effective in eliciting antiphonal long calls than either single whistles or single chirps (figure 33.2B). It is important to note that the level of antiphonal calling to single whistles or chirps was not significantly different from the level of calling back to white noise, suggesting that those response levels represent a baseline. In terms of overall long call production rate, whole call playbacks were also able to elicit more calls than single whistles or chirps. Response levels to whole calls versus sequences of chirps were equivalent, providing clues to the acoustic cues necessary in eliciting antiphonal calls by isolated tamarins. Overall, our data suggest that the most meaningful unit from the perspective of socially isolated receivers is the whole call. This represents an important first step in deciphering the perceptually salient features of the cotton-top tamarins' combination long call.

Future Directions

One of the central questions in the study of animal vocal communication concerns how the structure of acoustic signals relates to the behavior of receivers. Like learning any foreign communication system, one of the first problems ethologists confront is the difficulty of parsing acoustic signals into the relevant units of analysis—that is, units that are meaningful in terms of eliciting specific behaviors from the intended receivers (Chomsky and Halle 1968; Green and Marler 1979). For primate vocal behaviors, such investigations are only in their infancy, yet the implications for understanding the cognitive capacities and brain design of primates are tremendous (Ghazanfar and Hauser 1999).

Our research on cotton-top tamarin long calls suggests that the smallest unit of production is the syllable, but that the smallest unit of perception is the whole call. While our understanding of the units of perception and production is relatively limited at this point, new data will provide more insights into the meaningful units in tamarin vocal communication and the neural substrates for vocal behavior. At the perceptual level, for example, there are many behavioral experiments in both humans and other taxa demonstrating that a common strategy for signal identification and localization is sensitivity to two or more spectral or temporal components. Detection of such sound combinations often leads to individuals producing species-specific behaviors. In our case, tamarins antiphonally call more readily to combination long calls containing both the chirp and multiple whistles than to either syllable type alone.

One candidate neuronal mechanism that may underlie this behavioral selectivity is the nonlinear response of neurons to the combined elements of a signal compared with responses to the individual elements alone. Neurons that are combination sensitive to multisyllabic species-specific vocalizations are described extensively in the auditory systems of bats, frogs, and song-birds (Fuzessery and Feng 1983; Margoliash and Fortune 1992; O'Neill and Suga 1979). We predict that, like other organisms, primates with long, multisyllabic vocalizations will have neurons that are combination sensitive, particularly when the meaningful unit of perception is the whole call, not the individual syllables. In the present case, we predict that the central auditory system of tamarins should contain many neurons that are combination sensitive to both temporal and spectral components of the chirp-whistle sequence.

Acknowledgments

We would like to thank Marc Hauser for his support, guidance, and unfaltering enthusiasm in all aspects of our investigations. We would also like to thank R. Delgado, M. Hauser, V. Janik, D. Katz, and L. Santos for their excellent comments on earlier versions of this manuscript. This work was funded by Harvard University and by a grant from the National Institutes of Health (grant DC00377-02) to A.A.Ghazanfar and National Science Foundation grants (SBR-9602858 and 9357976) to M. Hauser.

References

Chapman, C. A. and Weary, D. M. (1990). Variability in spider monkey's vocalizations may provide basis for individual recognition. *American Journal of Primatology* 22: 279–284.

Chomsky, N. and Halle, M. (1968). *The Sound Patterns of English.* Cambridge, Mass.: MIT Press.

Clark Arcadi, A. (1996). Phrase structure of wild chimpanzee pant hoots: Patterns of production and interpopulation variability. *American Journal of Primatology* 29: 159–178.

Cleveland, J. and Snowdon, C. T. (1981). The complex vocal repertoire of the adult cotton-top tamarin, *Saguinus oedipus oedipus. Zeitschrift für Tierpsychologie* 58: 231–270.

Cynx, J. (1990). Experimental determination of a unit of song production in zebra finch (*Taeniopygia guttata*). *Journal of Comparative Psychology* 104: 3–10.

Fitch, W. T. (2000). The evolution of speech. *Trends in Cognitive Science* 4: 258–267.

Fuzessery, Z. M. and Feng, A. S. (1983). Mating call selectivity in the thalamus and midbrain of the leopard frog (*Rana p. pipiens*): Single and multiunit analyses. *Journal of Comparative Physiology A* 150: 333–344.

Ghazanfar, A. A. and Hauser, M. D. (1999). The neuroethology of primate vocal communication: Substrates for the evolution of speech. *Trends in Cognitive Sciences* 3: 377–384.

Ghazanfar, A. A., Flombaum, J. I., Miller, C. T., and Hauser, M. D. (2001). The units of perception in the antiphonal calling behavior of cotton-top tamarin (*Saguinus oedipus*): Playback experiments with long calls. *Journal of Comparative Physiology A* 187: 27–35.

Goustard, M. (1976). The vocalizations of hylobates. In *Gibbon and Siamang*, D. M. Rumbaugh, ed., pp. 135–166. Basel, Switzerland: S. Karger.

Green, S. and Marler, P. (1979). The analysis of animal communication. In *Social Behavior and Communication, Handbook of Behavioral Neurobiology.* Vol. 3, P. Marler and J. Vandenbergh, eds., pp. 73–158. New York: Plenum.

Hauser, M. D. (1996). *The Evolution of Communication.* Cambridge, Mass.: MIT press.

Hauser, M. D. and Andersson, K. (1994). Left hemisphere dominance for processing vocalizations in adult, but not infant rhesus monkeys. *Proceedings of the National Academy of Sciences, U.S.A.* 91: 3946–3948.

Hauser, M. D., Agnetta, B., and Perez, C. (1998). Orienting asymmetries in rhesus monkeys: Effect of time-domain changes on acoustic perception. *Animal Behaviour* 56: 41–47.

Lieberman, P. (1984). *The Biology and Evolution of Language.* Cambridge, Mass.: Harvard University Press.

Margoliash, D. and Fortune, E. S. (1992). Temporal and harmonic combination-sensitive neurons in the zebra finch's HVc. *Journal of Neuroscience* 12: 4309–4326.

Marler, P. (1968). Aggregation and dispersal: Two functions in primate communication. In *Primates:*

Studies in Adaptation and Variability, P. C. Jay, ed., pp. 420–438. New York: Holt, Rinehart, and Winston.

Marshall, A. J., Wrangham, R. W., and Arcadi, A. C. (1999). Does learning affect the structure of vocalizations in chimpanzees? *Animal Behaviour* 58: 825–830.

Miller, C. T., and Hauser, M. D. (in prep). What is the unit of production of cotton-top tamarin (*Saguinus oedipus*) long calls?

Mitani, J. C. (1985). Responses of gibbons (*Hylobates muelleri*) to self, neighbor, and stranger duets. *International Journal of Primatology* 6: 193–200.

Mitani, J. C. and Marler, P. (1989). A phonological analysis of male gibbon singing behavior. *Behaviour* 109: 20–45.

Mitani, J. and Nishida, T. (1993). Contexts and social correlates of long distance calling by male chimpanzees. *Animal Behaviour* 45: 735–746.

Narins, P. M. and Capranica, R. R. (1978). Communicative significance of the two-note call of the treefrog *Eleutherodactylus coqui. Journal of Comparative Physiology, A.* 127: 1–9.

Newman, J. D., Lieblich, A., Talmage-Riggs, G., and Symmes, D. (1978). Syllable classification and sequencing in twitter calls of squirrel monkeys (*Saimiri sciureus*). *Zeitschrift für Tierpsychologie* 47: 77–88.

O'Neill, W. E. and Suga, N. (1979). Target range-sensitive neurons in the auditory cortex of the mustache bat. *Science* 251: 565–568.

Podos, J., Peters, S., Rudnicky, T., Marler, P., and Nowicki, S. (1992). The organization of song repertoires in song sparrows: Themes and variations. *Ethology* 90: 89–106.

Quine, W. V. (1973). On the reasons for the indeterminacy of translation. *Journal of Philosophy* 12: 178–183.

Robinson, J. G. (1979). An analysis of the organization of vocal communication in the titi monkey *Callicebus moloch. Zeitschrift für Tierpsychologie* 49: 381–405.

Robinson, J. G. (1984). Syntactic structure in the vocalizations of wedge-capped capuchin monkeys, *Cebus olivaceus. Behaviour* 90: 46–79.

Ryan, M. J. and Rand, A. S. (1990). The sensory basis of sexual selection for complex calls in the Tungara frog, *Physalaemus pustulosus* (sexual selection for sensory exploitation). *Evolution* 44: 305–314.

Searcy, W. A., Nowicki, S., and Peters, S. (1999). Song types as fundamental units in vocal repertoires. *Animal Behaviour* 58: 37–44.

Teixidor, P. and Byrne, R. W. (1999). The "whinny" of spider monkeys: Individual recognition before situational meaning. *Behaviour* 136: 279–308.

Waser, P. M. (1977). Individual recognition, intragroup cohesion, and intergroup spacing: Evidence from sound playback to forest monkeys. *Behaviour* 60: 28–74.

Waser, P. M. (1982). The evolution of male loud calls among mangabeys and baboons. In *Primate Communication*, C. T. Snowdon, C. H. Brown, and M. R. Petersen, eds., pp. 117–143. New York: Cambridge University Press.

Weiss, D., Garibaldi, B., and Hauser, M. D. (in press). Individual recognition of cotton-top tamarin (*Saguinus oedipus oedipus*) long calls. *Journal of Comparative Psychology.*

34 Exploring the Cognitive World of the Bottlenosed Dolphin

Louis M. Herman

An exceptionally large brain, a high degree of sociability, and easy trainability make the bottlenosed dolphin (*Tursiops truncatus*) an ideal species for studying intellectual processes and potential. Accordingly, our long-term research program into dolphin cognition, now spanning some 30 years, with as many as 22 years of study of individual animals, has been directed toward the description and analysis of these processes and potential. The philosophy guiding this approach is that the intellectual potential of a long-lived, presumptively intelligent species (such as the dolphin) is best revealed through a long-term program of intensive, special education in a culture that values education. Since these conditions surely favor the emergence of the full flower of human intellect, can comparable conditions also reveal the intellectual potential of other targeted species? To this end, we have worked intensively with different individual animals, using a broad-brush, multilevel approach that includes studies of sensory processes, cognitive characteristics, and communication. These different areas of study have increased our understanding of the perceptual, cognitive, and social worlds of the dolphin.

The brain of the adult bottlenosed dolphin is about 25 percent heavier than the average adult human brain (Ridgway 1990). Inasmuch as larger mammals tend to have larger brains, a more meaningful metric is to compare actual brain size with that expected for the species' body size (i.e., relative brain size) (Jerison 1973). Measures of relative brain size place the bottlenosed dolphin, and two or three other closely related delphinid species, second only to humans and well above the great apes (Marino 1998; Ridgway and Tarpley 1996). The dolphin cortex has a degree of fissurization and a surface area exceeding that of the human brain, although its depth (thickness) (ca. 1.3–1.8 mm) is shallower than that of the human brain (ca. 3.0 mm) (Ridgway 1990). In addition, the size of the cerebellum relative to the total brain is significantly larger in the dolphin than in the human (Marino et al. 2000). Recent work has demonstrated that the cerebellum is involved in cognitive processing in addition to its role in motor control (Leiner et al. 1995; Fiez 1996).

Social living and social pressures may be major selection forces driving the evolution of intellect (Humphrey 1976; Herman 1980); among cetaceans, as well as among other highly social mammalian species, social living can lead to the development of new behaviors and traditions through cultural transmission mechanisms, such as imitation and teaching (Rendell and Whitehead, 2001). Bottlenosed dolphins, as well as many other species of toothed whales (odontocetes), live in complexly organized social units (e.g., Conner et al. 1992). To function effectively within these units, the young dolphin must undergo extensive learning about the conventions and rules of the society, about cooperative and collaborative activities, and about the identities and even personalities of group members and associates (Herman 1991). The protracted period of development and dependence of young dolphins on their mothers and other group members allows the time and opportunity for extensive social learning to take place.

Dolphins maintained in oceanaria or in research facilities can readily transfer their social awareness and responsivity to the human caregivers. If the caregivers acknowledge and give proper attention to the social nature of the dolphin, a strong interest in humans can develop that allows for a close and cooperative long-term working relationship between the two species. Under these conditions, it becomes possible to work with a dolphin for protracted periods within behavioral research or training paradigms (e.g., Defran and Pryor 1980).

At the end of the 1960s, when I began to work with bottlenosed dolphins, there was little substantive material on the cognitive abilities of this species, such as might be obtained from experimental studies using the rules for scientific evidence (a few exceptions were studies by Bastian 1967; Kellogg and Rice 1966; Pryor et al. 1969). Instead, there was at times relatively unrestrained speculation about such things as dolphin language, oral traditions, and philosophies, inferred from the large brain of this species (see e.g., Lilly 1961, 1967).

My goal then was to establish factual bases for inferences about intelligence and other cognitive traits of dolphins through experimental laboratory behavioral research. To do this, I would have to study not only intelligence, but also fundamental issues about dolphin sensory abilities and learning. In the late 1960s, research on dolphin sensory processes was almost entirely focused on the dolphin's auditory and vocal abilities, especially echolocation (see e.g., Busnel 1967; Au 1993). Because of findings of exceptional hearing capabilities, a highly developed echolocation system, and the large extent of the auditory cortex (Ridgway 1990), dolphins were generally regarded as strict auditory specialists, making their way through their world primarily through sound—either passively listening or actively echolocating. Vision was typically believed to be relatively poor, or to be a secondary sense of minor functional value compared with echolocation (for a brief review of this early view, see Kellogg and Rice 1966).

I began my work by mapping out some basic learning and memory abilities of dolphins, studying their sense of vision, and filling in some unstudied areas on hearing. We (my students, research colleagues, and I) used the preliminary phases of these studies to develop a better understanding of stimuli and procedures that would best reveal competencies. Later studies built on the procedural knowledge that we and the dolphins had acquired, as well as the accumulating declarative and conceptual knowledge gained by the dolphins. It was soon apparent that dolphins were highly trainable (Defran and Pryor 1980), but that teaching models were in many cases more efficient and more effective than training models based on traditional instrumental conditioning techniques.

Our teaching models employ techniques that may commonly be used with young children, such as showing and demonstrating. In dolphins this entails such things as touching the dolphin with the correct object of a pair, pointing toward the correct location for a response, or actually leading the dolphin toward the correct object or location. When introducing new problems, we take care that they are at an initial level likely to lead to success, and then increase the difficulty and complexity at a rate or in contexts that tend to promote overall success. For example, a new discrimination or transfer test will generally be presented in a probe fashion, embedded among earlier established discriminations or previously successful transfer tests. Hence, even if the probes are responded to incorrectly, successful responses to the familiar materials are highly likely and the overall success rate will be high. Under such conditions, the dolphins develop a positive attitude toward the introduction of new materials, often offering spontaneous responses, and seemingly anticipating success.

Using these procedures, we have carried out studies of sensory abilities, cross-modal matching, memory, conceptual processes, vocal and motor mimicry, language understanding, self-awareness, and the mental representations that may underlie performance in these various cognitive tasks. Some of our more important findings in these areas are reviewed here.

Visual Processes

The degree to which a species can process complex information through different sensory systems, and the extent to which these systems are integrated and can serve as interfaces to higher cognitive centers, provide some measure of the cognitive flexibility of the species. The eye of the dolphin has clearly undergone many progressive adaptations to the underwater world (Dawson 1980). Our study of the visual acuity of bottlenosed dolphins revealed that good resolution was maintained both in air and underwater (Herman et al. 1975) and was roughly equivalent to that of pinniped species and many nonprimate terrestrial mammals. Color vision is absent or weak, as it is in most nonprimate mammals (Madsen and Herman 1980). Furthermore, we found that the visual system can clearly serve as a valuable interface to higher cognitive centers that deal with concepts, abstractions, and representations (Herman 1990).

Sensory Integration

Our findings on cross-modal matching have shown that the dolphin's visual and echolocation senses are highly integrated and that each directly yields object-based percepts that easily translate across these senses (Pack and Herman 1995; Herman et al. 1998). These and other related findings reveal that the perceptual world of the dolphin is monitored and richly organized through both the auditory and the visual domains. Dolphins should no longer be regarded as strict auditory specialists, but as multimodal animals, interfacing with their world through vision as well as hearing and echolocation.

Memory and Concept Learning

Memory is the bedrock on which learning and other higher cognitive processes rest. We have carried out extensive studies of short-term memory in dolphins as well as their capability for rule learning and for the development of abstract concepts (Herman et al. 1993b, 1994). In keeping with our findings on good visual acuity, our studies have shown that immediate memory for things seen (Herman et al. 1989) and for sounds heard (Herman and Gordon 1974; Thompson and Herman 1977) are each of high fidelity and endurance. In the Thompson and Herman (1977)

study, the dolphin was able to reliably indicate whether a probe sound occurring at the end of lists of sounds ranging from two to as many as six different sounds was or was not a member of that list. A strong recency effect was found, with sounds early in the list remembered less well than sounds late in the list. In studies of concept learning, Herman and Arbeit (1973) demonstrated that a dolphin could learn a generalized win-stay, lose-shift rule in a two-alternative forced-choice test using pairings of novel sounds. The development and nearly errorless application of an identity-matching concept was also shown in the previously cited visual and auditory memory studies. A dolphin was also able to learn and apply a same-different rule to new pairings or triplets of objects (Herman et al. 1994; Mercado et al. 2000).

Imitation

Imitation is a complex skill and a demanding cognitive trait. However, the definition of imitation, and which animals can imitate, are currently matters of intense debate among researchers (see. e.g., Whiten and Ham 1992; Heyes 1993). If imitation is within the capabilities of a species, it can be an efficient mechanism for social learning.

Dolphins are capable of extensive vocal and behavioral mimicry, a seemingly unique ability among nonhuman animals. Dolphins have demonstrated motor imitation of other dolphins or of humans (Richards et al. 1984; Herman, in press). In these studies the dolphin understood imitation as a concept that could be applied to any newly modeled sound or behavior; furthermore, imitation occurred only if the experimenter requested it through an abstract symbolic sound or gesture.

Dolphins, as imitative generalists (Herman, in press), must be able to mentally represent the behaviors of others or of themselves in order to copy or reproduce those behaviors. When a dolphin imitates a human's motor acts, it must in some cases form analogies between its body im-

Figure 34.1
Motor mimicry of a human (graduate student Amy Miller) by the dolphin Elele. The behavior is not trained. Dolphins are excellent motor mimics of each other in their natural world and, remarkably, easily transfer this skill to imitate human motor acts. In doing so, the dolphin must relate its body image to the human's perceived body plan, creating analogies in some cases. In the image, the dolphin uses its raised tail as an analogy to the human's raised leg, as well as imitating the back bend of the human.

age and the human's body plan. For example, if the human raises a leg in the air, the dolphin will raise its tail (figure 34.1), and if the human waves his or her arms, the dolphin will wave its pectoral fins. The unquestionably extensive imitative abilities of the dolphin may derive from the naturally occurring highly synchronous or closely coordinated natural behaviors often seen among pairs or groups of dolphins. Synchrony may function to assist in tasks such as foraging and prey capture, but may also be an expression of social affiliation.

Language Learning

Semantics and syntax are considered the core attributes of any human natural language (Pavio and Begg 1981). Our studies of language comprehension have revealed capabilities in the dolphin for processing both semantic and syntactic information (Herman et al. 1984; Herman 1986; Herman and Uyeyama 1999). The primary syntactic device used in our language studies has been word order. The dolphin is capable of understanding that changes in word order change meaning. It can respond appropriately, for instance, to such semantic contrasts as *surfboard person fetch* (take the person to the surfboard) and *person surfboard fetch* (take the surfboard to the person).

In these language studies, the dolphin demonstrated an implicit representation and understanding of the grammatical structure of the language. For example, the language-trained dolphin Akeakamai was able to spontaneously understand logical extensions of a syntactic rule (Herman et al. 1984) and was able to extract a semantically and syntactically correct sequence from a longer anomalous sequence of language gestures given by a human (Herman et al. 1993a). To perform this extraction, the dolphin in some cases had to conjoin nonadjacent terms in the sequence. For example, the anomalous string glossed as *water speaker Frisbee fetch* violates a syntactic rule in that there is no rule that accommodates three object names in a row. However, embedded in this sequence are two semantically and syntactically correct three-item sequences, *water Frisbee fetch* (bring the Frisbee to the stream of water), and *speaker Frisbee fetch* (bring the Frisbee to the underwater speaker). In sequences of this type, the dolphin almost always extracted one or the other of the correct three-item sequences and operated on that implicit instruction. In theory, in responding to these anomalous sequences, the dolphin utilized its implicitly learned mental representation or schema of the grammar of the language to include not only word-order rules but also the semantic rules determining which items are transportable and which are not (neither the stream of water nor the underwater speaker affixed to the tank wall can be transported). No explicit training was given for these rules.

Representation

One of the issues in animal language studies is whether the symbols used to refer to objects or actions function as representations of those things; this is the problem of linguistic reference. For example, when a tutored ape uses a symbol for candy, does it understand that the symbol refers to or represents candy, or does it merely treat the symbol as a means to obtain candy (by using it on seeing candy present)? Much of the early ape language work failed to show that the symbols used were understood referentially (Savage-Rumbaugh 1986). The clearest indication that the dolphin Akeakamai understood the gestural symbols of her language referentially was her ability to report "yes" or "no" (by pressing one or another of two paddles) in response to gestural questions asking whether specific gesturally named objects were or were not present in her tank world (Herman and Forestell 1985). The ability to understand symbolic references to absent objects is one of the clearest indicants that the symbols represent the referent. In addition, the dolphin understood that if an experimenter pointed to a distal object, it was a reference to that object (Herman et al. 1999).

Television scenes are representations of the real world and, as humans, we respond to them as we might to the real world but understand that they are not the real world. A cat might respond to a television image of a moving bird in the same way it would respond in the real world, failing to discriminate between the representation and the real world. In other cases, the cat or other animals might simply ignore the television scene, seemingly failing to recognize that anything meaningful or relevant is occurring. The latter behavior has been reported, for example, for language-trained common chimpanzees, who only learned to respond appropriately to television scenes after long periods of watching their human companions responding (Savage-Rumbaugh 1986).

In contrast, all four of our dolphins, on the very first occasion that they were exposed to television, responded spontaneously and appropriately to televised images of people gesturing to

Figure 34.2
The dolphin Akeakamai watching the image of a person projected live on a television screen placed behind an underwater window. Akeakamai watches a sequence of gestures by the trainer expressed to her in a familiar gestural language. The sequence conveys an instruction to the dolphin, who then carries out the instruction as accurately and reliably as she does when instructions are conveyed to her in the real world. No training was involved in responding to TV images. Akeakamai responded accurately to TV instructions the first time she experienced them, which was also the first time she experienced television of any sort.

them (Herman et al. 1990). They responded in the same way as they did to live people, faithfully carrying out the gestural instructions conveyed by the image (figure 34.2). The dolphins understood, however, that the television scene was not the real world. For example, if the trainer tossed a ball in the air and then gestured to the dolphin to imitate the action, the dolphin did not attempt to retrieve the ball in the television scene, but used one in its real world.

Self-Awareness

Self-awareness is a multidimensioned concept that has usually been studied through the mirror self-recognition mark test (Gallup 1970). We chose to ask a different question about self-awareness: was a dolphin aware of its own recent behaviors (Mercado et al. 1998)? We taught the dolphin an abstract gesture, which we called "repeat." If this gesture occurred, it signaled the dolphin to do again what it just did; in essence to imitate its own behavior. A behavior was to be repeated only if that particular gesture was given. As an alternative to the repeat signal, some other gesture might be given that called for a behavior different from the one just executed. The demonstrated ability of the dolphin to reliably repeat or not repeat its previous behavior indicated that it maintained a mental representation of the behavior last performed and updated that as each new behavior was performed.

Conclusions

It seems clear that many of the studies we undertook would not have been possible, and many of the dolphin capabilities described would have gone unrevealed, without the implementation of the initial guiding educational philosophy. Immersion in a long-term program of intensive special education results in the accumulation of knowledge, concepts, rules, strategies, and a general level of intellectual sophistication that allows for the understanding and solution of a broad range of increasingly complex problems or tasks. Many of the later studies we carried out were not anticipated earlier (for example, interpretation of television scenes) because the groundwork was not yet in place, and the next step was not evident. The educational approach we used with the dolphin, a species with a life span stretching into the 40s or 50s, is not possible of course with short-lived species, but is applicable to such interesting species for cognitive investigation as elephants and the great apes. The work we have carried out with dolphins has expanded our understanding of the perceptual and cognitive world of this species, and certainly has demonstrated that the dolphin's reputation for intelligence is well earned.

Intelligence is of course a term with many definitions and interpretations, but I prefer to view it as flexibility of behavior (Herman and Pack 1994). By flexibility, I mean the ability to organize and carry out behaviors that are appropriate to new situations, new contexts, or new events, and that are not necessarily genetically determined or part of the species' naturally occurring repertoire of behaviors. Flexibility is demonstrated then by the animal's ability to go beyond the boundaries of its naturally occurring behaviors or the context of its natural world.

Perhaps the most daunting task facing potential investigators of dolphin cognition is to find a place where such studies can take place. Facilities such as our Kewalo Basin Marine Mammal Laboratory in Honolulu are rare, and opportunities for research may be limited mainly to oceanaria or marine parks. At such places, however, the investigator may be constrained by the competing uses of the animals for demonstrations or display. This situation may be improving, though, as the benefits of research, its educational value, and even its display value come to be appreciated by the managers of these oceanaria and parks. Given the availability of facilities, and the access to animals, there is almost an unlimited opportunity for new discoveries about dolphin cognition. Topics such as theory of mind, social awareness, imitation, productive language, interanimal communication, and much more, are relatively unstudied and await only the investigator and the opportunity.

Acknowledgments

Preparation of this paper was supported by grants from the Earthwatch Institute and The Dolphin Institute and by Grant IBN 0090744— from the National Science Foundation. I thank Adam Pack for helpful comments on the manuscript and the many graduate students, postdoctoral associates, and assistants who participated in and contributed to the various research projects described.

References

Au, W. W. L. (1993). *The Sonar of Dolphins.* New York: Springer-Verlag.

Bastian, J. (1967). The transmission of arbitrary environmental information between bottlenose dolphins. In *Animal Sonar Systems: Biology and Bionics*, R.-G. Busnel, ed., pp. 807–873. Jouy-en Josas, France: Laboratoire de Physiologie Acoustique.

Busnel, R.-G. (ed.) (1967). *Animal Sonar Systems: Biology and Bionics.* Jouy-en Josas, France: Laboratoire de Physiologie Acoustique.

Conner, R. A., Smolker, R. A., and Richards, A. F. (1992). Dolphin alliances and coalitions. In *Coalitions and Alliances in Humans and Other Animals*, A. H. Harcourt and F. B. M. de Waal, eds., pp. 415–443. Oxford: Oxford University Press.

Dawson, W. W. (1980). The cetacean eye. In *Cetacean Behavior: Mechanisms and Functions*, L. M. Herman, ed., pp. 53–100. New York: Wiley-Interscience.

Defran, R. H. and Pryor, K. (1980). The behavior and training of dolphins in captivity. In *Cetacean Behavior: Mechanisms and Functions*, L. M. Herman, ed., pp. 319–362. New York: Wiley-Interscience.

Fiez, J. A. (1996). Cerebellar contributions to cognition. *Neuron* 16: 13–15.

Gallup, G. G., Jr. (1970). Chimpanzees: Self-recognition. *Science* 167: 86–87.

Herman, L. M. (1980) Cognitive characteristics of dolphins. In *Cetacean Behavior: Mechanisms and Functions*, L. M. Herman, ed., pp. 363–429. New York: Wiley-Interscience.

Herman, L. M. (1986). Cognition and language competencies of bottlenosed dolphins. In *Dolphin Cognition and Behavior: A Comparative Approach*, R. J. Schusterman, J. Thomas, and F. G. Wood, eds., pp. 221–251. Hillsdale, N.J.: Lawrence Erlbaum Associates.

Herman, L. M. (1990). Cognitive performance of dolphins in visually guided tasks. In *Sensory Abilities of Cetaceans: Laboratory and Field Evidence*, J. A. Thomas and R. A. Kastelein, eds., pp. 455–462. New York: Plenum.

Herman, L. M. (1991). What the dolphin knows, or might know, in its natural world. In *Dolphin Societies: Discoveries and Puzzles*, K. Pryor and K. S. Norris, eds., pp. 349–364. Los Angeles: University of California Press.

Herman, L. M. (in press). Vocal, social, and self-imitation by bottlenosed dolphins. In *Imitation in Animals and Artifacts*, C. Nehaniv and K. Dautenhahn, eds., Cambridge, Mass.: MIT Press.

Herman, L. M. and Arbeit, W. R. (1973). Stimulus control and auditory discrimination learning sets in the bottlenosed dolphin. *Journal of the Experimental Analysis of Behavior* 19: 379–394.

Herman, L. M. and Forestell, P. H. (1985). Reporting presence or absence of named objects by a language-trained dolphin. *Neuroscience and Biobehavioral Reviews* 9: 667–691.

Herman, L. M. and Gordon, J. A. (1974). Auditory delayed matching in the bottlenosed dolphin. *Journal of the Experimental Analysis of Behavior* 21: 19–26.

Herman, L. M. and Pack, A. A. (1994). Animal intelligence: Historical perspectives and contemporary approaches. In *Encyclopedia of Human Intelligence*, R. Sternberg, ed., pp. 86–96. New York: Macmillan.

Herman, L. M. and Uyeyama, R. K. (1999). The dolphin's grammatical competency: Comments on Kako (1998). *Animal Learning & Behavior* 27: 18–23.

Herman, L. M., Peacock, M. F., Yunker, M. P., and Madsen, C. (1975). Bottlenosed dolphin: Double-slit pupil yields equivalent aerial and underwater diurnal acuity. *Science* 139: 650–652.

Herman, L. M., Richards, D. G., and Wolz, J. P. (1984). Comprehension of sentences by bottlenosed dolphins. *Cognition* 16: 129–219.

Herman, L. M., Hovancik, J. R., Gory, J. D., and Bradshaw, G. L. (1989). Generalization of visual matching by a bottlenosed dolphin (*Tursiops truncatus*): Evidence for invariance of cognitive performance with visual or auditory materials. *Journal of Experimental Psychology: Animal Behavior Processes* 15: 124–136.

Herman, L. M., Morrel-Samuels, P., and Pack, A. A. (1990). Bottlenosed dolphin and human recognition of veridical and degraded video displays of an artificial gestural language. *Journal of Experimental Psychology: General* 119: 215–230.

Herman, L. M., Kuczaj, S. III, and Holder, M. D. (1993a). Responses to anomalous gestural sequences by a language-trained dolphin: Evidence for processing of semantic relations and syntactic information. *Journal of Experimental Psychology: General* 122: 184–194.

Herman, L. M., Pack A. A., and Morrel-Samuels, P. (1993b). Representational and conceptual skills of dolphins. In *Language and Communication: Comparative Perspectives*, H. R. Roitblat, L. M. Herman, and P. Nachtigall, eds., pp. 273–298. Hillside, N.J.: Lawrence Erlbaum Associates.

Herman, L. M., Pack, A. A., and Wood, A. M. (1994). Bottlenosed dolphins can generalize rules and develop abstract concepts. *Marine Mammal Science* 10: 70–80.

Herman, L. M., Pack, A. A., and Hoffmann-Kuhnt, M. (1998). Seeing through sound: Dolphins perceive the spatial structure of objects through echolocation. *Journal of Comparative Psychology* 112: 292–305.

Herman, L. M., Abichandani, S. L., Elhajj, A. N., Herman, E. Y. K., Sanchez, J. L., and Pack, A. A. (1999). Dolphins (*Tursiops truncatus*) comprehend the referential character of the human pointing gesture. *Journal of Comparative Psychology* 113: 1–18.

Heyes, C. M. (1993). Imitation, culture and cognition. *Animal Behavior* 46: 999–1010.

Humphrey, N. K. (1976). The social function of intellect. In *Growing Points in Ethology*, P. P. G. Bateson and R. A. Hinde, eds., pp. 301–317. Cambridge: Cambridge University Press.

Jerison, H. J. (1973). *Evolution of the Brain and Intelligence*. New York: Academic Press.

Kellogg, W. N. and Rice, C. E. (1966). Visual discrimination and problem solving in a bottlenose dolphin. In *Whales, Dolphins, and Porpoises*, K. S. Norris, ed., pp. 731–754. Berkeley: University of California Press.

Leiner, H. C., Leiner, A. L., and Dow, R. S. (1995). The underestimated cerebellum. *Human Brain Mapping* 2: 244–254.

Lilly, J. C. (1961). *Man and Dolphin*. New York: Doubleday.

Lilly, J. C. (1967). *The mind of the Dolphin: A Nonhuman Intelligence*. New York: Doubleday.

Madsen, C. J. and Herman, L. M. (1980). Social and ecological correlates of vision and visual appearance. In *Cetacean Behavior: Mechanisms and Functions*, L. M. Herman, ed., pp. 101–147. New York: Wiley-Interscience.

Marino, L. (1998). A comparison of encephalization levels between odontocete cetaceans and anthropoid primates. *Brain, Behavior and Evolution* 51: 230–238.

Marino, L., Rilling, J. K., Lin, S. K., and Ridgway, S. H. (2000). Relative volume of the cerebellum in dolphins and comparison with anthropoid apes. *Brain, Behavior and Evolution* 56: 204–211.

Mercado, E. III, Murray, S. O., Uyeyama, R. K., Pack, A. A., and Herman, L. M. (1998). Memory for recent actions in the bottlenosed dolphin (*Tursiops truncatus*): Repetition of arbitrary behaviors using an abstract rule. *Animal Learning & Behavior* 26: 210–218.

Mercado, E. III, Killebrew, D. A., Pack, A. A., Macha, I. V. B., and Herman, L. M. (2000). Generalization of same-different classification abilities in bottlenosed dolphins. *Behavioural Processes* 50: 79–94.

Pack, A. A. and Herman, L. M. (1995). Sensory integration in the bottlenosed dolphin: Immediate recognition of complex shapes across the senses of echolocation and vision. *Journal of the Acoustical Society of America* 98: 722–733.

Pavio, A. and Begg, I. (1981). *Psychology of Language*. Englewood Cliffs, N.J.: Prentice-Hall.

Pryor, K., Haag, R., and O'Reilly, J. (1969). The creative porpoise: Training for novel behavior. *Journal of the Experimental Analysis of Behavior* 12: 653–661.

Rendell, L. and Whitehead, H. (2001). Culture in whales and dolphins. *Behavioral and Brain Sciences* 24: 309–382.

Richards, D. G., Wolz, J. P., and Herman, L. M. (1984). Vocal mimicry of computer-generated sounds and vocal labeling of objects by a bottlenosed dolphin, *Tursiops truncatus*. *Journal of Comparative Psychology* 98: 10–28.

Ridgway, S. H. (1990). The central nervous system of the bottlenose dolphin. In *The Bottlenose Dolphin*, S. Leatherwood and R. R. Reeves, eds., pp. 69–97. New York: Academic Press.

Ridgway, S. H. and Tarpley, R. J. (1996). Brain mass comparisons in Cetacea. In *Proceedings of the International Association of Aquatic Animal Medicine*, Vol. 2, pp. 55–57. Philadelphia: University of Pennsylvania Press.

Savage-Rumbaugh, E. S. (1986). *Ape Language: From Conditioned Response to Symbol*. New York: Columbia University Press.

Thompson, R. K. R. and Herman, L. M. (1977). Memory for lists of sounds by the bottlenosed dolphin: Convergence of memory processes with humans? *Science* 195: 501–503.

Whiten, A. and Ham, R. (1992). On the nature and evolution of imitation in the animal kingdom: Reappraisal of a century of research. *Advances in the Study of Behavior* 21: 239–283.

35 Chimpanzee Signing: Darwinian Realities and Cartesian Delusions

Roger S. Fouts, Mary Lee A. Jensvold, and Deborah H. Fouts

Darwinian Realities

Truly discontinuous, all-or-none phenomena must be rare in nature. Historically, the great discontinuities have turned out to be conceptual barriers rather than natural phenomena. They have been passed by and abandoned rather than broken through in the course of scientific progress. The sign language studies in chimpanzees have neither sought nor discovered a means of breathing humanity into the soul of a beast. They have assumed instead that there is no discontinuity between verbal behavior and the rest of human behavior or between human behavior and the rest of animal behavior—no barrier to be broken, no chasm to be bridged, only unknown territory to be explored. (R. Gardner et al. 1989, p. xvii)

Cross-Fostering

While chimpanzees (*Pan troglodytes*) have great difficulty adapting their vocalizations to human speech (Hayes and Hayes 1951; Hayes and Nissen 1971), they can freely move their hands, meaning that a gestural language is well suited to their abilities. R. A. and B. T. Gardner recognized this in their sign language studies with young chimpanzees. In 1966, the Gardners brought 10-month-old Washoe to the University of Nevada-Reno when they began their cross-fostering study. The Gardners described this approach as follows:

Cross-fostering a chimpanzee is very different from keeping one in a home as a pet. Many people keep pets in their homes. They may treat their pets very well, and they may love them dearly, but they do not treat them like children. True cross-fostering—treating the chimpanzee infant like a human child in all respects, in all living arrangements, 24 hours a day every day of the year—requires a rigorous experimental regime that has rarely been attempted. (R. A. Gardner and Gardner 1998, p. 292)

The Gardners and students in the cross-fostering project used only American Sign Language (ASL) in Washoe's presence (B. T. Gardner and Gardner 1971, 1974, 1989; R. A. Gardner and Gardner 1969).

In teaching sign language to Washoe [and other later cross-fosterlings] we imitated human parents teaching young children in a human home. We called attention to everyday events and objects that might interest the young chimpanzees, for example, THAT CHAIR, SEE PRETTY BIRD, MY HAT. We asked probing questions to check on communication, and we always tried to answer questions and to comply with requests. We expanded on fragmentary utterances using the fragments to teach and to probe. We also followed the parents of deaf children by using an especially simple and repetitive register of ASL and by making signs on the youngsters' bodies to capture their attention. (R. A. Gardner and Gardner 1998, p. 297)

In 1970 Washoe left Reno with companions Roger and Deborah Fouts for the Institute of Primate Studies (IPS) at the University of Oklahoma in Norman. The Gardners began a second cross-fostering project with four other infant chimpanzees. Moja, Pili, Tatu, and Dar were born in American laboratories and each arrived in Reno within a few days of birth. Moja arrived in November 1972 and cross-fostering continued for her until the winter of 1979 when she left for IPS. In 1980, Washoe and Moja moved with the Fouts to the Chimpanzee and Human Communication Institute (CHCI) on the campus of Central Washington University in Ellensburg, Washington. Tatu arrived in Reno in January 1976 and Dar in August 1976. Cross-fostering continued for Tatu and Dar until May 1981, when they left to join Washoe and Moja in Ellensburg. Pili arrived in Reno in November 1973 and died of leukemia in October 1975.

The size of the chimpanzees' vocabulary, their responses to *Wh-* questions (where, why, when etc.), number of utterances, proportion of phrases, variety of phrases, length of phrases, complexity of phrases, and inflection all grew

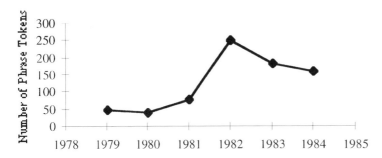

Figure 35.1
Number of phrase tokens.

throughout 5 years of cross-fostering (R. A. Gardner et al. 1992; B. T. Gardner and Gardner 1974, 1989, 1998). "Washoe, Moja, Pili, Tatu, and Dar signed to friends and strangers. They signed to each other and to themselves, to dogs and to cats, toys, tools, even to trees" (R. A. Gardner and Gardner 1989, p. 24). Signing was a robust behavior in the chimpanzees.

Cultural Transmission

At CHCI we continued to explore how the chimpanzees acquired signs and used them to communicate with humans and each other. The first of these studies began in 1978. In 1979, Washoe adopted a 10-month-old son, Loulis. To show that Loulis would learn signs from Washoe and other signing chimpanzees without human intervention, we restricted human signing when Loulis was present except for seven specific signs, WHO, WHAT, WHERE, WHICH, WANT, SIGN, and NAME. Humans used vocal English to communicate in his presence. Loulis began to sign in 7 days; at 15 months of age he combined signs; and at 73 months of age his vocabulary consisted of 51 signs (R. S. Fouts 1994; R. S. Fouts et al. 1982, 1989).

Human observers maintained written records of Loulis's signing and behavioral development. We used all of the records from his tenth month (the first month of the project) to his seventy-second month. From this record we plotted the growth of Loulis's phrases. A phrase is two or more different signs within two utterance boundaries. Utterance boundaries are defined by a pause marked by a relaxation of the hands within the signing area, or dropping the hands from the signing area altogether. The observer indicated utterance boundaries in the field records with a slash. Reiteration, where a sign is repeated for emphasis, did not meet the requirement for a phrase in that it did not contain two different signs.

Phrase tokens provide information on the frequency of all phrases that appeared in a year. YOU CHASE and CHASE YES are examples of two different phrase tokens. When Loulis signed ME ME GOOD GOOD once on March 1, 1984 and ME ME GOOD GOOD once on May 28, 1984, this was counted as two tokens. Figure 35.1 shows the total number of phrase tokens recorded for Loulis each year. Loulis's pattern is similar to that of Moja in figure 2 of B. T. Gardner and Gardner (1998).

We grouped phrase tokens into types according to the signs that they contained regardless of the order of the signs in the utterance. For example, all phrases that contained the signs THAT HURRY, HURRY THAT, and THAT THAT HURRY HURRY THAT THAT were the same phrase type containing the two signs HURRY and THAT. This provides information on the

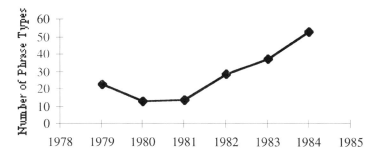

Figure 35.2
Number of phrase types: two sign phrases.

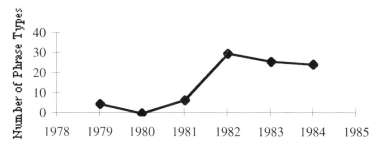

Figure 35.3
Number of phrase types: three or more sign phrases.

variety of phrases that Loulis produced. Figure 35.2 shows the total number of phrase types recorded for Loulis each year. After the third year of the project, Loulis showed a steady increase in the variety of his phrases. This pattern was similar to that of Moja, Tatu, Pili, and Dar in figure 3 of B. T. Gardner and Gardner (1998). Figure 35.3 shows the development of phrase types with three or more signs for Loulis. An example is HURRY YOU TICKLE. After the fourth year of the project, there was a sharp increase in the variety of Loulis's three or more sign phrases.

Loulis's phrase development paralleled that of the cross-fostered chimpanzees. Like human children, the development of phrases grew gradually in Loulis and the cross-fostered chimpanzees (B. T. Gardner and Gardner 1998). Loulis's

acquisition of phrases is particularly impressive since it occurred in the absence of human signing and his only models were other signing chimpanzees.

Remote Videotaping

In June 1984, the restriction on signing around Loulis ended and we turned our attention to recording Loulis's use of signs by remote videotaping, a technique that allowed the behavior of the chimpanzees to be recorded with no humans present. In the original method, three cameras were mounted in a chimpanzee enclosure, with each focused on part of the enclosure. Later a fourth camera was added. The cameras were attached to television monitors and to a videocassette recorder (VCR) in another room away from the chimpanzees. Only one camera recorded at a time and the VCR operator could control which camera recorded.

D. H. Fouts (1994) made 45 hours of remote videotape recordings to examine Loulis's interactions with Washoe, Moja, Tatu, and Dar. Loulis initiated 451 interactions, both signed and nonsigned, with the other chimpanzees. Forty percent (181) of those interactions were directed to his male peer, Dar. Loulis used 206 signs in his interactions and 114 of those were directed toward Dar. D. H. Fouts (1994) also reported 115 private signs that Loulis made when his face and body were not oriented toward another chimpanzee.

Loulis signed to the other chimpanzees and they signed to each other as well. A later study by Cianelli and Fouts (1998) found that the chimpanzees often used emphatically signed ASL signs during high-arousal interactions such as fights and active play. For example, after separating Dar and Loulis during a fight and with all the chimpanzees still screaming, Washoe signed COME HUG to Loulis. Loulis signed NO and continued to move away from Washoe. These results indicate that the chimpanzees' signing is very robust indeed and is a regular part of their interactions.

Bodamer (1987) looked for instances of private signing by the other chimpanzees in the videotapes recorded by D. H. Fouts (1994). He found 90 instances of private signing. These were signs made in the absence of interactive behaviors, such as looking toward another individual. He classified these into categories of private speech that humans use (Furrow 1984). We later recorded 56 more hours of remote videotape and found 368 instances of private signing (Bodamer et al. 1994). In both samples, one of the most common categories of signing was referential (59 percent in the 56-hour sample). In this category the chimpanzee signed about something present in the room, for example, naming the pictures in a magazine. The informative category, an utter-

ance that refers to an object or event that is not present, accounted for 12 percent in the 56-hour sample and 14 percent in the 45-hour sample. An example of an utterance in this category was when Washoe signed DEBBI to herself when Debbi was not present.

One category of private signing was imaginative and accounted for 17 instances in the 56 hours of remote videotaping. We later recorded 15 hours of remote videotape while the chimpanzees' enclosure was filled with toys. We found six instances of imaginary play. We classified these into categories of imaginary play that human children use (Matthews 1977). There were four instances of animation in which the chimpanzee treated an object as if it was alive. For example, Dar signed PEEKABOO to a stuffed bear. There were four instances of substitution in which the chimpanzee treated one object as if it were another. For example, Moja wore a shoe and signed SHOE. She then removed the shoe, put a purse on her foot, and zipped it up (Jensvold and Fouts 1993).

Williams (1995) used remote videotaping to examine the five chimpanzees' nighttime behavior. The chimpanzees were more active at night than we previously had assumed. There were even a few instances of signing in their sleep.

Conversational Context

While remote videotaping provides a way to discover what the chimpanzees do in the absence of humans, at other times we are interested in controlling variables and measuring the chimpanzees' responses within the context of their typical daily signed interactions with their human caregivers. This is the legacy of the Gardner cross-fostering project; they used rigorous methodology within the usual routine of the cross-fostering environment. In the Gardner experiments and in our own, the chimpanzees were free to leave the testing situation and to respond to their world with their full repertoire of behaviors. Typically in comparative psychology the experimenter tests the participant in an artificial environment in order to control all variables. However, this so greatly removes the participant from his or her natural environment that we often discover more about the intelligence of the experimenter than that of the participant. The following studies were all conducted during naturally occurring signing interactions between the chimpanzees and their human caregivers without compromising methodological controls.

The PCM system (B. T. Gardner et al. 1989) describes how a sign is formed, using place where the sign is made, configuration of the hand, and movement of the hand. During everyday activities such as cleaning, meals, and playtime, Davis (1995) introduced a distortion in some of her signs to measure the chimpanzees' response to the mispronunciations. The distortions always occurred on the place of the sign. Low distortions were made 1 to 4 inches from the standard form of the sign. Medium distortions were made 5 to 8 inches from the standard form of the sign. High distortions were made 9 to 12 inches from the standard form of the sign. For example, the standard form of the sign CRACKER is a fist hitting the elbow. In low distortion the fist hit the forearm; in medium distortion the fist hit the wrist; and in high distortion the fist hit the forehead. In response to low distortion messages, the chimpanzees restored the sign to its original form. When the distortion was high or medium, they typically did not respond. Like humans, the chimpanzees are tolerant of slight mispronunciations of signs. When the mispronunciation increased, the chimpanzees' responding decreased. This study used naturally occurring interactions with a human interlocutor to test the chimpanzees' perception of semantics. Other experiments tested pragmatic aspects of the chimpanzees' signed interactions with humans.

At the original CHCI facility, the chimpanzees had access to a suite of enclosures. One of the enclosures was across the hall from a human workroom. When a caregiver was in the workroom, the chimpanzees often came to the nearby enclosure to request objects or activities. They often made noises if the human was not looking at them. Bodamer and Gardner (2002) systematically studied these initiations. The interlocutor sat in the workroom with his back toward the chimpanzees' enclosure. When the chimpanzee made a noise, he turned and faced the chimpanzee immediately or after a 30-second delay. When the interlocutor was not facing the chimpanzees, they made noises, such as Bronx cheers, and rarely signed. The few times the chimpanzees signed, they used signs that made noise, such as DIRTY, in which the back of the hand hits the bottom of the jaw. Closed with force, this sign is noisy. In the delayed-response condition, the noises became louder and faster. Once the interlocutor faced the chimpanzees, they stopped making sounds and signed. Using a naturally occurring situation, this experiment showed that the chimpanzees initiate interactions and sign spontaneously.

In another test of conversational skill, the interlocutor used one of four types of probe: general questions, on-topic questions, off-topic questions, or negative statements (Jensvold and Gardner 2000). When the interlocutor asked a general question, the chimpanzees frequently expanded across turns, showing a persistence in their original topic and giving the interlocutor more information. When the interlocutor asked a relevant on-topic question, the chimpanzees responded with many incorporations and expansions. These responses are indicators of topic maintenance. When the interlocutor asked an off-topic question, the chimpanzees often failed to respond and when they did respond, they used few incorporations and expansions. When the interlocutor made a negative statement, Washoe and Dar often did not respond. The chimpanzees' responses were contingent and appropriate to the interlocutor's questions or statements and resembled patterns of conversation found in similar studies of human children.

By using rigorous methods that allow the chimpanzees to demonstrate their behaviors in a context-appropriate situation, sign language studies of chimpanzees show remarkable similarities between human and chimpanzee behaviors. These similarities support the biological reality that species differ by degree.

Future Research

We plan to continue to explore the rules governing the chimpanzees' conversations with each other and with humans. A recent new direction has been to examine the non-ASL gestures that the CHCI chimpanzees use to communicate with each other. We have already found evidence that they are using non-ASL gestures in the fashion of a dialect. At present we are expanding this research to the study of gestural dialects among free-living chimpanzee communities in Africa.

Cartesian Delusions

Nature, Mr. Allnutt, is what we were put in this world to rise above.
—Huston and Agee, *The African Queen*

The Darwinian view is very different from the Greek Platonic view and the more recent Cartesian view that holds that man is superior to all other beings, including women. Descartes held that a definite gap or difference in kind existed between man and the defective automata below him. Some scientists today continue to uphold the existence of such gaps in nature and accept the absence of evidence as evidence of absence. If a chimpanzee fails to perform like a human in a particular experiment, these scientists maintain that there are differences in kind between species and that there is a chasm between humans and the rest of nature.

Chimpanzees acquired the signs of ASL from humans and other chimpanzees. The chimpanzees used signs when conversing with each other, even when no humans were present. They used the signs to sign to themselves and in imaginary play about things that were present as well as not present. They initiated interactions with humans and appropriately adjusted their answers to variations in the interlocutor's signs and questions. Sign language studies fill some of the gaps between humans and the rest of nature that were created in the minds of philosophers and are maintained by human arrogance.

References

Bodamer, M. D. (1987). Chimpanzees Signing to Themselves. Unpublished master's thesis. Central Washington University, Ellensburg, Wash.

Bodamer, M. D. and Gardner, R. A. (2002). How cross-fostered chimpanzees initiate conversation. *Journal of Comparative Psychology* 116, in press.

Bodamer, M. D., Fouts, D. H., Fouts, R. S., and Jensvold, M. L. A. (1994). Functional analysis of chimpanzee (*Pan troglodytes*) private signing. *Human Evolution* 9: 281–296.

Cianelli, S. N. and Fouts, R. S. (1998). Chimpanzee to chimpanzee American Sign Language communication during high arousal interactions. *Human Evolution* 13: 147–159.

Davis, J. Q. (1995). The Perception of Distortions in the Signs of American Sign Language by a Group of Cross-Fostered Chimpanzees (*Pan troglodytes*). Unpublished master's thesis, Central Washington University, Ellensburg, Wash.

Fouts, D. H. (1994). The use of remote video recordings to study the use of American Sign Language by chimpanzees when no humans are present. In *The Ethological Roots of Culture*, R. A. Gardner, B. T. Gardner, B. Chiarelli, and F. X. Plooij, eds., pp. 271–284. The Netherlands: Kluwer.

Fouts, R. S. (1994). Transmission of human gestural language in a chimpanzee mother–infant relationship. In *The Ethological Roots of Culture*, R. A. Gardner, B. T. Gardner, B. Chiarelli, and F. X. Plooij, eds., pp. 257–270. The Netherlands: Kluwer.

Fouts, R. S., Hirsch, A. D., and Fouts, D. H. (1982). Cultural transmission of a human language in a chimpanzee mother–infant relationship. In *Psychobiological Perspectives: Child Nurturance*. Vol. 3, H. E. Fitzgerald, J. A. Mullins, and P. Page, eds., pp. 159–196. New York: Plenum.

Fouts, R. S., Fouts, D. H., and Van Cantfort, T. E. (1989). The infant Loulis learns signs from cross-fostered chimpanzees. In *Teaching Sign Language to Chimpanzees*, R. A. Gardner, B. T. Gardner, and T. Van Cantfort, eds., pp. 280–292. Albany: State University of New York Press.

Furrow, D. (1984). Social and private speech at two years. *Child Development* 55: 355–362.

Gardner, B. T. and Gardner, R. A. (1971). Two-way communication with an infant chimpanzee. In *Behavior of Nonhuman Primates*. Vol. 4, A. Schrier and F. Stollnitz, eds., pp. 117–184. New York: Academic Press.

Gardner, B. T. and Gardner, R. A. (1974). Comparing the early utterances of child and chimpanzee. In *Minnesota Symposium on Child Psychology*. Vol. 8, A. Pick, ed., pp. 3–23. Minneapolis: University of Minnesota Press.

Gardner, B. T. and Gardner, R. A. (1989). Cross-fostered chimpanzees: II. Modulation of meaning. In *Understanding Chimpanzees*, P. G. Heltene and L. A. Marquardt, eds., pp. 234–241. Cambridge, Mass.: Harvard University Press.

Gardner, B. T. and Gardner, R. A. (1998). Development of phrases in the early utterances of children and cross-fostered chimpanzees. *Human Evolution* 13: 161–188.

Gardner, B. T., Gardner, R. A., and Nichols, S. G. (1989). The shapes and uses of signs in a cross-fostering laboratory. In *Teaching Sign Language to Chimpanzees*, R. A. Gardner, B. T. Gardner, and T. E. Van Cantfort, eds., pp. 55–180. Albany: State University of New York Press.

Gardner, R. A. and Gardner, B. T. (1969). Teaching sign language to a chimpanzee. *Science* 165: 664–672.

Gardner, R. A. and Gardner, B. T. (1989). A cross-fostering laboratory. In *Teaching Sign Language to Chimpanzees*, R. A. Gardner, B. T. Gardner, and T. E. Van Cantfort, eds., pp. 1–28. Albany: State University of New York Press.

Gardner, R. A. and Gardner, B. T. (1998). *The Structure of Learning*. Mahwah, N.J.: Lawrence Erlbaum Associates.

Gardner, R. A., Gardner, B. T., and Van Cantfort, T. E. (1989). *Teaching Sign Language to Chimpanzees*. Albany: State University of New York Press.

Gardner, R. A., Van Cantfort, T. E., and Gardner, B. T. (1992). Categorical replies to categorical questions by cross-fostered chimpanzees. *American Journal of Psychology* 105: 25–57.

Hayes, K. J. and Hayes, C. (1951). The intellectual development of a home-raised chimpanzee. *Proceedings of the American Philosophical Society* 95: 105–109.

Hayes, K. J. and Nissen, C. H. (1971). Higher mental functions of a home-raised chimpanzee. In *Behavior of Nonhuman Primates*. Vol. 4, A. Schrier and F. Stollnitz, eds., pp. 59–115. New York: Academic Press.

Jensvold, M. L. A. and Fouts, R. S. (1993). Imaginary play in chimpanzees (*Pan troglodytes*). *Human Evolution* 8: 217–227.

Jensvold, M. L. A. and Gardner, R. A. (2000). Interactive use of sign language by cross-fostered chimpanzees (*Pan troglodytes*). *Journal of Comparative Psychology* 114: 335–346.

Matthews, W. S. (1977). Modes of transformation in the initiation of fantasy play. *Developmental Psychology* 13: 212–216.

Williams, K. (1995). Comprehensive Nighttime Activity Budgets of Captive Chimpanzees (*Pan troglodytes*). Unpublished master's thesis, Central Washington University, Ellensburg, Wash.

Michael Tomasello and Klaus Zuberbühler

One way to investigate animal cognition is to investigate acts of social communication. In the broadest definition of the term, many acts of communication do not involve complex cognitive processes, since they are simply involuntary behavioral reactions or emotional displays in social situations. However, especially in the case of primates, there are many acts of communication that clearly involve complex cognitive processes, since they are undoubtedly "flexible behavioral adaptations in which individual organisms make informed choices based on mental representations" (Tomasello and Call 1997, p. 12). Here we review—very schematically—what is currently known about the cognitive processes involved in primate communication. Based on the above definition, we focus on two key aspects: (1) the flexibility with which individuals use their communicative signals and (2) the mental representations that might underlie this use of signals. We review first the findings for vocal communication, which mostly involve monkeys, and then those for gestural communication, which mostly involve apes.

Vocal Communication

Primates vocalize to one another most often in the context of evolutionarily urgent events such as avoiding predators, defending against aggressors, traveling as a group, and discovering food. For the most part, all of the individuals of a given species use the same vocal signals, and no new vocal signals are invented by individuals. However, there is some flexibility in the precise manner in which a given call is produced. For example, rhesus monkeys can be trained to alter the duration of their "coo" calls (Sutton et al. 1973); newly paired pygmy marmosets tend to converge on a common structure in their "trill" calls (Snowdon and Elowson 1999); and subordinate cotton-top tamarins change the way they

produce food "chirps" as soon as they become dominant (Roush and Snowdon 1999).

Ontogenetically, there is evidence that in natural populations learning plays at least some role in determining the exact acoustic structure of vocalizations. For example, the coo calls of rhesus monkeys are acoustically more similar within than between matrilines (Hauser 1992), and a number of species display population-specific "dialects" in some vocal signals (e.g., saddle-backed tamarins, Hodun et al. 1981; chimpanzees, Mitani et al. 1992, and others). Experimental studies suggest, however, that such flexibility is limited. For example, in a cross-fostering experiment, Owren et al. (1992) found only limited modifications in the calls of two cross-fostered macaque species. The modifications that were found occurred in only a few limited contexts and involved only a subtle shift in some vocalization patterns and an increase in the frequency of use of calls already in the animals' repertoires. It is also relevant that squirrel monkeys produce most call types from soon after birth, even if they are reared in isolation (see Snowdon et al. 1997 for a review), and that hybrid gibbons (*Hylobates lar* and *H. pileatus*) produce songs with an acoustic structure morphologically intermediate between those of the two parent species (e.g., Geissmann 1984).

As with call morphology, call usage seems to have only limited flexibility, with learning playing a highly limited role. Thus in most cases calls are used in adultlike contexts from early in infancy, but then there is a learning phase in which more adultlike usage is fine tuned (Seyfarth and Cheney 1997). For example, infant vervet monkeys often make mistakes by giving an eagle alarm call in response to various moving things in the sky (Seyfarth and Cheney 1986), and they produce intergroup calls whenever they are distressed. Only later do they confine these to adultlike contexts (Hauser 1989). And young pigtail macaques are less precise when using agonistic screams than are adults (Gouzoules and Gouzoules 1989). Flexibility of use persists to some degree in adults. For example, in cotton-top tamarins, social and reproductive status can affect call usage (Roush and Snowdon 2000), and Japanese macaques show population-level differences in their use of food and contact calls (Green 1975; Sakura 1989).

An especially important type of flexibility concerns audience effects, in which an individual uses its vocal signals differently, depending on the social-communicative situation. Audience effects are especially important because they suggest that individuals are strategically modifying their use of a signal based on their current assessment of how it might affect potential recipients. For example, red-bellied tamarins produce food calls when they discover food, but the call rates depend on whether other group mates are present (Caine et al. 1995); male chimpanzees pant-hoot more frequently in traveling contexts when their alliance partners are nearby (Mitani and Nishida 1993); and vervet monkey females adjust the rate of alarm calls according to whether their own offspring are present, while males call more in the presence of females than males (Cheney and Seyfarth 1985). On the other hand, macaque females who were shown a predator, either in the presence of their offspring or alone, did not attempt to alert ignorant offspring more than knowledgeable ones, suggesting that audience effects may not be the result of callers assessing the knowledge of recipients (Cheney and Seyfarth 1990). Overall, there clearly are audience effects for some primate vocal signals, but the degree and consistency of the effects are modest and the underlying cognitive processes are not well understood.

Undoubtedly, primates display most flexibility in the way they perceive and understand vocal signals. The classic case for context-sensitive semanticity is vervet monkey alarm calls in which individuals respond to acoustically distinct calls with particular types of antipredator responses,

even in the absence of the predator (Seyfarth et al. 1980). [See Zuberbühler et al. (1997) and Zuberbühler (2000c) for comparable research with Diana monkeys and Campbell's monkeys, respectively; see Macedonia (1990) for research with prosimians.] The other classic case is rhesus monkey recruitment screams, which encode various aspects of the agonistic encounter, especially the relative rank of the attacker and the severity of the attack (Gouzoules et al. 1984). In this case responders base their reaction (e.g., aiding the attacked individual or not) on their understanding of who the aggressor is and what kind of aggression is occurring. Convincing evidence that recipients are indeed responding to the meaning (reference) of such calls, and not to such things as their emotional intensity or the like, comes from habituation experiments in which individuals habituated to some call show dishabituation only when the meaning (referent) of the call is changed (e.g., Cheney and Seyfarth 1988; Zuberbühler et al. 1999; Zuberbühler 2000c). It is interesting that there are no convincing observations of such "referential signals" in any ape species, the closest possibility being chimpanzees' differential use of food grunts for different amounts of food (Hauser and Wrangham 1987; Hauser et al. 1993).

As recipients, primates thus exhibit a fairly sophisticated understanding of the meanings associated with various calls, which possibly involves mental representations of the referential situation. Even more impressive, some species seem to be able to take into account the possible causes of an alarm call, rather than responding to the alarm calls directly (Zuberbühler 2000a,b). Thus, savannah baboons respond differentially to a played-back submissive grunt, depending on their understanding of who the caller is and whom she is giving the submissive grunt to (Cheney et al. 1995). Finally, the learning skills used in call comprehension show almost unlimited flexibility, since a number of primate species can learn to use effectively the calls of various other sympatric species, both primate and nonprimate. Thus,

Diana monkeys understand the meaning and underlying cause of other primate species' alarm calls (Zuberbühler 2000c), guinea fowl alarm calls (Zuberbühler 2000a), and different kinds of chimpanzee screams (Zuberbühler 2000b; see Hauser 1988, for similar findings with vervet monkeys).

In sum, although primate vocalizations seem to be under significant genetic control in their morphology and usage, there is also evidence that the callers have some limited flexibility. In particular, social factors may affect the structure of certain calls, and audience effects may influence the way calls are used in some circumstances. Current evidence suggests that primates possess the most flexibility in call comprehension. Many primate species comprehend vocal signals as indications of specific external events; this includes the calls of other species, which obviously must be individually learned. Moreover, the recipients take into account various kinds of causal relations in making the appropriate responses.

Gestural Communication

Primates communicate using manual and bodily gestures mainly in social contexts such as play, grooming, nursing, and during sexual and agonistic encounters. These are in general less evolutionarily urgent functions than those signaled by vocal communication. Unlike the case of vocal signals, however, there is good evidence that individuals of some primate species, especially apes, may invent new gestural signals as needed. Goodall (1986) and Tomasello et al. (1985), for example, report much variability in the gestures used by individual chimpanzees, including a number of idiosyncratic gestures used by single individuals only that could not have been either genetically determined or socially learned. Tanner and Byrne (1993, 1996) also report much individual variability in the gestures used by gorillas. The only systematic studies of monkeys'

gestural communication give no information about individual differences, but it does not seem likely that individuals invent gestures on their own (Maestripieri 1998).

Chimpanzees (and probably gorillas) learn their gestural signals via a process of ontogenetic ritualization (Tomasello 1996) in which two organisms shape one another's behavior in repeated instances of a social interaction. The general form of this type of learning is as follows: (1) Individual A performs behavior X. (2) Individual B reacts consistently with behavior Y. (3) Subsequently B anticipates A's performance of X, on the basis of its initial step, by performing Y. (4) Subsequently A anticipates B's anticipation and produces the initial step in a ritualized form (waiting for a response) in order to elicit Y. The main point is that a behavior that was not at first a communicative signal becomes one by virtue of the anticipations of the interactants over time. There is no evidence that chimpanzees, or any other ape species, acquire gestural signals by imitative learning (Tomasello and Call 1997). In the case of monkeys, de Waal and Johanowicz (1993) looked at the reconciliation gestures of juvenile rhesus and stumptail macaques who were co-housed for a period of 5 months and found that the nature of the gestures and displays for reconciliation remained unchanged for both species (even though some other behaviors did change).

With regard to flexibility of use, Tomasello et al. (1994, 1997) found that many chimpanzee gestures were used in multiple contexts, sometimes across widely divergent behavioral domains (an average of 2.5 gestures per individual were used for more than one function). Also, sometimes different gestures were used interchangeably in the same context toward the same end—and individuals sometimes performed these in rapid succession in the same context (e.g., initiating play first with a "poke-at" followed by an "arm raise"). By all accounts, the gestural signals of monkey species do not show this same degree of flexibility, but rather their gestural signals are as tightly tied to particular communicative situations as their vocal signals. However, in some instances both monkeys and apes have been observed to use some gestures in a way that suggests tactical deception, which—regardless of the appropriateness of this appellation—at least indicates that the human observer saw a gesture used outside its ordinary context (Whiten and Byrne 1988).

In terms of audience effects, Tomasello et al. (1994, 1997) found that chimpanzee juveniles give a visual signal to solicit play (e.g., "arm raise") only when the recipient is already oriented appropriately, but they use their most insistent attention-getter, a physical "poke-at," most often when the recipient is socially engaged with others. Tanner and Byrne (1993) reported that a female gorilla repeatedly used her hands to hide her play face from a potential partner, indicating some flexible control of the otherwise involuntary grimace, as well as a possible understanding of the role of visual attention in gestural communication (see also de Waal 1986, for a similar observation for chimpanzees). In an experimental setting, Call and Tomasello (1994) found that at least some orangutans also were sensitive to the gaze direction of their communicative partner. Kummer (1968) reported that before they set off foraging, male hamadryas baboons engage in "notifying behavior" in which they approach another individual and look directly into their face, presumably to make sure that the other is looking before the trek begins. Overall, audience effects are very clear in the case of ape gestures, and there may be such effects for monkey gestures as well.

Since the gestures of apes are produced so flexibly, flexible skills of gesture comprehension are required as well. Virtually no ape gestures are referential in the sense that they indicate an external entity (e.g., there is no pointing in the human fashion); they mostly concern the dyadic interactions among group mates. It is interesting for the process of comprehension that Tanner

and Byrne (1996) have reported on a number of gorilla gestures that seem to be iconic. That is, an adult male gorilla often seemed to indicate to a female playmate iconically, by using his arms or whole body, the direction in which he wanted her to move, the location he wanted her to go to, or the action he wanted her to perform (see also Savage-Rumbaugh et al. 1977, for bonobos). However, these might simply be normal ritualized gestures, with the iconicity being only in the eyes of the human; in fact, a role for iconicity in gorillas' and other apes' comprehension of gestures has not yet been demonstrated (Tomasello and Call 1997).

In sum, primate gestural communication shows more flexibility than primate vocal communication, perhaps because it concerns less evolutionarily urgent activities than those indicated most often with vocalizations. Apes create new gestures routinely and in general use many of their gestures quite flexibly; more research is needed with monkeys. Audience effects and flexibility of comprehension are routine with ape gestural communication, and again more research is needed with monkeys. However, neither ape nor monkey gestures seem to be used referentially, as are some monkey vocal signals.

Conclusion

By most definitions, cognition involves mental representation. Nonhuman primates mentally represent their worlds in many different ways, including in the domains of space, objects, categories, quantities, social relationships, and others (Tomasello and Call 1997). Primate communication also involves mental representation in some cases, most clearly in the representational vocalizations used by some monkey species. A major question in this respect is whether apes also use some representational signals, either vocal or gestural, in some as-yet to be discovered domain.

But the prototype of a cognitive process also involves the flexible use of these mental representations. Much primate communication, perhaps especially the gestural communication of apes, involves very flexible behavioral strategies. Ape gestures have flexible connections between signal and function, a clear sensitivity to audience, and at least some combinatorial possibilities, all of which rely on the flexible way in which gestural signals are ontogenetically ritualized. The most urgent question in this respect is whether some monkey species also use their gestural signals in highly flexible ways in some as-yet to be discovered behavioral domain.

Finally, there is also the question of how to characterize the communicative activities of human-raised apes who learn something resembling human language (e.g., Greenfield and Savage-Rumbaugh 1990; Savage-Rumbaugh et al. 1993). A review of this research is beyond the scope of this essay, but it is clear that there are a plethora of interesting but unanswered questions about these unique individuals.

References

Caine, N. G., Addington, R. L., and Windfelder, T. L. (1995). Factors affecting the rates of food calls given by red-bellied tamarins. *Animal Behaviour* 50: 53–60.

Call, J. and Tomasello, M. (1994). Production and comprehension of referential pointing by orangutans (*Pongo pygmaeus*). *Journal of Comparative Psychology* 108: 307–317.

Cheney, D. L. and Seyfarth, R. M. (1985). Vervet monkey alarm calls: Manipulation through shared information? *Behaviour* 94: 739–751.

Cheney, D. L. and Seyfarth, R. M. (1988). Assessment of meaning and the detection of unreliable signals by vervet monkeys. *Animal Behaviour* 36: 477–486.

Cheney, D. L. and Seyfarth, R. M. (1990). Attending to behaviour versus attending to knowledge: Examining monkey's attribution of mental states. *Animal Behaviour* 40: 742–753.

Cheney, D. L., Seyfarth, R. M., and Silk, J. B. (1995). The responses of female baboons (*Papio cynocephalus ursinus*) to anomalous social interactions: Evidence for causal reasoning? *Journal of Comparative Psychology* 109: 134–141.

de Waal, F. (1986). Deception in the natural communication of chimpanzees. In *Deception. Perspectives on Human and Nonhuman Deceit*, R. W. Mitchell and N. S. Thompson, eds., pp. 221–244. Albany: State University of New York Press.

de Waal, F. B. M. and Johanowicz, D. L. (1993). Modification of reconciliation behavior through social experience: An experiment with two macaque species. *Child Development* 64: 897–908.

Geissmann, T. (1984). Inheritance of song parameters in the gibbon song, analysed in 2 hybrid gibbons (*Hylobates pileatus* × *H. lar*). *Folia Primatologica* 24: 216–235.

Goodall, J. (1986). *The Chimpanzees of Gombe: Patterns of Behavior*. Cambridge, Mass.: Harvard University Press.

Gouzoules, H. and Gouzoules, S. (1989). Design features and developmental modification of pigtail macaque, *Macaca nemestrina*, agonistic screams. *Animal Behaviour* 32: 182–193.

Gouzoules, S., Gouzoules, H., and Marler, P. (1984). Rhesus monkey (*Macaca mulatta*) screams: Representational signalling in the recruitment of agonistic aid. *Animal Behaviour* 32: 182–193.

Green, S. (1975). Variation of vocal pattern with social situation in the Japanese monkey (*Macaca fuscata*): A field study. In *Primate Behavior, Developments in Field and Laboratory Research*. Vol. 2, L. A. Rosenblum, ed., pp. 1–102. New York: Academic Press.

Greenfield, P. M. and Savage-Rumbaugh, E. S. (1990). Grammatical combination in *Pan paniscus*: Processes of learning and invention in the evolution and development of language. In *"Language" and Intelligence in Monkeys and Apes. Comparative Developmental Perspectives*, S. T. Parker and K. R. Gibson, eds., pp. 540–578. Cambridge: Cambridge University Press.

Hauser, M. D. (1988). How infant vervet monkeys learn to recognize starling alarm calls: The role of experience. *Behaviour* 105: 187–201.

Hauser, M. D. (1989). Ontogenetic changes in the comprehension and production of vervet monkey (*Cercopithecus aethiops*) vocalizations. *Journal of Comparative Psychology* 103: 149–158.

Hauser, M. D. (1992). Articulatory and social factors influence the acoustic structure of rhesus monkey vocalizations: A learned mode of production? *Journal of the Acoustical Society of America* 91: 2175–2179.

Hauser, M. D. and Wrangham, R. W. (1987). Manipulation of food calls in captive chimpanzees: A preliminary report. *Folia Primatologica* 48: 24–35.

Hauser, M. D., Teixidor, P., Field, L., and Flaherty, R. (1993). Food-elicited calls in chimpanzees: Effects of food quantity and divisibility? *Animal Behaviour* 45: 817–819.

Hodun, A., Snowdon, C. T., and Soini, P. (1981). Subspecific variation in the long calls of the tamarin, *Saguinus fusicollis. Zeitschrift für Tierpsychologie* 57: 97–110.

Kummer, H. (1968). *Social Organization of Hamadryas Baboons.* Chicago: University of Chicago Press.

Macedonia, J. M. (1990). What is communicated in the antipredator calls of lemurs: Evidence from playback experiments with ring-tailed and ruffed lemurs. *Ethology* 86: 177–190.

Maestripieri, D. (1998). Primate social organization, vocabulary size, and communication dynamics: A comparative study of macaques. In *The Evolution of Language: Assessing the Evidence from Nonhuman Primates*, B. King, ed., pp. 89–112. Santa Fe, N.M.: School of American Research.

Mitani, J. C. and Nishida, T. (1993). Contexts and social correlates of long-distance calling by male chimpanzees. *Animal Behaviour* 45: 735–746.

Mitani, J. C., Hasegawa, T., Gros-Louis, J., Marler, P., and Byrne, R. (1992). Dialects in wild chimpanzees? *American Journal of Primatology* 27: 233–243.

Owren, M. J., Dieter, J. A., Seyfarth, R. M., and Cheney, D. L. (1992). Evidence of limited modification in the vocalizations of cross-fostered rhesus (*Macaca mulatta*) and Japanese (*M. fuscata*) macaques. In *Topics in Primatology. Human Origins*, T. Nishida, W. C. McGrew, P. Marler, M. Pickford, and F. B. M. de Waal, eds., pp. 257–270. Tokyo: University of Tokyo Press.

Roush, R. S. and Snowdon, C. T. (1999). The effects of social status on food-associated calling behaviour in captive cotton-top tamarins. *Animal Behaviour* 58: 1299–1305.

Roush, R. S. and Snowdon, C. T. (2000). Quality, quantity, distribution and audience effects on food calling in cotton-top tamarins. *Ethology* 106: 673–690.

Sakura, O. (1989). Variability in contact calls between troops of Japanese macaques: A possible case of neu-tral evolution of animal culture. *Animal Behaviour* 38: 900–902.

Savage-Rumbaugh, E. S., Wilkerson, B. J., and Bakeman, R. (1977). Spontaneous gestural communication among conspecifics in the pygmy chimpanzee (*Pan paniscus*). In *Progress in Ape Research*, G. H. Bourne, ed., pp. 97–116. New York: Academic Press.

Savage-Rumbaugh, E. S., Murphy, J., Sevcik, R. A., and Brakke, K. E. (1993). Language comprehension in ape and child. *Monographs of the Society for Research in Child Development* 58: 221.

Seyfarth, R. M. and Cheney, D. L. (1986). Vocal development in vervet monkeys. *Animal Behaviour* 34: 1640–1658.

Seyfarth, R. M. and Cheney, D. L. (1997). Behavioral mechanisms underlying vocal communication in nonhuman primates. *Animal Learning & Behavior* 25: 249–267.

Seyfarth, R. M., Cheney, D. L., and Marler, P. (1980). Vervet monkey alarm calls: Semantic communication in a free-ranging primate. *Animal Behaviour* 28: 1070–1094.

Snowdon, C. T. and Elowson, A. M. (1999). Pygmy marmosets modify call structure when paired. *Ethology* 105: 893–908.

Snowdon, C. T., Elowson, A. M., and Roush, R. S. (1997). Social influences on vocal development in New World primates. In *Social Influences on Vocal Development*, C. T. Snowdon and M. Hausberger, eds., pp. 234–248. Cambridge: Cambridge University Press.

Sutton, D., Larson, C., Taylor, E. M., and Lindeman, R. C. (1973). Vocalization in rhesus monkey: Conditionability. *Brain Research* 52: 225–231.

Tanner, J. E. and Byrne, R. W. (1993). Concealing facial evidence of mood. Perspective taking in a captive gorilla. *Primates* 34: 451–457.

Tanner, J. E. and Byrne, R. W. (1996). Representation of action through iconic gesture in a captive lowland gorilla. *Current Anthropology* 37: 162–173.

Tomasello, M. (1996). Do apes ape? In *Social Learning in Animals: The Roots of Culture*, B. Galef and C. Heyes, eds., pp. 319–346. San Diego, Calif.: Academic Press.

Tomasello, M. and Call, J. (1997). *Primate Cognition.* New York: Oxford University Press.

Tomasello, M., George, B., Kruger, A., Farrar, M., and Evans, A. (1985). The development of gestural communication in young chimpanzees. *Journal of Human Evolution* 14: 175–186.

Tomasello, M., Call, J., Nagell, K., Olguin, K., and Carpernter, M. (1994). The learning and use of gestural signals by young chimpanzees: A trans-generational study. *Primates* 35: 137–154.

Tomasello, M., Call, J., Warren, J., Frost, G. T., Carpenter, M., and Nagell, K. (1997). The ontogeny of chimpanzee gestural signals: A comparison across groups and generations. *Evolution of Communication* 1: 223–259.

Whiten, A. and Byrne, R. W. (1988). Tactical deception in primates. *Behavioral and Brain Sciences* 11: 233–273.

Zuberbühler, K. (2000a). Causal cognition in a nonhuman primate: Field playback experiments with Diana monkeys. *Cognition* 76: 195–207.

Zuberbühler, K. (2000b). Causal knowledge of predators' behaviour in wild Diana monkeys. *Animal Behaviour* 59: 209–220.

Zuberbühler, K. (2000c). Interspecific semantic communication in two forest monkeys. *Proceedings of the Royal Society of London B* 267: 713–718.

Zuberbühler, K., Noe, R., and Seyfarth, R. M. (1997). Diana monkey long-distance calls: Messages for conspecifics and predators. *Animal Behaviour* 53: 589–604.

Zuberbühler, K., Cheney, D. L., and Seyfarth, R. M. (1999). Conceptual semantics in a nonhuman primate. *Journal of Comparative Psychology* 113: 33–42.

Barbara Smuts

Most research investigating how communication may shed light on primate social cognition has focused on vocal rather than visual (gestural) signals (Tomasello and Call 1997). Since visual signals predominate when individuals relate "up close and personal," their study is especially useful for understanding how animals establish, maintain, and negotiate affiliative and cooperative relationships. Here I describe research on gestural communication in wild baboons and domestic dogs. "Gestures" include all nonvocal actions with potential communicative significance, including facial expressions; body postures; tail carriage; variations in gait and body carriage; and motions of limbs, muzzles, and other body parts that may or may not involve touching another animal. Most of this work relies on detailed analysis of videotaped interactions and only preliminary results are available.

Greetings among Male Olive Baboons

During a long-term study of male–female relationships in a large group of olive baboons (*Papio anubis*) near Gilgil, Kenya (Smuts 1999), I became intrigued by ritualized greetings between adult males in which one male would present his posterior to another and then allow him to handle or even mouth his genitals. Literally placing the source of one's future reproductive success in the hands of another male seemed like an act of extreme trust inconsistent with the highly aggressive nature of male baboons.

To address this paradoxical behavior, I teamed up with John Watanabe, a cultural anthropologist interested in ritual (Smuts and Watanabe 1990; Watanabe and Smuts 1999). Olive baboons live in female-bonded societies (Wrangham 1980), so males spend much of their adult lives with unrelated and initially unfamiliar others. Most male–male interactions involve mutual antagonism, including occasional severe wound-ing. Adult males virtually never groom or play with one another and participate in only two kinds of nonagonistic interactions with other males: greetings and alliances.

When one male baboon approaches another, the other usually threatens or avoids him. However, a subset of approaches is accompanied by an exaggerated gait and the "come hither" face (a striking expression with eyes narrowed and ears flattened back against the skull). Sometimes the approaching male also lip smacks, another friendly sign. These very distinctive signals invite the other male to greet, and, compared with routine approaches, the approached male is much less likely to move away. Typically, he will indicate acceptance of the greeting by reciprocating eye contact (in other contexts, eye contact constitutes a threat) and often by lip smacking and making the come hither face in return.

Upon completion of the approach, the males usually begin an exchange of gestures that typically involves one presenting his hindquarters while the other either grasps his hips, mounts him, touches his scrotum, or pulls his penis. Less often, one greeter nuzzles the other, or, very rarely, they embrace and play briefly (this is the only context in which I have seen adult males playing with one another). The gestures used during a single greeting most often entail asymmetrical roles, with one male taking the more active, dominant role (e.g., mounting). Occasionally, a mutual exchange occurs in which each one mounts the other in turn, or each touches the other's genitals simultaneously or in rapid succession. Immediately after the exchange, one (or occasionally both) of the males moves rapidly away using the same exaggerated gait characteristic of the approach. The entire sequence usually takes no more than a few seconds.

Initiation of a greeting never guaranteed its completion. Either male may break off at any time simply by moving away, and in our sample of 637 adult male greetings, nearly half of the time one male pulled away before completing the exchange. Occasionally (7 percent of our sample) attempts to greet ended in threats, chases, or fights. Remarkably, however, of the roughly 1000 greetings documented in this study and subsequent research by Smuts (see later discussion), not one resulted in a discernible injury. Thus these greetings stand in striking contrast to the agonism of virtually all other male baboon interactions.

The 12 fully adult males in our troop were either old males (non-natal males past their physical prime who had lived in the troop for at least 2 years, many for much longer) or young males (recently matured natal males or recently transferred males in physical prime). Based on the outcomes of dyadic agonistic encounters, all young males individually outranked all old males. Among baboons in general, higher-ranking males mate more often with estrous females than do lower-ranking males, but during our study—and typical of this larger population of olive baboons (Bulger 1993)—consort frequency and male rank were uncorrelated. Lower-ranking old males achieved unexpectedly high mating success, in part by forming coalitions with each other in which they jointly harassed a male in consort with a fertile female until a consort turnover occurred. All of these challenges targeted young males; we never saw old males challenge each other's consortships. After the turnover, the female almost always ended up with one of the old males involved in the coalition (Smuts 1999).

Although some of these coalitions probably represented mutualistic opportunism (Noe 1992), many of them entailed repeated alliances between the same partners, roughly equal access to the female, and inhibition of competition against alliance partners in other contexts; they thus appeared to involve reciprocal altruism (Packer 1977). To form such valuable coalitions, an older male must somehow cooperate with potential allies who are usually the very rivals whom he has long tried to harass, intimidate, bluff, and occasionally wound. How do these males convey to each other their readiness to establish cooperative relationships that rely on a degree of trust?

This is where male greetings, we hypothesized, play a decisive role. The patterns of greeting tended to reflect coalitional behavior (or its absence). Greetings between young males almost always displayed considerable tension, if they occurred at all. Young males had the lowest percentage of completed greetings (one-third) because each one attempted to adopt the dominant, active role while avoiding being in the subordinate, passive role. This resulted in mutual circling and failure to greet. Young males never formed coalitions with one another, and except for their attempted greetings, they mostly avoided each other and even avoided associating with the same females (Smuts 1999). In contrast, the old males who formed coalitions with each other tended to have relatively relaxed greetings, and they completed most (two-thirds) of them.

Several examples illustrate these patterns. The two highest-ranking males, at the time engaged in a tense standoff for dominance, repeatedly attempted to greet, but we never saw them succeed because neither would adopt the passive role. In contrast, between pairs of young males that had achieved a relatively stable dominance relationship, one was usually willing to accept the subordinate role during greetings. Repeated greetings between such pairs acknowledge (and perhaps reinforce) established dominance relationships with minimal risk of aggression. To the extent that both animals benefit from such acknowledgment, these greetings constitute a rudimentary form of cooperation.

One pair of old males, Alex and Boz, greeted more often than any other males. Unrelated but familiar to one another after 7 years of cohabitation in the same troop, they had the longest-standing, most consistent alliance of any pair of males and routinely helped each other take fertile females away from younger rivals. Unlike all other male pairs, neither tried to dominate the other, and they remained the only pair of males observed defending each other in fights with other males.

These two males' greetings showed a unique combination of features. First, they completed nearly all of their greetings. Second, almost half of their greetings included genital touching (the seemingly most risky and intimate form of greeting) compared with only 18 percent of greetings overall. Third, neither Alex nor Boz ever resisted having his genitals handled by the other, although in other dyads males often pulled away from attempts at genital touching. Finally, in contrast to the asymmetry of greeting roles characteristic of most male–male dyads, their greetings reflected near-perfect symmetry (out of a total of 43 gestural exchanges, Alex adopted the passive role in 22, the active role in 21). Indeed, whenever we saw them greet twice in rapid succession, the male who took the active role in the first greeting initiated the subsequent greeting by inviting his partner to adopt the active role. Their greetings thus paralleled their "turn-taking" in coalitions.

How exactly might greeting rituals foster cooperation? The distinctive approach that initiates a greeting seems to function as a kind of metacommunication (see Bekoff and Allen, chapter 53 in this volume) signifying that what follows is no ordinary approach. Thus, greetings appear to constitute a neutral ground that allows exploration of social roles through asymmetric, symmetric, or reciprocal actions. This exploration occurs in a context in which, because no resources are at stake, the chances of escalated aggression are minimized. The possibility of breaking off a greeting at any time without fear of retribution provides an incremental mechanism for testing another male's willingness to cooperate while minimizing one's own investment in the relationship. Because allowing another male to handle one's genitals presumably poses risks, it may function as an honest signal of willingness to cooperate. Males interested in establishing reliable alliances could thus be using the greetings as a way of expressing their good intentions in a world of otherwise suspicious, highly competitive individuals (Smuts and Watanabe 1990).

This hypothesis is indirectly supported by two other facts. First, the gestures used in greeting derive from two kinds of cooperative relationships: the mother–infant bond (lip smacking, embracing, and nuzzling) and sexual invitations and mating (presenting hindquarters, grasping hips, mounting, and genital contact). Second, when adult males solicit each other as allies in the heat of a contest for an estrous female, they employ telegraphic versions of the same gestures used during greetings to request and offer aid. In short, we hypothesize that male baboon greetings involve intentional communication designed to probe and evaluate another male's willingness (or lack of willingness) to cooperate, and that such communication allows potential allies to: (1) more wisely choose whom to solicit for aid and (2) more effectively coordinate actions during actual coalition formation.

Two kinds of evidence are needed to evaluate this hypothesis further. First, the relationships we identified between greeting behavior and alliance formation were correlational and do not demonstrate causation. Causation would be supported if we could show that changes in greeting behavior (such as a shift to more reciprocal roles) predict changes in alliance formation (such as an increase in reciprocal coalition formation). Unfortunately, the tendency for competing males to disappear into thick vegetation makes it difficult to obtain a large sample of alliances.

The second kind of evidence entails more fine-grained analysis of exactly what transpires during greetings and especially during successive greetings between pairs of males over periods of time. To this end, I videotaped over 400 greetings between adult male olive baboons at Gombe National Park, Tanzania. Preliminary analyses indicate that although most greetings conform to the basic formula described earlier, the precise form a particular gesture (e.g., presenting) takes, the choice and timing of gestures relative to what the partner is doing, and the way gestures are combined, vary between dyads and within individuals, depending on whom they are greeting. This microvariation supports our claim that baboons use greeting gestures flexibly and crea-tively in a way that suggests intentional commu-nication (Tomasello and Call 1997, p. 10). Future analyses will focus on patterns of gestural varia-tion within and across dyads in order to deter-mine whether some variations predict, or at least correlate with, specific aspects of relationships.

In addition to male–male greetings, the vid-eotaped sample also includes several hundred greetings involving all other possible combina-tions of age-sex classes. These greetings are cur-rently being analyzed to address questions about how greetings relate to alliance formation and friendship among females and between the sexes. We also aim to investigate the development of greeting behavior. Although the basic gestures used in greetings appear within the first few months (or even weeks) of life, young animals probably learn through experience how to use and combine these gestures in ways that help them achieve social goals.

Bekoff and Allen (chapter 53 in this volume) hypothesize that play fighting may also help animals to establish cooperative relationships. Male baboon greetings and play fighting have at least two unusual features in common. First, each involves "metacommunication" (e.g., the play-bow in canid play and the exaggerated gait and come-hither face in baboon greetings) sig-naling that what is about to transpire follows special rules. Second, each allows the opportunity for suspension of the behavioral asymmetries characteristic of routine interactions between animals of different ranks. Combined, these two features, which may be unique to play and greet-ings, create opportunities for rich communication about cooperative possibilities between the actors.

Play Fighting in Dogs

Intrigued by these similarities between greet-ings and play, my graduate student Erika Bauer and I have begun a quantitative study of self-handicapping and role reversal (see Bekoff and Allen, chapter 53 in this volume, for definitions)

during play fighting between pairs of domestic dogs. Like the baboon studies just described, this research analyzes videotaped sequences to allow detailed investigation of roles. Our sample in-cludes about 20 domestic dogs (adults or adoles-cents) in various partner combinations.

Dog play is a particularly interesting and suit-able arena in which to investigate variation in role asymmetry-symmetry during play because: (1) dogs play a lot; (2) outside of play, it is pos-sible to discern clear asymmetries in dominance status for most dyads (personal observation); and (3) they descend from wolves, a species char-acterized both by adult play and by a complex interplay of cooperation and competition.

Several patterns emerge from preliminary analyses. First, the degree of role reversal varies dramatically. For example, out of a sample of 12 different dyads, in one pair the dominant animal never adopted a subordinate role, whereas in another pair the dominant animal adopted the subordinate role 80 percent of the time (most dyads fell somewhere in between). Second, par-ticular individuals reversed roles more often with some partners than with others. Third, even within dyads, the degree of role reversal can vary considerably from one play bout to another. Fourth, in at least some dyads, dogs seem much more willing to reverse roles than the primates for whom quantitative data exist (such as rhesus macaques and squirrel monkeys; Biben 1998). Rhesus and squirrel monkeys have quite rigid dominance hierarchies compared with some other primates, such as chimpanzees (de Waal 1989). Dog and wolf dominance hierarchies (personal observation and Zimen 1982, respectively) appear more like those of chimps, in that subordinates are sometimes quite relaxed around dominants.

These comparisons raise the possibility that styles of play fighting may reflect species differ-ences in styles of dominance relationships. To evaluate this hypothesis, we plan to extend our study of role reversals during play to include chimpanzees, captive wolves, and other mam-malian species that exhibit different dominance styles. We also intend to analyze the videotapes for contingencies within and between play bouts that may influence the degree of role reversal. For example, if animal *A* shows little role reversal and her partner, *B*, loses interest in playing, does *A* then exhibit more role reversal as a strategy for inducing *B* to play more? Or, if one animal begins to challenge another's dominance rank, does he begin by decreasing his willingness to adopt the subordinate role during play (Pellis, chapter 52 in this volume)?

These sorts of questions establish common ground between our study of mammalian play and our research on baboon greetings. In each instance, we examine subtle behaviors detectable only through analysis of videotapes to find out what cues the animals emit and respond to, and how these cues may be used to negotiate concerns that extend beyond the immediate transaction to include alternative future paths for how the relationship may unfold. In our own species, such negotiations rely so much on verbal communication that we tend to assume that without words, animals cannot "talk" about the future. These studies represent one small step toward determining if this is true.

The Value of Videotaping Interactions

While videotaping greetings among baboons, whenever possible we also taped other interesting social phenomena, such as aggression, play, and formation of coalitions. Although this footage has not yet been analyzed systematically, casual perusal suggests that it will provide indisputable documentation of behaviors that may provide insights into social cognition. For example, one videotaped sequence shows two male baboons, *A* and *B*, allying against a third male, *C*, who is sitting minding his own business. *A* is a young alpha male, and *C* and *B* are young, recent immigrants of roughly equal rank. (Although coalitions among young males were rare at Gilgil, among the Gombe baboons, males used them to probe the fighting ability of other young males.) *A* and *B* jointly threaten *C* as they approach

him. *C*, who generally avoids all fights, at first ignores them. As *A* and *B* come closer, *C* gets up and begins to move away. Just as *C* leaves, *B*, who is on the far side of *A* from *C*, reaches out and apparently deliberately shoves *A* into *C*'s path. Because *C* is not looking at them, he does not see the shove, and from his perspective it must seem as if *A* suddenly lunged at him. *C* turns and attacks *A*, *A* flees, and *C* chases *A* for at least a hundred meters before they disappear into the bushes. *B* watches, unperturbed. With one well-timed shove, he managed to provoke aggression between *A* and *C*, perhaps increase tension between them, and gain potentially useful information about their relationship.

I was surprised when I saw this sequence on tape because I did not realize that baboons used such deliberate strategies to incite third-party aggression. Even had I detected this sequence in real time, it would constitute just one more anecdote whose details could (and should!) be questioned by people who were not there. However, because this sequence is on tape, we know exactly what happened, and different people can observe the sequence and decide whether they agree on its interpretation. In combination with dozens of other videotaped observations, events like this can comprise a database suitable for testing a hypothesis.

A second example of the value of videotaped observations concerns alliance formation in baboons. When I tried to record alliances on paper in the field, I often became frustrated and confused, because at one moment *A* and *B* were allied against *C* and then suddenly it seemed as if *B* and *C* were allied against *A*. I secretly thought that I was not a good enough observer until I examined such episodes on tape. I then discovered that, indeed, the baboons often change alliance partners so quickly that a temporary partnership could be missed in the blink of an eye. Even in slow motion, these shifts are hard to follow, but by watching a sequence over and over, one can eventually describe what happened.

The result resembles a choreographer's notation for a complex routine involving many dancers, and it is not obvious what it all means. The baboons probably know exactly what such rapid shifts signify, and I hope that intelligible patterns will emerge once a large enough number of episodes are analyzed. Such shifting partnerships are to my mind the single most striking attribute of baboon alliances, and yet they have received virtually no recognition in the literature, probably because they are so hard to follow unless they are recorded on videotape.

A third example concerns the astounding degree of responsiveness to a partner's cues during greetings. When videotaped greetings are observed in slow motion, it becomes immediately apparent that the two baboons respond to each other's subtlest shifts in movement and glances with split-second timing that cannot possibly be documented, or often even perceived, in real time. After watching hundreds of greetings in slow motion, I have become convinced that at least as important as *what* each baboon does is *how* they do it, and specifically, how finely coordinated their movements are with those of the other baboon.

In slow motion, some greetings look like the awkward steps of two people first learning to dance together, whereas others look like Fred Astaire and Ginger Rogers. Preliminary analyses indicate that the latter usually involve greetings between established friends, whereas the former involve greetings between individuals who are not friends (but who may be attempting to change that). If these results hold up, they suggest that coordinated movement may be one important way animals establish reliable, trusting relationships without words (Savage Rumbaugh et al. 1993). In fact, to the extent that such fine coordination depends on a major investment of time, this method of evaluating partners may be more reliable than mere verbal commitments.

These are just a few examples of the kinds of questions that can be addressed only through videotaped observations. Such observations, in my view, hold tremendous potential for delving much more deeply into the world of animal communication and thereby, animal social cognition.

References

Biben, M. (1998). Squirrel monkey playfighting: Making the case for a cognitive training function for play. In *Animal Play. Evolutionary, Comparative, and Ecological Perspectives*, M. Bekoff and J. A. Byers, eds., pp. 161–182. Cambridge: Cambridge University Press.

Bulger, J. B. (1993). Dominance rank and access to estrous females in male savanna baboons. *Behaviour* 127: 67–103.

de Waal, F. B. M. (1989). Dominance "style" and primate social organization. In *Comparative Socioecology. The Behavioural Ecology of Humans and Other Animals*, V. Standen and R. Foley, eds., pp. 243–264. Oxford: Oxford University Press.

Noe, R. (1992). Alliance formation among male baboons: Shopping for profitable partners. In *Coalitions and Alliances in Humans and Other Animals*, A. H. Harcourt and F. B. M. de Waal, eds., pp. 284–321. Oxford: Oxford University Press.

Packer, C. (1977). Reciprocal altruism in *Papio anubis*. *Nature (London)* 265: 441–443.

Savage-Rumbaugh, E. S., Murphy, J., Sevcik, R. A., Brakke K. E., Williams, S. L., and Rumbaugh, D. M. (1993). Language comprehension in ape and child. *Monographs of the Society for Research in Child Development* 58: v–221.

Smuts, B. B. (1999). *Sex and Friendship in Baboons*, 2nd. ed. Cambridge, Mass: Harvard University Press.

Smuts, B. B. and Watanabe, J. M. (1990). Social relationships and ritualized greetings in adult male baboons (*Papio cynocephalus anubis*). *International Journal of Primatology* 11: 147–172.

Tomasello, M. and Call, J. (1997). *Primate Cognition*. Oxford: Oxford University Press.

Watanabe, J. M. and Smuts, B. B. (1999). Explaining religion without explaining it away: Trust, truth, and the evolution of cooperation in Roy A. Rappaport's "The obvious aspects of ritual." *American Anthropologist* 101: 98–112.

Wrangham, R. W. (1980). An ecological model of female-bonded primate groups. *Behaviour* 75: 262–300.

Zimen, E. (1982). A wolf pack sociogram. In *Wolves of the World*, F. H. Harrington and P. C. Paquest, eds., pp. 282–322. Park Ridge, N.J.: Noyes.

38 Animal Vocal Communication: Say What?

Drew Rendall and Michael J. Owren

Researchers interested in animal cognition have sometimes viewed communication as a privileged source of insight into animal minds (Griffin 1976). This view is inspired in large part by analogy to human experience, where language both reflects and affects thought and so provides a window into the workings of our own minds. Not surprisingly, cognitively oriented research in animal vocal communication has also been influenced by other analogies to human language. Without a doubt, the sine qua non of language is meaning, communicated via arbitrarily structured words. Proceeding by analogy, researchers in animal communication have thus sought meaning in arbitrarily structured vocalizations, the overarching question being, "*What are they saying?*"

On the one hand, relying on linguistic analogy is perfectly natural. The concept of meaning is certainly familiar to us from language (even if it is sometimes difficult to pin down very precisely) and it makes sense to think that animal vocalizations, like human words, have meaning and are *about* things. On the other hand, this approach is a bit peculiar in that it uses a single, recent, and potentially highly derived system of communication (language) to model scores of other phylogenetically older and evidently simpler systems—thereby inverting scientific common sense. Furthermore, by shoehorning a potentially wide array of communicative phenomena into a single linguistic frame, the approach risks seriously underestimating the diversity of potential mechanisms and functions of animal communication.

For the past few years we have been pursuing a different approach (Owren and Rendall 1997, 2001; Rendall and Owren, in review). In keeping with basic ethological and evolutionary principles, we assume only that the function of communication must ultimately be to influence the behavior of others in ways that are, on average, beneficial to the signalers (and potentially, although not necessarily, also to the listeners). While such influence may be exerted through a variety of simple mechanisms, none need involve meaning per se. Consistent with other features of organismal biology, however, they probably do involve intimate connections between signal structure and function. The approach can perhaps be most simply summarized as emphasizing that it may not be so much *what* is said that matters, but rather *how* it is said, and *who* says it.

A Nonlinguistic Approach to Animal Communication

"How You Say It"—Direct Effects of Vocalizations on Listener Attention and Affect

"It's not *what* you say but *how* you say it!" The point of this familiar refrain is that information content may be less important than the manner of presentation. This principle is manifest in animal communication in the fact that for many taxa, certain kinds of sounds have direct and marked influences on listener behavior. One extreme but ubiquitous example is the acoustic startle reflex. This involuntary response is particularly triggered by abrupt (i.e., rapid onset) sounds, producing immediate attentional shifts and the interruption of ongoing activity. It also induces a host of basic nervous system responses, including stimulation of reticular formation nuclei in the brainstem that help to regulate overall brain activation. The phenomenon is thought to occur in every hearing species (Eaton 1984), demonstrating that sound can have direct access to low-level nervous-system mechanisms that guide behavior.

Other examples of sounds with direct effects on listener behavior are common. For instance, handlers and herders of various domesticated animals have long capitalized on the impact of sounds such as whistles, tongue clicks, and lip smacks to manage their charges (McConnell 1991). Here, rapidly repeated pulses and signals with dramatic frequency upsweeps are used to increase motor activity, while smooth, continuous signals with gradual, descending pitch help to decrease activity. Humans themselves are responsive to the same patterns, with frequency upsweeps being used to capture receiver attention and increase arousal in both infant-directed speech (Fernald 1992; Kaplan and Owren 1994; Papousek et al. 1991) and music (Schneider 1963), and arousal-reducing downsweeps having a soothing effect. Additional familiar examples in humans include the effects of fingernails scraping on chalkboards, infant crying, and contagious laughter, all of which have directly noxious or pleasant effects on listeners.

Among nonhuman primates, the taxon with which we are most familiar, numerous vocalizations bear the mark of being designed for similar direct effects. For example, across many species, alarm vocalizations to warn of predators are structurally similar and preserve acoustic features that are well suited to capturing and manipulating attention and arousal in listeners (reviewed in Owren and Rendall 2001; Rendall and Owren, in review). They are typically short, abrupt-onset, broadband calls that elicit immediate orienting responses and movements preparatory to flight. In fact, the same basic alarm call structure is evident in a range of other mammals (Owings and Morton 1998), and also in some birds (Marler 1955), suggesting that the direct acoustic effects on listener attention and affect may be highly conserved, most likely stemming from a function in ancestral vertebrates of identifying and localizing sounds so as to avoid predators and capture prey.

Developmental studies in primates have shown that such generalized startle responses to species' alarm calls are induced even in very young infants in the absence of significant experience with either the calls or predators (e.g., Seyfarth and Cheney 1986; Herzog and Hopf 1983, 1984), as expected from the operation of low-level, brainstem and subcortical structures associated with sound localization, orienting, and autonomic responding. Because attentional and affective systems are the scaffolding for learning, the acoustic properties of these calls most likely specifically facilitate association of the salient dimensions of predator encounters and functionally appropriate escape responses (Rendall and Owren, in review).

Such evolved auditory sensitivities to certain kinds of sound create the opportunity for senders to use vocalizations to engage others, thereby influencing the attention, arousal, and concomitant behavior of listeners in many contexts. Among nonhuman primates, one entire class of vocalizations that we have labeled "squeaks, shrieks, and screams" appears to capitalize on this potential. These sounds are numerous and diverse, and are present in every well-documented primate vocal repertoire, as well as those of many other mammals, birds, and amphibians. They are marked by sharp onsets, dramatic fluctuations in frequency and amplitude, and chaotic spectral structures—exactly the sorts of features that have direct impact on animal nervous systems (Kitko et al. 1999).

Sounds of this type are produced in especially large numbers by younger animals (Owren et al. 1993) and may represent the bulk of their vocal output. In fact, such calls ought to be of greatest value to youngsters, who have little power to influence the behavior of older and larger individuals in other ways. For example, while a frustrated weanling cannot force its mother to permit nursing or close physical contact, it can produce sounds whose acoustic features pick at the mother's attentional mechanisms, increase her arousal state, and with repetition may become quite aversive (Hammerschmidt et al. 1994; Todt 1988). Adults can be similarly impotent, notably when interacting with dominant individuals. Lower-ranking victims of aggression seldom offer much serious physical resistance, whether they are youngsters or adults, but can make themselves unappealing targets by screaming vociferously, producing loud, jarring bursts of broadband noise and piercing, high-frequency, tonal sounds in highly variable streams whose aversive qualities are difficult to habituate to.

"Who Says It"—Indirect Effects of Vocalizations Mediated Through Social Acuity

A second straightforward but effective way to influence others vocally is simply to advertise one's identity. This strategy is likely to be observed in highly social species where identity has a deterministic influence on the frequency and quality of social interactions. In many nonhuman primates, for instance, individual identity, kinship, and social rank are the pillars of complex social behavior. Who's who and doing what to whom is crucially important, and as a result the animals are attentive to each other's social identity and actions (Cheney et al. 1986).

Vocalizations in these species thus often contain conspicuous cues to caller identity, and listeners are demonstrably responsive to such cues. For example, acoustic analyses have revealed clear identity cues in the "contact" calls used to coordinate activity among dispersed group members in several species (reviewed in Rendall et al. 1998, 2000). For some species, field experiments involving the playback of these calls have confirmed that different group members can be distinguished by voice and that identity cues form the basis for differential responses by listeners (e.g., Mitani 1985; Snowdon and Cleveland 1980; Waser 1977). Among rhesus monkeys, for example, adult females are more responsive to the contact calls of adult female kin than to those of unrelated females in the group (Rendall et al. 1996). In many species, females are also highly responsive to the calls of their infants. For instance, baboon females discriminate the "lost" calls of their own infants from those of other infants and often search anxiously for their own infant if they hear it calling (Rendall et al. 2000).

Conspicuous cues to caller identity are also present in many close-range calls used to mediate face-to-face social interactions (Owren et al. 1993, 1997). Here again, work has confirmed that the identity cues are salient to listeners and influence their behavior. For example, among baboon females, the strongest determinant of listeners' responses to quiet, grunt vocalizations produced in various social contexts is the identity and social rank of the caller. Listeners are far more attentive to the calls of higher-ranking than lower-ranking females (Rendall et al. 1999).

In such face-to-face social interactions, conspicuous vocal cues to identity might seem wholly redundant, given that obvious cues to identity are available visually. However, vocal cues to identity actually have a special capacity to influence behavior in this context because they can be paired with behavioral acts that have significant consequences for others and become predictive of the acts (Owren and Rendall 1997, 2001). For example, in many social species, dominant individuals routinely antagonize more subordinate group members by combining conspicuous threat vocalizations with emotion-inducing aggressive acts such as hitting, biting, and chasing. The significance of the threat calls in predicting attack means that dominant animals can thereafter elicit learned fear and withdrawal from subordinates through the use of the vocalizations alone. Indeed, most conflicts are resolved in exactly this manner, with no aggressive exchange. Identity cues feature centrally here because, given the hierarchically structured social networks of these species, animals will routinely hear many such calls by others over the course of a day without being attacked. Hence, what makes the calls uniquely predictive in any given situation is who produced them.

Depending on one's social potency, then, simply announcing identity can be an effective way to influence others in a range of contexts. In fact, vocalizations are uniquely suited to exerting such influence because they are discrete stimuli that can be readily controlled by the sender and paired with emotion-inducing behavioral acts. They are also more difficult to ignore than signals in other modalities (simply turning one's head away effectively short-circuits a visual signal, but not a vocal one).

Not all calls are similarly imbued with conspicuous identity cues, however. For example, acoustic analyses and field playback experiments on scream vocalizations produced by victims of aggressive attack show that these calls contain few cues to caller identity (Rendall et al. 1998; Owren 2000). At first, this result might seem surprising because screams have traditionally been assumed to operate over longer distances to recruit aid from genetic relatives who may be out of sight (e.g., Gouzoules et al. 1984). However, in light of the foregoing points, the findings are actually quite sensible. Almost by definition, victims of aggressive attack are socially impotent relative to their attackers, and so their identity is unlikely to influence the balance of power when they are attacked. As a result, the most effective strategy for victims may be to capitalize on the direct effects of sound and be as obnoxious as possible. Loud, hypervariable, noisy screams physically preclude identity cueing, but their aversive properties may motivate attackers to relent, thereby "turning them off" (Owren and Rendall 1997; Rendall 1996; Rendall et al. 1998).

Implications for Mind

The alternative approach to animal vocal communication that we have taken proposes two simple ways in which vocalizations can influence listener behavior in a variety of contexts, and doubtless there are many other similarly simple ways. The two sorts of effects we propose stem directly from basic conditioning principles and in accordance with the widespread relevance of such principles appear to be broadly applicable across numerous types of vocalization and many species of nonhuman primates. In some cases they appear more broadly applicable to other animal taxa and other sensory domains (e.g.,

Endler and Basolo 1998; Ryan 1998). Both effects emphasize the intimacy of the connection between signal structure and function, in contrast to linguistic approaches, which, by analogy to human words, stress the arbitrary nature of signal design. Neither effect recommends use of the linguistic construct of meaning. Thus, vocalizations that serve their function either through direct effects on receiver psychology or through indirect effects mediated by a developed social acuity are not readily interpreted in terms of meaning. To reiterate, the significance of such signals lies not in *what* is said, but rather *who* says it and *how*. Hence, linguistic constructs, like meaning, may have limited application to animal vocalizations.

In fact, recent comparative research on intentionality points to the same conclusion. In human language, the meaning of words stems, not from their sound structure per se, but from the shared conceptual representations they conjure in the minds of speaker and listener alike. Effective communication via shared conceptual representations is made possible by the fact that speakers understand the relation between the words they use and the conceptual representations they instantiate, and they know that others share this understanding. Moreover, they know that others can be informed of phenomena about which they are unaware using words whose representational designata are nevertheless understood. Hence, speech is a deliberate attempt to invoke and sometimes manipulate shared conceptual representations in the minds of others, oftentimes with the goal to inform. The active invocation of such representations, and the sensitivity to others' mental states that is entailed, gives language use in humans the formal property of being intentional (as defined by Dennett 1983). Word meaning, then, is a derivative characteristic of human language that stems from its intentionality.

In contrast, communication in nonhuman primates is not similarly intentional. Recent reviews indicate that even the most cognitively sophisticated nonhuman primates do not use vocalizations as a deliberate attempt to inform others or otherwise modify their mental states (Cheney and Seyfarth 1996, 1998; Seyfarth and Cheney 1997; Rendall and Owren in review). In fact, callers prove to be surprisingly disengaged from their audience, often calling redundantly when all those around them are already aware of a situation, but failing to call in circumstances in which others are unaware of events and could profit from being informed. Evidently they do not appreciate the effect that their calls can have on *what* listeners know. These and additional related findings have led authors to conclude that the proximate cognitive mechanisms underlying human language and nonhuman primate vocal communication are in fact fundamentally different (Cheney and Seyfarth 1996). While the content, or meaning, of human word use hinges on intentional agents—*meaners* to mean what they say—nonhuman primate communication evidently does not. By extension then, we are left to wonder whether there can ever truly be *meaning* in primate vocalizations without there being a *meaner*?

Our response to this question is to suggest that the vocalizations of nonhuman primates (and probably those of many other animal taxa) function differently than the words of human language. Although the animals are clearly capable of forming rich, multidimensional representations of important events, invoking and manipulating such conceptual representations per se is not evidently a central feature of their communication. Instead, the vocalizations of nonhuman primates appear to function primarily through their influence on the affective systems that guide behavior. Note that our emphasis here is on the effect of signals on the affective systems that govern the behavior of receivers, rather than on the affective systems that govern the signal production of signalers, making our approach different in important respects from those of Smith (1977) and others (e.g., Lancaster 1975; Rowell and Hinde 1962).

In accordance with its focus on affective processes, our approach emphasizes subcortical systems like the brainstem and limbic structures that control attention, arousal, and affect, rather than higher-level cortical systems associated with conceptual representation and language comprehension in humans. While research on the neurobiology of vocal communication in primates has long pointed to the central role played by subcortical systems (reviewed in Deacon 1997; Hauser 1996; Jürgens 1998), work outside the laboratory has emphasized higher-level processes, if only by implication, as a result of analogy to human language. However, cortical processing in humans and nonhumans alike is necessarily shaped by the input received from lower-level brain centers. Therefore probing the role that cortically based cognition may play in any communication system depends on first understanding the processing that occurs at subcortical levels. We suggest that actively studying potential effects at these lower levels is likely to contribute to a more solid foundation for understanding the diversity of functions and mechanisms in animal communication, including those that may ultimately result in cortically based processes like the conceptual representations that underlie language use in humans.

Acknowledgments

We thank Sergio Pellis and John Vokey for many helpful discussions of topics related to this paper. We are also grateful for ongoing support from the Natural Sciences and Engineering Research Council of Canada, the University of Lethbridge, and Cornell University.

References

Cheney, D. L. and Seyfarth, R. M. (1996). Function and intention in the calls of non-human primates. In *Evolution of Social Behaviour Patterns in Primates and Man*, W. G. Runciman, J. Maynard Smith, and R. I.

M. Dunbar, eds., pp. 59–76. Oxford: Oxford University Press.

Cheney, D. L. and Seyfarth, R. M. (1998). Why animals don't have language. In *The Tanner Lectures on Human Values*, G. B. Pearson, ed., pp. 175–209. Salt Lake City: University of Utah Press.

Cheney, D. L., Seyfarth, R. M., and Smuts, B. B. (1986). Social relationships and social cognition in nonhuman primates. *Science* 234: 1361–1366.

Deacon, T. W. (1997). *The Symbolic Species*. New York: W. W. Norton.

Dennett, D. (1983). Intentional systems in cognitive ethology, The "Panglossian" paradigm defended. *Behavioral and Brain Sciences* 6: 343–355.

Eaton, R. C. (ed.) (1984). *Neural Mechanisms of Startle Behavior*. New York: Plenum.

Endler, J. A. and Basolo, A. L. (1998). Sensory ecology, receiver biases and sexual selection. *Trends in Ecology and Evolution* 13: 415–420.

Fernald, A. (1992). Meaningful melodies in mothers' speech to infants. In *Nonverbal Vocal Communication: Comparative and Developmental Approaches*, H. Papousek, U. Jürgens, and M. Papousek, eds., pp. 262–282. Cambridge: Cambridge University Press.

Gouzoules, S., Gouzoules, H., and Marler, P. (1984). Rhesus monkey (*Macaca mulatta*) screams: Representational signaling in the recruitment of agonistic aid. *Animal Behaviour* 32: 182–193.

Griffin, D. (1976). *The Question of Animal Awareness: Evolutionary Continuity of Mental Experience*. New York: Rockefeller University Press.

Hammerschmidt, K., Ansorge, V., Fischer, J., and Todt, D. (1994). Dusk calling in Barbary macaques (*Macaca sylvanus*): Demand for social shelter. *American Journal of Primatology* 32: 277–289.

Hauser, M. D. (1996). *The Evolution of Communication*. Cambridge, Mass.: Harvard University Press.

Herzog, M. and Hopf, S. (1983). Effects of species-specific vocalizations on the behaviour of surrogate-reared squirrel monkeys. *Behaviour* 86: 197–214.

Herzog, M. and Hopf, S. (1984). Behavioral responses to species-specific warning calls in infant squirrel monkeys reared in social isolation. *American Journal of Primatology* 7: 99–106.

Jürgens, U. (1998). Mammalian vocalization: With special reference to the squirrel monkey. *Naturwissenschaften* 85: 376–388.

Kaplan, P. S. and Owren, M. J. (1994). Dishabituation of visual attention in 4-month-olds by infant-directed frequency sweeps. *Infant Behavior and Development* 17: 47–358.

Kitko, R., Gesser, D., and Owren, M. J. (1999). Noisy screams of macaques may function to annoy conspecifics. *Journal of the Acoustical Society of America* 106: 2221.

Lancaster, J. B. (1975). *Primate Behavior and the Emergence of Human Culture*. New York: Holt, Rinehart and Winston.

Marler, P. (1955). Characteristics of some animal calls. *Nature* 176: 6–8.

McConnell, P. B. (1991). Lessons from animal trainers: The effect of acoustic structure on an animal's response. In *Perspectives in Ethology: Human Understanding and Animal Awareness*. Vol. 9, P. Bateson and P. Klopfer, eds., pp. 165–187. New York: Plenum.

Mitani, J. C. (1985). Sexual selection and adult male orangutan long calls. *Animal Behaviour* 33: 272–283.

Owings, D. H. and Morton, E. S. (1998). *Animal Vocal Communication: A New Approach*. Cambridge: Cambridge University Press.

Owren M. J. (2000). Spectral content and function of nonhuman primate vocalizations is shaped by both vocal-fold vibration and supra-laryngeal filtering characteristics. Paper presented at the *Fifth Seminar on Speech Production: Models and Data*. Kloster Seeon, Germany.

Owren, M. J. and Rendall, D. (1997). An affect-conditioning model of nonhuman primate vocal signaling. In *Perspectives in Ethology: Communication*. Vol. 12, D. H. Owings, M. D. Beecher, and N. S. Thompson, eds., pp. 299–346. New York: Plenum.

Owren, M. J. and Rendall, D. (2001). Sound on the rebound: Bringing form and function back to the forefront in understanding primate vocal signaling. *Evolutionary Anthropology* 10: 58–71.

Owren, M. J., Dieter, J. A., Seyfarth, R. M., and Cheney, D. L. (1993). Vocalizations of rhesus (*Macaca mulatta*) and Japanese (*M. fuscata*) macaques cross-fostered between species show evidence of only limited modification. *Developmental Psychobiology* 26: 389–406.

Owren, M. J., Cheney, D. L., and Seyfarth, R. M. (1997). The acoustic features of vowel-like grunt calls in chacma baboons (*Papio cynocephalus ursinus*). *Jour-*

nal of the Acoustical Society of America 101: 2951–2963.

Papousek, M., Papousek, H., and Symmes, D. (1991). The meanings of melodies in motherese in tone and stress languages. *Infant Behavior and Development* 14: 415–440.

Rendall, D. (1996). Social Communication and Vocal Recognition in Free-ranging Rhesus Monkeys (*Macaca mulatta*). Ph.D. dissertation, University of California at Davis.

Rendall, D. and Owren, M. J. (in review). No meaning required: Abandoning the quest for a language grail in animal communication. *Behavioral and Brain Sciences*.

Rendall, D., Rodman, P. S., and Emond, R. E. (1996). Vocal recognition of individuals and kin in free-ranging rhesus monkeys. *Animal Behaviour* 51: 1007–1015.

Rendall, D., Owren, M. J., and Rodman, P. S. (1998). The role of vocal tract filtering in identity cueing in rhesus monkey (*Macaca mulatta*) vocalizations. *Journal of the Acoustical Society of America* 103: 602–614.

Rendall D., Cheney D. L., Seyfarth R. M., and Owren, M. J. (1999). The meaning and function of grunt variants in baboons. *Animal Behaviour* 57: 583–592.

Rendall D., Cheney D. L., and Seyfarth R. M. (2000). Proximate factors mediating "contact" calls in adult female baboons and their infants. *Journal of Comparative Psychology* 114: 36–46.

Rowell, T. E. and Hinde, R. A. (1962). Vocal communication by the rhesus monkey (*Macaca mulatta*). *Proceedings of the Zoological Society of London* 138: 279–294.

Ryan, M. J. (1998). Sexual selection, receiver biases, and the evolution of sex differences. *Science* 281: 1999–2003.

Schneider, E. H. (1963). The rhythmic approach in music therapy. In *Music Therapy*. Vol. 12, E. H. Schneider, ed., pp. 71–97. Lawrence, Kan.: Allen.

Seyfarth, R. M. and Cheney, D. L. (1986). Vocal development in vervet monkeys. *Animal Behaviour* 34: 1640–1658.

Seyfarth, R. M. and Cheney, D. L. (1997). Behavioral mechanisms underlying vocal communication in non-human primates. *Animal Learning & Behavior* 25: 249–267.

Smith, W. J. (1977). *The Behavior of Communicating: An Ethological Approach*. Cambridge, Mass.: Harvard University Press.

Snowdon, C. T. and Cleveland, J. (1980). Individual recognition of contact calls by pygmy marmosets. *Animal Behaviour* 28: 717–727.

Todt, D. (1988). Serial calling as a mediator of interaction processes: Crying in primates. In *Primate Vocal Communication*, D. Todt, P. Goedeking, and D. Symmes, eds., pp. 88–107. Berlin: Springer-Verlag.

Waser, P. M. (1977). Individual recognition, intra-group cohesion and intergroup spacing: Evidence from sound playback to forest monkeys. *Behaviour* 60: 28–74.

Cracking the Code: Communication and Cognition in Birds

Christopher S. Evans

Since monkeys certainly understand much that is said to them by man, and when wild, utter signal-cries of danger to their fellows; and since fowls give distinct warnings for danger on the ground, or in the sky from hawks ... may not some unusually wise apelike animal have imitated the growl of a beast of prey, and thus told his fellow-monkeys the nature of the expected danger? This would have been a first step in the formation of a language.... When we treat of sexual selection we shall see that primeval man, or rather some early progenitor of man, probably first used his voice in producing true musical cadences, that is in singing, as do some of the gibbon-apes at the present day; and we may conclude from a widely-spread analogy, that this power would have been especially exerted during the courtship of the sexes,—would have expressed various emotions, such as love, jealousy, triumph,—and would have served as a challenge to rivals. (Darwin 1871, pp. 56–57)

Charles Darwin clearly believed that language had evolved from precursors in the natural signals of animals. As with so much of his writing, these passages anticipate recent research programs. He points out that monkeys and chickens have distinctive alarm calls for different kinds of danger, and goes on to suggest that language is the product of sexual selection. Darwin's argument is a case for continuity.

Over a hundred years later, this idea is still treated with considerable skepticism (e.g., Premack 1975; Luria 1982; Wallman 1992; Lieberman 1994). Critics typically take the Cartesian position that language is special, in the sense that all of its attributes are unique to humans. It follows that comparative studies should fail to reveal any comparable traits in nonhuman animals. These reservations are often summarized in two related assertions: First, that animal signals are simply a readout of emotional state and second, that their production is reflexive or involuntary. The resolution of this controversy is important because if Darwin was right, then we can use communication as a window on the minds of nonhuman animals. Evidence for continuity would also force us to re-think assumptions about the nature and extent of human uniqueness.

My research program focuses on the relationship between acoustic signaling and cognition in birds. I have adopted an ethological approach (Tinbergen 1963), choosing to study natural behavior of obvious functional importance. The techniques used include both controlled laboratory experiments to characterize mechanisms and studies of social groups under natural conditions to obtain insights about function. The theoretical assumption underpinning this work is that cognitive processes are adaptations in just the same way as physical structures.

Referential Signals

The first evidence that animal communication might be more complex than traditional models had anticipated came from Struhsaker's (1967) pioneering field studies of vervet monkeys (*Cercopithecus aethiops*). This work established that vervets have acoustically distinct alarm calls corresponding to their three principal classes of predator: eagles, leopards, and snakes. Seyfarth and Cheney followed up this work with playback experiments, convincingly demonstrating that calls are sufficient to evoke responses appropriate to the type of predator that had originally elicited the sound (Seyfarth et al. 1980). Macedonia's (1990) studies of ring-tailed lemurs (*Lemur catta*) provide similar evidence of predator class-specific alarm calls.

Vervets and lemurs have referential signals. In both species, identifiable external events reliably elicit a particular type of call and these signals are sufficient to evoke adaptive responses, even when contextual cues are unavailable. The strategy for exploring the characteristics of any system of referential signals involves mapping these relationships between eliciting conditions and signal structure, and between signal structure and receiver's response (Marler et al. 1992; Evans and Marler 1995; Evans 1997).

I am fascinated by the challenge of understanding the information encoded in animal signals. My subjects are golden Sebright chickens, an ornamental strain closely related to the ancestral red junglefowl (*Gallus gallus*). These birds have a large vocal repertoire (Collias 1987), which does not seem to have been altered by domestication. Most important, they have two distinct types of alarm call. A particular advantage of working with chickens is that they tolerate captivity well. It is thus possible to analyze quite precisely the conditions under which alarm calls are produced by presenting simulated predators and manipulating their characteristics.

The subjects in these laboratory experiments are males because aerial alarm calling is testosterone dependent (Gyger et al. 1988). Individual roosters are confined in a cage with a large video monitor supported on a frame overhead so that the screen is horizontal and facing downward (figure 1 in Evans and Marler 1992). Most of the time the monitor displays a blank white field, but periodically computer-generated animations of raptors can be triggered so that they fly across the screen. A second monitor at ground level allows footage of terrestrial predators to be presented. These video sequences evoke the full gamut of antipredator responses (figure 1 in Evans et al. 1993a). In early experiments, we showed the birds fast-moving hawk silhouettes overhead and an edited sequence of a walking raccoon (*Procyon lotor*) at ground level and found that there was an unambiguous relationship between alarm call and predator type with these prototypical stimuli. Only aerial alarm calls were evoked by the hawk animation and only ground alarm calls were evoked by the raccoon footage (Evans et al. 1993a).

One possible explanation for this finding was that alarm call type might simply reflect the spatial location of a stimulus. We assessed this idea by digitizing video sequences of a soaring hawk and of the raccoon and then editing them frame by frame to remove the background. This ensured that comparisons between the predator types would not be confounded by the settings in which they had been filmed. We then showed males the isolated hawk and raccoon, presenting each of them at ground level and overhead.

The results replicated our original demonstration that only aerial alarm calls are given in response to a raptor presented above and only ground alarm calls to a terrestrial predator in the normal position. The hawk at ground level and raccoon overhead both evoked a mixed response, but each of these stimuli was significantly less effective than the hawk overhead (figure 15.6 in Evans and Marler 1995). Evidently placing a terrestrial predator overhead is not sufficient for the birds to treat it as an aerial predator. On the other hand, spatial location clearly plays a role, because the number of calls elicited by both types of predator was reduced when they were in inappropriate positions. Unfortunately, this experiment has been misinterpreted as evidence that chicken alarm type is entirely determined by the location of the threat (Zuberbühler 2000). Neither the data nor our original description of this study (Evans and Marler 1995) logically supports such a conclusion.

Next, we turned to the question of specificity, concentrating on aerial alarm calls. What constitutes an adequate stimulus? Would any moving object do, or are alarm calls triggered only by potential predators? Experiments explored the importance of simple parameters such as apparent size, speed, and shape. We used computer-generated animations so that we could define stimulus attributes precisely; a single characteristic could be manipulated while all others were held constant. The speed continuum covered the full range that a bird would be likely to experience under natural conditions, from the leisurely movement of a vulture soaring on a thermal at one end, to the speed of an attacking harrier at the other. Similarly, the apparent size of hawk shapes (expressed in degrees subtended at the bird's eye) was varied from 1 to 8°, a range known to be associated with variation in alarm calling during natural encounters with potential predators (Gyger et al. 1987). Stimulus shape was manipulated by using a morphing algorithm to create a continuum from a disk to a realistic silhouette of a red-tailed hawk.

This series of experiments allowed us to define in some detail the characteristics of events that chickens respond to with aerial alarm calls. The most effective stimuli were found to be overhead (Evans and Marler 1995), large (apparent size > 4°; Evans et al. 1993b), fast-moving (apparent speed > 7.5 lengths/second; Evans et al. 1993b), and approximately bird shaped (Evans and Marler 1995).

Studies of signal production thus show that chickens have two distinct types of alarm call, each associated with a different class of predator, and that aerial alarms at least are evoked by a relatively specific subset of possible stimuli. What about signal receivers—are these sounds sufficient for selection of an appropriate response? To find out, we played both types of alarm call to isolated hens confined in a cage, with a small area of brush to provide cover. Hens responded to ground alarm calls by drawing themselves up into an unusually erect "alert" posture, becoming more active and scanning back and forth in the horizontal plane, as though trying to detect a threat approaching on the ground. When they heard aerial alarm calls, they reacted quite differently, running to cover and then crouching with their feathers sleeked and looking upward, precisely as though they were trying to detect an object moving overhead. Each of these responses is appropriate to the type of predator that originally elicited the call. We concluded that chicken alarm calls, like those of vervets and lemurs, are referential signals (Evans et al. 1993a).

Predators are not the only environmental events that chickens respond to with distinctive calls. Consider the social interaction depicted in figure 39.1. The male has found a food item and is producing characteristic pulsatile sounds, known traditionally as food calls (Collias and Joos 1953). A hen has responded by approaching and is now fixating closely on the fragment of grass in the male's beak. The male will typically then let the hen have the food, either by dropping it on the ground in front of her or by allowing her to take it directly from between his mandibles.

There are two types of explanation for this behavior. One possibility is that hens approach because food calls provide quite specific information about feeding opportunities (Marler et al. 1986). Alternatively, food calls might not really be "about" edible objects at all, but rather describe the subsequent behavior of the sender (Smith 1991), in which case hens would respond to these sounds because they predict a low probability of aggression. To distinguish between these accounts, we needed to study production of food calls. Are they dependent upon the availability of food, or are they just given by friendly males?

We used simple instrumental conditioning techniques to train roosters to peck a key for periodic deliveries of small food pellets (Evans and Marler 1994). During the first 2 minutes of each test session, the key was unlit and pecks delivered to it had no effect; then it was switched on, signaling that food was available. The results of this simple manipulation revealed that food calling is indeed dependent upon the presence of food items and cues reliably associated with them. When the key was first lit, the rate of food calling increased by an order of magnitude, then dropped slowly as the males become satiated. Tests with a hen confined in an adjacent cage also showed that there was no temporal relationship between food calling and courtship behavior (Evans and Marler 1994). These results demonstrated that there is a predictive relationship between food availability and production of food calls by males, which is incompatible with the idea that these sounds reveal only the subse-

Figure 39.1
Typical interaction between a food-calling male golden Sebright chicken and a hen. Note close inspection of the food item by the female.

quent behavior of the sender, although it is likely that this information is also encoded.

The next study considered the receiver's point of view. Studies of call production suggested that food calls might signal feeding opportunities, but in a natural interaction hens have lots of other information available to them. Roosters are not only calling, but also performing stereotyped 'tidbitting' movements in which they pick up and drop the food item repeatedly (figure 39.1). Females can clearly see these visual signals and also perhaps the food itself. To determine whether food calls alone are sufficient to explain the hens' responses, we needed to conduct playback experiments in which all of these other cues were stripped away.

Our first study compared food calls with ground alarm calls, which have similar acoustic characteristics. The second series of playbacks compared food calls with contact calls, which are also produced during affiliative social interac-

tions, and are hence matched for the subsequent behavior of the sender. Hens responded to recorded food calls by moving about the cage, pausing repeatedly to fixate on the ground in front of them. There was no such increase in substrate-searching behavior when ground alarm calls were played back. Even though these two sounds are structurally similar, they have qualitatively different effects on the hens' behavior, and these reflect the different circumstances under which the calls are produced (Evans and Evans 1999). Contact calls also had no effect on the duration of substrate searching. This result demonstrates that the effects of food calls are quite specific and suggests that the behavior of hens is mediated by the likely availability of food, rather than a desire to approach a nonaggressive companion (Evans and Evans 1999).

Taken together, these laboratory studies of alarm calling and food calling reveal a perhaps surprising degree of communicative complexity.

Other work, which I will not describe in detail here, shows that production of these calls is not simply dependent upon the eliciting stimulus (food or predators) but is also sensitive to the presence of appropriate receivers (Evans and Marler 1991, 1992, 1994). Such audience effects are consistent with a degree of volitional control.

Communication and Cognition

The acoustic signals of birds are thus substantially more sophisticated than most theorists have anticipated (Evans and Marler 1995; Evans 1997). The same is true of the alarm calls of primates (Cheney and Seyfarth 1990). Such signal systems do not, however, necessarily require us to postulate complex cognitive underpinnings; there are several alternative interpretations.

Consider vervets giving snake alarms to an approaching python. Their system, like those of chickens, has the following properties: One type of call (X) is reliably elicited by a specific class of environmental events (Y), and presentations of X are sufficient to evoke responses adaptive for dealing with Y.

How are we to interpret such observations? One possibility is that the alarm call is semantic, or representational—it tells companions specifically "There is a Y" (Cheney and Seyfarth 1990). But other interpretations cannot be excluded. Call X might reflect a Y-specific internal state ('snake fear'; Premack 1975; Lieberman 1994). Alternatively, it might not have an external referent "snake," but rather a behavioral referent that describes the subsequent actions of the sender ("I'm going to stand bipedally and peer into the grass"; Smith 1991). Finally, we have the interpretation favored by Baron-Cohen (1992) and Wallman (1992): Call X is instructive. It tells companions "Quick! Look down!"

Any claim for languagelike attributes in the natural signals of animals depends critically upon underlying cognitive processes. It requires that call X really be representational. Each of the other three possibilities produces the same be-

havioral output, but in these cases the parallel with language is illusory. This is why some theorists remain skeptical about the degree of continuity implied by the contemporary literature on referential signaling.

It has been challenging to distinguish among these models because the problem does not seem at first to be experimentally tractable. Some years ago, Joe Macedonia and I suggested that it might be best to adopt a neutral stance, acknowledging the constraints on interpretation of observations and playback experiments; studies like those that I have discussed so far demonstrate only that call X is functionally referential—receivers behave *as if* it predicts a class of environmental events (Macedonia and Evans 1993). Early attempts to grapple with this problem empirically in the vervet system (Cheney and Seyfarth 1988, 1990) provided only weak evidence for semantic properties (Evans 1997), but there is now agreement that the central issue is whether referential signals have their effects by evoking representations of the eliciting event.

Representation is a special word in psychology, cognitive science, and linguistics, so it is important to explain my use of it here. Consider the relationship between football jersey numbers and player identity, a simple example borrowed from Gallistel (1990). Mapping between these systems is systematic and constrained in two ways: First, assignment is one-to-one (each player can only wear a single number). Second, numbers are unique (the same number cannot be given to more than one player). These rules define a functioning isomorphism between the number system and the individual identity system.

There is a clear parallel between this type of mapping operation and the use of "representation" in neural networks, where it denotes the relationship between input values and a pattern of nodal activation. This general usage in psychology has a long history; it can be traced back at least as far as William James (1890), who speculated along the same lines about the relationship between stimuli and brain activity.

Gallistel (1990) calls representations of this type "nominal representations." He places them at the lowest level in a hierarchy of increasingly complex types of representation, characterized by the computational operations required to derive them. Gallistel considers nominal representations impoverished because the only operation that they can compute is *equals to* or *identity*. But such operations are functionally important. Animals frequently have to test for a match between stored information and a stimulus that they have encountered, and often the decisions made as a consequence of such processes really matter. For example, birds can learn to avoid insects with warning coloration after a single unpleasant experience. Each subsequent encounter is effectively a matching-to-sample task, requiring comparison with stored information about morphology.

In some systems we can be confident that adult behavior reflects the acquisition of representations during development. Oscine birds acquire a detailed model of song while they are still nestlings and reproduce it as adults (including all of the nuances of the local dialect) by matching their own output against this template. Exploration of the specialized systems necessary for this feat has been one of the most successful paradigms in neuroethology (Marler 1990; DeVoogd 1994). A similar point can be made about filial imprinting. Elegant experiments by Bateson, Horn, and their colleagues have shown that storage of information about an imprinting object (in nature this would be the mother's face) involves centers in the chick's left forebrain; this process is now understood at the cellular level (Horn 1990; Honey et al. 1995).

There is hence no doubt that animals have representations. Indeed, they may prove to be ubiquitous. What is contentious is whether they are involved in communication. To make this question specific, and thus testable, let us return to chicken food calls and ask how we might explain the characteristic substrate-searching response of a hen. There are two distinct possibilities. The call might evoke looking downward directly by triggering the appropriate motor pattern. Alternatively, food calls might stimulate retrieval of stored information about food, which then determines the hen's response. In the absence of additional evidence, we might prefer the reflexive account for reasons of parsimony.

To address this problem, we chose a strategy based upon classical work in associative learning which established that behavior is influenced by stored information about reinforcer properties (reviewed by Shettleworth 1998). The general approach in these studies was to explore the effects of manipulating one class of environmental events. Animals were trained to associate two distinct stimuli with two different types of reinforcer. Then one reinforcer, but not the other, was selectively devalued, either by pairing it with a toxin or by satiation. This caused a change in response that was specific to the stimulus associated with the devalued reinforcer, demonstrating that the animals had formed separate representations for each of the two rewards (Colwill and Rescorla 1985; Holland 1990; Hall 1996). Results of this kind are incompatible with the idea that responses to conditioned stimuli are entirely reflexive; they require us instead to postulate simple cognitive processes.

The logic of our design was closely analogous. We reasoned that if chicken food calls encode information about feeding opportunities, then responses evoked by playback should be affected by prior experience of food. Such a change in responsiveness should be highly specific; responses to another, similar, call type should be unaffected. Neither of these predictions is generated by alternative nonrepresentational models of call processing by the receiver.

Ground alarm calls were selected as a control because, as indicated, they closely match the acoustic structure of food calls, but have very different eliciting conditions. During test trials, calls were played back to hens either 3 minutes after delivery of a small quantity of food (three fresh corn kernels), or without such a pre-playback experience. The critical planned comparisons tested for an interaction between pre-playback experience and signal type.

The effects of manipulating prior access to food selectively changed responses to food calls, exactly as predicted by a representational model (Evans and Evans, in review). Playback of food calls evoked anticipatory feeding behavior, but only when this had not been preceded by a food delivery. In no-food trials, hens responded to food calls by searching the substrate significantly more than in trials with matched ground alarm calls. This difference was abolished by allowing a brief feeding bout before playback of the call. Our interpretation is that under these conditions, the signal provided no new information. In contrast, the response to ground alarm calls was entirely unaffected by prior experience of food. These results demonstrate that chicken food calls do not affect the behavior of companions by evoking motor patterns in a reflexive way. Rather, they stimulate access to a representation of food.

Recently completed experiments with alarm calls have produced precisely comparable results; responses are selectively changed by prior experience of the corresponding predator type. We conclude that chicken calls produce effects by evoking representations of a class of eliciting events. This finding should contribute to resolution of the debate about the meaning of referential signals. We can now confidently reject reflexive models, those that postulate only behavioral referents (Smith 1991), and those that view referential signals as imperative (Baron-Cohen 1992; Wallman 1992). The humble and much-maligned chicken thus has a remarkably sophisticated system. Its calls denote at least three classes of external objects. They are not involuntary exclamations, but are produced under particular social circumstances.

Clearly, representational signaling is not restricted to our closest primate relatives (see reviews by Hauser 1996; Shettleworth 1998). One of the major challenges for the future will now be to identify the ecological and social factors responsible for the evolution of such systems. The insights obtained will integrate studies of cognition more closely with those of function.

Acknowledgment

This research was supported by grants from the National Institutes of Health, the Australian Research Council, and Macquarie University.

References

Baron-Cohen, S. (1992). How monkeys do things with "words." *Behavioral and Brain Sciences* 15: 148–149.

Cheney, D. L. and Seyfarth, R. M. (1988). Assessment of meaning and the detection of unreliable signals by vervet monkeys. *Animal Behaviour* 36: 477–486.

Cheney, D. L. and Seyfarth, R. M. (1990). *How Monkeys See the World: Inside the Mind of Another Species.* Chicago: University of Chicago Press.

Collias, N. E. (1987). The vocal repertoire of the red junglefowl: A spectrographic classification and the code of communication. *Condor* 89: 510–524.

Collias, N. E. and Joos, M. (1953). The spectrographic analysis of sound signals of the domestic fowl. *Behaviour* 5: 175–188.

Colwill, R. M. and Rescorla, R. A. (1985). Post-conditioning devaluation of a reinforcer affects instrumental responding. *Journal of Experimental Psychology: Animal Behavior Processes* 11: 120–132.

Darwin, C. (1871). *The descent of Man and Selection in Relation to Sex.* London: J. Murray.

DeVoogd, T. J. (1994). The neural basis for the acquisition and production of bird song. In *Causal Mechanisms of Behavioural Development*, J. A. Hogan and J. J. Bolhuis, eds., pp. 49–81. Cambridge: Cambridge University Press.

Evans, C. S. (1997). Referential signals. *Perspectives in Ethology* 12: 99–143.

Evans, C. S. and Evans, L. (1999). Chicken food calls are functionally referential. *Animal Behaviour* 58: 307–319.

Evans, C. S. and Evans, L. (in review) Representational signalling in birds. *Proceedings of the Royal Society of London. Series B: Biological Sciences.*

Evans, C. S. and Marler, P. (1991). On the use of video images as social stimuli in birds: Audience effects on alarm calling. *Animal Behaviour* 41: 17–26.

Evans, C. S. and Marler, P. (1992). Female appearance as a factor in the responsiveness of male chickens during anti-predator behaviour and courtship. *Animal Behaviour* 43: 137–143.

Evans, C. S. and Marler, P. (1994). Food calling and audience effects in male chickens, *Gallus gallus*: Their relationships to food availability, courtship and social facilitation. *Animal Behaviour* 47: 1159–1170.

Evans, C. S. and Marler, P. (1995). Language and animal communication: Parallels and contrasts. In *Comparative Approaches to Cognitive Science*, H. L. Roitblat and J. Arcady-Meyer, eds., pp. 341–382. Cambridge, Mass.: MIT Press.

Evans, C. S., Evans, L., and Marler, P. (1993a). On the meaning of alarm calls: Functional reference in an avian vocal system. *Animal Behaviour* 46: 23–38.

Evans, C. S., Macedonia, J. M., and Marler, P. (1993b). Effects of apparent size and speed on the response of chickens, *Gallus gallus*, to computer-generated simulations of aerial predators. *Animal Behaviour* 46: 1–11.

Gallistel, C. R. (1990). *The Organization of Learning*. Cambridge, Mass.: MIT Press.

Gyger, M., Marler, P., and Pickert, R. (1987). Semantics of an avian alarm call system: The male domestic fowl, *Gallus domesticus*. *Behaviour* 102: 15–40.

Gyger, M., Karakashian, S. J., Dufty, A. M., and Marler, P. (1988). Alarm signals in birds: The role of testosterone. *Hormones and Behavior* 22: 305–314.

Hall, G. (1996). Learning about associatively activated stimulus representations: Implications for acquired equivalence and perceptual learning. *Animal Learning & Behavior* 24: 233–255.

Hauser, M. D. (1996). *The Evolution of Communication*. Cambridge, Mass.: MIT Press.

Holland, P. C. (1990). Event representation in Pavlovian conditioning: Image and action. *Cognition* 37: 105–131.

Honey, R. C., Horn, G., Bateson, P., and Walpole, M. (1995). Functionally distinct memories for imprinting stimuli: Behavioral and neural dissociations. *Behavioral Neuroscience* 109: 689–698.

Horn, G. (1990). Neural bases of recognition memory investigated through an analysis of imprinting. *Philosophical Transactions of the Royal Society of London, B* 329: 133–142.

James, W. (1890). *The Principles of Psychology*. New York: Henry Holt.

Lieberman, P. (1994). Human language and human uniqueness. *Language and Communication* 14: 87–95.

Luria, A. (1982). *Language and Cognition*. Cambridge, Mass.: MIT Press.

Macedonia, J. M. (1990). What is communicated in the antipredator calls of lemurs: Evidence from antipredator call playbacks to ringtailed and ruffed lemurs. *Ethology* 86: 177–190.

Macedonia, J. M. and Evans, C. S. (1993). Variation among mammalian alarm call systems and the problem of meaning in animal signals. *Ethology* 93: 177–197.

Marler, P. (1990). Song learning: The interface between behavior and neuroethology. *Philosophical Transactions of the Royal Society of London, B* 329: 109–114.

Marler, P., Dufty, A., and Pickert, R. (1986). Vocal communication in the domestic chicken: I. Does a sender communicate information about the quality of a food referent to a receiver? *Animal Behaviour* 34: 188–193.

Marler, P., Evans, C. S., and Hauser, M. D. (1992). Animal signals: Motivational, referential, or both? In *Nonverbal Vocal Communication: Comparative and Developmental Approaches*, H. Papousek, U. Jürgens, and M. Papousek, eds., pp. 66–86. Cambridge: Cambridge University Press.

Premack, D. (1975). On the origins of language. In *Handbook of Psychobiology*, M. S. Gazzaniga and C. B. Blackmore, eds., pp. 591–605. New York: Academic Press.

Seyfarth, R. M., Cheney, D. L., and Marler, P. (1980). Monkey responses to three different alarm calls: Evidence of predator classification and semantic communication. *Science* 210: 801–803.

Shettleworth, S. J. (1998). *Cognition, Evolution, and Behavior*. New York: Oxford University Press.

Smith, W. J. (1991). Animal communication and the study of cognition. In *Cognitive Ethology: The Minds of Other Animals*, C. A. Ristau, ed., pp. 209–230. Hillsdale, N.J.: Lawrence Erlbaum Associates.

Struhsaker, T. T. (1967). Auditory communication among vervet monkeys (*Cercopithecus aethiops*). In *Social Communication among Primates*, S. A. Altmann, ed., pp. 281–324. Chicago: University of Chicago Press.

Tinbergen, N. (1963). On aims and methods of ethology. *Zeitschrift für Tierpsychologie* 20: 410–433.

Wallman, J. (1992). *Aping Language*. Cambridge: Cambridge University Press.

Zuberbühler, K. (2000). Referential labelling in Diana monkeys. *Animal Behaviour* 59: 917–927.

IV SELF AND OTHER: THE EVOLUTION OF COGNITIVE COOPERATORS

Gordon G. Gallup, Jr., James R. Anderson, and Daniel J. Shillito

Can animals recognize themselves in mirrors? Gallup (1970) conducted an experimental test of this question using a relatively simple approach. Individually housed chimpanzees (*Pan troglodytes*) were confronted with a full-length mirror outside their cages for a period of 10 days. The chimpanzees initially reacted as if they were seeing another chimpanzee and engaged in a variety of social displays directed toward the reflection. These social responses waned after the first few days. Rather than continue to respond to the mirror as such, the chimpanzees began using the mirror to respond to themselves by engaging in mirror-mediated facial and bodily movements and self-directed responses such as grooming parts of the body only visible in the mirror. The transition from social to self-oriented responding gave the impression that the chimpanzees had learned to recognize themselves; i.e., that they had come to realize that their behavior was the source of the behavior being depicted in the mirror.

To assess this possibility Gallup devised the mark test. Each chimpanzee was anesthetized and while it was unconscious, a red mark was applied to the brow above one eye and the top half of the opposite ear. A nonodorous, nonirritant dye was used, so that upon recovery from the anesthetic the chimpanzees would have no knowledge of the marks. Observations in the absence of the mirror confirmed this because the chimpanzees rarely touched the marks. When the mirror was reinstated, however, the effect was dramatic. The chimpanzees looked at their reflection and guided their fingers to the marks on their faces that could only be seen in the mirror. In addition to touching the marks repeatedly and looking at their fingers, some even smelled their fingers.

Comparative Data on Self-Recognition

Gallup's (1970) initial study was comparative in the sense that the same procedures were carried out with three different species of monkeys (*Macaca*): stumptailed, rhesus, and cynomolgus macaques. The monkeys' initial reactions to the reflection were also social, but in contrast to the chimpanzees, the tendency to respond as if in the presence of another monkey persisted. Even after 3 weeks of mirror exposure, none of the monkeys showed any mirror-aided self-directed behaviors, nor did they use the mirror to investigate the marks during the mark test. The major implications of the study were not only that chimpanzees shared with humans the capacity for self-recognition, but that the capacity might be limited to those primates most closely related to humans, namely the great apes (family Pongidae).

In the three decades since this study, a substantial literature has accumulated concerning the phylogenetic distribution of self-recognition. Studies involving dozens of species and scores of individual prosimians and monkeys have been conducted to determine to what extent other primates might show any of the criterion behaviors for self-recognition (spontaneous mirror-guided self-exploration and mark-directed responses on the mark test). Primates tested for mirror-image reactions include lemurs and bushbabies (prosimians); squirrel monkeys and several species of marmosets, tamarins, and capuchin monkeys (New World monkeys); several species of baboons and guenons (African Old World monkeys); and numerous species of macaques (Asian Old World monkeys). Even gibbons (Asiatic apes of the family Hylobatidae) have been assessed.

Some studies have been carefully designed to replicate Gallup's original mirror exposure and mark test regime, while others have systematically manipulated other variables in an effort to increase the likelihood of getting monkeys to make the transition from a socially oriented to a self-oriented perception of their reflection. Experimental manipulations have ranged from giving monkeys many months or even years of almost continuous exposure to their reflections, starting mirror-image experience at a very young age, and providing mirrors of various shapes and sizes and at various angles and degrees of accessibility. Other attempts to find self-recognition have included training monkeys to attend explicitly to their reflections, training them to use the reflected environmental information as a cue to find otherwise invisible objects, or adapting marking procedures in an effort to improve the animals' responsiveness to the change in their normal physical appearance (Anderson 1984; Anderson and Gallup 1999; Gallup 1987, 1994).

The collective contribution of all of these studies is that we can now eliminate several potential explanations for the failure of most primates to show any signs of self-recognition. They do not fail because they need earlier or longer experience with mirrors or because they cannot grasp the correspondence between the reflected and the real environments. It is not a simple lack of attention or motivation that causes nonself-recognizing species to fail to show mirror-mediated mark-directed responses. A control procedure introduced by Gallup et al. (1980) shows that individuals who exhibit no responses to marks in critical body regions during mark tests will respond to body marks that are directly visible without a mirror, such as the wrist or the belly (see also Shillito et al. 1999). Nor do they fail to understand the source of the reflection because of an aversion to making eye contact with their image. When two mirrors are arranged at an angle so that eye-to-eye contact with the reflection is impossible, social responses may diminish, but they are not replaced by patterns of self-directed responding (Anderson and Roeder 1989; also Shillito et al. 1999).

The failure to find self-recognition in monkeys is not due to poor problem-solving skills. Although many species use tools (Beck 1980), experiments with tool-using macaque and capuchin monkeys have revealed no relationship between tool use and self-recognition (Anderson and Marchal 1994). Claims of incipient self-recognition in monkeys or protorecognition have appeared over the years, but in each case the evidence is flawed, inconclusive, or has proven impossible to replicate (see Anderson and Gallup 1999 for a general review). Along with the growing list of failures to find self-recognition in monkeys are studies that have shown similarities in the effectiveness of live conspecifics and mirror-image stimulation for eliciting a variety of social behaviors, reinforcing the view that monkeys perceive their reflection as some kind of conspecific (Anderson 1994).

Thus, Gallup's (1970) suggestion that self-recognition might be limited to the great apes has held up well. Numerous studies have confirmed self-recognition in chimpanzees and there is now information regarding the importance of early social experience (Gallup et al. 1971), along with data on developmental trends and individual differences (Lin et al. 1992; Povinelli et al. 1993). What about other species of great apes? Self-recognition in orangutans is well established (Suarez and Gallup 1981), and several individuals in two captive groups of bonobos have shown mirror-mediated self-exploration (Hyatt and Hopkins 1994; Walraven et al. 1995) indicative of self-recognition.

On the other hand, most studies have failed to find convincing evidence in gorillas, in spite of modifications to the original mirror and mark test procedure (Suarez and Gallup 1981; Shillito et al. 1999). There is one claim of positive evidence for a gorilla (Patterson and Cohn 1994) that has had extensive contact with humans from an early age. It has been hypothesized that under normal circumstances the capacity for self-recognition may not develop in gorillas, but that enculturation in the form of early and extensive rearing by humans may result in the formation of the critical neural connections required for the expression of this capacity (Povinelli 1994).

Attempts to Discredit the Mirror Test

The existence of such decisive phyletic differences has been disconcerting to researchers who are committed to the idea that species differences are a matter of degree rather than kind (Tomasello and Call 1997). They have adopted a number of tactics in an attempt to discredit the mirror test findings and salvage the intellectual continuity hypothesis. Some have tried to modify the criteria for self-recognition. For example, Hauser et al. (1995) claim that when the mark's salience was increased by applying it to species-significant morphological features (hair tufts on the subjects' heads), they found positive evidence of mirror self-recognition in cotton-top tamarins. Hauser et al. contend that increased viewing time during the mirror-present condition compared with the mirror-absent condition of the mark test constitutes evidence that the tamarins recognized their image. Given that monkeys view their reflections as conspecifics (see Anderson 1994 for a review), it is not surprising that Hauser's subjects exhibited an increase in viewing time when confronted with what appeared to be another monkey with bright purple hair tufts (especially given the importance of these tufts to the species). Furthermore, a review of Hauser's videotaped mirror tests reveals that the subjects' mark-directed responses were not attempts to investigate the marks. Indeed the monkeys do not even touch the marks. Rather, they appear to be scratching an irritation (in fact, they even used their hind legs to do this in a manner similar to a dog scratching at fleas). More important, Hauser and his colleagues now concede that despite several attempts, they have been unable to replicate their original findings of self-recognition in tamarins (Hauser et al. 2001).

Epstein and colleagues (1981) have argued that self-awareness does not exist and apparent self-aware acts are by-products of environmental contingencies. Epstein et al. used an extensive training regimen to engineer positive mark test results with pigeons. Not only have these findings proven difficult if not impossible to replicate (Thompson and Contie 1994), but the logic behind their approach is seriously flawed. In their zeal to discredit the mirror test, Epstein et al. (1981) missed the point. The mark test was devised as a means of verifying impressions based on observations of chimpanzees using mirrors to inspect and manipulate otherwise invisible body parts. Training an animal to respond to marks on its body, without collateral evidence of self-recognition, indicates more about the achievements of the researchers who designed the training procedures than any underlying ability of the animal (Gallup and Suarez 1986).

If training pigeons to peck at marks on their bodies is equivalent to what chimpanzees do, then as a by-product of such training, the pigeons ought to engage in other indicators of self-recognition, such as mirror-mediated self-preening behavior. They do not. Furthermore, instead of training the prerequisite responses and then administering the mark test, Epstein et al. trained their pigeons to emit the criterion response (i.e., pecking the mark). This is equivalent to teaching human subjects the correct answers on an IQ test; granted their scores would improve, but such results would reveal nothing about their cognitive abilities. In a recent article on the pitfalls of radical behaviorism, Frans de Waal describes the paper by Epstein et al. as "one of the greatest travesties in behavioral science" (de Waal 1999, p. B6).

More recently, Heyes (1994) championed an anesthesia hypothesis in still another attempt to discredit the mirror test. Heyes claimed that the increase shown by chimpanzees in mark-directed touches after the introduction of the mirror is an artifact of incomplete recovery from anesthesia. Heyes contended that during the mirror-absent condition of the mark test, the subjects are still feeling the effects of the anesthetic agent and consequently their ambient level of self-touching behavior is suppressed. As the subjects continue to recover from anesthesia during the subsequent mirror-present condition, their normal level of autogrooming returns, accounting for an increased number of coincidental mark touches during this period. In the same vein, Heyes argued that the species differences on the mirror test are due to differences in rates of nonspecific, ambient face touching, rather than cognitive distinctions between species.

However, had Heyes conducted a thorough review of the literature, it would have been clear from the outset that her concerns were untenable (Gallup et al. 1995). Most researchers wait until their subjects have fully recovered from the anesthesia (at times waiting more than 24 hours) before administering the mark test. In fact, some studies have surreptitiously marked chimpanzees without the use of anesthesia, using sham-marking protocols, and have obtained positive results on the mark test (e.g., Calhoun and Thompson 1988).

Furthermore, the chronology of the subjects' responses during the mirror test also shows that Heyes's argument is flawed. The anesthesia hypothesis predicts that subjects should exhibit a gradual increase in mark touches throughout the mirror-present condition as the anesthetic's inhibiting effects on motor responses diminish. However, the results from mirror tests do not follow this pattern (Povinelli et al. 1997). Rather, self-recognizing subjects exhibit a pronounced interest in their marks immediately after the mirror is introduced, but this interest is short lived. Once chimpanzees discover that the marks are inconsequential, their attention rapidly wanes (as indicated by decreased mark contacts).

The results from a recent experiment examining the temporal pattern and topography of chimpanzees' mark touches further invalidate Heyes's position. Povinelli and colleagues (1997) recorded the duration and number of times subjects touched their marked eyebrow and ear as well as their corresponding unmarked brow and ear. Contrary to Heyes, mark-directed behavior was conditional upon seeing themselves in the mirror, and chimpanzees focused their attention almost exclusively on their marked facial features rather than their unmarked eyebrow and ear. Thus mark touches clearly represent attempts to investigate the marks that can only be seen in the mirror, rather than random contacts brought on by increased activity levels.

Self-Recognition in Humans

In parallel with the comparative studies of nonhuman primates, developmental psychologists have tried to map the ontogeny of self-recognition in children and elucidate the relationship between self-recognition and other aspects of the developing sense of self. The modern era of self-recognition studies on humans began when Amsterdam (1972) observed the reactions of young infants and toddlers to a mirror and devised a less rigorous version of the mark test that involved applying rouge to the child's nose (see Gallup 1994 for a critique).

Children toward the end of the first year of life show mostly social responses to their reflection (e.g., smiling, vocalizing). These sometimes persist into the second year, when coy reactions and avoidance are also observed. Not until around 18 months do some infants use the reflection to investigate the mark on their nose, the majority doing so by 2 years of age. One study reported a positive relationship between performance on tasks related to self-recognition and tasks related to object permanence (Bertenthal and Fischer 1978), but age as a covariate can rarely be ruled out in such studies. In an attempt to establish more precise age norms, subsequent research has used video techniques to demonstrate the importance of contingency cues in the infant's developing understanding of its own image (Johnson 1982; for a review see Anderson 1984).

The existence of individual differences in self-recognition among chimpanzees is also well documented among humans. Mentally retarded persons sometimes fail to learn to recognize themselves in mirrors (Harris 1977). Among autistic children, self-recognition is developmentally delayed and can remain absent in as many as 30 percent of the cases (Spiker and Ricks 1984). People diagnosed with schizophrenia often experience a disassociation with their image in mirrors and come to respond to the reflection as if they were in the presence of another person (Gallup et al. in press). At the other end of the life-span, some people with Alzheimer's disease lose the ability to recognize themselves in mirrors (Biringer and Anderson 1993).

Self-Recognition, Self-Awareness, and Mental State Attribution

If the species and individual differences in self-recognition are real, are they important? Mirror self-recognition is an indicator of self-awareness (Gallup 1979). In its most rudimentary form, self-awareness is the ability to become the object of your own attention. When you see yourself in a mirror, you are literally the object of your own attention, but most organisms respond to themselves in mirrors as if confronted by another organism. The ability to correctly infer the identity of the image in the mirror requires a preexisting sense of self on the part of the organism making that inference. Without a sense of self, how would you know who you were seeing when confronted with your reflection in a mirror? Recent neuro-psychological evidence is highly consistent with the proposition that self-recognition taps into the ability to conceive of oneself. Patients with damage to the frontal cortex are not only impaired in their ability to recognize their own faces but also show corollary deficits in self-evaluation and autobiographical memory (Keenan and Wheeler in press).

As an extension of this line of reasoning, it has been argued that the ability to infer mental states in others (known as theory of mind, or mental state attribution) is a by-product of being self-aware (Gallup 1982). The rationale for postulating a connection between self-recognition and mental state attribution is simple. If you are self-aware, then you are in a position to use your experience to model comparable experiences in others. For example, when you see someone who is carrying a large box approach a door, you might be prompted to walk ahead of them and hold the door open. This is based on a set of inferences you make as a consequence of having found yourself in comparable situations. Based on the direction they are walking and their proximity to the door, you make an attribution about their intention to leave the building, coupled with a corollary attribution about their inability to open the door because their hands are full. Thus, making inferences about states of mind in others is a logical extension of your experience with your own mental states. Knowledge of self is an inductive springboard for an inferential knowledge of others.

According to this model, species that fail to recognize themselves in mirrors should likewise fail to show evidence of introspectively based social strategies. Because of their inability to take into account what other individuals may know, want, or intend to do, intentional instances of deception, gratitude, grudging, sympathy, and empathy should be absent in organisms that are not self-aware. Monkeys, for example, which fail to show self-recognition, seem incapable of taking into account mental states in other monkeys (Cheney and Seyfarth 1990). The same holds true for humans.

People who fail to recognize themselves in mirrors are often deficient in their ability to infer what other people are thinking. For instance, it is only after children learn to recognize themselves that evidence of being able to take into account what other people are feeling or seeing begins to become evident (Carruthers and Smith 1996). Both embarrassment and the use of personal pronouns, as well as the development of an autobiographical memory, also appear only after children show evidence of self-recognition (Howe and Courage 1997; Lewis et al. 1989). In contrast to normal children, autistic children (who show self-recognition impairment) are deficient in their ability to make inferences about what other people are thinking (Baron-Cohen 2000). Schizophrenics also show deficits in mental state attribution, and schizophrenic patients often respond to themselves in mirrors as though they were seeing other people (for details, see Gallup et al. in press).

Other developmental studies also bear on the hypothesis that self-recognition is related to the ability to infer mental states in others. Johnson (1982) found that mirror self-recognition in 18–24-month-olds was positively related to the likelihood of altruistic or prosocial intervention (e.g., helping, comforting) when the mother feigned distress. Asendorpf and Baudonniüre (1993) classified infants as self-recognizers or nonself-recognizers and paired them in the presence of objects. Long phases of synchronic imitation (both infants acting similarly on identical objects while monitoring each other) were most characteristic of dyads composed of self-recognizers, and the authors relate self-recognition to the development of perspective-taking skills.

Neuropsychological Correlates of Self-Recognition

There is growing evidence that self-recognition can be localized in the human brain. For instance, Keenan and colleagues (Keenan et al. 1999) found that there were no differences in the reaction time required to identify familiar faces as opposed to their own face when people responded with their right hand. However, when they were required to press keys with their left hand, the subjects identified their own faces faster than the faces of other people. As a consequence of contralateral control, the left-hand

advantage suggests that self-recognition is related to processing of information in the right cerebral hemisphere.

Keenan et al. (2001) report another intriguing case of hemispheric lateralization of self-recognition involving epileptic patients undergoing preoperative evaluation using the application of intracarotid amobarbital, known as the WADA test. These patients were shown a morphed picture consisting of their own face and that of a famous person while either the right or left hemisphere was anesthetized. Upon recovery from anesthesia, they remembered seeing their own faces when the left hemisphere was anesthetized, but only the faces of familiar people when the right hemisphere was anesthetized. Thus, it is possible to literally turn self-recognition on and off by selectively deactivating different parts of the brain.

There are also reports of brain-damaged patients who failed to recognize themselves in mirrors. Breen (1999) describes a patient with damage restricted to the right prefrontal cortex who could identify other people with a mirror, but insisted that his own reflection was not himself. Keenan and Wheeler (in press) review a number of other studies that implicate both right hemispheric lateralization and localization of self-recognition in the prefrontal cortex. In support of the idea that self-recognition is a by-product of self-awareness, other studies have shown that self-evaluation and autobiographical memories are also localized in the right prefrontal region.

The same part of the brain that appears to be responsible for self-recognition is also crucial for making inferences about what other people are thinking. For example, Stone and colleagues (1998) report that patients with right frontal cortex damage have difficulty representing mental states in other people. Likewise, Happe et al. (1999) found that patients with right hemisphere damage were impaired in their ability to interpret mental state attribution narratives, and failed to appreciate humor that requires understanding the mental state of different characters. More recently, Stuss et al. (2001) found that in contrast to patients with brain damage elsewhere, those with lesions restricted to the right frontal lobes were deficient in visual perspective-taking and in their ability to detect instances of deception.

Thus, the neuropsychological evidence bolsters the proposition that self-recognition is an indicator of self-awareness, and that mental state attribution is a by-product of self-awareness (Gallup 1982). The frontal cortex (particularly the right prefrontal cortex) appears to be involved in self-recognition, self-evaluation, episodic (autobiographical) memory, introspection, humor, and mental state attribution. Furthermore, deficits in mirror self-recognition and mental state attribution are characteristic of a number of psychiatric disorders (see Gallup et al. in press).

Finally, in light of what we now know about the neuroanatomical correlates of self-awareness in humans, it is important to return to a comparative perspective because preliminary data may shed new light on the gorilla's peculiar inability to recognize itself in a mirror. Relative to the other great apes that show self-recognition, gorillas appear to have a smaller, less well developed frontal cortex (Semendeferi 1999). Indeed, gorilla brains are not only smaller in areas that have been implicated in social intelligence, but they are also less structurally and anatomically lateralized than those of their chimpanzee and orangutan counterparts (LeMay and Geschwind 1975).

References

Amsterdam, B. (1972). Mirror self-image reactions before the age of two. *Developmental Psychobiology* 5: 297–305.

Anderson, J. R. (1984). Monkeys with mirrors: Some questions for primate psychology. *International Journal of Primatology* 5: 81–98.

Anderson, J. R. (1994). The monkey in the mirror: The strange conspecific. In *Self-Awareness in Animals and Humans: Developmental Perspectives*, R. W. Mitchell, S. T. Parker, and M. L. Boccia, eds., pp. 315–329. New York: Cambridge University Press.

Anderson, J. R. and Gallup, G. G., Jr. (1999). Self-recognition in non-human primates: Past and future challenges. In *Animal Models of Human Emotion and Cognition*, M. Haug and R. E. Whalen, eds., pp. 175–194. Washington D.C.: American Psychological Association.

Anderson, J. R. and Marchal, P. (1994). Capuchin monkeys and confrontations with mirrors. In *Current Primatology: Social Development, Learning and Behaviour*, J.-J. Roeder, B. Thierry, J. R. Anderson, and N. Herrenschmidt, eds., pp. 371–380. Strasbourg: University Louis Pasteur.

Anderson, J. R. and Roeder, J.-J. (1989). Responses of capuchin monkeys (*Cebus apella*) to different conditions of mirror stimulation. *Primates* 30: 581–587.

Asendorpf, J. B. and Baudonnüre, P.-M. (1993). Self-awareness and other-awareness: Mirror self-recognition and synchronic imitation among unfamiliar peers. *Developmental Psychobiology* 29: 88–95.

Baron-Cohen, S. (2000). The cognitive neuroscience of autism: Evolutionary approaches. In *The New Cognitive Neurosciences*, 2nd ed., M. Gazzaniga, ed., pp. 1249–1257. Cambridge, Mass.: MIT Press.

Beck, B. B. (1980). *Animal Tool Behavior: The Use and Manufacture of Tools by Animals*. New York: Garland.

Bertenthal, B. I. and Fischer, K. W. (1978). Development of self-recognition in the infant. *Developmental Psychology* 14: 44–50.

Biringer, F. and Anderson, J. R. (1993). Self-recognition in Alzheimer's disease: Use of mirror and video techniques and enrichment. In *Recent Advances in Aging Science*. Vol. 1, E. Beregi, I. A. Gergely, and K. Rajczi, eds., pp. 697–705. Bologna: Monduzzi Editore.

Breen, N. (1999). Misinterpreting the mirrored self. Paper presented at the annual meeting of the Association for the Scientific Study of Consciousness.

Calhoun, S. and Thompson, R. L. (1988). Long-term retention of self-recognition by chimpanzees. *American Journal of Primatology* 15: 361–365.

Carruthers, P. and Smith, P. K. (1996). *Theories of Theories of Mind*. Cambridge: Cambridge University Press.

Cheney, D. L. and Seyfarth, R. W. (1990). *How Monkeys See the World: Inside the Mind of Another Species*. Chicago: University of Chicago Press.

de Waal, F. B. M. (1999). The pitfalls of not knowing the whole animal. *Chronicle of Higher Education* 26: B4–6.

Epstein, R., Lanza, R. P., and Skinner, B. F. (1981). "Self-awareness" in the pigeon. *Science* 212: 695–696.

Gallup, G. G., Jr. (1970). Chimpanzees: Self-recognition. *Science* 167: 86–87.

Gallup, G. G., Jr. (1979). *Self-Recognition in Chimpanzees and Man: A Developmental and Comparative Perspective*. New York: Plenum.

Gallup, G. G., Jr. (1982). Self-awareness and the emergence of mind in primates. *American Journal of Primatology* 2: 237–248.

Gallup, G. G., Jr. (1987). Self-awareness. In *Comparative Primate Biology, Behavior, Cognition, and Motivation*. Vol. 2B, J. R. E. G. Mitchell, ed., pp. 3–16. New York: Liss.

Gallup, G. G., Jr. (1994). Self-recognition: Research strategies and experimental design. In *Self-Awareness in Animals and Humans: Developmental Perspectives*, S. T. Parker, R. W. Mitchell, and M. L. Boccia, eds., pp. 35–50. New York: Cambridge University Press.

Gallup, G. G., Jr. and Suarez, S. D. (1986). Self-awareness and the emergence of mind in humans and other primates. In *Psychological Perspectives on the Self*. Vol. 3, J. Suls and A. Greenwald, eds., pp. 3–26. Hillsdale, N.J.: Lawrence Erlbaum Associates.

Gallup, G. G., Jr., McClure, M. K., Hill, S. D., and Bundy, R. A. (1971). Capacity for self-recognition in differentially reared chimpanzees. *Psychological Record* 21: 69–74.

Gallup, G. G., Jr., Wallnau, L. B., and Suarez, S. D. (1980). Failure to find self-recognition in mother–infant and infant–infant rhesus monkey pairs. *Folia Primatologica* 33: 210–219.

Gallup, G. G., Jr., Povinelli, D. J., Suarez, S. D., Anderson, J. R., Lethmate, J., and Menzel, E.-W. J. (1995). Further reflections on self-recognition in primates. *Animal Behaviour* 50: 1525–1532.

Gallup, G. G., Jr., Anderson, J. R., and Platek, S. M. (in press). Self-awareness, social intelligence, and schizophrenia. In *The Self and Schizophrenia: A Neuropsychological Perspective*, A. S. David and T. Kircher, eds. Cambridge: Cambridge University Press.

Happe, F. G., Brownell, H., and Winner, E. (1999). Acquired "theory of mind" impairments following stroke. *Cognition* 70: 211–240.

Harris, L. P. (1977). Self-recognition among institutionalized profoundly retarded males: A replication. *Bulletin of the Psychonomic Society* 9: 43–44.

Hauser, M. D., Kralik, J., Botto-Mahan, C., Garrett, M., and Oser, J. (1995). Self-recognition in primates: Phylogeny and the salience of species-typical features. *Proceedings of the National Academy of Sciences, U.S.A.* 92: 10811–10814.

Hauser, M. D., Miller, C. T., Liu, K., and Gupta, R. (2001). Cotton-top tamarins (*Saguinus oedipus*) fail to show mirror-guided self-exploration. *American Journal of Primatology* 53: 131–137.

Heyes, C. M. (1994). Reflections on self-recognition in primates. *Animal Behaviour* 47: 909–919.

Howe, M. L. and Courage, M. L. (1997). The emergence and early development of autobiographical memory. *Psychological Review* 104: 499–523.

Hyatt, C. W. and Hopkins, W. D. (1994). Self-awareness in bonobos and chimpanzees: A comparative approach. In *Self-Awareness in Animals and Humans: Developmental Perspectives*, S. T. Parker, R. W. Mitchell, and M. L. Boccia, eds., pp. 248–253. New York: Cambridge University Press.

Johnson, D. B. (1982). Altruistic behavior and the development of the self in infants. *Merill-Palmer Quarterly* 28: 379–388.

Keenan, J. P. and Wheeler, M. (in press). The neuropsychology of self. In *The Self and Schizophrenia: A Neuropsychological Perspective*, A. S. David and T. Kircher, eds. Cambridge: Cambridge University Press.

Keenan, J. P., McCutcheon, B., Freund, S., Gallup, G. G., Jr., Sanders, G., and Pascual-Leone, A. (1999). Left-hand advantage in a self-face recognition task. *Neuropsychologia* 37: 1421–1425.

Keenan, J. P., Nelson, A., O'Connor, M., and Pascual-Leone, A. (2001). Self-recognition and the right hemisphere. *Nature* 409: 305.

LeMay, M. and Geschwind, N. (1975). Hemispheric differences in the brains of great apes. *Brain, Behavior, and Evolution* 11: 48–52.

Lewis, M., Sullivan, M. W., Stanger, C., and Weiss, M. (1989). Self-development and self-conscious emotions. *Child Development* 60: 146–156.

Lin, A. C., Bard, K. A., and Anderson, J. R. (1992). Development of self-recognition in chimpanzees (*Pan troglodytes*). *Journal of Comparative Psychology* 106: 120–127.

Patterson, F. G. P. and Cohn, R. H. (1994). Self-recognition and self-awareness in lowland gorillas. In *Self-Awareness in Animals and Humans: Developmental Perspectives*, S. T. Parker, R. W. Mitchell, and M. L. Boccia, eds., pp. 273–290. New York: Cambridge University Press.

Povinelli, D. J. (1994). How to create self-recognizing gorillas (but do not try it on macaques). In *Self-Awareness in Animals and Humans: Developmental Perspectives*, S. T. Parker, R. W. Mitchell, and M. L. Boccia, eds., pp. 291–300. New York: Cambridge University Press.

Povinelli, D. J., Rulf, A. B., Landau, K. R., and Bierschwale, D. T. (1993). Self-recognition in chimpanzees (*Pan troglodytes*): Distribution, ontogeny, and patterns of emergence. *Journal of Comparative Psychology* 107: 347–372.

Povinelli, D. J., Gallup, G. G., Jr., Eddy, T. J., Bierschwale, D. T., Engstrom, M. C., Perilloux, H. K., and Toxopeus, I. B. (1997). Chimpanzees recognize themselves in mirrors. *Animal Behaviour* 53: 1083–1088.

Semendeferi, K. (1999). The frontal lobes of the great apes with a focus on the gorilla and the orangutan. In *The Mentalities of Gorillas and Orangutans*, S. T. Parker, R. W. Mitchell, and H. L. Miles, eds., pp. 70–95. Cambridge: Cambridge University Press.

Shillito, D. J., Gallup, G. G., Jr., and Beck, B. B. (1999). Factors affecting mirror behavior in western lowland gorillas, *Gorilla gorilla. Animal Behavior* 57: 999–1004.

Spiker, D. and Ricks, M. (1984). Visual self-recognition in autistic children: Developmental relationships. *Child Development* 55: 214–225.

Stone, V. E., Baron-Cohen, S., and Knight, R. T. (1998). Frontal lobe contributions to theory of mind. *Journal of Cognitive Neuroscience* 10: 640–656.

Stuss, D. T., Gallup, G. G., Jr., and Alexander, M. P. (2001). The frontal lobes are necessary for theory of mind. *Brain* 124: 279–286.

Suarez, S. and Gallup, G. G., Jr. (1981). Self-recognition in chimpanzees and orangutans, but not gorillas. *Journal of Human Evolution* 10: 157–188.

Thompson, R. K. R. and Contie, C. L. (1994). *Further Reflections on Mirror Usage by Pigeons: Lessons from Winnie-the-Pooh and Pinocchio Too.* New York: Cambridge University Press.

Tomasello, M. and Call., J. (1997). *Primate Cognition.* Oxford: Oxford University Press.

Walraven, V., van Elsacker, L., and Verheyen, R. (1995). Reactions of a group of pygmy chimpanzees (*Pan paniscus*) to their mirror images: Evidence of self-recognition. *Primates* 36: 145–150.

41

When Traditional Methodologies Fail: Cognitive Studies of Great Apes

Robert W. Shumaker and Karyl B. Swartz

Since the classic work by Kohler (1925/1976), Yerkes and Yerkes (1929), and Tinklepaugh (1932), researchers have been developing methods to investigate the minds of apes. In this essay, we present a set of principles that we use in guiding our joint and separate research programs that address cognition in great apes. The general questions that guide our research address broad cognitive skills such as symbolic representation and the role of memory in serial processing. We are interested in understanding how apes extract and use information in solving problems. As we attempt to characterize cognition, we develop methods that guide the ape through experimental procedures to reveal the cognitive processes involved in performing a task. We come to our research from different backgrounds, but with similar perspectives and goals. Shumaker is an evolutionary biologist interested in primate behavior and perception. Swartz is an experimental psychologist with a comparative development emphasis.

In light of the evolutionary closeness of humans and the other great apes, one assumption underlying our research is continuity of cognitive ability. Our own experiences, as well as data from the human literature, are useful as a reference point for developing procedures for investigating cognition in great apes. Although this assumption of continuity provides a starting point, we formulate specific research questions and develop methodologies that are appropriate to the behavior of the species being studied. We study these animals because we are interested in understanding them in their own right.

It is a common misperception that the behavior of apes is "analogous to," "similar to," "like," or "equivalent to" young humans or humans with developmental delay. Although we can look at developmental parallels in the achievement of specific cognitive skills in apes and humans, we cannot make statements about overall ability.

Great apes are not the equivalent of preschool-age human children. For example, a $2\frac{1}{2}$-year-old human and a $2\frac{1}{2}$-year-old chimpanzee may both show the ability to recognize themselves in a mirror (Amsterdam 1972; Lewis and Brooks-Gunn 1979; Lin et al. 1992), but that does not imply the equivalence of other cognitive skills. Furthermore, Murofushi (1997) reported that an adult female chimpanzee has demonstrated the ability to numerically label random arrays of dots (ranging from one to six) faster than adult humans tested on the identical task. This does not imply that adults of both species share an overall equivalence in cognitive skill. The point is that a discussion of specific abilities generates meaningful comparisons, while a search for overall equivalence is unproductive.

Our empirical methodologies are diverse, depending on the cognitive process under investigation. While we frequently use classical and operant conditioning principles in our research design, the endpoint is not to document these forms of associative learning. Rather, we view these as techniques to initially convey information to the apes about the task at hand. Our intention is to create a situation that allows the ape to demonstrate the ontogeny of a particular cognitive skill or ability.

Our goal is to discover the nature of the internal psychological state and the series of cognitive operations that an animal successively passes through as a solution to a problem is achieved. Not only is it important to devise a task that will allow the ape to demonstrate the "how and why" of the solution, we also may find that some of our preexisting assumptions about the usefulness of a particular task, or the ability of a particular animal, may be challenged. What we may perceive as the most efficient means to elucidate the cognitive processes that lead to a solution may not be optimal, or could be completely in error. Complicating this problem is the inevitable presence of significant behavioral and cognitive differences among individual apes (Boysen 1994). The challenge for the experimenter is to balance task and subject variables in such a way that reliable data are collected and the integrity of the research agenda is maintained.

In our program, we attempt to address these potential complications through a combination of short- and long-term studies. The short-term studies focus on discrete questions based on a well-defined cognitive capacity and have a clearly defined end point. In contrast, the long-term studies have no defined concluding point and are based on a graduated set of cognitive skills. Examples of short-term studies include mirror self-recognition (Swartz and Evans 1991, 1997), observational learning (Shumaker et al. 1998), quantity judgment (Shumaker et al. 2001), two-choice discrimination learning set formation (Shumaker et al. in preparation), and habituation and responses to novelty (Swartz et al. in preparation). Examples of long-term studies include a language acquisition study (Shumaker 1997) and an investigation of serial learning and memory (Swartz et al. in preparation).

The value of embedding short-term studies in a longitudinal research plan is that these short, well-defined studies can inform us about certain abilities of the animals and hence can indicate whether the methodology developed for the research program is sound. Discovering relatively quickly that a particular ability is demonstrated by our subjects gives us more confidence in our overall design. That is, it gives us confidence that the basic skills required for demonstrating more complex cognitive abilities are present, and that we have developed a methodology that can reveal them. Over time, what we consider to be the more sophisticated task evolves into a set of basic tasks that serve as stepping-stones to more complex cognitive processes.

From the ape's perspective, the advantage to embedding short-term studies in a larger research plan is that the well-defined questions of short-term studies instruct them about the basic task, with smaller demands and greater opportunity for success. Most of the experimental tasks that we develop to ask cognitive questions of the apes are out of their range of experience and may initially appear confusing. By experiencing success along the way, the ape gradually becomes more sophisticated and better able to perform our experimental tasks. To address more concretely the issues raised here, we turn to some specific examples from our individual and joint research programs that illustrate our perspective.

When a Well-Established Methodology Fails: Mirror Self-Recognition by Gorillas

In 1970, Gallup presented the first observation of mirror self-recognition (MSR) in great apes. Four juvenile chimpanzees (*Pan troglodytes*) were each presented with a total of 80 hours of exposure to a mirror. Initially, each individual responded to the mirror image with social behavior, suggesting that it may have perceived the mirror image as an unfamiliar conspecific. Over the course of the mirror exposure phase of the study, the mirror-directed social behaviors waned, and mirror-guided self-directed behaviors appeared. These behaviors included such things as picking the teeth by using the mirror image to guide the hands, visually inspecting parts of the body that were otherwise visually inaccessible, and blowing bubbles with the mouth while visually inspecting the bubbles in the mirror. The presence of these mirror-guided self-directed behaviors was the first evidence that Gallup presented for MSR.

To obtain a more objective measure of self-recognition, Gallup devised a task that is commonly referred to as the "mark test." He anesthetized each animal and placed a red odorless mark on the forehead and opposite ear, locations that could not be seen by the chimpanzee without the mirror. Each ape was tested individually. When the chimpanzees recovered from the anesthesia, each was observed for a control period without the mirror present, and the number of mark-directed responses was counted. Only one mark touch occurred during the control period, compared with a total of 27 mark-directed responses across the four chimpanzees when the mirror was present. The finding that each chimpanzee touched the mark on his or her own body rather than attempting to touch the mark on the mirror image provided convincing evidence to support the conclusion that the animals showed MSR.

Over the past 30 years, there has been much controversy about the best methods to use to study MSR, as well as which species do or do not show MSR. The mark test continues to be the task most often used to show MSR (Gallup 1994; Gallup et al. chapter 40 in this volume; Swartz and Evans 1991, 1994), although some researchers have used the presence of self-directed behavior as the criterion for MSR (Povinelli et al. 1993). Important features of the mark test are that the subject be unaware of the placement of the mark, that anesthetized animals be fully conscious during the control period as well as when provided with the mirror, that subjects be provided with sufficient nonmarked mirror exposure prior to conducting the mark test, and that the animals be of the appropriate developmental age when tested [although the developmental course of MSR is not yet understood in any of the great apes; see Swartz (1998)].

To date, despite numerous experimental attempts, MSR has never been clearly demonstrated by monkeys (Anderson 1984; Swartz 1998). MSR appears to be a phenomenon limited to apes, including lesser apes (Ujhelyi et al. 2000). Some, but not all chimpanzees older than $2\frac{1}{2}$ years of age show evidence of MSR using the mark-test criterion (Lin et al. 1992; Swartz and Evans 1991, 1997). Orangutans (*Pongo pygmaeus*) appear to show MSR when tested under conditions similar to those used with chimpanzees (see Swartz et al. 1999). No mark test has been presented to a bonobo (*Pan paniscus*), but they have been reported to show self-directed behavior when presented with a mirror. While MSR is universally accepted for chimpanzees and orangutans, the presence of this ability in gorillas (*Gorilla gorilla gorilla*) is hotly debated and still unresolved (see Swartz et al. 1999).

Koko, a gorilla proficient in sign language, has demonstrated clear evidence that she recognizes herself in the mirror (Patterson and Cohn 1994). Of 23 gorillas tested for evidence of MSR (Shillito et al. 1999; see Swartz et al. 1999), only 6 individuals provided any evidence for this ability. Studies have been conducted with 6 orangutans, with 5 demonstrating MSR (see Swartz et al. 1999), and 163 chimpanzees have been tested, with 73 showing MSR (see Swartz et al. 1999). As in other areas of scientific inquiry, such as tool use, chimpanzees are overwhelmingly represented in the literature, making them the de facto standard for comparison (Beck 1982; Swartz et al. 1999). We believe that the use of a "chimpanzee standard" when investigating MSR is an excellent example of faulty expectations on the part of the experimenter. Specifically, all studies investigating MSR in gorillas have defined the phenomenon as it appears in chimpanzees and have limited their methodology to that which has been successful with chimpanzees.

The fact that gorillas have frequently failed the traditional mark test has at least two possible explanations. First, it could be that gorillas are simply incapable of understanding their own reflection in a mirror (Gallup 1994; Shillito et al. 1999). Second, it might be that the mark test is not an effective way to test gorillas and that a reasonable variant of the mark test that is more compatible with gorilla behavior would allow the expression of the capacity.

Koko and the other five gorillas cited in the literature (see Swartz et al. 1999) eliminate the first explanation as a valid possibility. To assess the validity of the second possibility, we modified the traditional mark test and administered this new procedure to Mopie, an adult male gorilla who had previously been reported to fail a traditional mark test (Shillito et al. 1999).

Our modification was designed to increase motivation to attend to the mirror image and to the location of the mark (Swartz and Shumaker in preparation). Mopie was provided with additional mirror exposure beyond what he had had previously (Shillito et al. 1999). Each day before receiving mirror exposure, Mopie was engaged by the experimenter in a trading interaction. The experimenter placed an adhesive-backed, red, 2-cm-diameter paper dot (of the type used to mark file folders) on various parts of Mopie's home enclosure, one at a time. If Mopie peeled the dot from the enclosure and gave it to the experimenter, he received a treat. Placement of dots began on the enclosure, but proceeded to body areas that were clearly visible to Mopie without the use of the mirror. On two or three trials each day that the dots were placed on his body, Mopie was asked to present his head and a dot was placed there above his eyes and removed before the end of the session by the experimenter.

When Mopie was reliably pulling each visible dot from his enclosure or his body and presenting it to the experimenter mark test were begun. One of the dots was left on his head for the mirror exposure period, although as in the previous sessions the experimenter behaved as if it had been removed. On the seventh mark test, Mopie removed the dot from his head, using the mirror to guide his hand. He then immediately presented the dot to the experimenter and received a reward. That mirror-guided dot-directed behavior occurred within two minutes after the mirror was presented. Two unsuccessful mark tests followed and further attempts to replicate this procedure were abandoned because Mopie was aware of the application of the dot, and checked his head by sweeping one hand across his brow (in a motion that was distinctly different from the mirror-guided removal of the dot performed previously) whenever it was touched in the trading session and before a mirror was available.

We then replaced the adhesive dots with the beam from a laser pointer to present a red dot on the enclosure and on Mopie's body. Each time that he touched the laser spot, Mopie received a reward. Once he was reliably touching each laser spot, we presented a laser spot on his head during mirror exposure. Mopie made no apparent attempt to touch this spot. Given his previous performance with the paper dot, we attempted another slight variation. Instead of shining the laser beam on his head, the dot was placed directly under his chin onto his chest and was invisible except in the mirror reflection. During this presentation, he looked directly at his reflection in the mirror and placed an index finger over the laser dot. He then extended his hand for a reward (Swartz and Shumaker, in preparation).

This slight modification of the traditional mark test provided motivation for the gorilla to find and remove the adhesive dot, or to find and touch the laser spot. Although it may have been the case that some tactile sensation was produced from the adhesive dot, Mopie did not remove the dot from his head until he looked in the mirror. He clearly used the mirror reflection to guide his hand during removal of the dot. Any potential problem in interpretation based on possible tactile cues was eliminated by the use of the laser pointer. We think that this is a clear example of how a traditional test, although powerful with chimpanzees, is not appropriate for gorillas. A modification of the procedure provided the opportunity for the gorilla to demonstrate an ability that was in question.

This example illustrates the value of short-term studies in the context of a long-range research program. Mopie's demonstration of a specific ability, MSR, not only provides strong evidence that he has the ability to perform this task, but also suggests an alternative method for exploring this phenomenon in other gorillas and possibly other species. Knowing that Mopie has the ability to show MSR allows us to develop a long-term study of related but more complex cognitive skills involved in mental state attribution, often termed "theory of mind" (Gallup 1982; Premack and Woodruff 1978).

When Nonexperimental Observations Conflict with Controlled Studies: Observational Learning by Great Apes

Observational learning, also frequently called "social learning," is the process in which an observer acquires a behavior by watching another individual, referred to as the demonstrator, perform that same behavior (Galef 1988). There are three different forms of observational learning. In order of complexity, they are social facilitation, stimulus enhancement, and imitation. Social facilitation refers to a situation in which a behavior appears to be "contagious." A flock of birds taking off one by one is an example. Stimulus enhancement describes a change in an observer's behavior toward some object in the environment as a result of seeing a demonstrator have a positive experience involving that object. If one monkey in a group uses a stick to obtain a favored food that is otherwise out of reach, and another monkey shows greater interest in holding and manipulating sticks as a result, stimulus enhancement would be a likely explanation for his increased attention toward sticks. Stimulus enhancement does not imply that the observer understands the full behavior, but is simply attracted to some object associated with a behavior. Imitation, the most sophisticated form of observational learning, describes a situation where an observer exactly replicates a new or unfamiliar behavior after watching a demonstrator perform that same behavior. A clear act of imitation implies that the observer has the mental ability to understand a situation from the demonstrator's point of view.

It is generally accepted that a variety of monkey species show evidence for social facilitation and stimulus enhancement, but not for imitation (Nishida 1986; Beck 1980; Visalberghi and Fragaszy 1990; Whiten and Ham 1992; Zuberbühler et al. 1996). The forms of observational learning that exist for great apes are less clear. The current debate is focused on whether great apes are capable of imitation. Observations of wild great apes argue for the existence of all three forms of observational learning (Goodall 1986; Boesch 1991). The behaviors that exist in different groups of wild great apes vary (Boesch 1994; McGrew 1994; Tomasello 1996) and are persistent at the local population level. As individuals move among groups, these behaviors spread rapidly and become fixed within a population (Huffman and Wrangham 1994; Wrangham and Goodall 1989). One likely explanation for this behavioral change is imitation between immigrating demonstrators and observers from the local population. Apes in the wild have also been reported to imitate behaviors that were performed by humans (Russon and Galdikas 1995).

In captivity, imitation has been documented for apes that have been subjects in language research experiments. These imitative acts have been reported to occur spontaneously as well as at the request of the experimenter (Hayes and Hayes 1952; Miles et al. 1996; Gardner and Gardner 1969; Savage-Rumbaugh and Lewin 1994).

However, the results from some captive studies that have been specifically designed to test great ape observational learning abilities conflict with these observations (Tomasello et al. 1987; Nagell et al. 1993; Call and Tomasello 1994, 1995). Much like the previously mentioned studies of MSR, we suggest that these investigations failed to consider the perspective of the apes that served as the subjects.

Nagell et al. (1993) presented 2-year-old humans and chimpanzees of varying ages with a tool task to assess their ability to imitate. Initially, both the human and ape subjects were given the opportunity to observe a human demonstrator use a rake to obtain an out-of-reach food reward. The demonstration was performed without error. When given the chance to try the task themselves, the humans clearly imitated the performance of the demonstrator. The chimpanzees did not imitate, but devised their own successful methods to obtain the food with the rake. The authors concluded that the humans clearly imitated, while the chimpanzees focused more on the results of the activity rather than the process that was used to acquire the reward. The same study was carried out with orangutans (Call and Tomasello 1994), yielding similar results, even when the study was modified to include an orangutan demonstrator.

Call and Tomasello (1995) further investigated the observational learning skills of orangutans and human children. Both groups observed a human demonstrator either push, pull, or rotate a rod that was attached to a box. A specific sequence of movements resulted in a food reward being delivered. The human demonstrator knew the sequence in advance and performed it in front of both groups. The humans reliably imitated, while the orangutans did not, even with an orangutan demonstrator. It is interesting that the humans were never tested using an ape demonstrator.

Shumaker et al. (1998) report that under a different methodology orangutans can clearly imitate the behavior of a demonstrator. In this task, three pairs of orangutans were tested on a match-to-sample task of increasing complexity that served as the beginning stage for a long-term study of language acquisition. Each pair had a dominant and subordinate individual. The dominant individual served as the demonstrator while the subordinate had the opportunity to observe all interactions between the experimenter and the demonstrator. In this task the demonstrator in each pair learned to label a specific food with its corresponding abstract symbol. During all phases of the task, the experimenter focused only on the demonstrator and provided no direct instruction, in any form, to the observer.

Three aspects of the results provide support for imitation by the observers. During the course of the study, two of the demonstrators briefly left the testing area on ten separate occasions. Each of these times, their observer immediately approached the testing apparatus and made a

selection from the array of symbols that were present. The observers did not require preliminary training to acquaint them with the "rules" of the task. They approached the experimental apparatus, pointed to a stimulus, and then extended their hand (or lips) toward the experimenter for a reward. The topography of the pointing response provided a second piece of evidence supporting imitation. The demonstrators had been trained to produce a clear pointing response using only one finger, a response form that is not typical of orangutans. None of the observers required training in this form; all responded with the single-finger discrete response they had observed in the demonstrators. Finally, of the ten selections made by the observers, nine were correct. The observers exactly replicated the novel behavior that had been exhibited by their demonstrator partner, clearly satisfying the definition of imitation.

Rather than a lack of ability to imitate, we propose that experimental design figures prominently in the explanation for the difference in results in the studies that have been cited. Nagell et al. (1993) and Call and Tomasello (1994, 1995) only used demonstrators (human or ape) that were competent at the experimental task. Therefore the observers only saw successful performances. In Call and Tomasello's (1995) study that utilized a vending apparatus operated by pushing, pulling, or rotating, the workings of the box were completely hidden from the ape. In addition, each ape in that study was given between 0.5 and 2.5 hours to interact with the test apparatus prior to actual data collection. During that time, no food reward was possible, even though the ape investigated the apparatus and could move the handle freely. It is entirely possible that during this habituation period the ape learned that no particular action resulted in a food reward.

Shumaker et al. (1998) used a significantly different methodology. Specifically, the demonstrator and observer were both completely naive at the beginning of the study. The observer was therefore able to see both mistakes and correct answers, rather than just accurate performance.

In addition, every aspect of the task was made obvious at all times, and every correct answer was always rewarded. We propose that these methodological differences had a significant impact on the performance of the apes.

In summary, the examples of MSR and observational learning cited in this essay illustrate the major influence that experimenter expectations and inappropriate methodology can have on the conclusions that are formed from otherwise well-executed and reasonable studies. We encourage the notion that whenever possible, experimenters must first consider the perspective of the subjects during the design of an experiment. We also encourage the implementation of short-term, direct studies as a means to assess the appropriateness of a methodology before embarking on long-term investigations. Overall, the natural behavior and inclinations of each species must be fully considered when conceptualizing, designing, and implementing a research program.

References

Amsterdam, B. (1972). Mirror image reactions before age two. *Developmental Psychobiology* 5: 297–305.

Anderson, J. R. (1984). Monkeys with mirrors: Some questions for primate psychology. *International Journal of Primatology* 5: 81–98.

Beck, B. B. (1980). *Animal Tool Behavior: The Use and Manufacture of Tools by Animals*. New York: Garland.

Beck, B. B. (1982). Chimpocentrism: Bias in cognitive ethology. *Journal of Human Evolution* 11: 3–17.

Boesch, C. (1991). Teaching in wild chimpanzees. *Animal Behaviour* 41: 530–532.

Boesch, C. (1994). Hunting strategies of Gombe and Taï chimpanzees. In *Chimpanzee Cultures*, R. W. Wrangham, W. C. McGrew, F. B. M. de Waal, and P. G. Heltne, eds., pp. 77–92. Cambridge, Mass.: Harvard University Press.

Boysen, S. T. (1994). Individual differences in the cognitive abilities of chimpanzees. In *Chimpanzee Cultures*, R. W. Wrangham, W. C. McGrew, F. B. M. de Waal, and P. G. Heltne, eds., pp. 335–350. Cambridge, Mass.: Harvard University Press.

Call, J. and Tomasello, M. (1994). The social learning of tool use by orangutans (*Pongo pygmaeus*). *Human Evolution* 9: 297–313.

Call, J. and Tomasello, M. (1995). Use of social information in the problem solving of orangutans (*Pongo pygmaeus*) and human children (*Homo sapiens*). *Journal of Comparative Psychology* 109: 308–320.

Galef, B. G., Jr. (1988). Imitation in animals: History, definitions, and interpretation of data from the psychological laboratory. In *Social Learning: Psychological and Biological Perspectives*, T. Zentall and B. Galef, eds., pp. 3–28. Hillsdale, N.J.: Lawrence Erlbaum Associates.

Gallup, G. G., Jr. (1970). Chimpanzee self-recognition. *Science* 167: 86–87.

Gallup, G. G., Jr. (1982). Self-awareness and the emergence of mind in primates. *American Journal of Primatology* 2: 237–248.

Gallup, G. G., Jr. (1994). Self-recognition: Research strategies and experimental design. In *Self-Awareness in Animals and Humans: Developmental Perspectives*, S. T. Parker, R. W. Mitchell, and M. L. Boccia, eds., pp. 35–50. Cambridge: Cambridge University Press.

Gardner, B. T. and Gardner, R. A. (1969). Teaching sign language to a chimpanzee. *Science* 165: 664–672.

Goodall, J. (1986). *The Chimpanzees of Gombe: Patterns of Behavior*. Cambridge, Mass.: Harvard University Press.

Hayes, K. J. and Hayes, C. (1952). Imitation in a home-raised chimpanzee. *Journal of Comparative Psychology* 45: 450–459.

Huffman, M. A. and Wrangham, R. W. (1994). Diversity of medicinal plant use by chimpanzees in the wild. In *Chimpanzee Cultures*, R. W. Wrangham, W. C. McGrew, F. B. M. de Waal, and P. G. Heltne, eds., pp. 129–148. Cambridge, Mass.: Harvard University Press.

Kohler, W. (1925/1976). *The Mentality of Apes*. Reprinted 1976. New York: Liveright.

Lewis, M. and Brooks-Gunn, J. (1979). *Social Cognition and the Acquisition of the Self*. New York: Plenum.

Lin, A. C., Bard, K. A., and Anderson, J. R. (1992). Development of self-recognition in chimpanzees (*Pan troglodytes*). *Journal of Comparative Psychology* 106: 120–127.

McGrew, W. C. (1994). Tools compared: The material of culture. In *Chimpanzee Cultures*, R. W. Wrangham,

W. C. McGrew, F. B. M. de Waal, and P. G. Heltne, eds., pp. 25–40. Cambridge, Mass.: Harvard University Press.

Miles, H. L., Mitchell, R. W., and Harper, S. E. (1996). Simon says: The development of imitation in an enculturated orangutan. In *Reaching into Thought: The Minds of the Great Apes*, A. Russon, K. Bard, and S. T. Parker, eds., pp. 278–299. Cambridge: Cambridge University Press.

Murofushi, K. (1997). Numerical matching behavior by a chimpanzee (*Pan troglodytes*): Subitizing and analogue magnitude estimation. *Japanese Psychological Research* 39: 14–153.

Nagell, K., Olguin, R. S., and Tomasello, M. (1993). Process of social learning in the tool use of chimpanzees (*Pan troglodytes*) and human children (*Homo sapiens*). *Journal of Comparative Psychology* 107: 174–186.

Nishida, T. (1986). Local traditions and cultural transmission. In *Primate Societies*, B. B. Smuts, D. L. Cheney, R. M. Seyfarth, R. W. Wrangham, and T. T. Struhsaker, eds., pp. 462–474. Chicago: University of Chicago Press.

Patterson, F. G. and Cohn, R. H. (1994). Self-recognition and self-awareness in lowland gorillas. In *Self-Awareness in Animals and Humans: Developmental Perspectives*, S. T. Parker, R. W. Mitchell, and M. L. Boccia, eds., pp. 273–290. New York: Cambridge University Press.

Povinelli, D. J., Rulf, A. B., Landau, K., and Bierschwale, D. (1993). Self-recognition in chimpanzees (*Pan troglodytes*): Distribution, ontogeny, and patterns of emergence. *Journal of Comparative Psychology* 107: 347–372.

Premack, D. and Woodruff, G. (1978). Does the chimpanzee have a theory of mind? *Behavioral and Brain Sciences* 1: 515–526.

Russon, A. and Galdikas, B. M. F. (1995). Constraints on great apes' imitation: Model and action selectivity in rehabilitant orangutan (*Pongo pygmaeus*) imitation. *Journal of Comparative Psychology* 109: 5–17.

Savage-Rumbaugh, S. and Lewin, R. (1994). *Kanzi: The Ape at the Brink of the Human Mind*. New York: Wiley.

Shillito, D. J., Gallup, G. G., Jr., and Beck, B. B. (1999). Factors affecting mirror behaviour in western lowland gorillas, *Gorilla gorilla*. *Animal Behaviour* 57: 999–1004.

Shumaker, R. W. (1997). Observational Learning in the Orang utan. Unpublished master's thesis, George Mason University, Fairfax, Va.

Shumaker, R. W., Beck, B. B., Brown, L., and Taub, S. (1998). Observational learning in orangutans [abstract]. *American Journal of Primatology* 45: 208.

Shumaker, R. W., Palkovich, A. M., Beck, B. B., Guagnano, G., and Morowitz, H. (2001). Magnitude discrimination and spontaneous ordination by the orangutan. *Journal of Comparative Psychology* 115: 385–391.

Shumaker, R. W., Swartz, K., and Smits, W. (in preparation). Individual differences in learning set formation by orang utans (*Pongo pygmaeus*).

Swartz, K. B. (1998). Self-recognition in nonhuman primates. In *Comparative Psychology: A Handbook*, G. Greenberg and M. Haraway, eds., pp. 849–855. New York: Garland.

Swartz, K. B. and Evans, S. (1991). Not all chimpanzees (*Pan troglodytes*) show self-recognition. *Primates* 32: 483–496.

Swartz, K. B. and Evans, S. (1994). Social and cognitive factors in chimpanzee and gorilla mirror behavior and self-recognition. In *Self-Awareness in Animals and Humans: Developmental Perspectives*, S. T. Parker, R. W. Mitchell, and M. L. Boccia, eds., pp. 189–206. Cambridge: Cambridge University Press.

Swartz, K. B. and Evans, S. (1997). Anthropomorphism, anecdotes, and mirrors. In *Anthropomorphism, Anecdotes, and Animals*, R. W. Mitchell, H. L. Miles, and N. Thompson, eds., pp. 296–306. Albany: State University of New York Press.

Swartz, K. B. and Shumaker, R. (in preparation). Evidence for mirror self-recognition by a western lowland gorilla (*Gorilla gorilla gorilla*).

Swartz, K. B., Sarauw, D., and Evans, S. (1999). Comparative aspects of mirror self-recognition in great apes. In *The Mentalities of Gorillas and Orangutans in Comparative Perspective*, S. T. Parker, R. W. Mitchell, and H. L. Miles, eds., pp. 283–294. Cambridge: Cambridge University Press.

Swartz, K. B., Shumaker, R., and Smits, W. (in preparation). Habituation and novelty responses by infant orang utans (*Pongo pygmaeus*).

Swartz, K. B., Himmanen, S. A., Wolf, W. R., Jr., Shumaker, R. W., Bond, M., and Harris, G. W. (in preparation). Serial list learning by an orang utan (*Pongo pygmaeus*).

Tinklepaugh, O. L. (1932). The multiple delayed reaction with chimpanzees and monkeys. *Journal of Comparative Psychology* 13: 207–243.

Tomasello, M. (1996). Do Apes ape? In *Social Learning in Animals: The Roots of Culture*, C. M. Heyes and B. G. Galef, Jr., eds., pp. 319–346. New York: Academic Press.

Tomasello, M., Davis-Dasilva, M., Camak, L., and Bard, K. (1987). Observational learning of tool-use by young chimpanzees. *Human Evolution* 2: 175–183.

Ujhelyi, M., Merker, B., Buk, P., and Geissmann, T. (2000). Observations on the behavior of gibbons (*Hylobates leucogenys, H. gabriellae,* and *H. lar*) in the presence of mirrors. *Journal of Comparative Psychology* 114: 253–262.

Visalberghi, E. and Fragaszy, D. (1990). Do Monkeys ape? In *"Language" and Intelligence in Monkeys and Apes: Comparative Developmental Perspectives*, S. Parker and K. Gibson, eds., pp. 247–273. Cambridge: Cambridge University Press.

Whiten, A. and Ham, R. (1992). On the nature and evolution of imitation in the animal kingdom: Reappraisal of a century of research. In *Advances in the Study of Behavior*. Vol. 21, P. J. B. Slater, J. S. Rosenblatt, C. Beer, and M. Milinski, eds., pp. 239–283. New York: Academic Press.

Wrangham, R. W. and Goodall, J. (1989). Chimpanzee use of medicinal leaves. In *Understanding Chimpanzees*, P. G. Heltne and L. A. Marquardt, eds., pp. 22–37. Cambridge, Mass.: Harvard University Press.

Yerkes, R. M. and Yerkes, A. W. (1929). *The Great Apes.* New Haven, Conn: Yale University Press.

Zuberbühler, K., Gygax, L., Harley, N., and Kummer, H. (1996). Stimulus enhancement and spread of a spontaneous tool use in a colony of long-tailed macaques. *Primates* 37: 1–12.

42 Kinesthetic-Visual Matching, Imitation, and Self-Recognition

Robert W. Mitchell

The chimpanzee Viki was raised in their home by Keith and Cathy Hayes (see C. Hayes 1951). Having been treated as a human child, Viki acted like one—she played with her human parents, ate at the table, had trouble with toilet training, loved going for rides in the car, understood spoken language, used sounds to communicate her desires, pretended with imaginary objects, deceived, enjoyed attending to the pages of picture books, and used household tools like her human parents. Her actions provided evidence for activities previously thought by many scientists to be largely if not exclusively human, most particularly her abilities for extensive bodily imitation and mirror self-recognition. Although there had been earlier indications that great apes could imitate bodily actions and recognize themselves in mirrors (see Mitchell 1999), Viki supplied the first experimental evidence that animals can imitate diverse actions, doing a good job of re-creating the Hayes' actions when asked to do the same thing. Viki also used mirrors to clean up her face, attempt to pull out her tooth, and play with putting on lipstick, and thus clearly recognized her image (K. J. Hayes and Hayes 1955; K. J. Hayes and Nissen 1971). How is one to explain Viki's abilities?

Little attempt at an explanation was made until several years later, when some primatologists came to believe that apes who recognized themselves in mirrors needed to have a remarkably well-developed self-concept to do so, including a complex knowledge of their internal states that was presumed to be useful in interpreting others' mental states. Although the basic idea that some sort of "self-concept" is required for self-recognition seems plausible (for how else can one recognize one's body?), the nature of the proposed self-concept was problematic—how exactly did this self-concept allow an animal to recognize itself in a mirror? For example, how does knowing that one is self-congratulatory or selfish, or that one experiences mental imagery,

or that others have mental states, lead one to recognize one's self in a mirror? Obviously one must have some knowledge of what one looks like—a very specific form of self-concept—but this raises a paradox: How can you know what you look like *before* you recognize yourself in the mirror? My answer to this question [or, more specifically, Guillaume's (1926/1971) answer], organized a whole field of research in a new way, tying self-recognition, bodily imitation and pretense, communication via simulation, recognition that one is being imitated, and mental planning together in one neat skill—kinesthetic-visual matching (Mitchell 1993a,b, 1994, 1997a,b, 2000, 2002).

Guillaume studied the development of imitation in children he knew well—his own—and came up with the idea that bodily imitation and self-recognition require matching between kinesthesis and vision. He argued that imitative development does not start with, but rather leads to, making matches between one's own and another's actions. According to Guillaume, initially children try to re-create the effects other people have on objects, and only gradually come to re-create other people's actions. By repeated attempts at re-creation, the child learns to match his or her own kinesthetic feelings to the visual actions of another, and also becomes aware of what he or she looks like to others and in a mirror. In this way, children develop a match between inner and outer experience—between their own subjective experiences and those of others.

Piaget (1941/1962) acknowledged Guillaume's idea that mature bodily imitation and mirror self-recognition derived from matching between kinesthesis and vision. However, he disagreed with Guillaume's depiction of the cause of development, arguing instead that the child imitates because he or she is interested in reproducing an action per se, not just its effects; imitation thereby indicates intelligent awareness of the similarity between his/her own and others' actions, based on a match between kinesthetic and visual experiences. Whichever of these hypotheses is correct [and it is unclear whether they are really as different as Piaget believed—see Mitchell (1993b)], the essential idea that kinesthetic-visual matching is necessary for self-recognition and any robust generalized imitation seems well taken. After such lucid solutions to the problem, Guillaume's and Piaget's hypotheses were (as so often happens in science) largely ignored. Not surprisingly, numerous other individuals "independently" came up with kinesthetic-visual matching as the basis for imitation and self-recognition [see Mitchell 1997a for historical overview; strangely, Povinelli (2000) not only claims priority for this idea, but previously claimed it to be false (Gallup and Povinelli 1993)].

Kinesthetic-visual matching as I envision it is the recognition of similarity between the feeling of one's own body's extent and movement (variously called "kinesthesis," "somasthesis," or "proprioception") and how it looks (vision). We (and other animals) know generally where our appendages and mouth are at any given time, even though we do not see them; we also feel where they are when we move. This kind of knowledge seems necessary to engage in actions. But we also have an intimate knowledge of the relationship between these nonvisual perceptual experiences of our body and its visual appearance, even though we do not usually see ourselves in action. We have an idea of what we look like when we act—a general (and imprecise) idea of the "outline" of our bodies, and the relative positions of each part. That is why animals such as apes can recognize their mirror image and learn to imitate—because they sense (or can learn) that the image in the mirror or of the other is like their own image. And this sense can lead them to test the similarity, as when the captive gorilla Muni used a mirror to examine himself: He "looked at the mirror with his head placed between his legs ... Later he stood on his hands, resting his feet on the mirror. Returning to a sitting position, he lifted one leg and looked at his reflection, inspecting the parts of him that he ordinarily could not see. He obviously recognized himself" (Riopelle et al. 1971, p. 88).

Kinesthetic-visual matching is remarkable and does not seem to exist in many species of animals. Here is an example. Several years ago Jim Anderson and I provided a long-tailed macaque (*Macaca fascicularis*) named Rodrique, a former pet of a primatologist, with a concave mirror and videotaped his activities. A concave mirror provides an image of whatever is in front of it, but the image is upside down and transposed. Rodrique engaged in some remarkable activities with this mirror. We observed him putting his hand into the mirror, moving his thumb and forefinger as if grooming; as well as putting his head into the mirror and sticking his tongue out, as if licking. Certainly these were odd behaviors, and suggested to us that Rodrique might have been playing with his image—sticking his tongue out, or moving his fingers, to see what happened. But because his actions seemed like grooming and licking, we looked into the mirror ourselves, and discovered that from the perspective of having your head somewhat into a concave mirror, an image of a three-dimensional upside-down head is visible inside. It appeared that what Rodrique was doing (or trying to do) was grooming and licking the upside-down head of a monkey! He never recognized, after many hours of experience with the concave mirror, that the upside-down monkey head was his own. [We also tested him by marking his head with a red mark, but he showed no interest in the mark after looking in a flat mirror, with which he had also had experience—see Mitchell and Anderson (1993).] Rodrique's actions are intriguing. He seemed content to interact with the phantom monkey as if it were a real monkey, apparently never recognizing that this monkey did the same thing he did, not caring that he was grooming and licking nothing tangible. It would seem that Rodrique did not have the ability to connect what he knew and felt about his own body with what he saw the other monkey doing—i.e., kinesthetic-visual matching.

Kinesthetic-visual matching does not, of course, work in isolation from other matching abilities that are widespread throughout mammals. For example, most mammals can recognize matches between two visual stimuli (e.g., Rodrique knew it was a monkey in the mirror), perhaps including even parts of their own body and those of other species members (Mitchell 1993a). Kinesthetic-visual matching works in combination with, and is clearly dependent on, an organism's other matching skills (Mitchell 1994). It would be surprising if an animal with kinesthetic-visual matching did not have visual-visual matching or other within-modality matching skills, and more surprising still if an organism was only able to use one of these matching skills at a time. Thus, assuming that kinesthetic-visual matching is essential to a generalized ability for bodily imitation does not, as Whiten (2000, p. 499) states, "neglect the possible role of visual feedback [i.e.] visual-visual matching." (Surprisingly, the potential visual-visual matching available from actively grooming the phantom monkey with his fingers did not afford Rodrique any self-recognition, perhaps because the reflected image was upside down and reversed, and transformed as his fingers moved closer to the reflected image.) There are, of course, alternative means besides kinesthetic-visual matching by which organisms might match their own and others' bodies. Haptic-kinesthetic matching might be used, for example, by blind organisms (or sighted ones at that), and perhaps auditory-kinesthetic matching (for those organisms with echolocation). But kinesthetic-visual matching seems the most likely possibility for most mammals, especially primates; other matches to kinesthesis besides the visual seem to require more active exploration to be detected.

Imagine organisms (like Rodrique) without kinesthetic-visual matching. Presumably they have within-modality matching skills, so that visual-visual or haptic-haptic or auditory-auditory matching is relatively easy for them. They also presumably have internal mental images of these modalities, so that they might be able to think or dream with these images; and they have kinesthesis, so they know where their body parts are and have a general notion of the outline of their body. They could reenact their own actions, for deceptive purposes or in play. They might even have cross-modal matching abilities, such as recognizing the analogical relation between a brighter light and a louder sound. But without connections between kinesthesis and vision, the organism would not be able to have a visual mental image of its own body and would not be able to connect the kinesthetic image it has of its own body with any visual image. Thus, such an organism could not translate from its bodily feelings to a visual mental image of itself (and thus could not have a visual mental image of itself). Such a creature might have dreams or thoughts in which it observed things, but it could never represent itself visually—it could only be an observer not visually represented. And without an ability to translate between its own bodily feelings and those of others, it could never recognize that others have bodily feelings and thoughts like its own. In fact, such a creature could not attribute its experiences to itself, for without an ability to attribute psychological states to others, one cannot attribute them to oneself (and vice versa).

This last idea is surprising, because many people think that we understand others because we analogously attribute our own experiences to them; that is, we judge from our knowledge of ourselves to knowledge of others. But this judging is problematic, as the philosopher P. F. Strawson pointed out, because it is unclear how one can attribute experiences *to* oneself without an ability to attribute experiences per se:

There is no sense in the idea of ascribing states of consciousness to oneself, or at all, unless the ascriber already knows how to ascribe at least some states of consciousness to others. So he cannot (or cannot generally) argue "from his own case" to conclusions about how to do this; for unless he already knows how to do this, he has no conception of *his own case*, or any *case* (i.e., any subject of experiences). Instead, he just has evidence that pain, etc., may be expected when a certain body is affected in certain ways and not when others are. (Strawson 1958/1964, p. 393)

Such an organism would experience pain, of course, but would not have a notion of itself as the possessor of that or any other experience (see Mitchell 2000, for elaboration).

An organism with all of the capacities of an organism like Rodrique, but with the addition of kinesthetic-visual matching, would have a great variety of understandings available to it based on matching between bodies. Such organisms would know (or at least have a general idea of) what they look like when they act, and would recognize that others' actions are possible actions for their own body. Thus, not only could they recognize that the visual image of their body in the mirror looks like what their body movements feel like (and thus infer that the image in the mirror is an image of their body), but they could also do what they see others do; that is, imitate others' actions. Such organisms could also recognize when they are being imitated, or pretend to be another by acting like them. And they could have some awareness that others have psychological experiences, in that they can recognize the bidirectionality of kinesthetic-visual matching. Not only do they feel like what another looks like, but they look like what another feels like. In addition, these organisms can represent themselves visually in thought, and be able to translate from the visual image of themselves to their own kinesthetic movements in order to act in relation to the visual image; that is, they can plan to do things. (Of course language offers another means of planning, without visual images.)

This theoretical orientation has led me to examine the evidence of self-recognition, imitation, recognition of being imitated, pretense, and planning in humans and nonhumans. The most consistently examined evidence concerns self-recognition and bodily imitation, two activities that are of course common in humans and which become mature at about 18 months of age or

earlier. In fact, generalized imitation and self-recognition appear to develop at about the same time in human children (Asendorpf and Baudonnière 1993; Hart and Fegley 1994; Asendorpf et al. 1996), which supports my prediction that they develop from the common skill of kinesthetic-visual matching (Mitchell 1993a).[1] Unfortunately, access to the sorts of animals likely to show similar skills—apes and other large-brained animals—is difficult, so to test my prediction I have had to look to the literature to see what other scientists have discovered.

At first the evidence was pretty clear—apes showed self-recognition and imitation, but most other species did not (see Mitchell 1993a, 1997b). Experimental testing of chimpanzees showed that they can use a mirror to self-recognize when a mark has been applied to their face (Gallup 1970), and a redoing of the Hayes's work with Viki showed that chimpanzees and orangutans can engage in generalized bodily imitation (Custance et al. 1995; Miles et al. 1996). [Imitation on demand does not seem to be present in other species (Mitchell 1993a; Mitchell and Anderson 1993), although not many species have been tested.] Superficially, the evidence seemed clear—at least some members of each ape species can self-recognize (Swartz et al. 1999) and imitate diverse bodily actions, so that both are potentially species-wide skills. Add the evidence that dolphins (*Tursiops* species) can engage in generalized bodily imitation (Tayler and Saayman 1973; Bauer and Johnson 1994) and self-recognition (Marten and Psarakos 1994; Reiss and Marino 2001), and it would appear that the evidence supports the kinesthetic-visual matching hypothesis. Another point in its favor is evidence that cotton-top tamarins show self-recognition (Hauser et al. 1995) and marmosets, a closely related species, show some skill at bodily imitation (Bugnyar and Huber 1997; Voelkl and Huber 2000; although see later discussion).

The problem is that the evidence is not all that conclusive. Other than the chimpanzee Viki, the gorilla Koko, and the orangutan Chantek—

all human-reared apes—there is little evidence (except in human children, for whom it is robust) of the co-occurrence of generalized bodily imitation and self-recognition in the same individual (Mitchell 1997b). [Unfortunately, many scientists seem content to use evidence that theoretically codependent activities occur in different members of the same species to support their belief that the same ability is responsible for both activities; see discussion by Mitchell (1993b).] Also, there seems to be evidence against the hypothesis; some animals (gibbons) that are not known to imitate appear to show a skill for self-recognition (Ujhelyi et al. 2000), although new research may indicate some skill at imitation. And some animals that are not known to self-recognize (macaques) every once in a while act in ways that are suggestive of imitation via kinesthetic-visual matching (see Mitchell 2002).

In addition, evidence of kinesthetic-visual matching in marmosets and tamarins is not convincing. Although some researchers characterized two tamarins' actions toward a mirror as looking at body parts that are not visible without a mirror and five tamarins' actions as wiping off a mark near their face while looking in the mirror (Hauser et al. 1995), alternative interpretations are possible (Anderson and Gallup 1997; cf. Hauser and Kralik 1997). For example, the behavioral evidence is rare and/or ambiguous in its support of the claim that the tamarins were using the mirror to look at their normally visually inaccessible body parts or the mark on the hair above their face. Similarly, some potential evidence of bodily imitation in marmosets (Bugnyar and Huber 1997) may result from visual-visual matching and/or chance similarities in handedness between observer and demonstrator marmosets (see Mitchell 2002), but the evidence is still suggestive. More recent evidence of marmoset imitation (Voelkl and Huber 2000) is also suggestive. Observer marmosets tended to open a film canister to obtain food using the same body configuration (by hand or mouth) that the demonstrator marmoset had used. Specifically, after

observing another marmoset open film canisters using its mouth, 4 out of 6 observer marmosets opened at least one canister (actually, 2, 6, 11, and 13) out of 14–15 with their mouths (which is infrequent among marmosets). By contrast, after observing another marmoset open film canisters using its hands, all 5 observer marmosets opened 14–15 canisters out of 15 with their hands (none used its mouth).

While these data are suggestive, the actions the demonstrator marmosets used may have been perceived by the observer marmosets as indicating something distinctive about the canisters they later opened. Based on the pictures of both actions on p. 197 of Voelkl and Huber's article, the method of opening by mouth suggests that the marmoset is attacking the lid of the canister, whereas the method of opening by hand suggests that the marmoset is exploring what is inside the canister. Thus it may be that the marmosets did not imitate the specific action of opening used by the demonstrator marmoset, but rather responded in a similar attitude toward the objects. Similarly, chimpanzees tended to react aggressively or with exploration toward a hidden object when another chimpanzee who had seen the object acted aggressively or with exploration (Menzel 1973). Imitation at the level of intentionally re-creating the actions of others seems unlikely in these circumstances.

Given that access to and control over great apes and dolphins is particularly difficult, my own penchant is to explore more directly whether animals who have never shown self-recognition can learn generalized imitation. (That is, I am looking for potentially disconfirmatory evidence.) As noted earlier, one attempt to teach the long-tailed macaque Rodrique to imitate my bodily actions failed (Mitchell and Anderson 1993). Recently, collaborator Ellen Furlong has been unsuccessful in teaching a Syke's guenon (*Cercopithecus mitis albogularis*) at April Truitt's Primate Rescue Center in Nicholasville, Kentucky to imitate her actions using a method, similar to that used with Rodrique, of rewarding the animal for scratching where she (the model) scratches. Of course if we can teach a monkey to engage in generalized bodily imitation, we can test it for self-recognition, thereby providing evidence for or against the kinesthetic-visual matching hypothesis. But given other researchers' interest in self-recognition and imitation, particularly those studying human children, we probably will not have to wait long for more information about the correlation between these two activities in the same individuals.

Note

1. Two more recent and well-thought-out empirical studies also show a developmental synchrony among self-recognition, generalized bodily imitation, and pretense based on body modeling (Nielsen 2001; Baudonnière et al. 2002) which supports the kinesthetic-visual matching hypothesis. Neither study acknowledges this, and Nielsen even argues that the occurrence of deferred imitation earlier than generalized bodily imitation and self-recognition provides evidence against the kinesthetic-visual matching hypothesis. It is important to clarify here that deferred imitation is not necessarily evidence of generalized bodily imitation. In my presentation of the kinesthetic-visual matching hypothesis, I explicitly acknowledged that deferred imitation occurred earlier than self-recognition (Mitchell 1993a, p. 303), and consequently suggested that a more generalized skill for bodily (including facial) matching is needed to indicate a kinesthetic-visual matching mature enough to support self-recognition (p. 303).

References

Anderson, J. and Gallup, G. G., Jr. (1997). Self-recognition in *Saguinus*? A critical essay. *Animal Behaviour* 54: 1563–1567.

Asendorpf, J. B. and Baudonnière, P. M. (1993). Self-awareness and other-awareness: Mirror self-recognition and synchronic imitation among unfamiliar peers. *Developmental Psychology* 29: 88–95.

Asendorpf, J. B., Warkentin, V., and Baudonnière, P. M. (1996). Self-awareness and other-awareness II: Mirror self-recognition, Social contingency awareness, and synchronic imitation. *Developmental Psychology* 32: 313–321.

Baudonnière, P.-M., Margules, S., Belkhenchir, S., Carn, G., Pèpe, F., and Warkentin, V. (2002). A longitudinal and cross-sectional study of the emergence of the symbolic function in children between 15 and 19 months of age: Pretend play, object permanence understanding, and self-recognition. In *Pretending and Imagination in Animals and Children*, R. W. Mitchell, ed., pp. 73–90. New York: Cambridge University Press.

Bauer, G. and Johnson, C. M. (1994). Trained motor imitation by bottlenose dolphins (*Tursiops truncatus*). *Perceptual and Motor Skills* 79: 1307–1315.

Bugnyar, T. and Huber, L. (1997). Push or pull: An experimental study on imitation in marmosets. *Animal Behaviour* 54: 817–831.

Custance, D. M., Whiten, A., and Bard, K. A. (1995). Can young chimpanzees (*Pan troglodytes*) imitate arbitrary actions? Hayes and Hayes (1952) revisited. *Behaviour* 132: 837–859.

Gallup, G. G., Jr. (1970). Chimpanzees: Self-recognition. *Science* 167: 86–87.

Gallup, G. G., Jr. and Povinelli, D. J. (1993). Mirror, mirror on the wall, which is the most heuristic theory of them all? A response to Mitchell. *New Ideas in Psychology* 11: 295–325.

Guillaume, P. (1926/1971). *Imitation in Children*, 2nd ed. Chicago: University of Chicago Press.

Hart, D. and Fegley, S. (1994). Social imitation and the emergence of a mental model of self. In *Self-Awareness in Animals and Humans*, S. T. Parker, R. W. Mitchell, and M. L. Boccia, eds., pp. 149–165. Cambridge: Cambridge University Press.

Hauser, M. D. and Kralik, J. (1997). Life beyond the mirror: A reply to Anderson and Gallup. *Animal Behaviour* 54: 1568–1571.

Hauser, M. D., Kralik, J., Botto-Mahan, C., Garrett, M., and Oser, J. (1995). Self-recognition in primates: Phylogeny and the salience of species-typical traits. *Proceedings of the National Academy of Sciences U.S.A.* 92: 10811–10814.

Hayes, C. (1951). *The Ape in Our House*. New York: Harper and Brothers.

Hayes, K. J. and Hayes, C. (1955). The cultural capacity of chimpanzee. In *The Non-human Primates and Human Evolution*, J. A. Gavan, ed., pp. 110–125. Detroit: Wayne State University Press.

Hayes, K. J. and Nissen, C. (1971). Higher mental functions of a home-raised chimpanzee. In *Behavior of Nonhuman Primates*. Vol. 4, A. M. Schrier and F. Stollnitz, eds., pp. 60–115. New York: Academic Press.

Marten, K. and Psarakos, S. (1994). Evidence of self-awareness in the bottlenose dolphin (*Tursiops truncatus*). In *Self-Awareness in Animals and Humans*, S. T. Parker, R. W. Mitchell, and M. L. Boccia, eds., pp. 361–379. Cambridge: Cambridge University Press.

Menzel, E. W., Jr. (1973). Leadership and communication in young chimpanzees. In *Precultural Primate Behavior*, E. W. Menzel, Jr., ed., pp. 192–225. Basel: S. Karger.

Miles, H. L., Mitchell, R. W., and Harper, S. (1996). Simon says: The development of imitation in an enculturated orangutan. In *Reaching into Thought: The Minds of the Great Apes*, A. Russon, K. Bard, and S. T. Parker, eds., pp. 278–299. New York: Cambridge University Press.

Mitchell, R. W. (1993a). Mental models of mirror self-recognition: Two theories. *New Ideas in Psychology* 11: 295–325.

Mitchell, R. W. (1993b). Recognizing one's self in a mirror? A reply to Gallup and Povinelli, de Lannoy, Anderson, and Byrne. *New Ideas in Psychology* 11: 351–377.

Mitchell, R. W. (1994). The evolution of primate cognition: Simulation, self-knowledge, and knowledge of other minds. In *Hominid Culture in Primate Perspective*, D. Quiatt and J. Itani, eds., pp. 177–232. Boulder, Col.: University Press of Colorado.

Mitchell, R. W. (1997a). Kinesthetic-visual matching and the self-concept as explanations of mirror self-recognition. *Journal for the Theory of Social Behavior* 27: 101–123.

Mitchell, R. W. (1997b). A comparison of the self-awareness and kinesthetic-visual matching theories of self-recognition: Autistic children and others. *Annals of the New York Academy of Sciences* 818: 39–62.

Mitchell, R. W. (1999). Scientific and popular conceptions of the psychology of great apes from the 1790s to the 1970s: Déjà vu all over again. *Primate Report* 53: 1–118.

Mitchell, R. W. (2000). A proposal for the development of a mental vocabulary, with special reference to pretense and false belief. In *Children's Reasoning and the Mind*, K. Riggs and P. Mitchell, eds., pp. 37–65. Hove, UK: Psychology Press.

Mitchell, R. W. (2002). Imitation as a perceptual process. In *Imitation in Animals and Artifacts*, C. L. Nehaniv and K. Dautenhahn, eds., pp. 441–469. Cambridge, Mass.: MIT Press.

Mitchell, R. W. and Anderson, J. (1993). Discrimination learning of scratching, but failure to obtain imitation and self-recognition in a long-tailed macaque. *Primates* 34: 301–309.

Nielsen, M. (2001). A longitudinal investigation of the emergence of mirror self-recognition, imitation and pretend play in human infants. Paper presented at International Primatological Society meeting. Adelaide, Australia.

Piaget, J. (1945/1962). *Play, Dreams, and Imitation in Childhood*. New York: W. W. Norton.

Povinelli, D. J. (2000). *Folk Physics for Apes: The Chimpanzee's Theory of How the World Works*. Oxford: Oxford University Press.

Reiss, D. and Marino, L. (2001). Mirror self-recognition in the bottlenose dolphin. A case of cognitive convergence. *Proceedings of the National Academy of Sciences. U.S.A.* 98: 5937–5942.

Riopelle, A. J., Nos, R., and Jonch, A. (1971). Situational determinants of dominance in captive young gorillas. *Proceedings of the 3rd International Congress of Primatology* 3: 86–91.

Strawson, P. F. (1958/1964). Persons. In *Essays in Philosophical Psychology*, D. F. Gustafson, ed., pp. 377–403. New York: Anchor Books.

Swartz, K. B., Sarauw, D., and Evans, S. (1999). Comparative aspects of mirror self-recognition in great apes. In *The Mentalities of Gorillas and Orangutans*, S. T. Parker, R. W. Mitchell, and H. L. Miles, eds., pp. 283–294. Cambridge: Cambridge University Press.

Tayler, C. K. and Saayman, G. S. (1973). Imitative behaviour by Indian Ocean bottlenose dolphins (*Tursiops aduncus*) in captivity. *Behaviour* 44: 286–298.

Ujhelyi, M., Buk, P., Merker, B., and Geissmann, T. (2000). Observations on the behavior of gibbons (*Hylobates leucogenys, H. gabriellae,* and *H. lar*) in the presence of mirrors. *Journal of Comparative Psychology* 114: 253–262.

Voelkl, B. and Huber, L. (2000). True imitation in marmosets. *Animal Behaviour* 60: 195–202.

Whiten, A. (2000). Primate culture and social learning. *Cognitive Science* 24: 477–508.

43 Darwin's Continuum and the Building Blocks of Deception

Güven Güzeldere, Eddy Nahmias, and Robert O. Deaner

Deception in nonhuman animals is one of the most fertile areas of research for pursuing philosophical questions in cognitive ethology, but it is an area rife with controversy (Byrne and Whiten 1988; Whiten and Byrne 1988.) The literature on deception is interwoven with questions and claims about intelligence, levels of representation, intentionality, and consciousness (Dennett 1983; Mitchell 1986; Perner 1991). Our aim in this essay is not to survey or evaluate the entire literature; rather, we use the phenomenon of deception as a case study, in a framework of some relevant conceptual distinctions, to illustrate a point about the graded continuity of mental phenomena across species.

Darwin stated: "If no organic being excepting man had possessed any mental power, or if his powers had been of a wholly different nature from those of the lower animals, then we should never have been able to convince ourselves that our high faculties had been gradually developed. But it can be shewn that there is no fundamental difference of this kind" (Darwin 1871/1936, p. 445). That is, while there may be notable differences among closely related living species because of rapid evolutionary changes and lost ancestral varieties, given the theory of natural selection, there should be no inexplicable jumps in cognitive abilities. Conversely, when faced with what seem to be differences *in kind* between related species' abilities, we should look for other cognitive abilities that may underlie these differences (cf. Allen and Bekoff 1997; Pinker 1994 and Deacon 1997 discuss language in this context from different perspectives).

With this in mind, we examine the traditional distinctions between levels of deception in animals and suggest some cognitive abilities that may serve as building blocks to account for the apparent gaps between these levels, focusing on the order of primates. We then describe an experiment that tested for deception in lemurs and that suggests directions for further empirical work in this area.

Three Categories of Deception

Deception may be broadly defined as an agent's producing or withholding an act or a signal so that it is misinterpreted by another to the advantage of the agent (see Mitchell 1986 and Hauser 1996; cf. Whiten and Byrne 1988; Hauser 1997). As such, deception always involves misrepresentation, and it takes (at least) two animals to make misrepresentation lead to deception. These animals may be of different species (e.g., predators and prey) or the same species, as in the case of "sexual mimicry" (Wickler 1965, 1968; see also Hockham 2000, 2001; and Weldon and Burghardt 2001, for a discussion of the link between mimicry and deception) or in instances where one monkey leads another way from a known food source (see later discussion).

However, if we are interested in determining what sorts of cognitive abilities are needed to carry out different types of deception, we should not begin by looking at the complexity of the misrepresentation. Sometimes the use of fairly simple signals to deceive (e.g., a misdirecting eye gaze) may require significant cognitive abilities, whereas there may be no cognition at all behind informationally rich displays of misrepresentation (e.g., butterfly wings with intricate patterns that make the wing ends look like the butterfly's head) (Robbins 1981). Camouflage and other "hardwired" displays count as deception in our broad first-pass definition, as long as we allow the production of signals to include the production (or even possession) of misleading appearances.

In our view, the relevant gradations in the cognitive abilities involved in deceptive signals or behaviors should be delineated by the historical processes that produced them, the flexibility with which they can be displayed, and the degree to which the deceiving animal understands how the deception achieves a desired outcome. Many discussions of animal deception suggest these distinctions by classifying deceptive behaviors in three basic categories, which are distinguished along the same lines as many other apparently intelligent behaviors: genetic, learned, or intentional (Mitchell 1986).

In the first category, there are the apparently "hardwired" deceptive behaviors, such as the Mantis shrimp that "pretends" it is ready to fight a competitor even though its shell is soft from molting (Caldwell 1986) or the nonpoisonous butterfly that mimics the coloration of poisonous butterflies to ward off blue jays, a common predator (Brower 1969). In both of these cases, the misrepresenting organism does not *choose* to misrepresent, with the intention to deceive, among other possible behaviors. In fact, what it does is simply what it cannot help doing. The behaviors or traits do not get acquired through experience or modified through a history of learning. Rather, they are produced by a history of selection for certain genetic "programs." As such, they are relatively inflexible in the face of environmental contingencies (although they may require environmental cues to develop). This type of deception roughly corresponds to what Dennett (1983) characterizes as "zero-order intentionality," since the animal does not seem to represent the goal of its deceptive behavior (the behavior does not involve beliefs and desires).

There is, nonetheless, a difference between these two cases. In the case of the butterfly, the misrepresenting appearance is displayed in individual butterflies whether or not blue jays are around (although the existence of predators provided the selective pressures for the evolution of the wing display). On the other hand, the misrepresenting display of the molting shrimp (the extension of its claws despite its inability to use them), even if it is also the result of selective pressures, is brought forth neither randomly nor all the time, but only on occasion—when there is

a threatening intraspecific competitor, and only one of a certain size (Steger and Caldwell 1983; Caldwell 1986). In this sense, the shrimp requires more resources to be a deceiver than the butterfly. The butterfly just needs to "be," whereas the shrimp needs a perceptual system to detect the appropriate stimuli so that its misrepresentational system can be triggered. Even though the shrimp does not *decide* to display its claw any more than the butterfly decides to maintain its wings' visual appearance, the shrimp has the elemental ingredients that are necessary for the learning involved in the next level of deception; for instance, a perceptual system and mechanisms to discriminate among various situations. As such, the butterfly and the Mantis shrimp occupy neighboring but nonetheless different gradations in Darwin's continuum (see Mitchell's 1986 distinction between levels I and II).

The second category of deceptive behavior is typically described as involving (merely) behavioristic learning. An animal's normal behavior is transferred to a different context because it is reinforced by the rewarding reactions of a target animal when the target animal responds to the behavior as if it indicated something that it does not. Put more simply, the deceptive animal learns to do something because it pays off, and it pays off because the target animal misrepresents the behavior (but the deceiver need not connect the payoff and the misrepresentation; that is, it need not recognize *why* its behavior is effective). For instance, a young monkey is threatened and calls out for help, and his mother comforts him by allowing him to suckle. He learns to associate calling out with getting milk, so when he is hungry, he calls out as if he is threatened. The behavior is deemed deceptive in that the mother reacts as if the call indicates a threat even when it does not.

Such learned behaviors, although they may be relatively inflexible once learned, are not produced directly by a history of natural selection, but rather by particular environmental contingencies, by a history of reinforcement. This type of deception is similar to Dennett's (1983) concept of "first-order intentionality," because the animal does represent the goal of its deceptive behavior described from the "intentional stance."

Notice that for both genetic and learned cases of deception, the deceiver does not intend its behavior (or signal) to be deceptive. Indeed, its behavior is deceptive only in the sense that we (human observers) describe it that way, and we describe it that way only because the behavior causes the target animal to represent its environment in a way that we know is mistaken. Both the deceiver and the deceived are just going about their business, reacting to perceptual information in the only way they know how. We might say that the deceiving animal produces deceptive behaviors just because the behaviors *have* worked (in its phylogenetic or ontogenetic history), rather than because the animal understands how they *will* work; that is, by causing the target to misrepresent its environment.

The third category of deceptive behavior is thus portrayed as the only one that really fits our ordinary concept of human deception. This category involves deception that can be described as intentional in the sense that the deceiver *intends* to misrepresent information to its target in order to achieve some kind of goal. The crucial point is that the animal understands *how* its behavior deceives its target; that is, the deceiver recognizes the relationship between the target's mental states (e.g., perceptions, beliefs, desires) and its behavior, and the deceiver manipulates those mental states accordingly. Because it involves misrepresenting information so that the target behaves in a particular manner, such deception is sometimes labeled as "reasoned" or "planned."

This third category of deception corresponds approximately to Dennett's (1983) notion of "second-order intentionality" because the deceiving agent represents the beliefs, desires, and intentions of the target, although it may involve higher levels of intentionality as well. For instance, a clever child who wants her pesky brother to leave her room tells him that it's time for his favorite television cartoon. She knows that because he desires to watch the show, then if he believes it is on, he will leave to watch it, satisfying her goal. She creates a false belief in her brother that combines with desires she knows he possesses in order to make him act in a way that he would not have acted without the false belief. Note that one may create a false belief in another's mind with only a limited understanding of the nature of belief. For instance, children pass false-belief tasks and begin intentionally deceiving around age 4, without yet being able to conceptualize and articulate how they are doing so (e.g., Perner 1991, chapter 8; cf. Flavell 1999, 2000).

Building Blocks of Deception

It is notoriously difficult to distinguish between these three categories of behavior—genetic, learned, and intentional—whether dealing with deceptive or other intelligent behaviors. This is in part because, experimentally, the evidence is difficult to gather and in part conceptually because they exist along a continuum without such clear, simplistic dividing lines. We have already suggested a possible building block—perceptual and discriminatory mechanisms—that may underlie the transition from genetic to learned deception. The transition from learned deception to intentional deception may have even more gradations. In fact, it is unlikely that any nonhumans, even the great apes, have a well-developed understanding of how intentional deception works, at least according to most definitions of intentional deception. The reason is quite straightforward. These definitions generally hold that intentional deception requires a theory of mind, the ability to understand that others have beliefs and desires, and there is little persuasive evidence that any nonhuman animals are capable of this (Perner 1991; Whiten 1997). Full-fledged intentional deception may even require language (notice how difficult it is to come up with examples of intentional deception that do not include language—we usually create false beliefs by *telling* lies).

Thus we believe future research should be directed toward identifying the kinds of deception that are intermediate between learned and intentional deception and specifying the cognitive abilities that could underlie such deception. Perhaps the first step in this endeavor is acknowledging that an animal may misdirect another's behavior and understand how it is achieving this goal without necessarily understanding the mental states of its target as mental states; that is, without having a theory of mind.

In fact, we think that it is crucial to delineate this capacity, and propose to call it "proto-understanding of agency," an ability that is more sophisticated than the mere perceptual identification of different elements in one's environment (food, mate, predator, etc.) but more elementary than the recognition of mental states in others. What is sufficient for the presence of proto-understanding of agency is some ability (which itself can occur in gradations) to recognize that certain things in one's environment, such as conspecifics, prey, or predators, can be manipulated in very specific ways—ways that do not apply to various other things, including many other animals, trees, rocks, and the like. What is emphasized is the recognition of another, not so much as a *thinker*, but as a *doer*. Remove this recognition and what you get is deception in the first or second category. This is not to suggest that intentional deception or proto-understanding of agency does not involve learning. They surely do; for instance, an animal may have to learn what class of things in its environment are agents rather than passive objects.

Put differently, in the first two categories the fact that the deceived party is an agent does figure in the story, but it makes no difference to the deceiving animal's behavior in genetic deception, and it underlies only discriminatory but still stimulus-bound and reinforced behavior in learned deception. Higher levels of deception, on the other hand, call for something extra, which is usually expressed in terms of the animal's under-

standing of the beliefs and desires of its target. But many types of flexible and forward-looking deception require only that the deceiver be able to predict the responses of the target, not necessarily that it understand the target's mental states.

What we call "proto-understanding of agency" is, then, roughly the capacity to have a rudimentary understanding of a causal relation between what another animal is perceiving at the moment (or perhaps even what it perceived a while ago) and how it is going to respond in a particular situation based on that perception (cf. Gomez 1991, Whiten 1997). It allows an animal to discriminate between agents, who do things in and to their environments, and objects, which do not act on their environments. This rudimentary understanding of causal agency can then be put to use in deceiving other animals, based on the recognition that a target animal, which is usually a conspecific, is *manipulable* in that its behavior, unlike the movements of inanimate objects, plants, and some other animals, will change depending on its perceptions.

In addition to proto-understanding of agency, we want to suggest a second building block for intentional deception: the ability to flexibly inhibit normal behavior that would result in immediate reward. This ability is important in allowing an animal to withhold informative signals or behavior. Although many animals can inhibit certain behaviors in a small set of contexts, we suggest that the flexibility of inhibition may be crucial in allowing certain types of deception to take place. Unfortunately, inhibition in naturalistic contexts has rarely been studied (cf. Diamond 1990, Hauser 2000), but it is likely that there are considerable variations. In the next section we describe one context where flexible inhibition may be crucial for deception.

A Test Case for Primate Deception: The Menzel Paradigm

We suggest that the abilities of proto-understanding of agency and flexible inhibition allow deceptive behaviors that go beyond stimulus-response learning, but need not involve the theory of mind abilities usually associated with intentional deception. By examining how different kinds of primates express these abilities in an experimental paradigm that tests for deception, we can begin to chart how a sophisticated cognitive ability such as (intentional) human deception is constituted of more elementary abilities found in other species.

To illustrate the relationship between proto-understanding of agency, inhibition, and deception, consider the Menzel paradigm, in which a subordinate animal is shown the location of a food cache and then the group (including a dominant animal) ignorant of the location is released with the subordinate (Menzel 1974; see also Hirata and Matsuzawa in review). When Menzel used this paradigm on chimpanzees, the dominant male followed the subordinate female to the food cache and took the food; he used her as a source of information. In the course of repeated trials, the subordinate began to take evasive actions; for instance, she would lead the group to a smaller food cache and then dart to the larger cache. But the dominant male soon ignored these tactics, and sometimes he even moved away from her, only to turn quickly to catch her heading for the food.

As described, each ape clearly recognizes, in some sense, how the other ape "works" (does things as an agent). At a minimum, they are able to predict the other's future behavior based on current perceptions and behavior. For example, the subordinate recognizes that if the dominant is attending to her movements and she goes to the food, she will lose it. He, in turn, recognizes that some of her movements are not reliable indicators of where the food is located, and he learns that she is more likely to reveal the food if she does not see him watching her.

The two key points in these experiments are that the apes' learning would require making complex and often novel connections between subtle bodily cues (e.g., gaze, body orientation)

and future behavior, and second, the end result—no what matter what the ontogeny—is that the apes apparently recognize each other as agents perceiving and acting on their environments. Furthermore, each ape, because of the presence of the other, *inhibits* presumably powerful motivations to act as it normally would. The subordinate avoids moving toward a known food source; the dominant redirects attention away from a reliable indicator of food (i.e., the subordinate's movement). These behaviors represent a flexible type of inhibition driven, not by the presence of an aversive stimuli, but by a reward to be obtained at a later time. They are necessary for the deception and the counterdeception to occur.

In sum, the deceiving ape's behavior may not represent intentional deception with a theory of mind, but it probably involves cognitive abilities that outstrip those required for many types of merely learned behaviors. At the very least, the deceiver's behavior can be considered intentional in the sense that it has an intermediate goal to modify the perceptions or experience of the target ape so that its behavior is then altered (see Hauser 1997).

What's in a Lemur's Mind?

The Menzel paradigm has been replicated in a species of Old World monkey (mangabeys) that used some of the same tactics employed by the chimps; notably the subordinate explored several different tactics and learned to take indirect routes to the food or to wait until the dominant animal was preoccupied before going to the food (Coussi-Korbel 1994). We should note that the mangabeys did not necessarily perform deceptive behaviors as complex as those of the chimps, and some experiments suggest differences in the abilities of monkeys and apes to inhibit and to understand agency. However, here we focus on the differences in cognitive abilities demonstrated by strepsirhine primates and those demonstrated

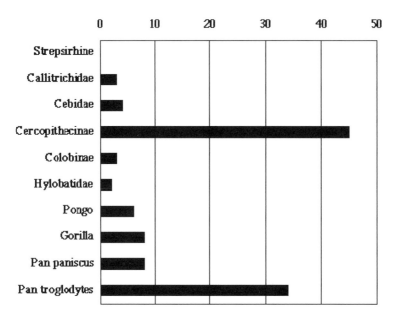

Figure 43.1
Taxonomic distribution of deception observations in primates (for details, see Byrne and Whiten 1992).

by both (some) monkeys and apes. The important point is that various types of deceptive behavior have been observed in most primate radiations, especially in baboons (cercopithecines) and chimpanzees, but no deception of any kind has been observed in any strepsirhine species (figure 43.1).

With this in mind, Robert Deaner replicated the Menzel paradigm with ringtailed lemurs at the Duke University Primate Center Durham, NC (Deaner 2000, in preparation).

In Deaner's setup (figure 43.2), a male ring-tailed lemur (the subordinate) was informed of the location of two grapes (the experimenter trained the lemurs to learn the location of the food by waving a towel in front of the correct location among eight possible sites), whereas a dominant female remained uninformed. Upon being released simultaneously into the foraging area, females regularly followed and displaced

the informed males and thus managed to obtain most of the grapes. Therefore the subordinate males had motivation to inhibit their movement toward the food location (just as the subordinate chimps and mangabeys did).

Even though this setup provided male ring-tailed lemurs with the opportunity to profit by deceiving females, deception was rare. For instance, in one representative dyad, a subordinate named Teres usually went directly to the site where he had observed the grapes hidden (as he did in baseline trials run alone), only to be followed by the dominant female, who took one or both of the grapes. Over many trials, however, Teres did employ several other tactics (table 43.1), and two of these even succeeded in deceiving the female and allowing Teres to gain both grapes without threat. Four times he departed from the trial shelter, waited for the female to move away from the baited site, and then pro-

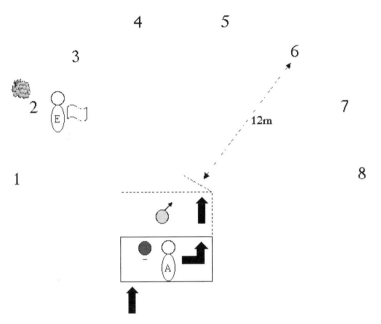

Figure 43.2
Experimental setup for study of ringtailed lemur deception. Male observes experimenter (E) waving towel to indicate location of grapes at feeding site 2. Meanwhile, female in trial shelter is monitored by assistant (A); after male is informed, female will be moved to the same part of the shelter as the male and the two lemurs will be released simultaneously.

ceeded to retrieve the grapes; twice he moved in the wrong direction, allowed the female to overtake him, and then moved to the correct site.

While these deceptive tactics were effective, Teres rarely used them, despite the fact that the female showed no indication of countering the deception by closely following Teres on the subsequent trial. Even more strikingly, across the 56 trials, and in subsequent trials with a second female, Teres did not begin using the deceptive tactics with increasing frequency. Thus there was no indication that he had learned what made the previous trial successful. In fact, in most dyads, after a male had previously been deceptive, in the next trial he usually proceeded directly to the baited site.

Although we cannot say for sure why male ringtailed lemurs do not practice deception more frequently in the Menzel paradigm, there are two likely possibilities. One is that, while the lemurs understand which behaviors are effective, they are unable to inhibit their tendency to go directly toward the food as soon as it is (apparently) accessible. This would explain why deceptive trials were frequently followed by direct dashes to the grapes. While an elementary understanding of agency (i.e., that behavior changes in response to what happens in one's perceived environment) is crucial to the ability to deceive, this understanding by itself, however sophisticated it may be, is ineffective in the absence of inhibition (whereas inhibition in the absence of understanding is

Table 43.1
Tactics employed by male ringtailed lemur, Teres, in a series of 56 dyadic trials with female, Adea

Description of Behavior	No. of Trials Used	Grapes/ Trial[a]	Tactic Classification
Depart immediately; move immediately in the correct direction; search baited site and nearby sites sequentially.	35	0.6 ± 0.7	Normal
Depart immediately; move immediately in the wrong direction; search unbaited site and nearby sites sequentially.	10	0.2 ± 0.6	Ineffective alternative
Wait in release cage; depart; move slowly to baited site.	2	0.5 ± 0.7	Ineffective alternative
Wait in release cage; depart; return to trial shelter for next trial (no search).	3	0.0 ± 0.0	Ineffective alternative
Depart immediately; wait (female moves in wrong direction); move to baited site.	4	2.0 ± 0.0	Effective alternative
Depart immediately; move immediately in wrong direction; stop at or before unbaited site (female searches unbaited site); move to baited site.	2	2.0 ± 0.0	Effective alternative

[a] Grapes per trial refers to the number of grapes Teres consumed per trial, out of a maximum of two.

useless). Supporting this possibility are data indicating that lemurs have more trouble than monkeys in an inhibition task where food must be obtained by initially pushing it away (Davis and Leary 1968).

Another possibility is that the lemurs cannot learn to associate their unusual behavior (i.e., their indirect route to the food) with a positive outcome (i.e., females moving away from the baited area). As we noted earlier, although associative learning is often thought of as a simple process, this is often not the case, especially when one must link, in particular contexts, events that rarely co-occur. In this case, the ability to associate the dominant's gaze and movements with her following behavior—that is, to have a proto-understanding of agency—could be extremely helpful and perhaps even necessary for deception.

Again there is some evidence for such a deficit in lemurs: Itakura (1996) and Anderson and Mitchell (1998) found that lemurs were less successful than monkeys and apes at co-orienting their visual gaze with humans, suggesting they did not recognize that gaze direction is indicative

of future behavior. Thus perhaps lemurs fail the Menzel task, whereas monkeys and apes pass it, because lemurs lack one or both of the building blocks for intentional deception—the abilities to inhibit certain behaviors and to recognize how their conspecifics, as agents, are *deceivable*.

Conclusion

What lessons can we draw from this study, in the context of questions about animal cognition? To say that the differences among chimps, monkeys, and lemurs on the Menzel task show that the lemurs are simply less *intelligent* tells us nothing about what specific cognitive abilities may explain the differences in behavior; it is to use the word "intelligent" as a placeholder for a more robust future explanation. Similarly, in many cases animals are said to deceive one another without a clear enough understanding of the cognitive abilities that allow the apparently deceptive behavior. We need more fine-grained concepts and experimental tasks to understand how different kinds of deceptive behavior are possible for different species.

In the experiments described it is difficult to distinguish a lack of understanding of agency from a lack of inhibition. Nonetheless, these two ingredients of intentional deception can be examined separately, and possibly other cognitive constituents of deception can be conceptually characterized and empirically tested. This kind of research would help reveal the underlying components of intelligence and allow us to map out more precisely Darwin's continuum in the mental landscape of nature.

References

Allen, C. and Bekoff, M. (1997). *Species of Mind: The Philosophy and Biology of Cognitive Ethology*. Cambridge, Mass.: MIT Press.

Anderson, J. and Mitchell, R. W. (1998). Macaques but not lemurs co-orient visually with humans. *Folia Primatologica* 70: 17–22.

Brower, L. P. (1969). Ecological chemistry. *Scientific American* 220 (2): 22–29.

Byrne, R. W. (1995). *The Thinking Ape*. New York: Oxford University Press.

Byrne, R. W. and Whiten, A. (1988). *Machiavellian Intelligence: Social Expertise and the Evolution of Intellect in Monkeys, Apes, and Humans*. Oxford: Clarendon Press.

Byrne, R. W. and Whiten, A. (1992). Cognitive evolution in primates: Evidence from tactical deception. *Man* 27: 609–627.

Caldwell, R. L. (1986). The deceptive use of reputation by stomatopods. In *Deception: Perspectives on Human and Nonhuman Deceit*, R. W. Mitchell and N. S. Thompson, eds., pp. 129–150 Albany: State University of New York Press.

Coussi-Korbel, S. (1994). Learning to outwit a competitor in mangabeys. *Journal of Comparative Psychology* 108: 164–171.

Darwin, C. (1871/1936). *The Descent of Man and Selection in Relation to Sex*. New York: Random House.

Davis, R. and Leary, R. (1968). Learning of detour problems by lemurs and seven species of monkeys. *Perceptual and Motor Skills* 27: 1031–1034.

Deacon, T. (1997). *The Symbolic Species: The Co-Evolution of Language and the Brain*. New York: W.W. Norton.

Deaner, R. (2000). An experimental study of deception in ringtailed lemurs (abstract). *American Journal of Physical Anthropology*, Suppl. 30: 135.

Deaner, R. (in preparation). An experimental study of deception in ringtailed lemurs. Duke University.

Dennett, D. C. (1983). Intentional systems in cognitive ethology: The Panglossian paradigm defended. *Behavioral and Brain Sciences* 6: 343–390.

Diamond, A. (1990). Developmental time course in human infants and infant monkeys, and the neural bases of inhibitory control in reaching. *Annals of the New York Academy of Science* 608: 637–676.

Flavell, J. H. (1999). Cognitive development: Children's knowledge about the mind. *Annual Review of Psychology* 50: 21–45.

Flavell, J. H. (2000). Development of children's knowledge about the mental world. *International Journal of Behavioral Development* 24 (1): 15–23.

Gomez, J.-C. (1991). Visual behavior as a window for reading the mind of others in primates. In *Natural Theories of Mind: Evolution, Development, and Simulation of Everyday Mindreading*, A. Whiten, ed., pp. 195–207. Oxford: Basil Blackwell.

Hauser, M. D. (1996). *The Evolution of Communication*. Cambridge, Mass.: MIT Press.

Hauser, M. D. (1997). Minding the behaviour of deception. In *Machiavellian Intelligence II: Extensions and Evaluations*, A. Whiten and R. W. Byrne, eds., pp. 112–143. Cambridge: Cambridge University Press.

Hauser, M. D. (2000). *Wild Minds: What Animals Really Think*. New York: Henry Holt.

Hirata, S. and Matsuzawa, T. (in review). Tactics to obtain a hidden food item in chimpanzee pairs: Implications for the understanding of others' knowledge.

Hockham, L. R. and Ritchie, M. G. (2000). Female secondary sexual characteristics: Appearance may be deceptive. *Trends in Ecology and Evolution* 15: 436–438.

Hockham, L. R. and Ritchie, M. G. (2001). Deception (mimicry): An integral component of sexual signals: Reply from Hockham and Ritchie. *Trends in Ecology and Evolution* 16: 228.

Itakura, S. (1996). An exploratory study of gaze-monitoring in nonhuman primates. *Japanese Psychological Research* 38: 174–180.

Menzel, E. W. (1974). A group of chimpanzees in a 1-acre field: Leadership and communication. In *Behavior of Nonhuman Primates*, A. M. Schrier and F. Stollintz, eds., pp. 83–153. New York: Academic Press.

Mitchell, R. W. (1986). A framework for discussing deception. In *Deception: Perspectives on Human and Nonhuman Deceit*, R. W. Mitchell and N. S. Thompson, eds., pp. 3–40. Albany: State University of New York Press.

Perner, J. (1991). *Understanding the Representational Mind*. Cambridge, Mass.: MIT Press.

Pinker, S. (1994). *The Language Instinct: How the Mind Creates Language*. New York: Willian Morrow.

Robbins, R. K. (1981). The "false head" hypothesis: Predation and wing pattern variation of lycaenid butterflies. *American Naturalist* 118: 770–775.

Steger, R. and Caldwell R. L. (1983). Intraspecific deception by bluffing: A defense strategy of newly molted stomatopods. *Science* 221: 558–560.

Weldon, P. J. and Burghardt, G. M. (2001). Deception (mimicry): An integral component of sexual signals. *Trends in Ecology and Evolution* 16: 228.

Whiten, A. (1997). The Machiavellian mindreader. In *Machiavellian Intelligence II: Extensions and Evaluations*, A. Whiten and R. W. Byrne, eds., pp. 144–173. Cambridge: Cambridge University Press.

Whiten, A. and Byrne, R. W. (1988). Tactical deception in primates. *Behavioral and Brain Sciences* 11: 233–244.

Wickler, W. (1965). Mimicry and the evolution of animal communication. *Nature* 208: 519–521.

Wickler, W. (1968). *Mimicry in Plants and Animals*. (R. D. Martin, trans.) New York: McGraw-Hill.

Brian Hare and Richard Wrangham

In the hope of understanding more about the origins of human social cognitive abilities such as teaching and deception, much attention has been centered on the question of whether chimpanzees, our closest extant relatives, understand others as psychological agents (Tomasello and Call 1997; Tomasello 1999). Currently, results from work with chimpanzees can be viewed as contradictory (see Heyes 1998; Hare 2001). One approach, which represents the majority of studies, has produced little compelling evidence that chimpanzees attribute psychological states of any kind (i.e., perceptions, attention, intentions, or beliefs) to others (see Heyes 1998; Tomasello and Call 1997, for reviews). Another approach, by contrast, finds evidence that chimpanzees can at least take another individual's visual perspective (Hare et al. 2000; Hare et al. 2001; Hirata and Matsuzawa 2001).

There are two possible resolutions to these findings. The first is empirical; further experimental work may show that one or both approaches are flawed. The second is conceptual; theoretical differences that lead to different experimental paradigms may account for the different findings. Given that we do not know the outcome of future studies, we focus here on the second potential resolution. We ask whether seemingly contradictory experimental outcomes can be attributed to the types of problems that the two different approaches suggest should be posed to test species.

Attempts to understand the evolution of cognitive abilities are typically framed by one of two kinds of hypothesis, here called the "general-purpose intelligence hypothesis" and the "adapted cognition hypothesis." We argue that in order to improve our ability to design and interpret experiments on social cognition, it is most constructive to consider the predictions and approaches of both these hypotheses. We illustrate our argument with investigations into chimpanzee and dog social cognition. The results of these studies are puzzling and potentially contradictory in relation to the general-purpose intelligence hypothesis alone, but they make sense in light of the adapted cognition hypothesis.

The General-Purpose Intelligence Hypothesis

Models of general-purpose intelligence attempt to explain the evolution of all-purpose cognitive mechanisms such as memory, categorization, learning, or reasoning that can vary between genetically canalized systems, such as those often observed in invertebrates (Gould and Gould 1986), and highly flexible, content-independent cognitive processes believed to underlie human intelligence. The variation in the flexibility of these general-purpose problem-solving abilities is commonly considered to be explained by increases in computing power during evolution, rather than being the result of an adaptation to any particular kind of problem (Gibson 1990). An increase in general-purpose intelligence might evolve, for example, as a physiological constraint is released, allowing an increase in brain size and computing power (Aiello and Wheeler 1995).

Therefore, the general-purpose intelligence hypothesis predicts (1) taxon-level (or brain-sized) differences in intelligence that are unrelated to obvious features of ecology or social behavior and (2) abilities that can be applied across contexts and have not been selected to solve any specific evolutionary problem(s). Thus paradigms that can be used across taxa and that are evolutionarily irrelevant offer the most experimental power.

Examples consistent with these predictions include evidence for mirror recognition in several hominid species, but not in other primates (Gallup 1982), and proposed taxon-level differences in reversal-learning tasks (Rumbaugh and Pate

1984a,b). In both cases, large-brained hominoid species, which evolved in radically different ecologies, outperform other primates in what are arguably evolutionarily irrelevant tasks.

The Adapted Cognition Hypothesis

In contrast, models of cognitive adaptation (e.g., domain specificity) assume that species evolve flexible mechanisms that function to solve particular kinds of ecological problems. Just as in the case of complex morphological or physiological adaptations, sophisticated cognitive adaptations are selected to solve the problems that an individual is most likely to encounter (Barkow et al. 1992). But cognitive adaptations are unlike other adaptations in that "evolution has relinquished its micromanagement of the behavioral interactions an organism has with its environment in deference to the individual and its judgment" (Tomasello and Call 1997, p. 9), allowing more flexible behavioral responses to ecological problems.

Thus the adapted cognition hypothesis predicts: (1) Since cognitive abilities have been selected to solve evolutionarily salient problems, a species' most flexible cognitive abilities will be demonstrated in the settings for which these abilities evolved. (2) Variation in cognitive abilities among species will be explained by the different kinds of ecological problems the species has faced in its evolutionary history. Accordingly, tests of the adapted cognition hypothesis should be designed to echo the behavioral and ecological problems of that species' evolutionary past.

The clearest examples of cognitive mechanisms being adaptations to a specific evolutionary history come from spatial tasks (see Hauser 2000 for a review). For example, bird species that cache food for later retrieval outperform those from closely related noncaching species. They also have a larger hippocampus, a structure involved in spatial memory (Balda and

Kamil 1989; Krebs et al. 1989). Clearly, ecological pressures favored caching ability, which selected for an enlarged hippocampus. As for experiments on animal social cognition, there seem to be few, if any, examples that support both of the predictions of the adapted cognition hypothesis.

Chimpanzee Behavior as a Guide to the Design of Experiments

In investigating the abilities of chimpanzees to attribute perceptions and beliefs to others, experimenters have used two approaches, utilizing different types of social problems. The classic approach utilizes a cooperative-communicative paradigm. In this paradigm, a human (and in one study a conspecific) shares information about the location of a monopolizable food resource that is hidden in a location the subject cannot see (Premack 1988; Povinelli et al. 1990, 1994, 1997, 1999; Tomasello et al. 1997; Call et al. 1998; Itakura and Tanaka 1998; Itakura et al. 1999; Call and Tomasello 1999; Call et al. 2000). In another version of the cooperative-communicative paradigm, a subject must signal to humans in some way in order to acquire a monopolizable piece of food (Premack 1988; Povinelli et al. 1992; Povinelli and Eddy 1996a,b; Reaux et al. 1999; Theall and Povinelli 1999). The alternative approach uses a competitive paradigm, in which socially housed chimpanzees compete with each other for food (Hare et al. 2000, 2001; Hirata and Matsuzawa 2001).

Given the different predictions of the general-purpose intelligence hypothesis and the adapted cognition hypothesis we outlined, it is vital to classify each of these experimental paradigms according to their evolutionary relevance. To do so, consider the frequency of cooperation and communication, and social competition, in the natural lives of chimpanzees.

Are there any natural analogs of the cooperative-communicative paradigm in wild chimpanzees? It is admittedly possible to imagine a natural event that mimics the social problem presented, where one individual needs information from a conspecific about hidden food. For example, imagine during a meat-eating episode that male *A* has a monkey carcass 15 m up in a tree. Next to him is a second male *B* who has no meat and is unsuccessfully begging for a piece from *A*. *A* sees one chunk of his meat, a leg perhaps, fall to the ground. As it falls, it bounces off a branch and flies leftward, so that only *A* sees where it lands. Can we expect *A* to share information about the location of the hidden food with *B*?

No. In Menzel's classic experiments in which a number of chimpanzees searched for hidden food, no informed individual ever spontaneously or overtly used communicative signals to indicate the location of hidden food to a conspecific (Menzel 1974). Chimpanzees systematically avoid sharing information about monopolizable food both in the wild (personal observation) and in captivity (Hauser and Wrangham 1987; Hauser et al. 1993). Because there appears to be no obvious parallel to this experimental setting in the wild, the cooperative-communicative paradigm is an excellent test of the general-purpose intelligence hypothesis since it appears to be so unnatural.

The social-competition paradigm, by contrast, presents chimpanzees with a problem they faced and solved on a daily basis over evolutionary time: outcompeting conspecifics for food. There are countless examples of food competition (see Goodall 1986). For example, lone individuals normally maintain an unchanged velocity as they approach a familiar fruit tree. However, when a party of chimpanzees approaches such a tree, they often break into a run at the last minute, and race to the preferred feeding spots (personal observation). In addition, there is much opportunity for the use of flexible strategies to out-compete competitors. For example, some chimpanzees can have greater knowledge than others about the location of food. Thus a party

approaching a fruit tree may include some who have fed from it daily for a week and others visiting for the first time. Therefore the knowledgeable chimpanzees could avoid informing ignorant individuals by using the best entrance routes or approaching the location with the highest density of fruit while they are unseen or while others are distracted. In addition, Goodall (1971) reports a number of observations in which a chimpanzee avoided food in the presence of ignorant but dominant competitors until they left. (In some cases they even proactively led the dominants away from the food.) In summary, chimpanzees regularly face problems of food competition, making it likely that their cognitive abilities function in part to solve these problems. The competitive paradigm is therefore a suitable choice for testing the adapted cognition hypothesis.

Testing the Adapted Cognition Hypothesis

Variation in Cognitive Ability According to Context

Experiments that utilize the cooperative-communicative approach have produced little compelling evidence for attribution of psychological states; these include tests of even the most basic social cognitive skills thought to be involved in human attribution (Povinelli 1996; Call et al. 2000). For example, in the object choice task, an experimenter hides food in one of two locations so that the subject knows the food is hidden, but does not know where. After baiting, the experimenter provides a social cue (eye direction, head direction, pointing, etc.) to indicate the location of the food to the subject. The question is whether the subject will use the cue provided by the human to find the hidden food. Most chimpanzees do not reliably use even the most obvious of cues, such as when the experimenter taps on the location of the food (Call et al. 1998, 2000). In addition, it takes many dozens of trials for chimpanzees to use an arbitrary and novel cue (such as a wooden block on the correct container) to find hidden food (Tomasello et al. 1997).

In contrast, experiments that utilize the competitive approach have produced promising results. Hare et al. (2000) baited an arena between the rooms of two chimpanzees with two pieces of food. As expected, in the control condition when two pieces of food were placed in view of both competitors, the dominant subject retrieved most of the food. If one piece of food was hidden from the dominant behind an occluder while the subordinate could see both pieces, as subordinates, the subjects preferred to retrieve the hidden piece of food that the dominant could not see. In contrast, if one piece was hidden from the subordinate behind an occluder, but the dominant could see both pieces, as dominants, the subjects preferred to retrieve the visible piece of food first to ensure that they obtained both pieces.

Hare et al. (2001) adapted the same competitive paradigm to test what chimpanzees know about what others have and have not seen in the immediate past in a number of situations. In experimental conditions, subordinates saw where food was hidden and that their dominant competitor did not. In control conditions, subordinates saw where food was hidden as the dominant watched. Subordinates preferred to retrieve food in experimental conditions. In addition, they refused to even approach more often in control conditions. Finally, in corroboration, Hirata and Matsuzawa (2001) found that during competition over hidden food, chimpanzees tend to use the behavior of competitors when they have seen food hidden, but ignore the behavior of ignorant individuals who have not seen food hidden.

Taken together, these findings demonstrate that chimpanzees know what conspecifics do and do not see, and furthermore, that they use this knowledge to develop flexible behavioral strategies in a number of different competitive situations. Most important, the combined results of

these two approaches are consistent with the first prediction of the adapted cognition hypothesis. Chimpanzees' social cognitive skills seem to be most flexible and sophisticated when solving the problems that this species faced during their evolutionary history.

Variation in Cognitive Ability According to Evolutionary History

The second prediction of the adapted cognition hypothesis is that different species will have different social cognitive abilities, which are correlated with the problems they faced in their evolutionary history. To take an example contrasting sharply with chimpanzees, dogs have evolved to cooperate and communicate with humans (Scott and Fuller 1967; Serpell and Barrett 1996). Accordingly, we expect that this species will show skills in the cooperative-communicative tasks that chimpanzees do not.

In the object choice task, dogs clearly can use a number of social cues provided by humans to find hidden food, including distal pointing and gaze direction (Miklosi et al. 1998; Mckinney and Sambrook 2000). But how flexibly can they use these social cues? Hare et al. (1998) and Hare and Tomasello (1999) ruled out the possibility that dogs were simply using the direction of their informant's movements as a cue. In addition, Hare and Tomasello (1999) found that whether it was an unfamiliar human or another dog, the identity of the informant had no effect on performance. Finally, unlike chimpanzees, adult dogs and puppies immediately used a novel and arbitrary cue to find hidden food (Agnetta et al. 2000).

Meanwhile wolves, the direct ancestors of dogs, who have not been selected to engage in cooperative-communicative interactions with humans, do not use social cues such as tapping, pointing, or head movement provided by an experimenter (Agnetta et al. 2000; Hare et al. submitted). Thus, the abilities of domestic dogs relative to primates and wolves in the object choice task supports the second prediction of the adapted cognition hypothesis that variation in social cognitive abilities is explained by variation in the problems presented in the past by different social-ecological contexts.

An Integrated Approach for Studying Social Cognition

The general-purpose intelligence hypothesis and the adapted cognition hypothesis are not necessarily mutually exclusive. For example, Tomasello and Call (1997) proposed that the ability of primates to understand tertiary relationships evolved first in the social domain and was later extended to the physical domain as well. Nevertheless since the two hypotheses have different implications, it is important to consider them separately.

First, if the adapted cognition hypothesis is right, the relationship of the experimental paradigm to the test species' evolutionary history affects the interpretation of results. Thus if a cognitive ability is found to be absent in an ecologically irrelevant experimental context, the result represents a good test of the predictions of the general-purpose intelligence hypothesis, but not of the adapted cognition hypothesis. Further experiments would be needed to test the predictions of the adapted cognition hypothesis.

Second, the adapted cognition hypothesis helps predict where cognitive flexibility and sophistication are most likely to reside (Barkow et al. 1992). If tests designed on this basis are successful, novel (ecologically irrelevant) problems can be introduced to test the general-purpose intelligence hypothesis. For example, to test for coordinated cooperation, Hare (2001) suggested an experiment in which two chimpanzees must work together in a physical task to outcompete a third, since the majority of cooperative interactions occur in attempts to outcompete third parties (Harcourt and de Waal 1992). If individuals showed coordination in this paradigm, then they could subsequently be tested in a novel noncompetitive version of the same task.

Unfortunately, there are no truly objective methods for operationalizing the evolutionary relevance of different paradigms. This makes careful consideration of a test species' ecology even more important in experimental designs and suggests that solving this problem is a priority for experimentalists (Hare 2001).

Finally, the relative importance of the two hypotheses in explaining the evolution of social cognition will be most easily assessed by testing a wider variety of species. Only by considering a wide range of species can one expect to find variation in cognitive abilities that corresponds with ecological differences (Burghardt and Gittleman 1990). For example, tamarins (Callitrichidae) have evolved as cooperative breeders, with adults being aided by juveniles in raising offspring. Tamarin helpers catch prey and can even give a food call that summons the young, to whom they give the food (Goldizen 1987). This suggests that unlike other primates, tamarins might demonstrate more flexibility in cooperative paradigms than in competitive paradigms (at least in those that involve food).

Experiments designed to test such ideas will therefore require comparative psychologists to work closely with behavioral ecologists. The results should clarify the evolutionary relationship between general-purpose intelligence and adapted cognition, both of which appear to play an important role in mammalian cognition. They will also raise fascinating questions about the mechanisms that limit a species' ability to solve social-cognitive problems that are posed in evolutionarily novel contexts.

Acknowledgments

We wish to thank the editors for inviting us to contribute. In addition, we wish to thank them for their valuable comments and Marc Hauser, Sonya Kahlenberg, Matthew McIntyre, Cory

Miller, Katarina Mucha, Michael Tomasello, and Michael Wilson for comments provided on an earlier version of this manuscript.

References

Agnetta, B., Hare, B., and Tomasello, M. (2000). Cues to food location that domestic dogs (*Canis familiaris*) of different ages do and do not use. *Animal Cognition* 3: 107–112.

Aiello, L. and Wheeler, P. (1995). The expensive tissue hypothesis. *Current Anthropology* 36: 199–221.

Balda, R. and Kamil, A. (1989). A comparative study of cache recovery by three corvid species. *Animal Behaviour* 38: 486–495.

Barkow, J., Cosmides, L., and Tooby, J. (1992). *The Adapted Mind: Evolutionary Psychology and the Generation of Culture*. New York: Oxford University Press.

Burghardt, G. M. and Gittleman, J. G. (1990) Comparative and phylogenetic analyses: New wine, old bottles. In *Interpretation and Explanation in the Study of Behavior*. Vol. 2, *Comparative Perspectives*, M. Bekoff and D. Jamieson, eds., pp. 192–225. Boulder, Col.: Westview Press.

Call, J. and Tomasello, M. (1999). A nonverbal false belief task: The performance of children and great apes. *Child Development* 70: 381–395.

Call, J., Hare, B., and Tomasello, M. (1998). Chimpanzee gaze following in an object-choice task. *Animal Cognition* 1: 89–99.

Call, J., Agnetta, B., and Tomasello, M. (2000). Social cues that chimpanzees do and do not use to find hidden objects. *Animal Cognition* 3: 23–34.

Gallup, G. (1982). Self-awareness and the emergence of mind in primates. *American Journal of Primatology* 2: 237–248.

Gibson, K. (1990). New perspectives on instincts and intelligences: Brain size and the emergence of hierarchical mental constructional skills. In *"Language" and Intelligence in Monkeys and Apes: Comparative Developmental Perspectives*, S. P. Parker, and K. R. Gibson, eds., pp. 97–128. Cambridge: Cambridge University Press.

Goldizen, A. (1987). Tamarins and marmosets: Communal care of offspring. In *Primate Societies*, B. B. Smuts, D. L. Cheney, R. M. Seyfarth, R. W. Wrangham, and T. T. Struhsaker, eds., pp. 34–43. Chicago: University of Chicago Press.

Goodall, J. (1971). *In the Shadow of Man*. Boston: Houghton Mifflin.

Goodall, J. (1986). *The Chimpanzees of Gombe*. Cambridge, Mass.: Harvard University Press.

Gould, S. and Gould, C. (1986). Invertebrate intelligence. In *Animal Intelligence: Insights into the Animal Mind*, R. Hoage and L. Goldman, eds., pp. 21–36. Washington, D.C.: Smithsonian Institution Press.

Harcourt, A. and de Waal, F. (1992). *Coalitions and Alliances in Humans and Other Animals*. New York: Oxford University Press.

Hare, B. (2001). Can competitive paradigms increase the validity of experiments on primate social cognition? *Animal Cognition* 4: 269–280.

Hare, B., Brown, M., Williamson, C., and Tomasello, M. (submitted). The domestication of social cognition in dogs.

Hare, B. and Tomasello, M. (1999). Domestic dogs (*Canis familiaris*) use human and conspecific social cues to locate hidden food. *Journal of Comparative Psychology* 113: 173–77.

Hare, B., Call, J., and Tomasello, M. (1998). Communication of food location between human and dog (*Canis familiaris*). *Evolution of Communication* 2: 137–159.

Hare, B., Call, J., Agnetta, B., and Tomasello, M. (2000). Chimpanzees know what conspecifics do and do not see. *Animal Behaviour* 59: 771–785.

Hare, B., Call, J., and Tomasello, M. (2001). Do chimpanzees know what conspecifics know? *Animal Behaviour* 61: 139–151.

Hauser, M. (2000). *Wild Minds*. New York: Henry Holt.

Hauser, M. and Wrangham, R. (1987). Manipulation of food calls in captive chimpanzees: A preliminary report. *Folia Primatologica* 48: 207–210.

Hauser, M., Teixidor, P., Field, L., and Flaherty, R. (1993). Food-elicited calls in chimpanzees: Effects of food quantity and divisibility. *Animal Behaviour* 45: 817–819.

Heyes, C. (1998). Theory of mind in nonhuman primates. *Behavioral and Brain Sciences* 21: 101–148.

Hirata, S. and Matsuzawa, T. (2001). Tactics to obtain a hidden food item in chimpanzee pairs (*Pan troglodytes*). *Animal Cognition* DOI 10.1007/S100710100084.

Itakura, S. and Tanaka, M. (1998). Use of experimenter-given cues during object-choice tasks by chimpanzees (*Pan troglodytes*), an orangutan (*Pongo pygmaeus*), and human infants (*Homo sapiens*). *Journal of Comparative Psychology* 112: 119–126.

Itakura, S., Agnetta, B., Hare, B., and Tomasello, M. (1999). Chimpanzees use human and conspecific social cues to locate hidden food. *Developmental Science* 2: 448–456.

Krebs, J., Sherry, S., Healy, V., Perry, V., and Vaccarino, A. (1989). Hippocampal specialization of food-storing in birds. *Proceedings of the National Academy of Sciences U.S.A.* 86: 1388–1392.

McKinley, J. and Sambrook, T. (2000). Use of human-given cues by domestic dogs (*Canis familiaris*) and horses (*Equus caballus*) *Animal Cognition* 3: 13–22.

Menzel, E. (1974). A group of young chimpanzees in a one-acre field. In *Behaviour of Nonhuman Primates*, A. M. Schrier and F. Stollnitz, eds., pp. 83–153. San Diego: Academic Press.

Miklosi, A., Polgardi, R., Topal, J., and Csanyi, V. (1998). Use of experimenter-given cues in dogs. *Animal Cognition* 1: 113–122.

Povinelli, D. (1996). Chimpanzee theory of mind?: The long road to strong inference. In *Theories of Theories of Mind*, P. Carruthers and P. Smith, eds., pp. 330–343. Cambridge: Cambridge University Press.

Povinelli, D. and Eddy, T. (1996a). What young chimpanzees know about seeing. *Monographs of the Society for Research on Child Development* 61 (2, Serial No. 247): 1–191.

Povinelli, D. and Eddy, T. (1996b). Factors influencing young chimpanzees' (*Pan troglodytes*) recognition of attention. *Journal of Comparative Psychology* 110: 336–345.

Povinelli, D., Nelson, K., and Boysen, S. (1990). Inferences about guessing and knowing by chimpanzees (*Pan troglodytes*). *Journal of Comparative Psychology* 104: 203–210.

Povinelli, D., Nelson, K., and Boysen, S. (1992). Comprehension of role reversal in chimpanzees: Evidence of empathy? *Animal Behaviour* 43: 633–640.

Povinelli, D., Rulf, A., and Bierschwale, D. (1994). Absence of knowledge attribution and self-recognition in young chimpanzees (*Pan troglodytes*). *Journal of Comparative Psychology* 108: 74–80.

Povinelli, D., Reaux, J., Bierschwale, D., and Allain, A. (1997). Exploitation of pointing as a referential gesture in young children, but not adolescent chimpanzees. *Cognitive Development* 12: 327–365.

Povinelli, D., Bierschwale, D., and Cech, C. (1999). Comprehension of seeing as a referential act in young children but not juvenile chimpanzees. *British Journal of Developmental Psychology* 17: 37–60.

Premack, D. (1988). "Does the chimpanzee have a theory of mind?" revisited. In *Machiavellian Intelligence. Social Expertise and the Evolution of Intellect in Monkeys, Apes, and Humans*, R. W. Byrne and A. Whiten, eds., pp. 160–179. New York: Oxford University Press.

Reaux, J., Theall, L., and Povinelli, D. (1999). A longitudinal investigation of chimpanzees' understanding of visual perception. *Child Development* 70: 215–290.

Rumbaugh, D. and Pate, J. (1984a). The evolution of cognition in primates: A comparative perspective. In *Animal Cognition*, H. L. Roitblat, T. Bever, and H. Terrace, eds., pp. 569–587. Hillsdale, N.J.: Lawrence Erlbaum Associates.

Rumbaugh, D. and Pate, J. (1984b). Primates' learning by levels. In *Behavioral Evolution and Integrative Levels*, G. Greenberg and E. Tobach, eds., pp. 221–240. Hillsdale, N.J.: Lawrence Erlbaum Associates.

Scott, P. and Fuller, J. (1967). *Genetics and the Social Behavior of the Dog*. Chicago: University of Chicago Press.

Serpell, J. and Barrett, P. (eds.) (1995). *The Domestic Dog: Its Evolution, Behaviour and Interactions with People*. Cambridge: Cambridge University Press.

Theall, L. and Povinelli, D. (1999). Do chimpanzees tailor their gestural signals to fit the attentional states of others? *Animal Cognition* 2: 207–214.

Tomasello, M. (1999). *The Cultural Origins of Human Cognition*. Cambridge, Mass.: Harvard University Press.

Tomasello, M. and Call, J. (1997). *Primate Cognition*. Oxford: Oxford University Press.

Tomasello, M., Call, J., and Gluckman, A. (1997). Comprehension of novel communicative signs by apes and human children. *Child Development* 68: 1067–1080.

45 Field Studies of Social Cognition in Spotted Hyenas

Kay E. Holekamp and Anne L. Engh

Considered in relation to body size, the brains of primates are relatively large and complex compared with those of other animals, including most nonprimate mammals (Macphail 1982; Harvey and Krebs 1990). In particular, the primate neocortex is very large in relation to the rest of the brain (Barton and Dunbar 1997). Furthermore, primates appear to be endowed with cognitive abilities that are superior to, and qualitatively different from, those observed in most other mammals (reviewed in Byrne and Whiten 1988; Harcourt and de Waal 1992; Tomasello and Call 1997).

Two different types of selection pressures have been hypothesized to favor the evolution of large brains and great intelligence (as defined by Kamil 1987) in primates. The first hypothesis suggests that intelligence has been favored in primates by selection pressures associated with complexity in the physical environment, particularly that confronted when navigating through a three-dimensional arboreal world (e.g., Povinelli and Preuss 1995) or when finding and obtaining food (e.g., Milton 1981). The second hypothesis suggests instead that the key selection pressures have been imposed by complexity associated with the labile social behavior of conspecific group members (Byrne and Whiten 1988). Predictions of this social complexity hypothesis have now been confirmed in a number of Old World primate species, suggesting that the evolution of intelligence has been more strongly influenced by social pressures than by nonsocial aspects of the environment (reviewed in Byrne 1994; Tomasello and Call 1997). Unfortunately the generality of this hypothesis is severely limited by the current dearth of information about social cognition in animals other than primates (Harcourt and de Waal 1992). In fact, most work in this area has focused exclusively on cercopithecine primates and great apes. The social complexity hypothesis, however, predicts that nonprimate animals that share with primates most salient features of their social life and resource distribution should possess many of the same features of social intelligence found in monkeys and apes.

Like most primates, many mammalian carnivores live in permanent, complex social groups that contain both males and females from multiple, overlapping generations. Gregarious carnivores also engage in a variety of behaviors, such as cooperative hunts of large vertebrate prey, which have prompted many observers to claim that these predators must possess extraordinary intellectual powers (e.g., Guggisberg 1962). The size of the carnivore neocortex is positively correlated with group size (Dunbar and Bever 1998). However, the cognitive abilities of carnivores have seldom been the subject of systematic study, and they are currently very poorly understood (e.g., Byrne 1994).

In our own research we are examining the cognitive mechanisms underlying social behavior and communication in one gregarious carnivore, the spotted hyena (*Crocuta crocuta*). Our ultimate objective is to determine whether hyenas exhibit some of the same cognitive abilities that are observed in primates. Evidence for the existence of shared cognitive abilities would suggest convergent evolution in these two distantly related taxa and would strongly support the social complexity hypothesis. In contrast, the failure to obtain such evidence would suggest that the social complexity hypothesis should be either rejected or revised. We use three different methods to investigate the cognitive abilities of gregarious carnivores: (1) comparative analysis based on literature review, (2) field experiments in the natural habitat, and (3) controlled observations of free-living hyenas. Here we summarize our work to date using each of these methods, although we hope readers will bear in mind the fact that much of it is still in progress.

Comparative Analysis of Group Travel

Here we sought to understand whether coordinated movements by gregarious carnivores reveal the operation of complex mental abilities. Like many primates (e.g., Boinski and Garber 2000), gregarious carnivores frequently travel with other members of their social groups. Group hunts by lions, spotted hyenas, and other carnivores often appear to involve intelligent coordination and division of labor among hunters (e.g., Guggisberg 1962; Stander 1992a,b). Studies of carnivore predatory behavior reveal that their group hunts represent more complexly organized phenomena than mere opportunistic grabs at prey. However, although myriad observers have claimed that the group hunting activity of large carnivores requires the operation of humanlike mental processes, coordinated hunting behavior by lions, hyenas, and social canids can in fact be most parsimoniously explained by the operation of a few simple mental rules of thumb, such as "Move wherever you need to in order to keep the selected prey animal between you and another hunter" (Holekamp et al. 2000). Currently there is no evidence that gregarious carnivores use mental algorithms more complex than simple rules of thumb to surround and capture prey. Disproving the rules of thumb hypothesis will require experimental evidence, not only that individual carnivores monitor both their prey and their fellow hunters (e.g., Stander 1992b), but also that they accurately anticipate the behavior of the latter based on knowledge of their goals.

Another form of group travel that reveals the mental operations of gregarious carnivores takes place when young cubs or pups are moved from one den site to another (Holekamp et al. 2000). Whereas group hunts make carnivores look very smart, den moves make them appear very stupid. When gregarious carnivores move infants (N) to a new den, they always make at least one extra trip ($N + 1$) to the old den, suggesting that instead of being able to count, during den moves

carnivores operate according to a rule of thumb that says "Revisit the old den until you find no more of your infants there." Thus the mental processes required to initiate and maintain seemingly complex movements need not be particularly sophisticated in carnivores. Interestingly, it appears that much coordinated group travel in primates, including cooperative hunting behavior by chimpanzees (e.g., Boesch and Boesch 1989), might also be explained effectively by these same simple rules of thumb.

Field Experiments and Controlled Observations of Spotted Hyenas

Many primates demonstrate the ability to form mental representations of tertiary, or third-party, relationships among conspecific group members (Tomasello and Call 1997). These involve interactions and relationships in which the observer is not directly involved. For example, adult male hamadryas baboons make adaptive decisions about whether to challenge a rival male based on their assessment of the strength of the social bond between the rival and his female associates (Bacchman and Kummer 1980). Tomasello and Call (1997) hypothesize that an ability to understand third-party relationships is unique to primates, and furthermore, that this distinguishes their mental abilities from those of all other animals. Indeed, the ability to recognize third-party relationships has not been documented in nonprimate mammals, even in species living in complex, stable societies such as elephants (Moss 1988), dolphins (Connor et al. 1992), or lions (Packer 1994). However, few published studies of these other gregarious mammals have focused on the cognitive abilities of these animals, so it remains unclear whether the absence of evidence that any of these nonprimate mammals can recognize tertiary relationships actually means that this ability is also absent. Our own field work has attempted to determine whether a primate-like ability to recognize tertiary relationships is exhibited by the spotted hyena, a carnivore whose social life is remarkably similar in most respects to that of many Old World primates.

The Social Lives of Hyenas and Old World Monkeys

Both spotted hyenas and Old World monkeys usually live in permanent social groups that commonly contain multiple adult males and multiple matrilines of adult female kin with offspring. Males leave their natal groups, whereas females are usually philopatric (Cheney and Seyfarth 1983; Smale et al. 1997). Adults of both taxa can be ranked in a linear dominance hierarchy based on outcomes of agonistic interactions, and priority of access to resources varies with social rank (Tilson and Hamilton 1984; Andelman 1985). In both taxa, members of the same matriline occupy adjacent rank positions in the group's hierarchy, and female dominance relations are extremely stable across a variety of contexts and over extended periods. Juvenile hyenas of both sexes acquire ranks immediately below those of their mothers (Holekamp and Smale 1991), and they do this via the same associative learning mechanisms as those documented in cercopithecine primates (Horrocks and Hunte 1983; Engh et al. 2000). Thus social status is often not determined by size, strength, or fighting ability in either of these taxa, as it is in many mammals (e.g., Clutton-Brock et al. 1982). Instead, juvenile monkeys and hyenas both learn during early life that they can dominate individuals ranked lower than their mothers. In both monkeys and hyenas, kin associate more closely than do non-kin, and individuals direct affiliative behavior more frequently toward kin than non-kin (Seyfarth 1980; East et al. 1993; Holekamp et al. 1997).

In both taxa, high-ranking animals are preferred over lower-ranking individuals as social companions (Seyfarth, 1980; Holekamp et al. 1997). Furthermore, the patterns of greeting behavior in *Crocuta* follow primate patterns of social grooming (East et al. 1993), in which individuals prefer to direct affiliative behavior toward high-ranking non-kin (Seyfarth and Cheney 1984). This suggests that hyenas, like monkeys, may recognize that some group members are more valuable social partners than others. Finally, triadic and more complex interactions (e.g., coalitions) appear to play important roles in both maintenance and reversals of social rank in free-living *Crocuta* (Zabel et al. 1992; Smale et al. 1993), as they do in many cercopithecine primates (Walters 1980; Datta 1986; Chapais 1992).

Playback Experiments on Vocal Recognition

The similarities between the social lives of spotted hyenas and Old World primates present an ideal opportunity for comparative analysis of their social knowledge. In our first attack on this problem (Holekamp et al. 1999), we conducted a series of playback experiments with hyenas in the field and compared the hyenas' behavioral responses during these experiments with those observed earlier in vervet monkeys (Cheney and Seyfarth 1990) and in other cercopithecine primates (e.g., Gouzoules et al. 1995).

In the vervet experiments (Cheney and Seyfarth 1980), the distress scream of a juvenile was played through a hidden loudspeaker to groups of females that contained the juvenile's mother as well as other adult breeding females (controls). The mothers responded more strongly to screams than the control females, and the controls often responded by looking at the mother without any apparent cues from the mother herself. Thus vervets behaved as though they recognized the close associative relationships existing among group members unrelated to themselves (Cheney and Seyfarth 1980, 1990).

We used playback experiments to determine whether hyenas are capable of identifying individual conspecifics on the basis of their long-distance "whoop" vocalizations and whether they recognize tertiary relationships among non-kin members of their clan, as occurs in vervets. Although the results obtained during our playback experiments showed clearly that hyenas can recognize the whoops of specific individuals, they yielded no evidence that hyenas can recognize third-party relationships (Holekamp et al. 1999). In contrast to vervets, after the playback, the control females in our experiments were no more likely to look at the mother of the whooping cub than at other control females. Although this result suggests that spotted hyenas do not recognize third-party relationships, it remains possible that they do recognize them, but that their orientation responses during playback experiments are poor indicators of this ability. If this alternative hypothesis is correct, then other social situations, in which actions based on social knowledge will unambiguously improve fitness outcomes for the actor, should reveal that hyenas can indeed recognize third-party relationships. Therefore we recently turned our attention to these other types of social situations.

Coalition Formation

Coalitions play an important role in the acquisition and maintenance of social rank in spotted hyenas (Zabel et al. 1992; Smale et al. 1993; Engh et al. 2000). When aggression between two hyenas escalates, one or more others may join the skirmish by forming a coalition with the attacker against the target individual. Typically, animals joining to form coalitions are all dominant to the victim. Thus when attempting to displace a larger subordinate animal from food, a hyena might benefit, for example, by delaying its attack until the arrival of a potential coalitionary ally who is higher ranking than the target animal.

We are currently analyzing changes in rates of aggression directed against subordinates after the arrival of hyenas of high and intermediate rank in order to learn more about hyenas' abilities to recognize tertiary relationships (Engh et al. in preparation). If hyenas increase their rates of aggression only after higher-ranking hyenas arrive on the scene, then they may be following a

simple rule of thumb, such as "Only attack a larger subordinate when another individual is present who is higher ranking than yourself." Alternatively, if the attack rate also increases following the arrival of an individual who is dominant to the victim but subordinate to the attacker, then the attacking hyena must recognize the relative ranks of the other two individuals. In the latter case, the hyenas would be indicating to us that they can indeed recognize tertiary relationships. Although we are still in the early stages of this study, the preliminary results suggest that hyenas increase their rates of aggression only when animals higher ranking than themselves arrive, indicating once again that they are probably using simple rules of thumb to make decisions, rather than knowledge of complex relationships.

Reconciliation and Redirected Aggression

Affiliative gestures functioning to repair social relationships damaged during a fight are called "reconciliation behaviors" (de Waal 1993). Reconciliation occurs in many primates during friendly reunions between former opponents shortly after aggressive conflicts (reviewed by Aurlei and de Waal 2000). Similarly, spotted hyenas appear to reconcile after 10 to 15 percent of fights, often by the loser lifting its leg in a friendly "greeting" gesture directed at its former opponent (Hofer and East 2000; Wahaj et al. 2001). After fights, Old World primates are known to reconcile, not only with their former opponents, but also with the kin of former opponents (e.g., Cheney and Seyfarth 1989), indicating that the conciliatory monkeys recognize those tertiary relationships. In contrast, we rarely observe hyenas reconciling with any animals but their former opponents. This suggests that even though the ability to recognize tertiary relationships under these circumstances might well enhance their fitness, hyenas do not behave as though they possess this ability.

In many primates, an animal that has been involved in a fight will redirect aggression and threaten a third, previously uninvolved individual (e.g., Cheney and Seyfarth 1989). Often such redirected aggression is directed toward a close relative of the prior opponent. Thus patterns of redirected aggression in primates resemble those seen in kin-based reconciliation, and both indicate that monkeys can recognize third-party relationships. Although we have not yet begun work on redirected aggression in spotted hyenas, it is very common in this species, and we plan in the near future to use these interactions to ask once again whether hyenas can recognize third-party relationships.

Conclusions

Much of our work is still in progress, but the initial results from our studies of social cognition in hyenas suggest that even though these predators live in the same types of complex, stable groups as many primates do, they lack some of the mental abilities common in monkeys and apes. If spotted hyenas cannot recognize third-party social relationships, then features of the social environment common to both primates and spotted hyenas may be necessary, but cannot be sufficient, to account for the evolution of the primates' ability to recognize relationships among unrelated group members. Our data to date suggest that hyenas may accomplish many of the same social feats as monkeys do by following simple rules. If it is confirmed in our ongoing studies, the finding that hyenas cannot recognize third-party relationships will suggest that the social complexity hypothesis should be modified to focus only on those aspects of primate social life that are not shared with gregarious carnivores.

Acknowledgments

This work was supported by National Science Foundation grant IBN9906445. We thank L. Smale for insightful comments on an earlier draft of this paper.

References

Andelman, S. J. (1985). Ecology and Reproductive Strategies of Vervet Monkeys (*Cercopithecus aethiops*) in Amboseli National Park, Kenya. Ph.D. thesis, University of Washington, Seattle.

Aurlei, F. and de Waal, F. B. M. (2000). *Natural Conflict Resolution*. Berkeley: University of California Press.

Bacchman, C. and Kummer, H. (1980). Male assessment of female choice in hamadryas baboons. *Behavioral Ecology and Sociobiology* 6: 315–321.

Barton, R. A. and Dunbar, R. I. M. (1997). Evolution of the social brain. In *Machiavellian Intelligence II: Extensions and Evaluations*, A. Whiten and R. W. Byrne, eds., pp. 240–263. Cambridge: Cambridge University Press.

Boesch, C. and Boesch, H. (1989). Hunting behaviour of wild chimpanzees in the Taï National Park. *American Journal of Physical Anthropology* 78: 547–573.

Boinksi, S. and Garber, P. (eds.) (2000). *On the Move: How and Why Animals Travel in Groups*. Chicago: University of Chicago Press.

Byrne, R. W. (1994). The evolution of intelligence. In *Behaviour and Evolution*, P. J. B. Slater and T. R. Halliday, eds., pp. 223–265. Cambridge: Cambridge University Press.

Byrne, R. W. and Whiten, A. (eds.) (1988). *Machiavellian Intelligence*. Oxford: Clarendon Press.

Chapais, B. (1992). The role of alliances in the social inheritance of rank among female primates. In *Coalitions and Alliances in Humans and Other Animals*, A. H. Harcourt and F. B. M. de Waal, eds., pp. 29–60. Oxford: Oxford University Press.

Cheney, D. L. and Seyfarth, R. M. (1980). Vocal recognition in free-ranging vervet monkeys. *Animal Behaviour* 28: 362–367.

Cheney, D. L. and Seyfarth, R. M. (1983). Nonrandom dispersal in free-ranging vervet monkeys: Social and genetic consequences. *American Naturalist* 122: 392–412.

Cheney, D. L. and Seyfarth, R. M. (1989). Redirected aggression and reconciliation among vervet monkeys, *Cercopithecus aethiops*. *Behaviour* 110: 258–275.

Cheney, D. L. and Seyfarth, R. M. (1990). *How Monkeys See the World*. Chicago: University of Chicago Press.

Clutton-Brock, T. H., Guiness, F. E., and Albon, S. D. (1982). *Red Deer: Behavior and Ecology of Two Sexes*. Chicago: University of Chicago Press.

Connor, R. C., Smolker, R. A., and Richards, A. F. (1992). Dolphin alliances and coalitions. In *Coalitions and Alliances in Humans and Other Animals*, A. H. Harcourt and F. B. M. de Waal, eds., pp. 415–443. Oxford: Oxford University Press.

Datta, S. B. (1986). The role of alliances in the acquisition of rank. In *Primate Ontogeny, Cognition and Social Behavior*, G. J. Else and P. C. Lee, eds., pp. 219–225. Cambridge: Cambridge University Press.

de Waal, F. B. M. (1993). Reconciliation among primates: A review of empirical evidence and unresolved issues. In *Primate Social Relationships*, W. A. Mason and S. P. Mendoza, eds., pp. 111–114. New York: State University of New York Press.

Dunbar, R. I. M. and Bever, J. (1998). Neocortex size predicts group size in carnivores and some insectivores. *Ethology* 104: 695–708.

East, M. L., Hofer, H., and Wickler, W. (1993). The erect "penis" as a flag of submission in a female-dominated society: Greetings in Serengeti spotted hyenas. *Behavioral Ecology and Sociobiology* 33: 355–370.

Engh, A. L., Esch, K., Smale, L., and Holekamp, K. E. (2000). Mechanisms of maternal rank "inheritance" in the spotted hyaena, *Crocuta crocuta. Animal Behaviour* 60: 323–332.

Engh, A. L., Siebert, E., Jensen, A. D., and Holekamp, K. E. (in preparation). Do spotted hyenas use allies strategically?

Gouzoules, S., Gouzoules, H., and Marler, P. (1995). Representational signaling in non-human primate vocal communication. In *Current Topics in Primate Vocal Communication*, E. Zimmerman, J. D. Newman, and U. Jürgens, eds., pp. 235–252. New York: Plenum.

Guggisberg, C. A. W. (1962). *Simba*. London: Bailey Brothers and Swinfen.

Harcourt, A. H. and de Waal, F. B. M. (eds.) (1992). *Coalitions and Alliances in Humans and Other Animals*. Oxford: Oxford University Press.

Harvey, P. H. and Krebs, J. R. (1990). Comparing brains. *Science* 249: 140–146.

Hofer, H. and East, M. L. (2000). Conflict management in female-dominated spotted hyenas. In *Natural Conflict Resolution*, F. Aureli and F. B. M. de Waal, eds., pp. 232–234. Berkeley: University of California Press.

Holekamp, K. E. and Smale, L. (1991). Rank acquisition during mammalian social development: The "inheritance" of maternal rank. *American Zoologist* 31: 306–317.

Holekamp, K. E., Cooper, S. M., Katona, C. I., Berry, N. A., Frank, L. G., and Smale. L. (1997). Patterns of association among female spotted hyenas (*Crocuta crocuta*). *Journal of Mammalogy* 78: 55–64.

Holekamp, K. E., Boydston, E. E., Szykman, M., Graham, I. Nutt, K. J., Birch, S., Piskiel, A., and Singh, M. (1999). Vocal recognition in the spotted hyaena and its possible implications regarding the evolution of intelligence. *Animal Behaviour* 58: 383–395.

Holekamp, K. E., Boydston, E. E., and Smale, L. (2000). Group travel in social carnivores. In *On the Move: How and Why Animals Travel in Groups*, S. Boinksi and P. Garber, eds., pp. 587–627. Chicago: University of Chicago Press.

Horrocks, J. and Hunte, W. (1983). Maternal rank and offspring rank in vervet monkeys: An appraisal of the mechanisms of rank acquisition. *Animal Behaviour* 31: 772–782.

Kamil, A. C. (1987). A synthetic approach to the study of animal intelligence. *Nebraska Symposium on Motivation* 7: 257–308.

Macphail, E. M. (1982). *Brain and Intelligence in Vertebrates*. Oxford: Clarendon Press.

Milton, K. (1981). Distribution patterns of tropical plant foods as an evolutionary stimulus to primate mental development. *American Anthropologist* 83: 534–548.

Moss, C. J. (1988). *Elephant Memories*. Boston: Houghton Mifflin.

Packer, C. (1994). *Into Africa*. Chicago: University of Chicago Press.

Povinelli, D. J. and Preuss, T. M. (1995). Theory of mind: Evolutionary history of a cognitive specialization. *Trends in Neuroscience* 18: 418–424.

Seyfarth, R. M. (1980). The distribution of grooming and related behaviors among adult female vervet monkeys. *Animal Behaviour* 28: 798–813.

Seyfarth, R. M. and Cheney, D. L. (1984). Grooming, alliances, and reciprocal altruism in vervet monkeys. *Nature* 308: 541–543.

Smale, L., Frank, L. G., and Holekamp, K. E. (1993). Ontogeny of dominance in free-living spotted hyenas: Juvenile rank relations with adults. *Animal Behaviour* 46: 467–477.

Smale, L., Nunes, S., and Holekamp, K. E. (1997). Sexually dimorphic dispersal in mammals: Patterns, causes and consequences. *Advances in the Study of Behavior* 26: 181–250.

Stander, P. E. (1992a). Foraging dynamics of lions in a semi-arid environment. *Canadian Journal of Zoology* 70: 8–21.

Stander, P. E. (1992b). Cooperative hunting in lions: The role of the individual. *Behavioral Ecology and Sociobiology* 29: 445–454.

Tilson, R. T. and Hamilton, W. J. (1984). Social dominance and feeding patterns of spotted hyaenas. *Animal Behaviour* 32: 715–724.

Tomasello, M. and Call. J. (1997). *Primate Cognition*. Oxford: Oxford University Press.

Wahaj, S., Guze, K., and Holekamp, K. E. (2001). Reconciliation in the spotted hyena (*Crocuta crocuta*). *Ethology* 107: 1057–1074.

Walters, J. (1980). Interventions and the development of dominance relationships in female baboons. *Folia Primatologica* 34: 61–89.

Zabel, C. J., Glickman, S. E., Frank, L. G., Woodmansee, K. B., and Keppel, G. (1992). Coalition formation in a colony of prepubertal spotted hyaenas. In *Coalitions and Alliances in Humans and Other Animals*, A. H. Harcourt and F. B. M. de Waal, eds., pp. 113–135. Oxford: Oxford University Press.

Robert M. Seyfarth and Dorothy L. Cheney

Monkeys and apes live in complex social groups and must master a formidable calculus if they are to survive and reproduce. Baboons' groups, for example, are often composed of 80 or more individuals drawn from 8 or 9 matrilineal families arranged in a linear dominance rank order. What sort of intelligence is required to navigate this social landscape? How do individuals acquire information about their companions, and how do they store it in memory? Such questions are interesting because understanding social relationships and predicting behavior may have been the most complex problems faced by our ancestors during periods when the human brain increased most dramatically in size. Some authors have even suggested that human intelligence evolved largely because selection favored individuals who could solve complex social problems (Jolly 1966; Humphrey 1976; Cosmides and Tooby 1994).

Research on the mechanisms that underlie primate social intelligence is, however, fraught with problems, largely because the behavior of monkeys and apes can be explained equally well in many different ways. Does a baboon that apparently knows the matrilineal kin relations of others have a "social concept," as some have argued (e.g., Dasser 1988), or has the baboon simply learned to link individual *A1* with individual *A2* through a relatively simple process like associative conditioning, as others believe (e.g., Thompson 1995)? At present, the preferred explanation often depends as much upon the scientist's mind as upon any objective understanding of the baboon's.

Our research is conducted in the field, among nonhuman primates living in their natural social groups. We use long-term observations and videotaped playback experiments to address the question: What must a monkey know, and how must its knowledge be structured, in order to account for its social behavior? Our goal is to develop models of social intelligence that account for existing behavior and that explain why, during the course of primate evolution, some cognitive strategies have been favored over others.

Knowledge of Other Animals' Kin Relations

East African vervet monkeys (*Cercopithecus aethiops*) live in groups of 8–30 individuals. Females remain throughout their lives in the group where they were born. When males are 5–6 years of age, they emigrate to a neighboring group. Adult females and their offspring can be arranged in a linear dominance hierarchy, with offspring ranking immediately below their mothers. The stable core of a vervet social group is thus a hierarchy of matrilineal families (Cheney and Seyfarth 1990).

Most friendly interactions, such as grooming and the formation of aggressive alliances, occur within families (reviewed in Cheney and Seyfarth 1990). Clearly, individuals distinguish their own matrilineal relatives from all others because their behavior toward them is so different. There is also evidence, however, that vervets recognize the close associates of other group members (Cheney and Seyfarth 1986; Harcourt 1988). For example, a vervet who has been involved in an aggressive interaction with a particular opponent will often soon afterward threaten a close relative of the opponent. Knowledge of other individuals' social relationships can only be obtained by attending to interactions in which one is not involved and making the appropriate inferences.

Studying vervet monkeys in Amboseli National Park, Kenya, we taperecorded the vocalizations given by known individuals in social interactions with one another. In a series of playback experiments, we then played the distress scream of a juvenile to a group of three adult females, one of whom was the juvenile's mother.

The females' responses were filmed. As expected, mothers looked toward the loudspeaker for longer durations than did control females. Even before she had responded, however, a significant number of control females looked at the mother. They behaved as if they recognized the link between a particular vocalization, a specific juvenile, and a particular adult female (Cheney and Seyfarth 1980, 1982).

In a more recent study of baboons (*Papio cynocephalus ursinus*) in the Okavango Delta of Botswana, two unrelated females heard a sequence of calls that mimicked an aggressive interaction between two other members of their group. The first sequence mimicked a fight between two individuals who were both unrelated to either female. Neither subject responded. The second sequence mimicked a fight between a relative of the dominant subject and another, unrelated individual. The subordinate responded by looking at the dominant. The third sequence mimicked a fight between one of the dominant's and one of the subordinate's relatives. Both females responded by looking at each other (Cheney and Seyfarth 1999). Moreover, after hearing this third sequence, the two females were more likely to be involved in an agonistic interaction than after hearing either of the two other sequences. Apparently, baboon females view their social groups, not just in terms of the individuals that comprise them, but also in terms of a network of social relationships in which certain individuals are linked with several others (for further discussion, see Seyfarth and Cheney in press).

Knowledge of Other Animals' Dominance Ranks

Along with matrilineal kinship, linear, transitive dominance relations are a pervasive feature of social behavior in groups of Old World monkeys. A rank order might emerge because individuals can recognize the transitive dominance relations that exist among others: *C* knows that *A* is dominant to *B*. Alternatively, monkeys might simply recognize who is dominant or subordinate to themselves. In the latter case, a transitive, linear hierarchy would be an incidental outcome of paired interactions. The hierarchy would be a product of the human mind, not the minds of the monkeys themselves.

There is evidence, however, that monkeys do recognize the rank relations that exist among others. For example, dominant female baboons often grunt to mothers with infants as they approach the mothers and attempt to handle their infants. The grunts seem to function to facilitate social interactions by appeasing anxious mothers, because an approach accompanied by a grunt is significantly more likely to lead to friendly interaction than an approach without a grunt (Cheney et al. 1995b).

Occasionally, however, a mother will utter a submissive call, or "fear bark," as a dominant female approaches. Fear barks unambiguously indicate subordination; they are never given to lower-ranking females. To test whether baboons recognize that only a more dominant animal can cause another individual to give a fear bark, we played to adult females a causally inconsistent call sequence in which a lower-ranking female apparently grunted to a higher-ranking female and the higher-ranking female apparently responded with fear barks. As a control, the subjects heard the same sequence of grunts and fear barks made causally consistent by the inclusion of additional grunts from a third female who was dominant to both of the others. For example, if the inconsistent sequence was composed of female 6's grunts followed by female 2's fear barks, the corresponding consistent sequence might begin with female 1's grunts, followed by female 6's grunts and ending with female 2's fear barks. The subjects responded significantly more strongly to the causally inconsistent sequences, suggesting that they recognized not only the identities of different signalers but also the rank relations that existed among others in their group (Cheney et al. 1995a).

Evolution

How does an individual benefit from knowing the relations that exist among others? Current hypotheses stress the importance of triadic alliances, which occur whenever two or more individuals join together in directing aggression against a third. While alliances occur in many species, only primates (and perhaps dolphins, see Connor et al. 2000) appear to be strategic in their choice of alliance partners (Harcourt 1988). In macaques, for example, males consistently solicit allies who outrank both themselves and their opponents (Silk 1999). If alliances play an important role in competitive interactions, and if alliances only succeed when the solicitor recruits an ally who outranks and will not join his opponent, then solicitors must know the relative ranks and kin relations of all possible allies and opponents. In other words, they must know about the relations that exist among others.

Underlying Mechanisms

Humans readily divide social companions into groups and organize these groups within a hierarchical structure. There is, however, no a priori reason to believe that the same mental operations underlie social knowledge in monkeys and apes. Several authors, for example, have argued that primate social behavior can be explained by relatively simple processes of associative learning and conditioning (Heyes 1994; Thompson 1995).

Schusterman and Kastak (1993, 1998; see also chapter 28 in this volume) taught a California sea lion, Rio, to group seemingly arbitrary visual stimuli into equivalence classes. She learned to associate, for example, *A1*, *A2*, and *A3*, even though they shared no physical features. Next, Rio was rewarded for selecting stimulus *A1* over stimulus *B1*. Finally, Rio was tested to determine whether she had begun to treat all *A* stimuli as equivalent to each other and all *B* stimuli as equivalent to each other, at least insofar as they followed the rule *if A1 > B1 then An > Bn*. Rio performed significantly above chance.

The authors suggest that the kind of equivalence judgments demonstrated by Rio constitute a general learning process that underlies the recognition of social relationships in many species. Thus, for example, a baboon or vervet monkey learns to group members of the same matriline together because they share a history of common association and functional relations. And when one monkey, upon hearing a juvenile scream, responds by looking at the juvenile's mother, she does so because members of the same matriline have effectively become "interchangeable" (Schusterman and Kastak 1998).

There is no doubt that associative processes provide a powerful and often accurate means for animals to assess the relationships that exist among different stimuli, including members of their own species. Indeed, it seems unlikely that a monkey could form a concept such as "closely bonded" without attending to social interactions and forming associations between one individual and another. To some extent, learning about other individuals' social relationships is by definition dependent on some form of conditioning. However, before concluding that all primate social knowledge can be explained on the basis of learned contingencies (Heyes 1994), we note several ways in which equivalence class relations fail to capture the complexity of primate social relations.

Equivalence classes are typically based on a single underlying association, such as spatial or temporal juxtaposition. By contrast, no single behavioral measure is either necessary or sufficient to define the association among individuals in a primate matriline. A mother and her infant son interact in ways very different from those of two subadult male brothers, yet all four may be recognized by others as part of the same kin group.

While the stimuli that make up an equivalence class are mutually substitutable (Schusterman and Kastak 1998), the individuals that form a matrilineal kin group are linked in more variable and less predictable ways. If infant baboon *A1* and juvenile baboon *A2* both associate at high rates with the same adult female and she associates with an adult male "friend" (Smuts 1985; Palombit et al. 1997), it would be correct to assume that the male is closely allied to the infant but incorrect to assume that he is equally closely allied to the juvenile.

Further complicating matters, individual primates belong to multiple classes simultaneously. An adult female baboon, for example, belongs to a matrilineal kin group, associates with one or more adult males, holds a particular dominance rank, and may be weakly or strongly linked to other females outside her matriline. The natural situation is considerably more complex than laboratory studies of equivalence classes.

Next, consider the problem of training. In Schusterman and Kastak's experiment, as in many other studies, the subject was first presented with stimuli that had links to one another (*A1*, *A2*, *A3*) and then rewarded for choosing stimuli from one class over those from another (*A1* > *B1*). Thus trained, the subject generalized her knowledge so that when presented with any other *AB* stimulus pair, she always chose *A*. Speaking conservatively, these results only tell us that when presented with certain stimuli and rewarded for following a particular rule with a subset, a sea lion will generalize the rule and apply it to all the other members of that subset. The experiment does not tell us whether, in the absence of training and reward, the sea lion would naturally recognize this particular rule, or if she did recognize it, whether she would apply it generally beyond her immediate experience.

The distinction between learning that is rewarded in the laboratory and learning that occurs in the wild is important, because any intervention by humans that selectively rewards one kind of learning over another potentially distorts an animal's natural method of acquiring and storing information. For example, pigeons trained to match-to-sample with just a few stimuli are not able to transfer their behavior to novel stimuli, although monkeys and chimpanzees do so easily. However, pigeons do learn to match similar stimuli if they are trained with hundreds of exemplars over hundreds of trials. Apparently, although pigeons can acquire the abstract concept same-different, they seem predisposed to attend to absolute stimulus properties and to form item-specific associations (Wright et al. 1988; see also Wasserman et al. 1995; Shettleworth 1998). Extensive training by humans, therefore, changes the ways in which pigeons classify stimuli.

Similarly, Tomasello and colleagues compared the performance of chimpanzees raised by humans (but without language training), chimpanzees raised by their own mothers, and 2-year-old children. Human-reared chimpanzees showed more imitation (Tomasello et al. 1993), more joint attention, and were more likely to use gestures to direct a demonstrator's attention (Carpenter et al. 1995) than chimpanzees raised by their own mothers. In another study, chimpanzees that had been trained to use tokens as symbols were able to solve match-to-sample tasks that required them to judge relations between relations. Naive chimpanzees could perceive these relations, but their knowledge seemed to remain tacit (Premack 1983; Thompson and Oden 1995). Mere exposure to humans, therefore, alters chimpanzees' problem-solving skills.

Finally, consider the magnitude of the problem. In Schusterman's and Kastak's experiment, Rio learned a total of 180 dyadic (two-item) comparisons. This is roughly equivalent to the number of different dyadic comparisons—but not the number of triads—that confront a monkey in a group of 14 individuals. Most primates, however, live in much larger groups. In a group of 80 animals, each individual confronts 3160 different dyadic combinations and 82,160 different triadic combinations. In other words, free-ranging monkeys and apes face problems in learning and memory that are not just quantitatively but also qualitatively different from those presented in the typical laboratory experiment (Seyfarth and Cheney 2001).

What might constitute an adaptive solution to the social demands placed on monkeys? When asked to remember long strings of items, humans learn the string faster and remember it better if some kind of "rule" allows them to group items into "chunks" (Miller 1956; Tulving 1962). Similarly, when asked to remember the location of specific food types in a radial maze, rats act as if they have organized locations into groups according to the food type they contain (Dallal and Meck 1990; Macuda and Roberts 1995). Faced with the problem of remembering a large complex dataset, then, both humans and rats are predisposed to search for statistical regularities in the data (Seyfarth and Cheney 2001). They do so naturally and without reinforcement. Why should monkeys be any different?

In our current research we are testing the hypothesis that baboons organize knowledge about their social companions into a two-level, nested hierarchy based on matrilineal kinship and rank. Given that baboons respond strongly to evidence of a rank reversal between two individuals, we are using field playback experiments to test whether the subjects' responses are stronger if the purported rank reversal involves two adjacently ranked individuals in different matrilines compared with two adjacently ranked individuals in the same matriline (Seyfarth and Cheney in press).

Conclusion

To survive and reproduce, a monkey must be able to predict the behavior of others. In nonhuman primate groups, where alliances are common, prediction demands that a monkey learn and remember all of its opponents' dyadic and triadic relations. The task is similar to the problems faced by humans and rats in memory experiments. In response to these pressures, we

suggest that nonhuman primates are innately predisposed to group other individuals into hierarchical classes, both for ease of recall and to facilitate predictions of behavior. The formation of hierarchical classes is an adaptive mental strategy, shaped by natural selection.

References

Carpenter, M., Tomasello, M., and Savage-Rumbaugh, E. S. (1995). Joint attention and imitative learning in children, chimpanzees, and enculturated chimpanzees. *Social Development* 4: 217–237.

Cheney, D. L. and Seyfarth, R. M. (1980). Vocal recognition in free-ranging vervet monkeys. *Animal Behaviour* 28: 362–367.

Cheney, D. L. and Seyfarth, R. M. (1982). Recognition of individuals within and between groups of free-ranging vervet monkeys. *American Zoologist* 22: 519–529.

Cheney, D. L. and Seyfarth, R. M. (1986). The recognition of social alliances among vervet monkeys. *Animal Behaviour* 34: 1722–1731.

Cheney, D. L. and Seyfarth, R. M. (1990). *How Monkeys See the World: Inside the Mind of Another Species.* Chicago: University of Chicago Press.

Cheney, D. L. and Seyfarth, R. M. (1999). Recognition of other individuals' social relationships by female baboons. *Animal Behaviour* 58: 67–75.

Cheney, D. L., Seyfarth, R. M., and Silk, J. B. (1995a). The responses of female baboons (*Papio cynocephalus ursinus*) to anomalous social interactions: Evidence for causal reasoning? *Journal of Comparative Psychology* 109: 134–141.

Cheney, D. L., Seyfarth, R. M., and Silk, J. B. (1995b). The role of grunts in reconciling opponents and facilitating interactions among adult female baboons. *Animal Behaviour* 50: 249–257.

Connor, R. C., Read, A. J., and Wrangham, R. W. (2000). Male reproductive strategies and social bonds. In *Cetacean Societies: Field Studies of Dolphins and Whales*, J. Mann, R. C. Connor, and P. L. Tyack, eds., pp. 247–269. Chicago: University of Chicago Press.

Cosmides, L. and Tooby, J. (1994). Origins of domain specificity: The evolution of functional organization. In *Mapping the Mind: Domain Specificity in Cognition and Culture*, L. A. Hirschfeld and S. A. Gelman, eds., pp. 85–116. Cambridge: Cambridge University Press.

Dallal, N. and Meck, W. (1990). Hierarchical structures: Chunking by food type facilitates spatial memory. *Journal of Experimental Psychology: Animal Behavior Processes* 16: 69–84.

Dasser, V. (1988). A social concept in Java monkeys. *Animal Behaviour* 36: 225–230.

Harcourt, A. H. (1988). Alliances in contests and social intelligence. In *Machiavellian Intelligence: Social Expertise and the Evolution of Intellect in Monkeys, Apes, and Humans*, R. W. Byrne and A. Whiten, eds., pp. 132–152. Oxford: Oxford University Press.

Heyes, C. M. (1994). Social cognition in primates. In *Animal Learning and Cognition*, N. J. Mackintosh, ed., pp. 281–305. New York: Academic Press.

Humphrey, N. K. (1976). The social function of intellect. In *Growing Points in Ethology*, P. Bateson and R. A. Hinde, eds., pp. 303–313. Cambridge: Cambridge University Press.

Jolly, A. (1966). Lemur social behavior and primate intelligence. *Science* 153: 501–506.

Macuda, T. and Roberts, W. A. (1995). Further evidence for hierarchical chunking in rat spatial memory. *Journal of Experimental Psychology: Animal Behavior Processes* 21: 20–32.

Miller, G. A. (1956). The magical number seven, plus or minus two: Some limits on our capacity for processing information. *Psychological Review* 63: 81–97.

Palombit, R. A., Seyfarth, R. M., and Cheney, D. L. (1997). The adaptive value of "friendships" to female baboons: Experimental and observational evidence. *Animal Behaviour* 54: 599–614.

Premack, D. (1983). The codes of man and beast. *Behavioral and Brain Sciences* 6: 125–167.

Schusterman, R. J. and Kastak, D. A. (1993). A California sea lion (*Zalophus californianus*) is capable of forming equivalence relations. *Psychological Record* 43: 823–839.

Schusterman, R. J. and Kastak, D. A. (1998). Functional equivalence in a California sea lion: Relevance to animal social and communicative interactions. *Animal Behaviour* 55: 1087–1095.

Seyfarth, R. M. and Cheney, D. L. (2001). Cognitive strategies and the representation of social relationships by monkeys. In *Evolutionary Psychology and Motiva-*

tion, Nebraska Symposium on Motivation. Vol. 48, J. A. French, A. C. Kamil, and D. W. Leger, eds., pp. 145–178. Lincoln: University of Nebraska Press.

Seyfarth, R. M. and Cheney, D. L. (in press). Hierarchical structure in the social knowledge of monkeys. In *Animal Social Complexity*, F. B. M. de Waal and P. L. Tyack, eds. Cambridge, Mass.: Harvard University Press.

Shettleworth, S. (1998). *Cognition, Evolution, and Behaviour*. Oxford: Oxford University Press.

Silk, J. B. (1999). Male bonnet macaques use information about third-party rank relationships to recruit allies. *Animal Behaviour* 58: 45–51.

Smuts, B. (1985). *Sex and Friendship in Baboons*. Chicago: Aldine.

Thompson, R. K. R. (1995). Natural and relational concepts in animals. In *Comparative Approaches to Cognitive Science*, H. Roitblat and J. A. Meyer, eds., pp. 175–224. Cambridge, Mass.: MIT Press.

Thompson, R. K. and Oden, D. L. (1995). A profound disparity revisited: Perception and judgment of abstract identity relations by chimpanzees, human infants, and monkeys. *Behavioural Processes* 35: 149–161.

Tomasello, M., Savage-Rumbaugh, E. S., and Kruger, A. C. (1993). Imitative learning of actions on objects by children, chimpanzees, and enculturated chimpanzees. *Child Development* 64: 1688–1705.

Tulving, E. (1962). Subjective organization in the free recall of "unrelated" words. *Psychological Review* 69: 344–354.

Wasserman, E. A., Hugart, J. A., and Kirkpatrick-Steger, K. (1995). Pigeons show same-different conceptualization after training with complex visual stimuli. *Journal of Experimental Psychology: Animal Behavior Processes* 21: 248–252.

Wright, A., Cook, R., and Rivera, J. (1988). Concept learning by pigeons: Matching to sample with trial-unique video picture stimuli. *Animal Learning & Behavior* 16: 436–444.

47 From the Field to the Laboratory and Back Again: Culture and "Social Mind" in Primates[1]

Andrew Whiten

As I was preparing to leave school for university, I recall expressing an interest in studying the mind; psychology perhaps, or even philosophy. This was an idea that clearly troubled my teachers. I think it was my biology teacher who announced that all psychologists were themselves pretty loony, so I would do much better to get a solid scientific foundation in a biological subject like zoology, then decide what to do next. And that is what I did, with the perhaps inevitable final intellectual destination of the sciences of the mind. Although I think my biology teacher was wrong about psychologists (well, mostly!), following his advice had the wonderful benefit that by the time I came to study the mind I was fully steeped in the principles of evolutionary biology.

Of the many likely consequences of this decision, I should highlight two here. One is the working hypothesis guiding my research—that the mind will only be understood as a biological adaptation, shaped by evolutionary processes to deal with a certain set of ecological challenges and opportunities. Second, intimately linked to this hypothesis is the methodological principle that a research program will yield important insights only if it is grounded in the animal's behavior in its natural habitat.

This perspective was greatly shaped by Lorenz, Tinbergen, and von Frisch, whose groundbreaking work led them to the 1973 Nobel Prize. Always building on a foundation of good natural history, they used a great variety of complementary methods, including systematic and quantitative observation, and experimentation in both field and laboratory. This combination was enormously productive and, given the way it was grounded in nature, achieved insights that no other approach could.

In contrast, turning to my own discipline, it seems all too easy to classify many contemporary primatologists as *either* field researchers *or* laboratory workers; as *either* observers *or* experimenters. Among the notable exceptions, who could be seen as emulating the catholic yet integrated methodological programs of Tinbergen and company, are such figures as Kummer, Matsuzawa, Cheney, and Seyfarth.

In this essay I illustrate our own efforts to this end in relation to research on social learning and culture. However, this is but one component of a larger topic we are studying, which is beyond the scope of this brief essay and therefore is sketched only in outline here. Its scope is essentially the "social mind." The Machiavellian intelligence hypothesis is that the intelligence of monkeys and apes is an adaptation, not so much to the challenges of the physical world (like finding food), as to the problems encountered in negotiating a particularly complex, natural, social world (see Byrne and Whiten 1988, including the foundational articles of Humphrey 1976, and others; Whiten 1999, 2000a; Whiten and Byrne 1997). This hypothesis has come to be supported by a series of studies that find measures of encephalization to be more closely related to indices of social complexity than to those concerning activities like finding food (Barton and Dunbar 1997).

Examples of our efforts to elucidate components of this social complexity include work on tactical deception (Whiten and Byrne 1988), management of conflict and reconciliation (Castles and Whiten 1998), and "mind reading" (or a "theory of mind"; Whiten 1998a, 2000b). This research reveals primate minds to be populated with a variety of sophisticated cognitive mechanisms that deal with a diversity of specifically social and complex problems in their daily lives. Learning from others (social learning) is a further aspect of this elaborate social mind. It is a phenomenon appropriately included in the scope of Machiavellian intelligence insofar as it is yet another way in which an animal can exploit to its benefit important information unique to its social environment.

Social Learning and Culture

Baboons survive in habitats that are not conducive to most other primates by exploiting a great
diversity of food types that can be difficult to
find and/or process (Whiten et al. 1991). For
example, adult olive baboons (*Papio anubis*) I
studied in Kenya in the late 1980s would yank
up clumps of a large sedge, *Mariscus*, and deftly
extract the small edible parts through a hierarchically organized sequence of actions (Whiten
1988). The clump would be held in one hand
while the other hand grasped a tuft of stems,
quickly twisting it around to detach it and bring
it to the mouth to bite off (and spit out) the
sheath; then the white, nutritious stem base was
revealed and could be bitten off; the next tuft was
twisted off and the process was repeated, until all
the tufts were dealt with. Juveniles did not use
this structured technique, but did observe their
elders doing it and later came to use it themselves. So is this an example of social learning?

In the field, it seems difficult, perhaps impossible, to tell whether it is. Primates develop slowly.
Behavior like the slick, adult style of *Mariscus*
processing typically emerges over periods of
weeks or months. There seems to be no way to
discriminate between the roles of individual and
social learning, let alone to differentiate between
alternative social learning processes. For example, maybe the juveniles learn by copying the
method from their elders (imitation); or perhaps
their attention is merely drawn to the sedge
(stimulus enhancement) and they work out the
optimal way to eat it through trial and error.
Typically, in the field, an ethologist sees the focal
subjects only at intervals, between which all
kinds of experience may be shaping them in ways
the scientist cannot see.

I therefore decided that to study the social
learning process, and particularly to probe
underlying cognitive processes, an experimental
approach would be necessary, although of course
it should be designed to incorporate the kinds of
problems observed in the field. The result was
what Debbie Custance and I called an "artificial
fruit"; an object with an edible core that, like
Mariscus, requires a series of manipulations
to remove or disable inedible components or
"defenses" (figure 47.1). The specific design of
our first fruit had two important aspects to it.
First, each of three different kinds of defense
could be removed using one of two alternative
methods; for example, twisting and pulling out a
pair of sticks wedging the fruit closed, or poking
them through. This meant that in the experiment, two samples of subjects would each see
only one of the two alternatives, and we could
then assess the extent to which the difference was
copied into the subjects' own subsequent actions.
The other aim of the design was to mimic aspects
of complex natural foraging like *Mariscus* eating; accordingly, the fruit had several different
kinds of defense that had to be dealt with sequentially, and repetitions or cycles of actions
were also built in. Note that this was not a tool-
using task. Most primates are not in fact great
tool users and we wanted to present tasks more
like those that all primates perform, to permit a
program of comparative studies.

As things turned out, the first subjects that
became available for these studies were young
chimpanzees (Whiten et al. 1996), not baboons.
In addition to the chimpanzees, we tested young
children because research on imitation inevitably
begs for comparison with a species (our own)
that is assumed to be strongly imitative ("*Homo
imitans*"; Meltzoff 1988). We found that the
children conformed to this expectation and typically produced high-fidelity copies of the particular actions they had seen used. The chimpanzees'
response was more likely to be a rougher, more
selective copy of the action they witnessed. They
were more ready to ignore what a model did if
they themselves could see a way to take a short-
cut. However, to varying extents the chimpanzees
did incorporate various aspects of the behavior
they had seen, such as twisting the wedged sticks;

A

Figure 47.1
(A) Working with "artificial fruit." A juvenile chimpanzee attempts to open an artificial fruit by pulling and twisting out a pair of bolts that jam the top shut. The subjects in an experiment will have seen the bolts either pulled out or poked through, so their imitative tendency can later be measured. Also holding the top shut is a handle that can be either turned or pulled up to release it; this is held in place by a pin that must be removed first by another pair of alternative actions. (B) Recording chimpanzee attempts on a "fruit." A young chimpanzee on Ngamba island, Uganda, is busy opening an artificial fruit and already has both bolts removed.

this was not actually necessary to remove the sticks and so reinforces our conclusion that our naturalistic experiment showed chimpanzees learning to apply a particular technique by imitating what they saw another individual doing.

Other parts of this research program delineate a number of different aspects of the social learning process in chimpanzees, such as the acquisition of information on the sequential structure of actions (Whiten 1998b). They also extend to a number of other species of apes and monkeys, some aspects of which we will return to later in this essay. The above summary, however, is sufficient to illustrate that the basic experimental approach is able to identify social learning in a way that has so far proved impossible for animals in the wild, who slowly and naturally perfect similar kinds of behavior. With this knowledge in mind, let us return to Africa.

Charting the Cultural Variation of Wild Chimpanzees

A few years ago, because of what we discovered through the experimental program, I was invited to write a review that also extended to the topic of culture (or tradition) in primates (Whiten 2000c). In the course of this I tried to establish just how much we knew about cultural variation in the species for which it seems most extensive—chimpanzees. This was both exciting and frustrating: exciting because, as long-term studies have accumulated, chimpanzee researchers have been able to construct larger and larger charts of the putative cultural variation across Africa (McGrew 1992; Boesch 1996); but this literature is a bit frustrating for several reasons. One is simply that this work tends to have been done by afficionados of chimpanzee research, who have not fussed too much with defining types of behavior that are still a little mysterious to the rest of us, so it is not so easy to cross-reference one expert's tables with another's. More important, the collating work has been done on the basis of

published records. This is not too satisfactory because this source is likely to be incomplete in both positive and (just as crucial) negative records at any one site; also, the frequency of a behavior pattern, which is clearly relevant to whether the behavior can be considered cultural, may not have been reported.

I felt a different approach was needed. Together with Christophe Boesch, I approached the research directors of the most long-term field sites and we agreed to proceed systematically through two phases of research. In the first, we drew up a list of behavior categories, published or not, that researchers suspected might be cultural variants. Full descriptions were set down and agreed upon (see Whiten et al. 2001).[2] With nine international contributors, this process was much more complex than one might guess! It generated the rather extraordinary total of 65 candidate cultural behaviors (a tribute in itself to the inventiveness of chimpanzees). The field directors and their staff then categorized each behavior pattern according to whether it was customary (typical for their site), habitual (less than customary but still consistent with social transmission), merely present, or absent, either with or without an obvious ecological explanation.

Collating all the information, we checked for behavior patterns that met our criteria for cultural variants: those customary or habitual in at least one community, yet absent without ecological explanation in at least one other. In this way we identified no less than 39 cultural variations (Whiten et al. 1999), a figure far exceeding that reported for any other species except our own (indeed, previous studies of traditions among animals have typically identified variation in just a single behavior). Although some behavior patterns occur in more than one community, each community has its own unique suite of variants, so that if we know enough about a chimpanzee, we can locate it geographically on the basis of its cultural repertoire, as we might do for people (Whiten and Boesch 2001) (see figure 47.2).

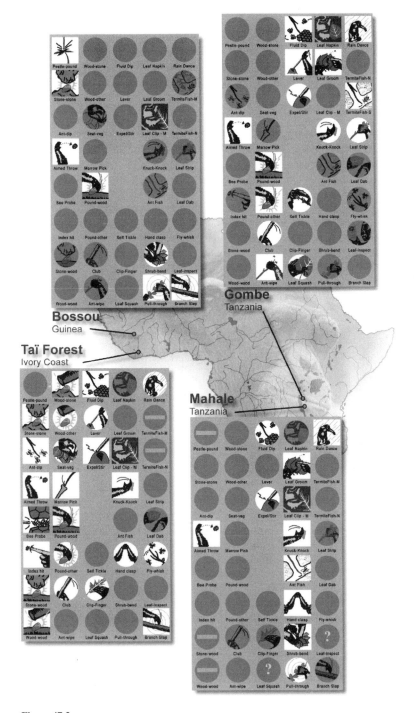

Figure 47.2
Chimpanzee cultures. Variation in behavior patterns of chimpanzees at four long-term study sites. Square, customary; circular, habitual; faded circular, present; blank, absent; bar, ecological reason for absence; question mark, unknown. From an ongoing collaborative project (see Whiten et al. 1999, 2001).

But are the variants really all cultural? Several kinds of information might be thought relevant to judging this, including the background experimental evidence that chimpanzees are social learners (and in some circumstances, imitators); that juveniles devote serious effort to scrutinizing adults performing these behaviors; that ecology is not a plausible explanation (for example, the same materials are available whether the behavior is customary or absent); and similarly, that a genetic difference is not plausible (for example, variation occurs between nearby communities of the same subspecies).

Take the case of ant dipping, which in the Taï Forest preserve in the Ivory Coast involves using a short stick to pick up a few driver ants and transfer them directly to the mouth, but at Gombe in Tanzania involves a much more efficient trick of swiping a hand along a long wand to gain a large ball of ants. Juveniles watch these activities intently; ecological explanations for the different methods seem unlikely since ants and sticks are available at each site; and a genetic cause seems unlikely, for the behavior does not even exist at Mahale, a short distance from Gombe. But note I say "unlikely." Can we prove the sticks and ants are really similar? Perhaps not. What would really clinch the cultural status of these behaviors is if we moved some chimps from Taï and Mahale to Gombe and they adopted the local technique. Since for ethical reasons this is not likely to happen, the importance of using relevant experiments to complement what the field research appears to be telling us, even if the experiments are not done in the wild, is clear.

Wider Horizons: The Comparative Cognitive Ethology of Social Learning

In this essay I have focused on what we are learning about chimpanzees. However, the two different kinds of method I described have been developed very much with comparative research in mind. The approach described earlier for the cross-cultural survey could, we hope, be used as a kind of template for other taxa of animals where multiple field studies are accumulating reliable data on the absence of behavior patterns in certain populations. Such data already exist for other primates, but further candidates include the topical cases of capuchins (Fragaszy and Perry, in press) and whales and dolphins, the cultural complexity of which has recently been debated (Rendell and Whitehead in press).

In the case of the experimental approaches, comparative research has already been extended to other primates (e.g., Custance et al. 1999; Stoinski et al. 2001) and nonprimates (e.g., Huber et al., 2001). Here, however, we soon encountered a problem in designing a task that could be used to make valid comparisons. We found, for example, that marmosets were unable to open an appropriately scaled-down version of the artificial fruit, either spontaneously or with a variety of training efforts. A case like this forces us to take a different approach. Accordingly, we created an artificial fruit appropriate to the marmosets' capabilities (Caldwell et al. 1999).

Perhaps this should lead us to explicitly recognize two different comparative approaches, one in which different tasks appropriate to each species are created (facilitating a comparison of social learning that takes account of differences in other underlying abilities) and another in which the same (appropriately scaled) task is given to all species (making for a more direct comparison of what they actually do acquire, whatever the underlying explanation). Probably it will be the complementary application of both these approaches—in conjunction with the very different methods described above—that has the greatest potential for allowing us to obtain an insightful, comparative, cognitive ethology of culture and social learning.

Acknowledgments

I am indebted to Gordon Burghardt for comments on an earlier draft of this essay.

Notes

1. With due respect to Kummer (1984).

2. The Chimpanzee Cultures Website (http://chimp.st-and.ac.uk/cultures) incorporates a searchable, illustrated database covering all the chimpanzee behaviors surveyed and their distribution.

References

Barton, R. A. and Dunbar, R. I. M. (1997). Evolution of the social brain. In *Machiavellian Intelligence II: Evaluations and Extensions*, A. Whiten and R. W. Byrne, eds., pp. 240–263. Cambridge: Cambridge University Press.

Boesch, C. (1996). The emergence of cultures among wild chimpanzees. In *Evolution of Social Behavior Patterns in Primates and Man*, W. G. Runciman, J. Maynard-Smith, and R. I. M. Dunbar, eds., pp. 251–268. New York: Oxford University Press.

Byrne, R. W. and Whiten, A. (eds.) (1988). *Machiavellian Intelligence: Social Expertise and the Evolution of Intellect in Monkeys, Apes and Humans*. Oxford: Oxford University Press.

Caldwell, C., Whiten, A., and Morris, K. (1999). Observational learning in the marmoset monkey, *Callithrix jacchus*. In *Proceedings of the Artificial Intelligence and Simulation of Behaviour Convention (AISB), Symposium on Imitation in Animals and Artifacts*, pp. 27–31. Edinburgh, Scotland: The Society for the Study of Artificial Intelligence and the Simulation of Behavior.

Castles, D. L. and Whiten, A. (1998). Post-conflict behaviour of wild olive baboons. I. Reconciliation, redirection and consolation. *Ethology* 104: 126–147.

Custance, D. M., Whiten, A., and Fredman, T. (1999). Social learning of "artifical fruit" processing in capuchin monkeys (*Cebus apella*). *Journal of Comparative Psychology* 113: 13–23.

Huber, L., Rechberger, S., and Taborsky, M. (2001). Social learning affects object exploration and manipulation in keas (*Nestor notabilis*). *Animal Behaviour* 62: 945–954.

Humphrey, N. K. (1976). The social function of intellect. In *Growing Points in Ethology*, P. P. G. Bateson and R. A. Hinde, eds., pp. 303–313. Cambridge: Cambridge University Press. Reprinted in Byrne and Whiten (1988).

Fragaszy, D. M. and Perry, S. (eds.) (in press). *The Biology of Traditions: Models and Evidence*. Cambridge: Cambridge University Press.

Kummer, H. (1984). From laboratory to desert and back: A social system of hamadryas baboons. *Animal Behaviour* 32: 965–971.

McGrew, W. C. (1992). *Chimpanzee Material Culture: Implications for Human Evolution*. Cambridge: Cambridge University Press.

Meltzoff, A. N. (1988). The human infant as *Homo imitans*. In *Social Learning: Psychological and Biological Perspectives*, T. Zentall and B. Galef, eds., pp. 319–341. Hillsdale, N.J.: Lawrence Erlbaum Associates.

Rendell, L. and Whitehead, H. (2001). Culture in whales and dolphins. *Behavioral and Brain Sciences* 24: 309–382.

Stoinski, T. S., Wrate, J. L., Ure, N., and Whiten, A. (2001). Imitative learning by captive western lowland gorillas (*Gorilla gorilla gorilla*) in a simulated food-processing task. *Journal of Comparative Psychology* 115: 272–281.

Whiten, A. (1988). Acquisition of foraging techniques in infant olive baboons. Paper presented at the XII Congress of the International Primatological Society.

Whiten, A. (1998a). Evolutionary and developmental origins of the mindreading system. In *Piaget, Evolution and Development*, J. Langer and M. Killen, eds., pp. 73–99. Hillsdale, N.J.: Lawrence Erlbaum Associates.

Whiten, A. (1998b). Imitation of the sequential structure of actions by chimpanzees (*Pan troglodytes*). *Journal of Comparative Psychology* 112: 270–281.

Whiten, A. (1999). The Machiavellian intelligence hypothesis. In *MIT Encyclopedia of the Cognitive Sciences*, R. Wilson and F. Keil, eds., pp. 495–497. Cambridge, Mass.: MIT Press.

Whiten, A. (2000a). Social complexity and social intelligence. In *The Nature of Intelligence*. Novartis Symposium 233: 185–201. Chichester, UK: Wiley.

Whiten, A. (2000b). Chimpanzee cognition and the question of mental re-representation. In *Meta-*

representations, D. Sperber, ed., pp. 139–167. Oxford: Oxford University Press.

Whiten, A. (2000c). Primate culture and social learning. *Cognitive Science* (Special Issue on Primate Cognition) 24: 477–508.

Whiten, A. (2001). Theory of mind in non-verbal apes? Conceptual issues and the critical experiments. In *Philosophy Supplement 49; Naturalism, Evolution and Mind*: 199–233.

Whiten, A. and Boesch, C. (2001). The cultures of chimpanzees. *Scientific American* 284: 48–55.

Whiten, A. and Byrne, R. W. (1988). Tactical deception in primates. *Behavioral and Brain Sciences* 11: 233–266.

Whiten, A. and Byrne, R. W. (eds.) (1997). *Machiavellian Intelligence II: Evaluations and Extensions.* Cambridge: Cambridge University Press.

Whiten, A., Byrne, R. W., Barton, R. A., Waterman, P. G., and Henzi, S. P. (1991). Dietary and foraging strategies of baboons. *Philosophical Transactions of the Royal Society B* 334: 27–35.

Whiten, A., Custance, D. M., Gomez, J.-C., Teixidor, P., and Bard, K. A. (1996). Imitative learning of artificial fruit processing in children (*Homo sapiens*) and chimpanzees (*Pan troglodytes*). *Journal of Comparative Psychology* 110: 3–14.

Whiten, A., Goodall, J., McGrew, W. C., Nishida, T., Reynolds, V., Sugiyama, Y., Tutin, C. E. G., Wrangham, R. W., and Boesch, C. (1999). Cultures in chimpanzees. *Nature* 399: 682–685.

Whiten, A., Goodall, J., McGrew, W. C., Nishida, T., Reynolds, V., Sugiyama, Y., Tutin, C. E. G., Wrangham, R. W., and Boesch, C. (2001). Charting cultural variation in chimpanzees. *Behaviour* 138: 1489–1525.

Richard W. Byrne

Probably, researchers from a greater variety of disciplines study the nonhuman primates than any other group of animals. Primates interest ecologists, zoologists, medical researchers, geneticists, anthropologists, and psychologists, like myself. Since nonhuman primates belong to the same order of mammals as ourselves, their cognitive processes are more likely than those of any other species of animal to be relevant to understanding the remote origins of the human mind.

One might think that this truth has long been generally accepted. Far from it; even today, some would dispute that primate behavior had *any* relevance to the vexed issue of the human mind (Macphail 1998). And in the days when what we would now call "evolutionary psychology" was called "comparative psychology," psychology did not do itself any favors by its choice of species. Typically, the comparative psychologist studied only a few species: the laboratory (white) strain of the rat; occasionally the ring-tailed lemur; more often the rhesus monkey and the chimpanzee. It was hard to escape the impression of a natural scale, with each living species a sort of model of an earlier stage in the evolution of the more advanced forms; indeed, for some practitioners, that really seems to have been their underlying theory (for details of this history, see Burghardt 1973; Burghardt and Gittleman 1990). In reality, of course, evolution seldom produces a linear progression. Yet, because comparative psychologists only had access to a restricted range of species handily available in captivity, even those who did not think in linear terms had difficulty convincing anyone that their theories did not retain the long-discredited logic of progressive evolution. Comparative psychology became something of a Cinderella subject in psychology.

Its change in fortune, and the subtle transformation to a properly evolutionary psychology, came with the huge burgeoning of primate field studies in the 1970s and 1980s. Fieldwork was done for many different purposes, but the growth in knowledge it has produced has at last allowed a genuinely comparative database to be built up. Field primatology began with a few isolated studies of "glamorous" or easy-to-watch species: chimpanzees, baboons, the species of macaque that live as commensals in Japan and India, those South American monkeys that happened to be marooned on Barro Colorado Island by the waters of the Panama Canal. It has grown to the present state in which virtually every branching point on the tree of primate phylogeny has at least one detailed study in the wild, and in some cases every species in a group has been studied.

Theoretical methods have also advanced over the same period, from the early days of two-species or two-population comparisons (see examples in Sussman 1979), to the modern use of quantitative comparisons performed as phylogenetic contrasts (to remove concerns of pseudoreplication resulting from possible phylogenetic inertia) across the whole order (e.g., Barton and Dunbar 1997). It is at last possible to focus clearly on the central questions:

• When did a particular cognitive trait enter the human lineage?

• What was its original adaptive function? (And has it been retained for the same reason, or is it now valuable for some different purpose?)

• What is the cognitive basis for the trait, and how does its organization relate to other mental capacities?

In order to illustrate how these questions may be approached, I use some recent studies of monkeys and great apes.

It would be relatively straightforward to establish when a trait originated if its presence or absence could be clearly identified in living species. Unfortunately, definitive evidence of the *absence* of a cognitive trait is often difficult to obtain, and we may have to be content with a surrogate measure and a residual level of uncertainty.

Monkeys and apes have long been known to show social manipulations that appear complex and clever to human observers: third-party support to win resources, ruses that rely on deception, long-term nurturing of friendships and reciprocal collaboration, targeted choice of allies and repair of disrupted relationships, and so on (see papers in Byrne and Whiten 1988; Harcourt and de Waal 1992). In contrast, people who study lemurs and lorises, or indeed most other mammals, report nothing very similar. Moreover, the simian primates (monkeys and apes) have unusually large brains for animals of their size (Jerison 1973; Passingham 1981); lemurs and lorises, on the other hand, have brains of more typical size for mammals. This difference is principally expressed in neocortical volume, and there is a direct relationship between neocortical volume and the amount of "clever-looking" behaviors that researchers observe. That applies to deception, to innovation, and to tool use (Byrne 1996b; Reader and Laland 2001). At least in the case of deception, this is not a by-product of the greater opportunities for researchers to see an interesting behavior if they watch a larger social group because the effect is independent of group size. In modern primate phylogeny, which is based on the pattern of differences in species' DNA, the monkeys and apes form a monophyletic clade; that is, they are a group descended from a single ancestor species. Because some fossils are available for calibrating the evolutionary tree revealed by molecular study, we can approximately date this ancestor species as living 30 million years ago. For these reasons, we know that the mental capacity to use other social individuals in a manipulative, clever-seeming fashion, including quite elaborate cooperation and the use of deceptive tactics, has a rather ancient origin in the human lineage (Byrne 1995, 2000).

The answers to functional questions are always more open to debate than matters of dating. No modern monkey is "equivalent" to the ancestors of monkeys and apes living of 30 million years age; there is no model of this extinct form. We cannot therefore study the original function of the enlarged simian neocortex. It is evident that all modern simians benefit socially from the cooperation and competition that their clever-looking behavior allows, but would they perhaps cope perfectly well without it if they could not afford large brains? In metabolic terms, brain tissue is the most "expensive" tissue in the entire body (Aiello and Wheeler 1995; Armstrong 1983); and uniquely, brain tissue remorselessly requires a constant energy supply or it deteriorates. Other things being equal, having a *smaller* brain is a good thing, so some positive advantage of brain enlargement must exist in every case where we find species with relatively large brains.

In fact, there is evidence that a large neocortex confers social benefits on modern monkeys and apes. The average group size in which they live is well predicted by the degree of their neocortical enlargement (Barton and Dunbar 1997; Dunbar 1992). On the other hand, measures of environmental complexity, such as range size and the distance of a day's journey (when corrected for the body size of the species concerned), do not correlate with neocortex size. Furthermore, in Old World monkeys and apes, which use grooming to build up friendly social relationships, the typical group size also predicts the amount of grooming seen; in a large group, more grooming is apparently necessary.

Although all these relationships are correlational, the associations between neocortex size and both a method of building up social relationships (grooming) and the frequency of use of a social tactic (deception), encourage the Machiavellian intelligence hypothesis: that an important selective pressure on the evolution of intelligence has been social complexity (Byrne 1996a; Humphrey 1976; Jolly 1966). The fact that the group sizes of modern primate species relate to their neocortex volumes suggests that social complexity may set an approximate upper limit on group sizes because of the demands that it places on the limited neocortex tissue available. Over longer time scales, pressure to live in ever-larger groups is felt as positive selection for an enlarged neocortex.

The benefits of a large neocortex—the underlying cognitive basis of monkey and ape social sophistication—are not easily determined. It is tempting, but it may be utterly wrong to assume that an animal that works over many months to build up a friendly relationship with another has some idea of the effect its behavior is having on the mind of the other. ("If I scratch his back often, he'll like me," rather than simply "If I scratch his back often, he'll probably scratch mine one day.") We also readily assume that an animal that uses a trick that relies on successful deception to gain some special resource actually meant to do so (i.e., it planned the effect of its actions in advance). That is, we assume the agent realizes that by producing a false belief in its victim, it may have risked losing a friend or gaining an enemy. The alternative explanation is that such behavior is a more prosaic mixture of genetic predispositions and rapid learning; often this is more likely.

Baboons, for example, are famous for building up long-term friendships that result in the benefits of social support on later occasions (Smuts 1983; Strum 1983). When these apparently important relationships are threatened by conflict, the baboons show behavior described as reconciliation (Castles et al. 1999; de Waal and van Roosmalen 1979). In this, friendly acts are actually more likely after a fight than before; however, as far as we know, all baboons will show these behaviors under the appropriate circumstances. They apparently do not have to be learned or deduced. So it is entirely possible that the baboon is genetically equipped with tendencies to direct affiliative acts to high-ranking members of its social group and to respond positively to affiliation shown by others, both at the time and afterward.

The underlying cognition needed to make these traits "pay" is rather simple. Baboons must recognize others as individuals and must be able to categorize those individuals by their dominance rank. They must also be able to categorize individuals by their past history of showing affiliative or aggressive acts toward themselves; for this, a single dimension would suffice, which we might call "self-friendliness." The genetic tendency of working to increase the level of self-friendliness in dominant members of the groups would automatically produce much of the social engineering we know from Old World monkeys like the baboons. Even reconciliation would follow, provided the animals are equipped with a tendency to show affiliation to dominant individuals whose self-friendliness has recently decreased. Simple rules of this sort would rapidly pay in evolutionary currency and are exactly what we would expect to evolve in highly social species.

Some of the "clever" behaviors of baboons, such as deception or innovation, are by no means universal; they are relatively rare, and each case is idiosyncratically different (Byrne and Whiten 1985). These are hallmarks of learning or deduction, but once again, they may not require any deep analysis of the social situation. Consider, for instance, a baboon that leaped to its hind legs and scanned the distance, for all the world as if it had seen a predator or an unexpected incursion of another baboon troop, at precisely the moment when it was being attacked by a dominant. Convenient? Indeed so; the attack was aborted, and no baboon troop or predator ever materialized. Yet this tactic may depend on no more than rapid learning. All that is required is a little history, in which perhaps the same baboon once actually did see a predator at a time when it was losing a fight and as a result was not thrashed (Byrne 1997). Avoidance of pain can function as a reward, making any preceding behaviors more probable in future similar cases. In this case, the preceding behaviors were leaping to the hind legs and scanning the distance. Learning must be rapid, but social insight may be lacking.

Innovations, likewise, are impressive and memorable if they happen to be beneficial and become enshrined traditions, but most primate researchers have seen oddities of behavior or fads that are pointless and simply die out. It is therefore quite possible that successful innovation depends on no more than rapid learning and a bit of luck, but lacks insightful understanding of the mechanism of the benefit conferred (Kummer and Goodall 1985). For example, Mike, one of the Gombe chimpanzees, discovered that banging empty kerosene cans together could help his rise in social dominance. Yet this device was not used by others at Gombe, suggesting that the chimpanzees were unable to understand the mechanism of Mike's good fortune.

Researchers have to be very cautious, then, in attributing to nonhuman primates the ability to understand social behavior or how things work in the mechanistic way that adult humans understand them. Rapid learning in social circumstances, a good memory for individuals and their different characteristics, and some simple genetic tendencies, can explain much that has impressed observers as intelligent behavior in simian primates. Is that all there is to be discovered? I believe not, but my confidence comes from the study of something rather less glamorous than social manipulation: feeding.

For most species of primates—monkeys, lemurs, and lorises—the challenge of feeding is largely a matter of getting to food (Byrne 1999a). Food may be scarce, dispersed, or concealed, but once it is at hand, eating is simple. [Interesting exceptions to this generalization occur in some lemurs, for instance, the aye-aye (*Daubentonia*), which uses echolocation to detect grubs in rotting wood and specialized teeth and finger adaptations to get them out (Erickson 1991); and the bamboo lemurs (*Hapalemur*), which use a highly specialized manual technique to eat the giant grasses they subsist on (Stafford et al. 1993).] However, in the great apes, remarkable feeding techniques are found that closely match the physical problems presented by local food sources. In chimpanzees, tool-making

and tool-using traditions vary from site to site (McGrew et al. 1979), and in gorillas, complex manual techniques are found that are specific to a number of different plants found only in the range of a few dozen groups of gorillas (Byrne and Byrne 1991, 1993; Byrne et al. 2001). Evidently great apes are able to learn elaborate techniques to solve manual problems.

A number of facts support the hypothesis that great apes can learn manual techniques by imitating the underlying structure of a behavior, filling in the details of execution in the most convenient way, often by trial and error (Byrne 1993; Byrne and Russon 1998). Although they are efficient and highly standardized in overall form in the local population, the gorillas' techniques are not the only, or the simplest, ways of obtaining the plants. One chimpanzee tool-using technique actually exists in two variants at different sites, even though the ant species, and the twigs used to make the tools, are available at each site. At Gombe, Tanzania, ants are collected on a large, stripped wand and scooped into the mouth in a ball (McGrew 1974); at Taï, the Ivory Coast, ant dipping is done with one hand, using a shorter stick that often has a frayed end (Boesch and Boesch 1990; see also Whiten, chapter 47 in this volume). The second method is less efficient, yet the Ivoirean chimpanzees have not discovered a better way. In both gorillas and chimpanzees, injuries from snares often maim the hands of exploring, curious infants and young juveniles. Surprisingly, individuals can survive with highly disabling hand injuries, even though they rely on complex manual techniques to feed on some important resources. Rather than growing up to acquire novel techniques that are specialized for making the most of the remaining manual function, these animals learn the same methods as able-bodied individuals and work around their own disabilities by using other limbs, fingers, chin, or branches to carry out the same process (Stokes and Byrne 2001).

To learn the organization of behavior imitatively, it is first necessary to "see" that organization—to go beyond the surface level in which behavior consists of fluid movement and reach the underlying structure. Judging by the great apes' skills, that will include recognizing the modular grouping as well as the linear sequence of actions, the coordination of the two hands (and sometimes mouth as well) used in different, complementary roles, and the use of some modules as subroutines in the service of the overall routine. As with the social skills of monkeys, this may not imply a deep understanding of mechanism. All these aspects of behavior produce distinctive traces in behavior, provided the observer can repeatedly watch the skill in action (Byrne 1999b). On the other hand, the ability to see the underlying modularity, hierarchical organization, and coordination of effectors is an essential starting point for understanding cause and effect and the purposes that lie behind an action. So it may be that the great apes' efforts at eating their more awkward foods will give important clues to understanding the evolutionary origins of the human capacity to understand causes and intentions.

References

Aiello, L. and Wheeler, P. (1995). The expensive tissue hypothesis. *Current Anthropology* 36: 199–221.

Armstrong, E. (1983). Metabolism and relative brain size. *Science* 220: 1302–1304.

Barton, R. and Dunbar, R. I. M. (1997). Evolution of the social brain. In *Machiavellian Intelligence II: Extensions and Evaluations*, A. Whiten and R. W. Byrne, eds., pp. 240–263. Cambridge: Cambridge University Press.

Boesch, C. and Boesch, H. (1990). Tool use and tool making in wild chimpanzees. *Folia Primatologica* 54: 86–99.

Burghardt, G. M. (1973). Instinct and innate behavior: Toward an ethological psychology. In *The Study of Behavior: Learning, Motivation, Emotion and Instinct*, J. A. Nevin and G. S. Reynolds, eds., pp. 322–400. Glenview Ill.: Scott, Foresman.

Burghardt, G. M. and Gittleman, J. G. (1990). Comparative and phylogenetic analyses: New wine, old bottles. In *Interpretation and Explanation in the Study of Behavior*. Vol. 2, *Comparative Perspectives*, M. Bekoff and D. Jamieson, eds., pp. 192–225. Boulder, Col.: Westview Press.

Byrne, R. W. (1993). Hierarchical levels of imitation. Commentary on M. Tomasello, A. C. Kruger, and H. H. Ratner, "Cultural learning." *Behavioral and Brain Sciences* 16: 516–517.

Byrne, R. W. (1995). *The Thinking Ape: Evolutionary Origins of Intelligence*. Oxford: Oxford University Press.

Byrne, R. W. (1996a). Machiavellian intelligence. *Evolutionary Anthropology* 5: 172–180.

Byrne, R. W. (1996b). Relating brain size to intelligence in primates. In *Modelling the Early Human Mind*, P. A. Mellars and K. R. Gibson, eds., pp. 49–56. Macdonald Institute for Archaeological Research, Cambridge.

Byrne, R. W. (1997). What's the use of anecdotes? Attempts to distinguish psychological mechanisms in primate tactical deception. In *Anthropomorphism, Anecdotes, and Animals: The Emperor's New Clothes?* R. W. Mitchell, N. S. Thompson, and L. Miles, eds., pp. 134–150. Albany: State University of New York Press.

Byrne, R. W. (1999a). Cognition in great ape ecology. Skill-learning ability opens up foraging opportunities. *Symposia of the Zoological Society of London* 72: 333–350.

Byrne, R. W. (1999b). Imitation without intentionality. Using string parsing to copy the organization of behaviour. *Animal Cognition* 2: 63–72.

Byrne, R. W. (2000). The evolution of primate cognition. *Cognitive Science* 24: 543–570.

Byrne, R. W. and Byrne, J. M. E. (1991). Hand preferences in the skilled gathering tasks of mountain gorillas (*Gorilla g. beringei*). *Cortex* 27: 521–546.

Byrne, R. W. and Byrne, J. M. E. (1993). Complex leaf-gathering skills of mountain gorillas (*Gorilla g. beringei*): Variability and standardization. *American Journal of Primatology* 31: 241–261.

Byrne, R. W. and Russon, A. E. (1998). Learning by imitation: A hierarchical approach. *Behavioral and Brain Sciences* 21: 667–721.

Byrne, R. W. and Whiten, A. (1985). Tactical deception of familiar individuals in baboons (*Papio ursinus*). *Animal Behaviour* 33: 669–673.

Byrne, R. W. and Whiten, A. (1988). *Machiavellian Intelligence: Social Expertise and the Evolution of*

Intellect in Monkeys, Apes and Humans. Oxford: Clarendon Press.

Byrne, R. W., Corp, N., and Byrne, J. M. E. (2001). Estimating the complexity of animal behaviour: How mountain gorillas eat thistles. *Behaviour* 138: 525–557.

Castles, D. L., Whiten, A., and Aureli, F. (1999). Social anxiety, relationships and self-directed behaviour among wild female olive baboons. *Animal Behaviour* 58: 1207–1215.

de Waal, F. and van Roosmalen, A. (1979). Reconciliation and consolation among chimpanzees. *Behavioral Ecology and Sociobiology* 5: 55–56.

Dunbar, R. I. M. (1992). Neocortex size as a constraint on group size in primates. *Journal of Human Evolution* 20: 469–493.

Erickson, C. (1991). Percussive foraging in the aye-aye, *Daubentonia madagascariensis. Animal Behaviour* 41: 793–801.

Harcourt, A. H. and de Waal, F. B. (eds.). (1992). *Coalitions and Alliances in Humans and Other Animals.* Oxford: Oxford University Press.

Humphrey, N. K. (1976). The social function of intellect. In *Growing Points in Ethology*, P. P. G. Bateson and R. A. Hinde, eds., pp. 303–317. Cambridge: Cambridge University Press.

Jerison, H. J. (1973). *Evolution of the Brain and Intelligence.* New York: Academic Press.

Jolly, A. (1966). Lemur social behavior and primate intelligence. *Science* 153: 501–506.

Kummer, H. and Goodall, J. (1985). Conditions of innovative behaviour in primates. *Philosophical Transactions of the Royal Society of London B* 308: 203–214.

Macphail, E. M. (1998). *The Evolution of Consciousness.* Oxford: Oxford University Press.

McGrew, W. C. (1974). Tool use by wild chimpanzees feeding on driver ants. *Journal of Human Evolution* 3: 501–508.

McGrew, W. C., Tutin, C. E. G., and Baldwin, P. J. (1979). Chimpanzees, tools, and termites: Cross cultural comparison of Senegal, Tanzania, and Rio Muni. *Man* 14: 185–214.

Passingham, R. E. (1981). Primate specializations in brain and intelligence. *Symposia of the Zoological Society of London* 46: 361–388.

Reader, S. M. and Laland, K. N. (2001). Brain size and intelligence: Comparative studies of innovation, tool use and social learning across the non-human primates. Talk given at the *18th Congress of the International Primatological Society. Adelaide 7–12 January 2001.*

Smuts, B. B. (1983). Special relationships between adult male and female olive baboons: Selective advantages. In *Primate Social Relationships*, R. A. Hinde, ed., pp. 262–266. Oxford: Blackwell.

Stafford, D. K., Milliken, G. W., and Ward, J. P. (1993). Patterns of hand and mouth lateral biases in bamboo leaf shoot feeding and simple food reaching in the gentle lemur (*Hapalemur griseus*). *American Journal of Primatology* 29: 195–207.

Stokes, E. J. and Byrne, R. W. (2001). Cognitive capacities for behavioural flexibility in wild chimpanzees (*Pan troglodytes*): The effect of snare injury on complex manual food processing. *Animal Cognition* 4: 11–28.

Strum, S. C. (1983). Use of females by male olive baboons (*Papio anubis*). *American Journal of Primatology* 5: 93–109.

Sussman, R. W. (ed.). (1979). *Primate Ecology. Problem-Oriented Field Studies.* New York: Wiley.

49 How Smart Does a Hunter Need to Be?

Craig B. Stanford

The belief that hunting animals tend to be clever is very deeply held in our culture. Our images of predators, from lions to eagles, portray them in positions of power, both physical and intellectual. Whether there is truth in this portrayal depends very much on one's definition of intelligence. If we employ two commonly used criteria—that intelligent animals typically use their cognitive abilities for environmental problem-solving and that being smart allows them to adjust quickly and frequently to novel situations—we find that most species of predators are not necessarily very intelligent. Many employ a highly evolved set of weapons and a few, such as lions and wolves, also use cooperation to kill their prey. Without the weapons of teeth and claws, even the most cooperative wolf hunt would be unsuccessful.

Humans are an exception. The idea that hunters are smart comes mainly from human subsistence hunting, in which outwitting one's quarry is more important than possessing the best weapon. Hunting for a living is still practiced by some traditional foraging societies, and in these societies meat is by far the favored food source (Cordain et al. 2000). The notion that hunting places a natural selection pressure on the evolution of intelligence acquired a bad name during the 1970s and 1980s, owing to a now-infamous body of theory often labeled "Man the Hunter." Sherwood Washburn and Chet Lancaster (1968) hypothesized a crucial role for hunting in the evolution of the human intellect because of the natural selection pressure placed upon coordination and communication during the hunt. This placed the evolution of the human mind in the brain of males, who hunted, rather than females, who tended not to hunt. In the early 1970s, other anthropologists pointed out that Man the Hunter neglected the role of the human female in the evolutionary process, citing data from a variety of traditional societies showing that women are responsible for procuring most of the protein

calories for the family group. In spite of the attention paid to male hunting behavior, these critics claimed that gathering by females was nutritionally more important. These criticisms led to a dismissal of theories about hunting and the early human diet (Tanner and Zihlmann 1976).

Through the 1980s, theories of early human foraging behavior focused mainly on scavenging rather than hunting (Blumenschine 1987). In the past decade, the pendulum has swung back to the importance of hunting, in part owing to field data on chimpanzee behavior. The primacy of high-quality foods such as meat is again at the center of hypothesized links between hunting and brain size (see Kaplan et al. 2000).

There is one nonhuman animal that is both our close relative and a predator in social groups, much like traditional foraging people. Chimpanzees are related closely enough to humans and also cognitively similar enough to them to suppose that the evolutionary pressures on their encephalization may have been similar to our own. In order to learn more about the role of cognition in hunting behavior, I turned to the predatory behavior of wild chimpanzees. In the early 1960s, when Jane Goodall began her now famous study of the chimpanzees of Gombe National Park, Tanzania, it was thought that chimpanzees were strictly vegetarian. Today, hunting by chimpanzees at Gombe has been well documented (Teleki 1973; Goodall 1986; Stanford 1998), and hunting patterns have been reported from most other sites in Africa where chimpanzees have been studied. These include Mahale National Park in Tanzania (Uehara et al. 1992), Kibale National Park in Uganda (Mitani and Watts 1999), and Taï National Park in the Ivory Coast (Boesch and Boesch 1989; Boesch 1994).

Chimpanzee society is described as fission-fusion because there is little cohesive group structure apart from mothers and their infants; instead, temporary subgroupings called "parties" come together and separate throughout the day. These parties vary in size in relation to the abundance and distribution of the food supply and the presence of estrous females, who serve as a magnet for males. Thus the size and membership of hunting parties vary greatly, from one to thirty-five. The hunting abilities of the party members as well as the number of hunters present can thus influence when a party hunts as well as whether it will succeed in catching a colobus.

Chimpanzee Predatory Behavior

After four decades of research on chimpanzees at Gombe, we know a great deal about their predatory patterns. A community of chimpanzees may kill and eat more than a hundred small- and medium-sized animals such as monkeys, wild pigs, and small antelopes each year. The most important vertebrate prey species in their diet, however, is the red colobus monkey. At Gombe, red colobus account for more than 80 percent of mammalian prey. Infant and juvenile colobus are caught in greater proportion than their availability (Stanford et al. 1994a); 75 percent of all colobus killed are immature.

Chimpanzees are largely fruit eaters, and meat consumption takes up only about 3 percent of the time they spend eating overall, which is less than in nearly all human societies. Adult and adolescent males do most of the hunting, making about 90 percent of the kills. Females also hunt, although more often they receive a share of meat from the male who either captured the meat or stole it from the captor. Although lone chimpanzees, both male and female, sometimes hunt, most often hunts are social. In other hunting species, cooperation among hunters yields greater success rates. In both Gombe and in the Taï Forest in the Ivory Coast, there is a strong positive relationship between the number of hunters and the odds of a successful hunt (Boesch and Boesch 1989; Stanford et al. 1994b). Although

most successful hunts result in a kill of a single colobus monkey, in some hunts from two to seven colobus may be killed.

In her early years of research, Jane Goodall (1986) noted that the Gombe chimpanzees tend to hunt in binges, during which they would hunt almost daily and kill large numbers of monkeys and other prey. The explanation for such binges has always been unclear. My own work focused on the causes for such spurts in hunting frequency, with unexpected results (Stanford 1998). The explanation for sudden changes in frequency seems to be related to whatever factors promote hunting itself; when such factors are present to a high degree or for an extended period of time, frequent hunting occurs. For example, the most intense hunting binge we have seen occurred in the dry season of 1990. From late June through early September, a period of 68 days, the chimpanzees were observed to kill 71 colobus monkeys in 47 hunts. It is important to note that this is the observed total; the actual total that includes hunts at which no human observer was present may be one-third greater. During this time the chimpanzees may have killed more than 10 percent of the entire colobus population within their hunting range (Stanford 1998).

Hunting and Intelligence

Every researcher has his or her own revealing anecdotes about some clever tactic employed by a hawk or a leopard to catch prey. It appears, however, that chimpanzees respond to myriad hunting scenarios just as they respond to other social situations, with highly flexible and context-dependent tactics for achieving their goal. I have seen male chimpanzees corner male colobus monkeys on narrow tree branches, then grab the limb with their hands and whip it up and down until the colobus were forced to leap off, allowing the hunters to rush in and capture babies from the group. I once saw the male chimpanzee Frodo attack a colobus group huddled in the top of a palm tree. He approached from beneath, then pulled a palm frond down, creating a temporary bridge over which the group tried to escape the tree. Frodo's hand holding the frond remained unseen until the last moment, when he lunged for (and missed) a mother colobus and her baby crossing to safety. Frodo appeared to set a trap for the colobus, one that nearly worked.

Both humans and chimpanzees are omnivores, eating a diet that is high in plant foods. The important decisions about when to eat meat are based on the nutritional costs and benefits of obtaining prey compared with the essential nutrients that the food provides relative to plants. However, social influences such as party size and composition seem to play an important role in mediating hunting behavior as well. A major goal of my research was understanding when and why chimpanzees decide to hunt colobus monkeys rather than forage for fruits, even though the hunt involves the risk of injury from colobus canine teeth and a substantial risk of failure to catch anything.

In his study of Gombe chimpanzee predatory behavior in the 1960s, Geza Teleki (1973) considered hunting to have a strong social basis. Other early researchers had said that hunting by chimpanzees might be a form of social display, in which a male chimp tries to show his prowess to other members of the community (Kortlandt 1972). In the 1970s, Richard Wrangham conducted the first systematic study of chimpanzee behavioral ecology at Gombe and concluded that predation by chimps was nutritionally based, but that some aspects of hunting behavior were not well explained by nutritional needs alone.

More recently, Toshisada Nishida and his colleagues in the Mahale Mountains chimpanzee research project reported that the alpha male there, Ntilogi, used colobus carcasses for political gain, withholding meat from rivals and doling it out to allies (Nishida et al. 1992). William McGrew (1992) has shown that those female Gombe chimps who receive generous shares of meat after a kill have more surviving offspring, indicating a reproductive benefit tied to meat eating.

My own preconception was that hunting must be nutritionally based. After all, meat from monkeys and other prey would be a package of protein, fat, and calories hard to equal from any plant food. I therefore examined the relationship between the odds of success and the amount of meat available with different numbers of hunters in relation to each hunter's expected payoff in meat obtained. When are the time, energy, and risk (in other words, the costs) involved in hunting worth the potential benefits, and therefore when should a chimp decide to join or not join a hunting party? And how do the costs of hunting compare with the costs and benefits of foraging for plant foods?

The results were surprising. I expected that as the number of hunters increased, the amount of meat available for each hunter would also increase. This would have explained the social nature of hunting by Gombe chimpanzees. If the amount of meat available per hunter declined with increasing hunting party size (because each hunter got smaller portions as party size increased), then it would be a better investment of time and energy to hunt alone rather than join a party. The success rate of lone hunters was only about 30 percent, while that of parties with ten or more hunters was nearly 100 percent. However, there was no relationship between the number of hunters and the amount of meat available per capita. This is because even though the likelihood of success increases with more hunters in the party, the most frequently caught prey animal is a 1-kg baby colobus monkey. Whether it is shared among four hunters or fourteen, such a small package of meat does not provide anyone with much food or incentive to hunt. The decision to join a hunting party therefore is based on some calculation of expected returns by potential hunters.

Whether intelligence is an important factor in hunting tactics by chimpanzees is a key issue because of the putative importance of hunting to the evolution of human intelligence. If intelligent hunters succeed more often than less intelligent hunters, and if the capture of meat has some survival value to the animals, then natural selection should have favored those chimpanzees who employed clever hunting tactics. If females prefer to mate with the best hunters, perhaps because the females benefit nutritionally from gifts of meat, then hunting performance would be subject to sexual selection as well.

Since there is little evidence that chimpanzees set off each morning with the intention of finding meat, optimizing their foraging routes to take advantage of the likely locations of colobus monkeys is probably not part of a chimpanzee hunting strategy. Many animal species, from bumblebees to hummingbirds, forage efficiently without any semblance of higher intelligence, since natural selection may program the ability to optimize travel routes to locate a maximum number of quality food sources per unit of area. A leopard needs extraordinarily evolved sensory capabilities and the weaponry to stalk and kill prey, but perhaps not the higher cognitive function that lies at the basis of the evolution of the human brain.

In many mammalian predators, however, hunting tactics and the expected behavior of the prey species must be learned. This is true for solitary hunters like leopards as well as for social hunters like lions and wolves. It reaches a peak in humans in traditional hunting and gathering societies, who spend years learning to be good hunters. Some anthropologists have even argued that the long human maturation period is evolutionarily related to the need to learn the skills for obtaining the highest quality foods—namely meat (Kaplan et al. 2000). Skills that might be important to making a kill, such as the flexibility to respond strategically to rapidly changing circumstances, should also be selected for and enhanced.

Conclusions

There are many reasons why chimpanzees hunt, and they vary according to season, group com-

position, and individual personalities (Stanford 1998). Future research in this area should be able to establish further the why and wherefore of hunting and sharing. Although most researchers (e.g., Boesch 1994) have drawn comparisons between chimpanzee hunting behavior and that of social carnivores such as wolves and lions, much more apt comparisons are to be found with human hunter-gatherers. In both humans and chimpanzees, meat is only a part of the diet, and decisions on whether to hunt must be made on an hourly basis. People forage for meat and also gather plant foods. Chimpanzees forage mainly for ripe fruit and hunt opportunistically when they happen to encounter prey. Their meat-sharing patterns are also more systematic and more nepotistic than those that researchers see in wild baboons, capuchin monkeys, and any other nonhuman primate. Whether intelligence is crucial to hunting and sharing in humans seems indisputable; its role among chimpanzees is likely to be only marginally less so.

References

Blumenschine, R. J. (1987). Characteristics of an early hominid scavenging niche. *Current Anthropology* 28: 383–407.

Boesch, C. (1994). Cooperative hunting in wild chimpanzees. *Animal Behaviour* 48: 653–667.

Boesch, C. and Boesch, H. (1989). Hunting behavior of wild chimpanzees in the Taï National Park. *American Journal of Physical Anthropology* 78: 547–573.

Cordain, L., Miller, J. B., Eaton, S. B., Mann, N., Holt, S. H. A., and Speth, J. D. (2000). Plant to animal subsistence rations and macronutrient energy estimations in world wide hunter-gatherer diets. *American Journal of Clinical Nutrition* 71: 682–692.

Goodall, J. (1986). *The Chimpanzees of Gombe: Patterns of Behavior*. Cambridge, Mass.: Harvard University Press.

Kaplan, H., Hill, K., Lancaster, J., and Hurtado, A. M. (2000). A theory of human life history evolution: Diet, intelligence, and longevity. *Evolutionary Anthropology* 9: 156–185.

Kortlandt, A. (1972). *New Perspectives on Ape and Human Evolution.* Amsterdam: Stichting Voor Psychobiologie.

McGrew, W. C. (1992). *Chimpanzee Material Culture.* Cambridge: Cambridge University Press.

Mitani, J. C. and Watts, D. (1999). Demographic influences on the hunting behavior of chimpanzees. *American Journal of Physical Anthropology* 109: 439–454.

Nishida, T., Hasegawa, T., Hayaki, H., Takahata, Y., and Uehara, S. (1992). Meat-sharing as a coalition strategy by an alpha male chimpanzee. In *Topics in Primatology.* Vol. I, *Human Origins,* T. Nishida, W. C. McGrew, P. Marler, and F. B. M. de Waal, M. Pickford, eds., pp. 159–174. Tokyo: University of Tokyo Press.

Stanford, C. B. (1998). *Chimpanzee and Red Colobus: The Ecology of Predator and Prey.* Cambridge, Mass.: Harvard University Press.

Stanford, C. B., Wallis, J., Matama, H., and Goodall, J. (1994a). Patterns of predation by chimpanzees on red colobus monkeys in Gombe National Park, Tanzania, 1982–1991. *American Journal of Physical Anthropology* 94: 213–228.

Stanford, C. B., Wallis, J., Mpongo, E., and Goodall, J. (1994b). Hunting decisions in wild chimpanzees. *Behaviour* 131: 1–20.

Tanner, N. M. and Zihlmann, A. L. (1976). Women in evolution part 1: Innovation and selection in human origins. *Signs: Journal of Women, Culture, and Society* 1: 585–608.

Teleki, G. (1973). *The Predatory Behavior of Wild Chimpanzees.* Lewisburg, Pa.: Bucknell University Press.

Uehara, S., Nishida, T., Hamai, M., Hasegawa, T., Hayaki, H., Huffman, M., Kawanaka, K., Kobayoshi, S., Mitani, J., Takahata, Y., Takasaki, H., and Tsukahara, T. (1992). Characteristics of predation by the chimpanzees in the Mahale Mountains National Park, Tanzania. In *Topics in Primatology.* Vol. 1, *Human Origins,* T. Nishida, W. C. McGrew, P. Marler, M. Pickford, and F. B. M. de Waal, eds., pp. 143–158. Tokyo: University of Tokyo Press.

Washburn, S. L. and Lancaster, J. B. (1968). The evolution of hunting. In *Man the Hunter,* R. Lee and I. DeVore, eds., pp. 293–303. Chicago: Aldine.

Wrangham, R. W. (1975). Behavioral biology of chimpanzees in Gombe National Park, Tanzania. Ph.D. Thesis, Cambridge University.

50 Insight from Capuchin Monkey Studies: Ingredients of, Recipes for, and Flaws in Capuchins' Success

Elisabetta Visalberghi

Many years ago, while passing in front of a group of capuchin monkeys (*Cebus apella*) at the Rome zoo, I was lucky to see an adult male pounding an unshelled peanut with a boiled potato. Why was he behaving like this? Peanuts can be opened easily and, moreover, boiled potatoes are soft. Could it have been that he was so fond of pounding with tools that he kept doing so even in circumstances in which the function would be sheer fun, rather than trying to achieve an impossible goal?

The fact that capuchins were doing something smart in a silly way, or something silly in a smart way, struck my interest. I was fascinated and overwhelmed by them, as many other scientists have been before me (Erasmus Darwin and Konrad Lorenz, to cite only two very famous ones). Capuchins are the right species for an enthusiastic, rational, and skeptical person like me. The delicate balance between chance and necessity, doing and understanding, ingredients and outcome has indeed fueled all of my research since then. What I have done is demonstrate on the one hand, how successful capuchins are in solving problems and on the other, how relatively little they understand of what they do.

My research has focused on sorting out the ingredients of, the recipes for, and the flaws in their success. Here I focus on capuchins' success in using tools and solving a task cooperatively and discuss how their behavioral traits (e.g., interest in objects, combinatorial activities, associative learning, and chance in the case of tool use; manipulative tendencies, high interindividual tolerance, associative learning, and chance in the case of cooperation) foster their success and how their cognitive capacities limit what they are able to understand about the conditions necessary for success.

Tool Use

The use of tools may enable or increase the exploitation of resources, such as foods that are difficult to obtain through direct action with hands or teeth. Our contemporary fascination with tool use in nonhuman species reflects a profound appreciation of the importance of tools in our own species. There is no doubt that tools have enabled humans to diversify their way of life and to exploit resources not available to other primates. Apart from the issue of intelligence, tool use is of interest to biologists because it is a means by which an individual can extend what it can do or where it can live. In the wild, the use of tools is widespread in chimpanzees, but is observed less often in other apes. However, great apes and several species of monkeys use tools readily in captivity (Tomasello and Call 1997). Among monkeys, capuchins stand out as masters of tool use. Although in natural settings they rarely use tools, in captivity, capuchins, like the apes, readily and spontaneously use tools in a large variety of circumstances (Visalberghi 1990; Anderson 1996).

Biologists and psychologists have speculated widely as to how nonhuman animals arrive at the efficient and sometimes elegant and skillful use of objects as tools. One common notion is that one individual learns to use a tool by observing another, as often happens in humans. In nonhuman primates, however, recent research has shown that imitation as we typically think of it (watching and then reproducing novel actions) plays a limited role (great apes) or none (monkeys) in learning a new tool-using behavior (Visalberghi and Fragaszy 1990).

The normal way nonhuman primates acquire a new tool-using skill involves both social influences and individual discovery (Fragaszy and Visalberghi 1989). For example, we presented capuchin monkeys with a food (applesauce) source inside a container. The food could only be obtained through openings too small for the monkeys' hands, but large enough to allow the insertion of sticks or straws, both of which were available to them. The monkeys showed immediate interest in the food and tried to get at it in every possible way. For the capuchin monkeys, one possible way was to poke into the container with an object at hand. This behavior is often seen during normal exploration, even when food is not involved. If the object happens to be the right size, shape, and strength, and if it is inserted deeply enough, then the monkey will succeed in getting the food. Once a monkey has accidentally succeeded this way (and here is where "chance" comes in), it is likely to repeat the actions and associate them with success. Other monkeys in the vicinity may observe this behavior closely and even obtain some food for themselves.

Although social circumstances can increase an individual's interest in the tool, the container, and its environs, there is no evidence to date that capuchins can learn the specific details of how to use a tool from watching others. It is clear, however, that social influences can restrict an individual's access to a site of particular interest to other group members and thus decrease the probability that the individual will discover how to solve the problem. Moreover, those individuals that reliably succeed in obtaining food from others may not become tool users themselves. Overall, the ingredients that make an individual a likely tool user range from exploratory and manipulative tendencies, to persistence in trying regardless of failure, to the immediate social context and individual social relationships. The notion that individual "intelligence" is the major factor accounting for success is not supported experimentally.

Why do captive capuchins (and other primate species) use tools more in captivity than in natural settings? We can dismiss the possibility that captive animals are somehow "smarter" than their wild counterparts. Rather, it seems that the combination of abundant free time for play and exploration, little space, and few and appropriate objects and surfaces, all make discovery of tool use more likely than in nature. When a researcher places something in the cage that provides an interesting new focus for activity, the monkey's or ape's natural propensity to explore the object is sufficient for the occasional accidental combination of actions and objects that produces desirable consequences (Fragaszy and Adams-Curtis 1991). The captive conditions seem to channel the individual's propensities in a way that makes the discovery of tool use more likely.

Although insight or comprehension is not required for using a tool, if it is present, it increases the effectiveness of tool use. The more an individual understands about the functioning of a tool, the more effectively he or she can use it or change it to improve its effectiveness. Furthermore, understanding allows an individual to deal with variations of a problem without having to resort to trial and error. These features are indeed part of what is special about the way in which humans manufacture and use their tools.

In the past, we demonstrated that capuchins learn to use a stick to push an object out of a horizontal transparent tube. Then we wondered whether this meant that they understood the properties of the tool and the relation between their action and the outcome; that is, that the stick moves the reward and pushes it out of the tube. To test this, we varied a property of the tube; we placed a trap in the middle of it so that the tube was now shaped like a T with a very short stem (figure 50.1). To push something out of this tube, the monkeys had to avoid moving the object in the direction of the trap. When faced with the task, capuchins that were otherwise proficient in pushing things out of tubes

Figure 50.1
The trap-tube task. In order to solve the task, the subject must insert the stick into the side of the tube from which it can push the reward out of the tube and not into the trap. Depending on the side in which the subject inserts the tool, it can either push the reward into the trap or push the reward out of the tube and obtain it. (A) Insertion in the wrong side of the tube. A reward lost in a previous trial is already inside the trap. (B) A capuchin monkey has inserted the stick into the correct side of the tube. Note that the reward is on the left side of the trap.

suddenly performed at chance level (Visalberghi and Limongelli 1996). One monkey eventually improved, but further testing showed that she was simply introducing the stick into the side of the tube farthest from the reward, without taking the position of the trap into account.

When chimpanzees and 3–4-year-old children were presented with the same task, some of the chimpanzees and almost all the children behaved in ways that suggested an understanding of the requirements of the task (Visalberghi and Limongelli 1996). When moving the reward with the stick, the successful individuals took into account the presence of the trap (or learned to do it) and the position of the reward in relation to the trap. These findings show that different primate species may have different levels of understanding of the same problem. We can say that their tool recipe has different ingredients and flaws. Moreover, if we limit ourselves to a description of the final outcome (i.e., whether they succeed in doing the task), we are going to miss those processes that lead to success and those that in other circumstances could prevent it.

The fact that capuchins are very good at using tools, though limited in their understanding of what they do, provided us with the opportunity to appreciate the ingredients needed for tool use to occur and those that, if absent, would produce flaws. And since the latter are visible only if ad hoc experiments are carried out, we also learned that laboratory experiments are fundamental to an assessment of cognitive abilities.

Cooperation

Cooperation is a topic that can benefit greatly from capuchin studies. Cooperation increases individual success and has fitness implications; most ethologists have studied cooperation from an evolutionary perspective (for a review see Dugatkin 1997) and very few have been interested in the proximal mechanisms fostering cooperation, which can be studied well in the laboratory.

Cooperation may arise because individuals are "programmed" to cooperate or as a cognitive adaptation to overcome situations in which an individual alone would not be as successful as two or more individuals acting together. In the latter case, and especially in primates, where cooperation has been considered more cognitively advanced than in other animals, you would expect individuals to take into consideration how, when, and where their common actions can be successful. Yet when primates engage in most of the so-called cooperative behaviors (e.g., forming coalitions and alliances, grooming, defending against predators, group hunting), it is very difficult to assess exactly what each individual must do to achieve the goal, what goal is being pursued by each individual, what each individual knows, and which variables it takes into account when acting. In these naturally occurring events, the scientist has little information on and no control over the many variables involved and therefore cannot assess the role played by each individual in promoting or preventing the achievement of the (supposed) goal.

This unsatisfactory condition has prompted experimental investigations on apes (Chalmeau 1994; for a review see Visalberghi 1997) and capuchin monkeys in our laboratory in Rome (Chalmeau et al. 1997; Visalberghi et al. 2000) and elsewhere (Mendres and de Waal 2000). The task that could be solved only through cooperation consisted of an apparatus that required both partners to pull a handle simultaneously in order for both to be rewarded. The handles were less than a meter apart so that the capuchins could closely monitor what the companion was doing. In particular, we expected them to learn to pay attention to where the other monkey was (far from the handle or close to it) and to what the other monkey was doing (about to pull or do something else that would affect the handle) (figure 50.2).

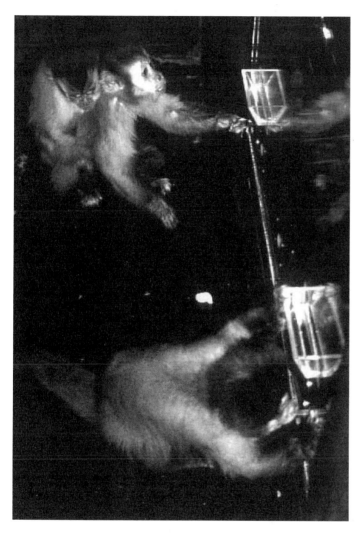

Figure 50.2
To solve the cooperation task, both individuals should pull their handles. The monkey in the front pulls by using its mouth instead of its hand, the monkey in the back has an infant.

Our capuchins did very well; they succeeded in the cooperation task by learning to take the spatial position of the partner into account (for a similar finding see Mendres and de Waal 2000). However, they did not consider the partner's behavior, that is, whether or not it was pulling. In short, the capuchins cooperated while lacking comprehension of some of the conditions for success; for example, they kept pulling (hundreds and hundreds of times!) even when their companion was meters away from the apparatus or when the companion was not pulling at all. The ingredients for their success were high interindividual tolerance (when chimpanzees faced the same task, some individuals were so afraid of their partner that they rarely, or never, pulled the handle) and the fact that they learned to pull more when the companion was present in the area close to the handle.

What is it then that capuchins are lacking? And in what circumstances would we expect them to be unable to cooperate? Cognitively grounded cooperation would call for a capuchin to pull (or learn to pull) when the partner is pulling or when the partner is at least close to the handle and about to pull the handle. Since they do not seem to understand that their own pulling is effective only when the companion is also pulling, if we assume that each subject can pull the handle only ten times, we predict little success for the capuchins. Conversely, a human being, or a child above a certain age, will make an effort to synchronize pulling. This is what full-fledged cooperation is all about: having a goal and a companion and performing the action(s) for achieving the goal that produce greater chances of success.

Another example might explain this point further; when faced with the problem of having to move a stone too heavy for one monkey in order to be able to get at a reward underneath it, macaques sometimes solve the task (Petit et al. 1992). However, most of the time each individual tries to move the stone by pushing it in a different or opposite direction from that in which the companion is pushing. In short, they show no cooperation, but since they perform so many pushing actions, a few times the macaques will, by pure chance, show "coproduction" of pushing actions. Yet, since the macaques do not understand "why" their actions were sometimes effective and sometimes not, they are unable to repeat the desired outcome.

When cooperation relies upon cognitive capacities that allow an appreciation of what is going on, it becomes a fundamental step in evolution and strongly increases behavioral flexibility. In contrast, when animals have a poor understanding of what they do, cooperation is more a matter of chance (capuchins) or of genes (ants, honeybees, etc.).

The idea of observing what capuchins do, how they learned it, and what they understand of what they do is still in my mind. The research now in progress is aimed at trying to sort out how capuchins learn to determine what they eat and the extent to which their individual learning is socially biased.

References

Anderson, J. R. (1996). Chimpanzees and capuchin monkeys: Comparative cognition. In *Reaching into Thought. The Minds of the Great Apes*, A. Russon, K. Bard, and S. Parker, eds., pp. 23–56. Cambridge: Cambridge University Press.

Chalmeau, R. (1994). Do chimpanzees cooperate in a learning task? *Primates* 35: 385–392.

Chalmeau, R., Visalberghi, E., and Gallo, A. (1997). Capuchin monkeys, *Cebus apella*, fail to understand a cooperative task. *Animal Behaviour* 54: 1215–1225.

Dugatkin, L. A. (1997). *Cooperation Among Animals. An Evolutionary Perspective*. New York: Oxford University Press.

Fragaszy, D. M. and Adams-Curtis, L. (1991). Generative aspects of manipulation in tufted capuchin monkeys (*Cebus apella*). *Journal of Comparative Psychology* 105: 387–397.

Fragaszy, D. and Visalberghi, E. (1989). Social influences on the acquisition and use of tools in tufted capuchin monkeys (*Cebus apella*). *Journal of Comparative Psychology* 103: 159–170.

Mendres, K. A. and de Waal, F. B. M. (2000). Capuchins do cooperate: The advantage of an intuitive task. *Animal Behaviour* 60: 532–529.

Petit, O., Desportes, C., and Thierry, B. (1992). Differential probability of "coproduction" in two species of macaque (*Macaca tonkeana, M. mulatta*). *Ethology* 90: 107–120.

Tomasello, M. and Call, J. (1997). *Primate Cognition*. New York: Oxford University Press.

Visalberghi, E. (1990). Tool use in *Cebus. Folia Primatologica* 54: 146–154.

Visalberghi, E. (1997). Success and understanding in cognitive tasks: A comparison between *Cebus apella* and *Pan troglodytes. International Journal of Primatology* 18: 811–830.

Visalberghi, E. and Fragaszy, D. (1990). Do monkeys ape? In *"Language" and Intelligence in Monkeys and Apes*, S. Parker and K. Gibson, eds., pp. 247–273. Cambridge: Cambridge University Press.

Visalberghi, E. and Limongelli, L. (1996). Acting and understanding: Tool use revisited through the minds of capuchin monkeys. In *Reaching into Thought. The Minds of the Great Apes*, A. Russon, K. Bard, and S. Parker, eds., pp. 57–79. Cambridge: Cambridge University Press.

Visalberghi, E., Pellegrini Quarantotti, B., and Tranchida, F. (2000). Solving a cooperation task without taking into account the partner's behavior. The case of capuchin monkeys (*Cebus apella*). *Journal of Comparative Psychology* 114: 297–301.

51 A Cognitive Approach to the Study of Animal Cooperation

Lee Alan Dugatkin and Michael S. Alfieri

The study of cooperative behavior in animals has played a central role in the field of behavioral ecology (Dugatkin 1997). Moving away from early views that cooperation either permeated the animal world (Kropotkin 1908) or was completely lacking in nonhumans (Huxley 1888), modern behavioral ecologists are attempting to build a sound theoretical framework for understanding the evolution of cooperation (Hamilton 1964; E. O. Wilson 1975; Axelrod and Hamilton 1981; Brown 1983; Dugatkin 1997; Sober and Wilson 1998). Progress in building such models has been good, and this is reflected in the hundreds of controlled studies of cooperation in animals that can be found in the primary literature (Dugatkin 1997).

In the past, one of us (L.A.D.) has argued that there are four paths to cooperation (Dugatkin et al. 1992; Mesterton-Gibbons and Dugatkin 1992; Dugatkin 1997). These paths can go by slightly different names, but are generally referred to as reciprocity, by-product mutualism, group selection, and kin selection (Mesterton-Gibbons and Dugatkin 1992; Dugatkin et al. 1992). As with any attempt to build an umbrella theory, objections to slicing the cooperation pie into four pieces certainly exist.

Here our goal is not to settle these arguments, but rather to present the case that splitting cooperation into the four paths mentioned here is quite useful in terms of understanding the different cognitive prerequisites for various forms of cooperation to take place in animals. To accomplish this goal, we outline each of these four paths to cooperation; at the same time we describe what each path requires and how they differ in terms of recognition of individuals and memory of specific events (the two cognitive variables we focus on).

Before proceeding any further, it might be useful to briefly touch on the relationship between recognition of individuals and memory of specific events. One important facet of this relationship is that recognition of individuals can exist in the absence of memory of specific events, and vice versa. For example, I may recognize you by some mosaic of facial features, but not necessarily remember anything about what you have done. Conversely, I may be able to remember that someone did X to me, without recalling who it was who actually undertook X.

The notion of recognition of individuals itself is not without problems. It could be argued that it is simply one case of what might be thought of as "category" recognition (Barnard and Burk 1979; Dugatkin and Sih 1995, 1998). A great deal of empirical work in animals has shown that animals can distinguish things based on size (e.g., large versus small), color (e.g., red versus blue), etc. If animals are able to apply enough of these categories to other individuals, eventually individual recognition will come about (Barnard and Burk 1979). As we will soon see, some paths to cooperation require a simple form of category recognition, while others require recognition of individuals per se.

Reciprocity

Trivers (1971) suggested that one path to cooperative behavior among humans and nonhumans alike is reciprocity. Under Trivers's formulation, reciprocity evolves when the (potentially) minor cost that one individual pays to help another individual in need is repaid when the recipient returns the favor (reciprocated) some time in the future. Such a system, however, is subject to cheating (= not cooperating = defecting) because the greatest "payoff" attainable in such scenarios goes to the recipient of a cooperative action, who then can fail to reciprocate in turn.

To formalize the evolution of reciprocity, behavioral ecologists employ the prisoner's dilemma game (Axelrod 1984; Dugatkin and Reeve 1998). Using mathematical formulas and computer simulations, Axelrod and Hamilton (1981) examined the success of an array of strategies (behavioral rules) in the iterated prisoner's dilemma game. They found that if the probability of meeting a given partner in the future was above some critical threshold, then a strategy called "tit for tat" (TFT; first created by Anatol Rapport) was a robust solution to the iterated prisoner's dilemma. TFT instructs a player to cooperate on the initial encounter with a partner and to subsequently copy the partner's last move. As such, TFT initially cooperates and then defects on defectors and cooperates with cooperators.

From a cognitive perspective what does TFT require in terms of recognition of individuals and memory? The answer, not surprisingly, depends on the ecology and population biology of the animal group being studied. In relatively rare circumstances, individuals involved in some sort of reciprocal interaction will have the same partner for all such actions. This might be due to individuals being somehow physically linked. Another scenario that might produce life-long partners is a scarcity of potential partners. Here partners are not physically bound together, but the lack of potential new partners in effect creates long-term interaction. When interactants are trapped with a given partner, all that is required to play TFT is the memory of a specific event (cooperation or cheating). Recognition of an individual, or for that matter recognition of any category, is not required.

A more realistic scenario for TFT involves individuals who are free to change partners and do so readily (Dugatkin 1997). In this case, TFT requires individuals to remember specific events (cooperation or cheating) and to recognize individuals. To play TFT when partners are swapped, you must do what your partner did on the last move, and that means an individual must recognize who they are paired with at any given moment.

While it may come as no surprise that primates possess the cognitive prerequisites to play TFT in such scenarios (see de Waal 1996), it is important to note that "simpler" animals such as guppies and even aquatic polychaete worms seem capable of playing TFT when trapped in a prisoner's dilemma (Sella 1985, 1988; see Dugatkin 1997 for a review). For example, *Ophryotrocha diadema* is a hermaphroditic worm that appears to engage in the "egg swapping" that is seen in many deep-sea fish (Fischer 1988). Individual worms pair up and take turns contributing eggs and sperm to matings. It is not known how these worms keep track of one another's behavior, yet "cheating" (failing to provide eggs at the appropriate turn) is rare comprising approximately 8 percent of interactions. (Sella and Lorenzi 2000). In fact, most of the time it is the availability of a new potential mate with riper eggs, and not cheating by one's partner that causes the dissolution of polychaete pairings. The fact that worms are capable of such behavior suggests that reciprocity and partner fidelity may be rooted deep in evolutionary time.

Kin Selection

Perhaps the most well-known path to cooperation is kin selection. Both Darwin (1859) and Haldane (1932) recognized that individuals were more prone to cooperate with relatives, but this idea was not formalized until the late W. D. Hamilton came up with his inclusive fitness or kin-selection models (Hamilton 1964). The heart of inclusive fitness models is that they modify prior formulations by considering the effect of a gene, not only on the individual that bears it, but on others as well; most important, those sharing genes that are identical by descent. "Hamilton's rule" (i.e., $rb-c > 0$) states that cooperation should be more common among kin because by helping kin, individuals are helping copies of their own genes, which just happen to reside in their blood relatives.

What recognition and memory requirements must be met for kin-selected cooperation to evolve? Again, the answer depends on the demography and biology of the population under study. At one extreme, individuals may find themselves always interacting with kin and only kin during some stage of their lives. During that time, a rule such as "treat everyone as if they were kin" would be favored, and this would in turn remove any memory or individual recognition requirements from kin-selected cooperation. While this sort of rule of thumb generally works, it is susceptible to parasitization. For example, cowbirds parasitize other species of birds by laying their eggs in their nests (Ortega 1998). The victims of such duplicity almost always treat cowbird young as their own, doling out all sorts of parental attention. The parents in such nests have adopted the "if it is in my nest treat it as offspring" rule, which generally works but is susceptible to cowbirdlike cheating.

In most cases of kin-related cooperation, individuals encounter both kin and non-kin on a normal basis. In such scenarios, individuals only need to be able to distinguish kin from non-kin (Reeve 1989). While there is some debate about exactly how animals actually recognize kin (Grafen 1990; Brown and Eklund 1994), such discriminatory abilities have been demonstrated many times (Fletcher and Michener 1987; Hepper 1991; Crozier and Pamilo 1996; Holmes and Matio 1998). For kin selection when individuals encounter both kin and non-kin alike, categorical recognition is needed, but neither recognition of individuals nor memory is required. One way to see this is to recognize that while kinship is an important variable in animal social interactions, animals typically do not use conditional strategic rules like "Do x when someone does y; otherwise do z" (rules that require memory and recognition of individuals) when interacting with relatives. The more common rule of thumb in kin-selected cooperation is if that you are detected as a relative, you are treated one way; if not, you are treated another way.

Group Selection

Group selection has a long and controversial history within evolutionary biology (D. S. Wilson 1983; D. S. Wilson and Sober 1994; Sober and Wilson 1998). Here we focus on "modern" group selection or what is often referred to as trait-group selection. In trait-group models, natural selection operates at two levels—within groups and between groups. Within groups, cooperators are always at a disadvantage because they pay a cost that cheaters do not. However, it is possible that the productivity of groups may be positively related to the number of cooperators (the between-group component of trait-group selection). In these models, cooperation evolves if the within-group cost incurred is offset by some between-group benefit, so that cooperative groups are more productive than selfish groups (D. S. Wilson 1980; Sober and Wilson 1998). For such group-level benefits to be manifest, groups must differ in the frequency of cooperators within them, and must be able to "export" the productivity associated with cooperation.

At the most basic level, group-selected cooperation requires neither recognition nor memory. For example, it can be shown that some forms of cooperation can evolve by group selection if groups are formed randomly and individuals have no memory or recognition abilities (D. S. Wilson 1980). That being said, group-selected cooperation would be favored if cooperators were able to identify and interact with other cooperators (Eshel and Cavalli-Sforza 1982; Peck 1993; D. S. Wilson and Dugatkin 1997; Roberts and Sherratt 1998). This would require cooperators to categorize others as "cooperator" or "cheater," but they would not necessarily possess the ability to recognize individual (such abilities, however, would facilitate the evolution of cooperation to an even greater extent). In-group biasing, in which individuals show a strong tendency to favor those in their group, is one possible outgrowth of group-selected cooperation when categorization ("same group," "different group") is possible. To see how powerful in-group biasing can be, consider two examples—one from chimps and one from humans.

Warfare is often defined as large-scale, open hostility between groups, in which both sides in the conflict use lethal force against the other (Boehm 1992). According to this definition, chimpanzees do not engage in war, but between-group interactions in chimpanzees do resemble the raiding behavior so common among many tribes of humans (Boehm 1992, 1999; Wrang-ham 1999). During raids, all-male chimpanzee patrol groups often travel into areas that abut their territorial boundaries (Bygott 1979; Nish-ida 1979; Goodall 1986) and move about in a wary fashion (Goodall 1986). These raids often involve the killing of a small number of members of the raided group and the capture of females. Occasionally raiding parties from two groups will meet one another. Rather than all-out aggression, both groups often engage in hostile vocalizations and then withdraw (Goodall 1986). However, when two raiding parties meet, violence sometimes erupts, resulting in the extinction of one group. For example, Nishida et al. (1985) provide evidence that raiding behavior in the Mahale Mountains of Tanzania resulted in a larger group extinguishing a smaller group of chimps.

Although the costs and benefits of raiding behavior are not known, it appears to be a good candidate for cooperation via group selection. Between-group selection favors such raids (because they most likely benefit all group members, not just the raiding party participants), yet if raiding is dangerous, selection within groups should favor cheating—i.e., letting others do the raiding, but continuing to reap the benefits.

In-group biasing can be even more powerful (and scary) in humans because it rears its head in even the most trivial decision-making processes. Consider Tajfel's (1970) study of in-group biasing in English teenagers in the late 1960s and early 1970s. In this study, 64 boys who attended the same school were asked to estimate the number of dots flashed on a wall. The subjects were then told that they fell into one of two groups: those that overestimated the number of dots or those that underestimated this number.

Once they learned whether they were dot over- or underestimators, each subject was placed in a room by himself with a series of forms. The forms asked the subjects how they would divide up monetary rewards and penalties between two other boys in the study. The pairs from which a subject could choose were made up of either two individuals from his own group, two from the other group, or one from each group.

The results were striking. When asked to divide up rewards and punishments between two individuals from the same group—either the group the subject belonged to or the one that he did not—rewards and punishments were split fairly. If, however, the choice was between someone from the subject's own group (for example, the dot overestimators) and an individual from the other group (the dot underestimators), the subjects consistently favored members of their own group, despite having no information on the actual identity of who was in either group. Simply knowing that others overestimated dots as they did, even if they never met the overestimators, was sufficient to cause an unequal distribution of monetary rewards and punishments.

It should be noted that a preference for those in one's own group is not restricted to primates. A preference for familiar individuals independent of relatedness has been demonstrated in several species of centrachid fish (Brown and Colgan 1986; Dugatkin and Wilson 1992) and guppies (Magurran et al. 1994; Griffiths and Magurran 1999).

By-product Mutualism

Cooperation via by-product mutualism occurs when animals live in "harsh" environments in which there is an immediate cost or penalty for not acting cooperatively (West-Eberhard 1975; Brown 1983; Mesterton-Gibbons and Dugatkin 1992; Connor 1995). Put simply, in harsh environments, the immediate net benefit of cooperating outweighs that of cheating. When this is not the case, in so-called "mild" environments, cooperation by by-product mutualism is not favored by natural selection.

While relatedness plays a role in the hunting behavior of lions, in many ways cooperative hunting in lions exemplifies by-product mutualism. Lions hunt in groups when they are stalking large predators that could not be taken by a single hunter, but work alone when hunting smaller prey (Scheel and Packer 1991). In the case of lions, large prey items constitute a harsh environment, while smaller prey items fall under the rubric of mild environments.

By-product mutualism adds a new twist to the cognitive prerequisites needed for different forms of cooperation to evolve. In by-product mutualism, neither memory nor recognition of individuals is necessary, and while recognition of categories is needed, it is not the same form of recognition of categories as in other types of cooperation. In this case, individuals need to categorize the environment they are in, rather than something specific about another individual. For example, in the case of lions, individual lions need not categorize other lions, but rather assign the environment to either the "mild" or "harsh" category.

Closing Thoughts

The evolution of cooperation continues to be an area of active research interest in behavioral ecology, yet few attempts have been made to explicitly link work on cooperation to animal cognition. Here we have suggested that one way to begin such work is by examining memory and recognition requirements and abilities with respect to four different types of cooperation. Depending on the specific type of cooperation under study, and the specific demographics and structure of the population under investigation, very different cognitive abilities are required for cooperation to be feasible.

To date, very few studies have examined recognition and memory in the context of cooperation (Dugatkin 1997). This may in part be because much of the modern work on cooperation is intimately linked to game theory models of social behavior (Maynard Smith 1982). These models focus on evolutionarily stable strategies and payoff matrices (Hammerstein 1998) and not on the cognitive prerequisites of cooperation. That is, game theory models predict what sort of behavioral strategy can evolve in a particular social environment without much regard for the cognitive requirements. This is not to say that disclaimers such as "only certain types of animals are likely to be cognitively sophisticated enough to . . ." are lacking in game theory papers on cooperation, but rather that detailed studies of recognition, memory, and cooperation are few and far between (see Dugatkin 1997 for a review).

We hope that this essay will spur the studies needed to truly understand the no-doubt complicated relationship between cooperation and cognition. For many reasons this will be a daunting task, even using the framework we have developed. Many instances of cooperation do not fall neatly into one of the four paths to cooperation, but instead combine aspects of more than one path (Dugatkin 1997). For example, work on blood sharing in vampire bats is often cited as a classic case of cooperation by reciprocity in animals (Wilkinson 1984). Yet it turns out that the vampire bats swapping blood meals are also related (Wilkinson 1987). This makes the vampire case no less interesting, but does raise the question of what sort of cognitive abilities we might expect when reciprocity is tied to relatedness.

Even if we could map out all possible combinations of the four paths to cooperation and make predictions with respect to recognition and memory, there would be hurdles to overcome. Work in comparative psychology as well as evolutionary psychology demonstrates that experiments on memory and recognition in animals are fraught with design problems and often require a huge number of controls to shed any light on the details of animal cognition (Balda et al. 1998; Dukas 1998; Shettleworth 1998). Yet rather than view these conceptual and experimental problems as stumbling blocks that make the study of cooperation and cognition impossible, we prefer to think of them as fascinating problems in and of themselves, and part of the tribulations associated with studying anything of real importance in science.

References

Axelrod, R. (1984). *The Evolution of Cooperation.* New York: Basic Books.

Axelrod, R. and Hamilton, W. D. (1981). The evolution of cooperation. *Science* 211: 1390–1396.

Balda, R., Pepperberg, I., and Kamil, A. (eds.) (1998). *Animal Cognition in Nature.* San Diego: Academic Press.

Barnard, C. and Burk, T. (1979). Dominance hierarchies and the evolution of individual recognition. *Journal of Theoretical Biology* 81: 65–73.

Boehm, C. (1992). Segmentary warfare and management of conflict: A comparison of East African chimpanzees and patrilineal-patrilocal humans. In *Coalitions and Alliances in Humans and Other Animals*, A. Harcourt and F. B. M. de Waal, eds., pp. 137–173. Oxford: Oxford University Press.

Boehm, C. (1999). *Hierarchy in the Forest.* Cambridge, Mass.: Harvard University Press.

Brown, J. L. (1983). Cooperation—A biologist's dilemma. In *Advances in the Study of Behavior*, J. S. Rosenblatt, ed., pp. 1–37. New York: Academic Press.

Brown, J. and Colgan, P. (1986). Individual and species recognition in centrachid fishes: Evidence and hypotheses. *Behavioral Ecology and Sociobiology* 19: 373–379.

Brown, J. L. and Eklund, A. (1994). Kin recognition and the major histocompatibility complex—An integrative review. *American Naturalist* 143: 435–461.

Bygott, J. D. (1979). Agonistic behavior, dominance and social structure in wild chimpanzees of the Gombe National Park. In *The Great Apes*, D. A. Hamburg and E. R. McCown, eds., pp. 405–427. Menlo Park, Calif.: Benjamin/Cummings.

Connor, R. C. (1995). The benefits of mutualism: A conceptual framework. *Biological Review* 1–31.

Crozier, R. H. and Pamilo, P. (1996). *Evolution of Social Insect Colonies: Sex Allocation and Kin Selection.* Oxford: Oxford University Press.

Darwin, C. (1859). *On the Origin of Species.* London: J. Murray.

de Waal, F. (1996). *Good Natured.* Cambridge, Mass.: Harvard University Press.

Dugatkin, L. A. (1997). *Cooperation Among Animals: An Evolutionary Perspective.* New York: Oxford University Press.

Dugatkin, L. A. and Reeve, H. K. (eds.) (1998). *Game Theory and Animal Behavior.* Oxford: Oxford University Press.

Dugatkin, L. A. and Sih, A. (1995). Behavioral ecology and the study of partner choice. *Ethology* 99: 265–277.

Dugatkin, L. A. and Sih, A. (1998). Evolutionary ecology of partner choice. In *Cognitive Ecology*, R. Dukas, ed., pp. 379–403. Chicago: University of Chicago Press.

Dugatkin, L. A. and Wilson, D. S. (1992). The prerequisites of strategic behavior in the bluegill sunfish. *Animal Behaviour* 44: 223–230.

Dugatkin, L. A., Mesterton-Gibbons, M., and Houston, A. I. (1992). Beyond the prisoner's dilemma: Towards models to discriminate among mechanisms of cooperation in nature. *Trends in Ecology and Evolution* 7: 202–205.

Dukas, R. (ed.) (1998). *Cognitive Ecology.* Chicago: University of Chicago Press.

Eshel, I. and Cavalli-Sforza, L. L. (1982). Assortment of encounters and the evolution of cooperation. *Proceedings of the National Academy of Sciences, U.S.A.* 79: 1331–1335.

Fischer, E. (1988). Simultaneous hermaphroditism, tit-for tat, and the evolutionary stability of social systems. *Ethology and Sociobiology* 9: 119–136.

Fletcher, D. and Michener, C. (eds.) (1987). *Kin Recognition in Animals.* New York: Wiley.

Goodall, J. (1986). *The Chimpanzees of Gombe: Patterns of Behavior.* Cambridge, Mass.: Harvard University Press.

Grafen, A. (1990). Do animals really recognize kin? *Animal Behaviour* 39: 42–55.

Griffiths, S. X. and Magurran, A. E. (1999). Schooling decisions in guppies (*Poecilia reticulata*) are based on familiarity rather than kin recognition by phenotype matching. *Behavioral Ecology and Sociobiology* 45: 437–445.

Haldane, J. B. S. (1932). *The Causes of Evolution.* London: Longmans Green.

Hamilton, W. D. (1964). The genetical evolution of social behaviour. I and II. *Journal of Theoretical Biology* 7: 1–52.

Hepper, P. G. (ed.) (1991). *Kin Recognition.* Cambridge: Cambridge University Press.

Holmes, W. and Mateo, J. (1998). How mothers influence the development of litter-mate preferences in Belding's ground squirrels. *Animal Behaviour* 55: 1555–1570.

Huxley, T. H. (1888). *The Struggle for Existence: A Programme.* New York: Nineteenth Century Magazine.

Kropotkin, P. (1908). *Mutual Aid.* London: William Heinemann.

Magurran, A. E., Seghers, B., Shaw, P., and Carvalho, G. (1994). Schooling preferences for familiar fish in the guppy, *Poecilia reticulata. Journal of Fish Biology* 45: 401–406.

Maynard Smith, J. (1982) . *Evolution and the Theory of Games.* Cambridge: Cambridge University Press.

Mesterton-Gibbons, M. and Dugatkin, L. A. (1992). Cooperation among unrelated individuals: Evolutionary factors. *Quarterly Review of Biology* 67: 267–281.

Nishida, T. (1979). The social structure of chimpanzees of the Mahale mountains. In *The Great Apes*, D. A. Hamburg and E. R. McCown, eds., pp. 73–121. Menlo Park, Calif.: Benjamin/Cummings.

Nishida, T., Hiraiwa-Hasegawa, M., Hasegawa, T., and Takahata, Y. (1985). Group extinction and female transfer in wild chimpanzees in the Mahale National Park, Tanzania. *Zeitschrift für Tierpsychologie* 67: 284–301.

Ortega, C. (1998). *Cowbirds and Other Brood Parasites.* Tucson: University of Arizona Press.

Peck, J. R. (1993). Friendship and the evolution of cooperation. *Journal of Theoretical Biology* 162: 195–228.

Reeve, H. K. (1989). The evolution of conspecific acceptance thresholds. *American Naturalist* 133: 407–435.

Roberts, G. and Sherratt, T. (1998). Development of cooperative relationships through increasing investment. *Nature* 394: 175–179.

Scheel, D. and Packer, C. (1991). Group hunting behaviour of lions: A search for cooperation. *Animal Behaviour* 41: 697–709.

Sella, G. (1985). Reciprocal egg trading and brood care in a hermaphroditic polychaete worm. *Animal Behaviour* 33: 938–944.

Sella, G. (1988). Reciprocation, reproductive success and safeguards against cheating in a hermaphroditic polychaete worm. *Biological Bulletin* 175: 212–217.

Sella, G. and Lorenzi, G. (2000). Partner fidelity and egg reciprocation in the simultaneously hermaphroditic polychaete worm *Ophryotrocha diadema. Behavioral Ecology* 11: 260–264.

Shettleworth, S. (1998). *Cognition, Evolution and Behavior.* New York: Oxford University Press.

Sober, E. and Wilson, D. S. (1998). *Unto Others.* Cambridge, Mass.: Harvard University Press.

Tajfel, H. (1970). Experiments in intergroup discrimination. *Scientific American* 223: 96–102.

Trivers, R. L. (1971) The evolution of reciprocal altruism. *Quarterly Review of Biology* 46: 189–226.

West-Eberhard, M. J. (1975). The evolution of social behavior by kin selection. *Quarterly Review of Biology* 50: 1–35.

Wilkinson, G. (1984). Reciprocal food sharing in vampire bats. *Nature* 308: 181–184.

Wilkinson, G. (1987). Reciprocal altruism in bats and other mammals. *Ethology and Sociobiology* 9: 85–100.

Wilson, D. S. (1980). *The Natural Selection of Populations and Communities.* Menlo Park, Calif.: Benjamin/Cummings.

Wilson, D. S. (1983). The group selection controversy: History and current status. *Annual Review of Ecological Systems* 14: 159–187.

Wilson, D. S. and Dugatkin, L. A. (1997). Group selection and assortative interactions. *American Naturalist* 139: 336–351.

Wilson, D. S. and Sober, E. (1994). Re-introducing group selection to the human behavioral sciences. *Behavioral and Brain Sciences* 17: 585–654.

Wilson, E. O. (1975) *Sociobiology: The New Synthesis.* Cambridge, Mass.: Harvard University Press.

Wrangham, R. (1999). Evolution of coalitionary killing. *Yearbook of Physical Anthropology* 42: 1–30.

Sergio M. Pellis

Play is usually thought of as a phenomenon of childhood (Burghardt 1998; Power 2000). In some species, however, playful modes of behavior are retained into adulthood (Pellis and Iwaniuk 1999b). Most often, adult–adult play involves play fighting (Aldis 1975; Fagen 1981). Unlike play fighting in juveniles, where the evidence for its functions has been difficult to discern, play among postpubertal individuals is more clearly associated with fitness-enhancing consequences. The literature shows that when it occurs in subadults and adults, play fighting is used in two general contexts—social bonding and social testing (Pellis and Pellis 1996; Pellis and Iwaniuk 2000). That is, play fighting is a tool that can be used to assess and manipulate conspecifics (Breuggeman 1978). A comparison of species with and without this tool offers some insight into the value of such play.

While an adult male mouse or rat will attack a male intruder (R. J. Blanchard and Blanchard 1994), there is a striking species difference when they encounter one another in a neutral arena. Mice follow one of two options: They aggressively attack the opponent or ignore him (Brain 1981; R. J. Blanchard et al. 1979). Rats have a third option: They engage in playful fighting akin to that of juveniles (Smith et al. 1999).[1] Such play fighting may lead to the establishment of a dominance relationship as in colonies (Pellis and Pellis 1992; Pellis et al. 1993). If such play fails to resolve the relationship, the encounter may escalate into a serious fight (Smith et al. 1998, 1999). In rats, unlike mice, such play fighting leads to a social world with more shades of gray, and so a greater demand for more sophisticated information processing (Whishaw et al. 2001). Therefore, play fighting among adults can be used as a window into social cognition.

Two questions arise from the finding that adults use play fighting to assess and manipulate others (Breuggeman 1978; Pellis and Iwaniuk 2000). What kind of information about the other animal can be acquired via play, and what are the structural properties of play fighting that make it a suitable means of assessment and social manipulation? Studies in my laboratory of play fighting among adult rats (*Rattus norvegicus*) illustrate some of the possibilities.

Play and Social Information

Within a colony, adult male rats form a dominance hierarchy (D. C. Blanchard and Blanchard 1990; Calhoun 1963; Flannery and Lore 1977). When two unfamiliar male rats confront one another in a neutral arena, the home status of the unfamiliar opponents affects the pattern of play and aggression that ensues (Smith et al. 1999). All males initiate play with a subordinate less often than with a dominant, and are more likely to evade the playful contact of a subordinate. However, they are more likely to engage a dominant in more prolonged physical contact. That is, during these encounters, the unfamiliar animals appear to recognize each other's respective home-colony status. What is not known is whether the unfamiliar pairmates actually need to engage in play fighting to make that determination, or whether nonplayful cues can provide such information. Rats are known to recognize dominant males by olfactory cues (Brown 1985). Indeed, before play fighting begins, rats engage in mutual anogenital investigation. It is also possible that visual cues may be involved (Calhoun 1963), such as a "macho" swagger (Dittman 1992) or hypermasculine body proportions (Karen Dean, personal communication). Recent studies in my laboratory have shown that whatever these cues are, rats can make these judgments at a distance.

Under laboratory conditions, rats are typically maintained on a nutritious but boring diet of processed rodent chow. Therefore, when they are offered a fat-rich and delectable treat such as a sunflower seed, they take it readily. However, they must first husk the seed, a task that takes them several days of practice before they achieve a high level of proficiency. When it is done properly, a rat can split the husk neatly into two halves (Whishaw et al. 1998). My postdoctoral research associate, Karen Dean, used seed husking to develop a sensitive test of social knowledge. Once this task is fully learned, both dominant and subordinate male rats maintain a high level of proficiency regardless of whether a dominant or a subordinate is present on the other side of a wire mesh partition in the test enclosure. However, when they are placed as an intruder in someone else's home cage, dominants and subordinates differ markedly in how they husk the seeds.

The dominant rat will continue to split his seeds neatly whether the dominant or subordinate resident is sitting on the opposite side of the partition. In contrast, the subordinate will perform normally if he is sitting next to the subordinate resident, but will shred the seed to pieces when the dominant resident is sitting next to him. That is, the subordinate's performance severely deteriorates in the presence of an unfamiliar dominant. It is important to note that in this test paradigm, the rats do not press against the wire partition and investigate one another; rather, they sit facing the experimenter, at least a body length away from each other. Therefore direct contact is not needed for a rat to recognize the relative status of the animal sitting next to it. Given that the resident's home cage is not washed before testing, the general odor of both residents, dominant and subordinate, must pervade the whole enclosure; this suggests that the intruders are using a combination of olfactory and visual cues to evaluate the neighboring animal (see Pellis et al. 1996). Whatever means they are using, they are doing so without having to interact physically. This suggests that when strangers meet, play is not necessary to assess the status of the opponent. Rather, play may be used to assess other features of the opponent. The play that occurs between colony members offers a clue as to what that assessment may involve.

In their home colony, the subordinates initiate more playful encounters with the dominant than they do with each other (Pellis et al. 1993). Furthermore, when they are playfully contacted by a dominant, they are more likely to roll over onto their backs, as they did as juveniles (Pellis and Pellis 1987). When contacted by another subordinate, they are more likely to remain standing and to push against the attacker with their flank (Pellis et al. 1993), as is typical of a dominant male (Pellis and Pellis 1991, 1992). In the absence of dominance relationships, all postpubertal males are more likely to stand and push than to roll over to a supine position (Pellis and Pellis 1990; Smith et al. 1996, 1998, 1999). Subordinates' frequent soliciting of playful contact with dominants, and their juvenilelike response to those dominants, suggests that the play fighting is used as a means of maintaining "friendly" relations with the dominant (Pellis et al. 1993). However, not all subordinates are equally obsequious.

The greater the dominance asymmetry between pairmates, the more juvenilelike the playful responses by the subordinate (Pellis and Pellis 1992). Furthermore, it is those pairmates that are least asymmetrical in their play relations that are the most likely to escalate the playful encounter into a serious fight (Pellis and Pellis 1991). Similarly, when male rats that are unfamiliar with each other are placed in a neutral arena, the dominant–dominant combinations are the ones most likely to escalate into serious fights (Smith et al. 1999). Close inspection of those escalations suggests that the play fighting preceding the serious fight is rougher. These observations suggest that when animals are testing their opponent's ability to maintain or gain a position of superior dominance, the play can escalate into a quasi-aggressive intensity. Therefore there are two extremes in the style of play available to an individual: a gentler form, seen when a subordinate is "sucking up" to a dominant, and a rougher form, seen when one rat is probing another for weaknesses. These differences in play intensity can be explicitly converted into formal rules of engagement, and so provide a basis for judging how such play can be used to assess and manipulate a partner.

The Rules Underlying Play Fighting

For play fighting to remain playful, it needs to follow the 50:50 rule (Aldis 1975; Altmann 1962). That is, both pairmates have to win close to 50 percent of the playful encounters. To achieve this, the rules of attack and defense differ from those in serious fighting (Pellis and Pellis 1998a). When an attack is launched during a serious fight, the attacking animal has to guard against retaliation from the opponent. To do so, the attacker typically incorporates some defensive tactic into its attack; this limits the defender's opportunity to counterattack (Pellis 1997). Similarly, when defending itself against a serious attack, the defender uses an intensity of defense that reduces the likelihood of a successful penetration by the attacker (Pellis 1997). In contrast, during play fighting, the attacker does not typically incorporate defensive maneuvers into its attack; this facilitates successful counterattacks by the defender (Pellis and Pellis 1998a). Also, when defending itself against a playful attack, the defender uses an intensity of defense that is lower than that in serious fights; this increases the likelihood of a successful contact by the attacker (Pellis and Pellis 1998a). Therefore, in play fighting, the tactics of attack and defense are decoupled to ensure that both animals get to contact their partner successfully. Following such a rule structure in serious fighting would be suicidal, since an opponent could seize upon an unguarded moment with great severity (Geist 1971).

An examination of the instances where play fights escalate into serious fights reveals that one of the partners, either when attacking or when defending, shifts the rule structure from the playful to the serious mode. This typically leads to the other animal switching from playful attacks to serious attacks (Pellis and Pellis 1998a). Nonetheless, bending the 50:50 rule provides a means of using play fighting for assessment and manipulation. For example, when a subordinate rat uses play fighting for social bonding with the dominant male in the colony, he bends the 50:50 rule in the dominant's favor. However, when a subordinate uses play fighting to probe the dominant for weakness, he bends the rule in his own favor. In the second case, the subordinate can assess how much of a deviation from equality the dominant will tolerate before retaliating aggressively. In such a scenario, the subordinate may follow a simple rule of thumb—if the dominant tolerates this deviation away from the 50:50 rule, then escalate further until the status is reversed; but if the dominant starts to respond forcefully, then back down before the encounter escalates into a serious fight.[2]

Species Comparisons

As noted earlier, play fighting is used by adults in two functional contexts: social bonding and social testing. These can be further subdivided, with social bonding including courtship and sexual and nonsexual pair bonding. Similarly, social testing can include some form of jostling for social status within an established group or evaluating the social potential of a stranger. In species that play as adults, there is considerable variation as to whether play fighting is used in all or only some of these contexts (Pellis and Iwaniuk 1999b, 2000). Some comparisons of related species illustrate this diversity.

Both male rats and hamsters can engage in playful fights as a precursor to copulation (personal observation). However, rats, but not hamsters, also use play fighting for bonding with dominants in their home colony, and for testing dominance relationships with both colony mates and strangers (Pellis et al. 1993; Pellis and Pellis 1993; Smith et al. 1999). Whereas both the slow loris (Erhlich and Musicant 1975) and the greater galago (Erhlich 1977) use play fighting within their colonies for social affiliation, only the galago uses play fighting in encounters with unfamiliar animals in a neutral arena (Newell 1971). Similarly, while subadult spotted hyenas appear to use play fighting for affiliation and hence integration into the clan (Drea et al. 1996), subadult brown hyenas preferentially engage the adults in play fights, indicating that they are probing for a position in the clan's dominance hierarchy (Mills 1990). What are needed are comparative studies that can explicitly evaluate the possible causal mechanisms that have generated this diversity (Pellis and Iwaniuk 1999b, 2000). Unfortunately, given the lack of information on most species, this is difficult to do at present.

An insight into why some species have chosen play as a solution for particular kinds of social problems would greatly enhance our ability to characterize the cognitive mechanisms involved. For example, comparative analyses of adult–adult play in primates indicate that both the sexual and nonsexual use of play is more likely in species with social systems that lead to lower levels of contact and familiarity among social partners (Pellis and Iwaniuk 1999b, 2000). Furthermore, it seems to be most common in species with relatively impoverished repertoires of signals useful for communication at a distance (Pellis and Iwaniuk 2000). Touch, which is necessary for play fighting to occur, may be a crucial means of evaluating a social fellow you cannot be sure about. Indeed, such information may also be valuable for individuals that know each other well. For example, following an argument, a "no" from your significant other in response to the question "Are you still mad at me?" may or may not reflect their true feelings. A touch on the shoulder when the question is asked obtains more honest information. Subtle perturbations in

rehearsed play routines between well-acquainted individuals may provide a means of evaluating changes in the relationship (Wolf 1984).

Conclusions

Several researchers have begun to analyze play in juveniles from a cognitive perspective (e.g., Allen and Bekoff 1997; Biben 1998; Thompson 1998). Two difficulties with this have emerged. First, there are not only superficial species differences in the content of the behavioral repertoire used in play, but there are also deep organizational differences (Pellis 1993; Pellis and Iwaniuk 1999a). Because of this, it is difficult to generalize from in-depth studies of single species. Therefore, broader comparative studies are needed even though these are more difficult to conduct (e.g., Lewis 2000; Parker and McKinney 1999). Second, as noted earlier, the fitness-enhancing outcomes of play by juveniles have been elusive. It is thus difficult to evaluate the variability present in play. The variability may reflect an inability to follow a plan or it may reflect adaptive adjustments that ensure a particular outcome is obtained. Clear end points would aid greatly in distinguishing between these possibilities. Shifting the focus to play between adults would help with both problems. The distribution of play in adulthood is more restricted and its content is less diverse than is the case for juveniles. Also, the play fighting present in postpubertal animals has more clearly discernible fitness-enhancing outcomes than is the case for play in juveniles. The use of an adult's perspective on play may also afford us an unexpected benefit—that of reexamining childhood play.

If the cognitive skills required for adults to engage in manipulative play fighting are taken as a developmental end point, then the play occurring in childhood can be reexamined for evidence that it is structured to enhance the development of those skills. Two lines of converging evidence support such a possibility. Even though social deprivation studies are limited by difficulties in ensuring that only certain experiences are restricted (Bekoff 1976), species comparisons reveal some intriguing differences. Deprivation of social play in the juvenile phase produces severe cognitive deficits in rats, whereas comparable experiments on other laboratory rodents do not (Einon et al. 1978, 1981). Such comparisons suggest that while rats need juvenile play to develop cognitive skills, other species, even though they engage in play as juveniles, do not.

Evidence from my laboratory has shown that play fighting in muroid rodents resembles species-typical patterns of adult precopulatory behavior (Pellis 1993). The resemblance is in both the body targets attacked and in the defensive maneuvers used to block attacks. During post-weaning development, playful interactions involve the use of behavior patterns at frequencies typical of adult sexual encounters (Pellis and Pellis 1998b). This is not the case for rats. In adult sexual encounters, most of the female's defensive maneuvers involve evasion, whereas in play, most involve turning to face the attacker. Whereas the former limits body contact, the latter enhances it. Indeed, changes in tactics occurring at the onset of the juvenile phase and at puberty (Pellis and Pellis 1990, 1997a) ensure that the body contact in play is further enhanced and exaggerated. That is, the species of rodent that has a pattern of juvenile play most different from the adult behavior being mimicked uses play fighting as an adult social strategy and is the one most adversely affected if deprived of play as a juvenile. This offers the opportunity to link specific childhood experiences to particular sociocognitive skills in adulthood (Pellis et al. 1999).

Acknowledgments

I thank Vivien C. Pellis and the editors for their thoughtful comments. The research from my laboratory was supported by grants from the Natural Sciences and Engineering Council (Canada) and the Harry Frank Guggenheim Foundation.

Notes

1. Whereas playful attacks in rats involve nosing the nape of the partner, agonistic attacks involve bites directed at the lower dorsum or the face (R. J. Blanchard et al. 1977; Pellis and Pellis 1987; Siviy and Panksepp 1987).

2. Some species have affiliative signals that demonstrate playful intention. When available, such signals may be able to increase the flexibility of the playful-serious gradient because they can be used to diffuse unwanted escalations (see Bekoff 1995; Pellis and Pellis 1996, 1997b).

References

Aldis, O. (1975). *Play Fighting*. New York: Academic Press.

Allen, C. and Bekoff, M. (1997). *Species of Mind*. Cambridge, Mass.: MIT Press.

Altmann, S. A. (1962). Social behavior of anthropoid primates: Analysis of recent concepts. In *Roots of Behavior*, E. L. Bliss, ed., pp. 277–285. New York: Harper and Brothers.

Bekoff, M. (1976). The social deprivation paradigm: Who's being deprived of what? *Developmental Psychobiology* 9: 499–500.

Bekoff, M. (1995). Play signals as punctuation: The structure of social play in canids. *Behaviour* 132: 419–429.

Biben, M. (1998). Squirrel monkey playfighting: Making the case for a cognitive training function for play. In *Animal Play. Evolutionary, Comparative, and Ecological Perspectives*, M. Bekoff and J. A. Byers, eds., pp. 161–182. Cambridge: Cambridge University Press.

Blanchard, D. C. and Blanchard, R. J. (1990). The colony model of aggression and defense. In *Contemporary Issues in Comparative Psychology*, D. A. Dewsbury, ed., pp. 410–430. Sunderland, Mass.: Sinauer Associates.

Blanchard, R. J. and Blanchard, D. C. (1994). Environmental targets and modeling of animal aggression. In *Ethology and Psychopharmacology*, S. J. Cooper and C. A. Hendrie, eds., pp. 133–157. New York: Wiley.

Blanchard, R. J., Blanchard, D. C., Takahashi, T., and Kelly, M. J. (1977). Attack and defense behaviour in the albino rat. *Animal Behaviour* 25: 6222–6634.

Blanchard, R. J., O'Connell, V., and Blanchard, D. C. (1979). Attack and defense behaviors in the albino mouse. *Aggressive Behavior* 5: 341–352.

Brain, P. F. (1981). Differentiating types of attack and defense in rodents. In *Multidisciplinary Approaches to Aggression Research*, P. F. Brain and D. Benton, eds., pp. 53–77. Amsterdam: Elsevier/North-Holland Biomedical Press.

Breuggeman, J. A. (1978). The function of adult play in free-ranging *Macaca mulatta*. In *Social Play in Primates*, E. O. Smith, ed., pp. 169–192. New York: Academic Press.

Brown, R. E. (1985). The rodents II: Suborder Myomorpha. In *Social Odours in Mammals*. Vol. 1, R. E. Brown and D. W. MacDonald, eds., pp. 345–457. Oxford: Clarendon Press.

Burghardt, G. M. (1998). Play. In *Comparative Psychology: A Handbook*, G. Greenberg and M. Harraway, eds., pp. 757–767. New York: Garland.

Calhoun, J. B. (1963). *The Ecology and Sociology of the Norway Rat*. Washington, D.C.: U.S. Department of Health, Education, and Welfare, Public Health Service.

Dittman, R. W. (1992). Body positions and movement patterns in female patients with congenital adrenal hyperplasia. *Hormones and Behavior* 26: 441–456.

Drea, C. M., Hawk, J. E., and Glickman, S. E. (1996). Aggression decreases as play emerges in infant spotted hyaenas: Preparation for joining the clan. *Animal Behaviour* 51: 1323–1336.

Einon, D., Morgan, M. J., and Kibbler, C. C. (1978). Brief periods of socialization and later behavior in the rat. *Developmental Psychobiology* 11: 213–225.

Einon, D., Humphreys, A. P., Chivers, S. M., Field, S., and Naylor, V. (1981). Isolation has permanent effects upon the behavior of the rat, but not mouse, gerbil, or guinea pig. *Developmental Psychobiology* 14: 343–355.

Erhlich, A. (1977). Social and individual behaviors in captive greater galagos. *Behaviour* 63: 192–214.

Erhlich, J. F. and Musicant, A. (1975). Social and individual behaviors in captive slow lorises. *Behaviour* 60: 195–220.

Fagen, R. (1981). *Animal Play Behavior*. New York: Oxford University Press.

Flannelly, K. and Lore, R. (1977). Observations of the subterranean activity of domesticated and wild rats (*Rattus norvegicus*): A descriptive study. *Psychological Record* 2: 315–329.

Geist, V. (1971). *Mountain Sheep*. Chicago: University of Chicago Press.

Lewis, K. P. (2000). A comparative study of primate play behaviour: Implications for the study of cognition. *Folia Primatologica* 71: 417–421.

Mills, M. G. L. (1990). *Kalahari Hyaenas. Comparative Behavioural Biology of Two Species*. London: Unwin Hyman.

Newell, T. G. (1971). Social encounters in two prosimian species: *Galago crassicaudatus* and *Nycticebus coucang*. *Psychonomic Society* 2: 128–130.

Parker, S. T. and McKinney, M. L. (1999). *Origins of Intelligence*. Baltimore, Md: Johns Hopkins University Press.

Pellis, S. M. (1993). Sex and the evolution of play fighting: A review and model based on the behavior of muroid rodents. *Play Theory and Research* 1: 55–75.

Pellis, S. M. (1997). Targets and tactics: The analysis of moment-to-moment decision making in animal combat. *Aggressive Behavior* 23: 107–129.

Pellis, S. M. and Iwaniuk, A. N. (1999a). The roles of phylogeny and sociality in the evolution of social play in muroid rodents. *Animal Behaviour* 58: 361–373.

Pellis, S. M. and Iwaniuk, A. N. (1999b). The problem of adult play fighting: A comparative analysis of play and courtship in primates. *Ethology* 105: 783–806.

Pellis, S. M. and Iwaniuk, A. N. (2000). Adult-adult play in primates: Comparative analyses of its origin, distribution and evolution. *Ethology* 106: 1083–1104.

Pellis, S. M. and Pellis, V. C. (1987). Play-fighting differs from serious fighting in both target of attack and tactics of fighting in the laboratory rat *Rattus norvegicus*). *Aggressive Behavior* 13: 227–242.

Pellis, S. M. and Pellis, V. C. (1990). Differential rates of attack, defense and counterattack during the developmental decrease in play fighting by male and female rats. *Developmental Psychobiology* 23: 215–231.

Pellis, S. M. and Pellis, V. C. (1991). Role reversal changes during the ontogeny of play fighting in male rats: Attack versus defense. *Aggressive Behavior* 17: 179–189.

Pellis, S. M. and Pellis, V. C. (1992). Juvenilized play fighting in subordinate male rats. *Aggressive Behavior* 18: 449–457.

Pellis, S. M. and Pellis, V. C. (1993). Influence of dominance on the development of play fighting in pairs of male Syrian golden hamsters (*Mesocricetus auratus*). *Aggressive Behavior* 19: 293–302.

Pellis, S. M. and Pellis, V. C. (1996). On knowing it's only play: The role of play signals in play fighting. *Aggression and Violent Behavior* 1: 249–268.

Pellis, S. M. and Pellis, V. C. (1997a). The prejuvenile onset of play fighting in laboratory rats (*Rattus norvegicus*). *Developmental Psychobiology* 31: 193–205.

Pellis, S. M. and Pellis, V. C. (1997b). Targets, tactics, and the open mouth face during play fighting in three species of primates. *Aggressive Behavior* 23: 41–57.

Pellis, S. M. and Pellis, V. C. (1998a). The structure–function interface in the analysis of play fighting. In *Animal Play. Evolutionary, Comparative, and Ecological Perspectives*, M. Bekoff and J. A. Byers, eds., pp. 115–140. Cambridge: Cambridge University Press.

Pellis, S. M. and Pellis, V. C. (1998b). Play fighting of rats in comparative perspective: A schema for neurobehavioral analyses. *Neuroscience and Biobehavioral Reviews* 23: 87–101.

Pellis, S. M., Pellis, V. C., and McKenna, M. M. (1993). Some subordinates are more equal than others: Play fighting amongst adult subordinate male rats. *Aggressive Behavior* 19: 385–393.

Pellis, S. M., McKenna, M. M., Field, E. F., Pellis, V. C., Prusky, G. T., and Whishaw, I. Q. (1996). Uses of vision by rats in play fighting and other close-quarter social interactions. *Physiology and Behavior* 59: 905–913.

Pellis, S. M., Field, E. F., and Whishaw, I. Q. (1999). The development of a sex-differentiated defensive motor pattern in rats: A possible role for juvenile experience. *Developmental Psychobiology* 35: 156–164.

Power, T. G. (2000). *Play and Exploration in Children and Animals*. Mahwah, N.J.: Lawrence Erlbaum Associates.

Siviy, S. M. and Panksepp, J. (1987). Sensory modulation of juvenile play in rats. *Developmental Psychobiology* 20: 39–55.

Smith, L. K., Field, E. F., Forgie, M. L., and Pellis, S. M. (1996). Dominance and age-related changes in the play fighting of intact and post-weaning castrated male rats (*Rattus norvegicus*). *Aggressive Behavior* 22: 215–226.

Smith, L. K., Forgie, M. L., and Pellis, S. M. (1998). Mechanisms underlying the absence of the pubertal shift in the playful defense of female rats. *Developmental Psychobiology* 33: 147–156.

Smith, L. K., Fantella, S.-L. N., and Pellis, S. M. (1999). Playful defensive responses in adult male rats depend on the status of the unfamiliar opponent. *Aggressive Behaviour* 25: 141–152.

Thompson, K. V. (1998). Self-assessment in juvenile play. In *Animal Play. Evolutionary, Comparative, and Ecological Perspectives*, M. Bekoff and J. A. Byers, eds., pp. 183–204. Cambridge: Cambridge University Press.

Whishaw, I. Q., Sarna, J. R., and Pellis, S. M. (1998). Evidence for rodent-common and species-typical limb and digit use in eating, derived from a comparative analysis of ten rodents. *Behavioural Brain Research* 96: 79–91.

Whishaw, I. Q., Metz, G. A. S., Kolb, B., and Pellis, S. M. (2001). Accelerated nervous system development contributes to behavioral efficiency in the laboratory mouse: A behavioral review and theoretical proposal. *Developmental Psychobiology* 39: 151–170.

Wolf, D. P. (1984). Repertoire, style and format: Notions worth borrowing from children's play. In *Play in Animals and Humans*, P. K. Smith, ed., pp. 175–193. Oxford: Basil Blackwell.

53 The Evolution of Social Play: Interdisciplinary Analyses of Cognitive Processes

Marc Bekoff and Colin Allen

The Value of Social Play

Progress in understanding animal cognition requires interdisciplinary collaboration among biologists, psychologists, cognitive scientists, neuroscientists, and philosophers. In our own case, as a biologist and a philosopher, our work has combined empirical and conceptual studies of social play. In this essay we describe how our personal interests have contributed to our cooperative efforts.

Our work is rooted in a series of long-term empirical studies of social play. When Marc Bekoff decided to study social play for his doctoral research, many people told him that it was a waste of time, for it was impossible to define and many others had tried to study it and failed. While this provided the perfect challenge for a graduate student who had the full support of his advisor, Michael W. Fox, Marc frankly thought that his research on play would end when he received his degree. He was very wrong indeed.

Social play is a fascinating topic because it combines many elements of cooperation, communication, and learning (see Pellis, chapter 52 in this volume), as well as providing a possible prototype for the evolution of morality (playing fair). According to the social intelligence hypothesis, intelligence is an adaptation for social living (Jolly 1966; Humphrey 1976; Byrne and Whiten 1988), which suggests that social play might be an excellent domain for the investigation of cognitive abilities in a variety of animal species (see also Power 2000; Burghardt 2002).

Marc's main interests have been in what animals do when they play (the structure of play), the development of play, how animals communicate their intentions to play, and what possible functions play may serve—why play has evolved. The animals he has studied in depth are all members of the family Canidae—domestic dogs, coyotes, wolves, foxes, and hybrids. The pri-

mary approach has been to take detailed notes while observing animals playing, along with videotaping them for later analysis. These early efforts clearly showed that there were species differences in social play as well as significant individual differences, even among littermates. Individual differences were especially apparent for animals of different social ranks. Sex differences were few. Marc also came to realize that there was something unique about how animals communicated their intentions to engage in or to continue social play, and became interested in Gregory Bateson's (Bateson 1955) ideas about metacommunication—communication about communication.

Species differences in social play among canids permit one to study in more detail how play is communicated. First it was shown that a specific play signal, the bow, was highly stereotyped and necessarily so. Detailed measures of the duration and form of bows (Bekoff 1977a) showed that they clearly were a ritualized action, the result of which is the performance of a clear and unambiguous signal. There was very little variability in bows; this made sense in that when canids and other animals play, they use behavior patterns from various other contexts, namely, predation, aggression, and reproduction, and individuals need to know that "this is play and not attempts at predation, aggression, or reproduction."

Ethology Meets Philosophy

Work on conceptual issues has always been carried on alongside Marc's empirical work. Bekoff and Byers (1981) formalized a well-received working definition of play. Subsequently Marc worked with philosopher Dale Jamieson, who put him in touch with Colin Allen, leading to collaboration on a series of articles on conceptual issues in play, bridging the gap between ethological and philosophical inquiries (Bekoff and Allen 1992, 1998; Allen and Bekoff 1994, 1997).

Philosophical interest in ethology often centers on why it may or may not be scientifically useful and important for ethologists to ascribe meaning and content to animal communication (Dennett 1987), and how to characterize concepts in non-human animals. When Colin met Marc Hauser, Dorothy Cheney, and Robert Seyfarth at the University of California, Los Angeles, he rapidly became fascinated by the problems of describing animal concepts (Allen and Hauser 1991; Allen 1999) and specifying what the vocalizations of vervet monkeys mean (Cheney and Seyfarth 1990; Allen 1992a; Allen and Saidel 1998). Saying that a call means "Leopard coming!" potentially tells us about the function of that call, but articulating the correct relationship between meaning and biological function is a far from trivial task; indeed it is a task that some philosophers dismissed as hopeless, albeit on questionable grounds (Allen 1992b; Allen and Bekoff 1995, 1997). These interests converged with Marc's interests on the questions of how to define play (Allen and Bekoff 1994) and how to describe the function and content of the signals used during play (Bekoff and Allen 1992).

While we all seem to be able to recognize play when we see it, the problem of defining play is illustrated by the fact that many behavioral biologists have been tempted to define it in terms of what it is *not*—it is not aggression, predation, or reproduction—rather than what it *is*. Despite expressing doubt about the merits of providing a definition of play, Bekoff and Byers (1981) went ahead and attempted to give one. Rather than use this short essay to go into the details of definitions, we refer readers to Bekoff and Allen (1998) for a summary. However, the position we take is that working definitions are just that— *working* definitions (see also Fagen 1981 and Burghardt 2002). Such definitions are bound to be imperfect in the absence of the empirical research needed to refine them. Thus we criticized Rosenberg's (1990) effort to call the biological

study of play into question on the basis of his definition of play, which required the players to possess a concept of pretense, which in turn required second-order intentionality—a restrictive definition that would rule out playing even in young human children (Allen and Bekoff 1994). Rosenberg's views were based upon arguments about the impossibility of accurately specifying the meaning of animal concepts; identical arguments were criticized by Allen (1992b) and Allen and Bekoff (1994).

Although the difficulties of specifying meaning should not be understated, assigning meanings to animal signals can help us to understand their functions (however, see Rendall and Owren, chapter 38 in this volume, for a contrary view). For instance, because the behavior patterns seen during social play also occur during aggression, predation, or sexual behavior, the signals exchanged during play might be characterized as telling others "I want to play," "This is still play no matter what I am going to do to you," or "This is still play regardless of what I just did to you." These interpretations raise questions that have not yet been answered, such as whether canids or other animals are capable of attributing intentional states to each other during play— perhaps a limited application of a "theory of mind" in a specific domain (Allen and Bekoff 1997, chapter 6).

Nevertheless, conceiving the meaning of play signals in this way prompted a study that re-analyzed the production of play bows during play sequences by infant canids (domestic dogs, wolves, and coyotes) (Bekoff 1995). It was found that play bows were used nonrandomly, especially when biting accompanied by rapid side-to-side shaking of the head was performed. This kind of biting takes place during serious aggressive and predatory encounters and can easily be misinterpreted if its meaning is not modified by a play signal.

Individuals also engage in role reversing and self-handicapping (Bekoff and Allen 1998) to maintain social play. Each can serve to reduce asymmetries between the interacting animals and foster the reciprocity that is needed for play to occur. Self-handicapping occurs when an individual performs a behavior that might compromise her on himself. For example, a coyote might not bite her play partner as hard as she can, or she might not play as vigorously as she can. Watson and Croft (1996) found that red-necked wallabies adjusted their play to the age of their partner. When a partner was younger, the older animal adopted a defensive, flat-footed posture, and pawing rather than sparring occurred. In addition, the older player was more tolerant of its partner's tactics and took the initiative in prolonging interactions.

Role reversing occurs when a dominant animal performs an action during play that would not normally occur during real aggression. For example, a dominant animal might voluntarily not roll over on his back during fighting, but would do so while playing. In some instances role reversing and self-handicapping might occur together. For example, a dominant individual might roll over while playing with a subordinate animal and inhibit the intensity of a bite. From a functional perspective, self-handicapping and role reversing, similar to using specific play invitation signals or altering behavioral sequences, might signal an individual's intention to continue to play.

The Meaning of Play Signals

There are likely to be questions about whether information about intentions is really being signaled during play sequences. Even if signals indicate to a human observer an individual's intention to play, it is a separate question whether they do so to a conspecific playmate. Here our work has been guided by Ruth Millikan's (1984) ideas about signal content. Bekoff and Allen (1992) argue that play bows meet the definition of an "intentional icon," the basic kind of meaningful signal in Millikan's theory, because play

bows have the function of conveying information about intentions to play partners. Unlike other approaches to a theory of mind, Millikan's approach allows play signals to have this role even if canids, or other animals, lack a fully general capacity for reasoning about the mental states of others (Allen and Bekoff 1997). Such a functional approach to meaning can seem less natural than the more familiar approaches to meaning based on folk psychology (e.g., Dennett 1987), which tend to assume fully conscious, rational processing of meanings. However, these differing approaches may in fact be complementary, representing different forms of explanation of animal behavior (Allen, in press).

Whether the animals are psychologically aware of these meanings requires an investigation of their cognitive abilities with respect to signal function. Such a study might be attempted in connection with how they handle erroneous or false signals such as those that might occur during deception (Allen and Hauser 1993; Allen and Bekoff 1997). However, in the specific context of play, such studies are likely to be very difficult. There is little evidence that play signals are used to deceive others in canids or other species. Cheaters are unlikely to be chosen as play partners because others can simply refuse to play with them and choose others, and limited data on captive and wild infant coyotes show that cheaters have difficulty getting other young coyotes to play (Bekoff, personal observations). It is also not known if individuals select play partners based on what they have observed during play by others.

Neurobiological Bases of Sharing Intentions

It is useful to ask how a play bow (or other action) might provide the recipient with information about the sender's intentions. Perhaps an individual's experiences with play can promote learning about the intentions of others. Perhaps it is possible that the recipient shares the intentions (beliefs, desires) of the sender based on the recipient's own prior experiences of situations in which he or she performed play bows.

Recent research suggests a neurobiological basis for sharing intentions. "Mirror neurons," found in macaques, fire when a monkey executes an action and also when the monkey observes the same action being performed by another monkey (Gallese et al., chapter 56 in this volume). Frith and Frith (1999) report the results of neural imaging studies in humans that suggest a neural basis for one form of social intelligence: understanding others' mental states (mental state attribution). More comparative data are needed to determine if mirror neurons (or their functional equivalents) are found in other taxa and if they might actually play a role in the sharing of intentions between individuals engaged in an ongoing social interaction such as play. Neuroimaging studies will also be useful.

Why Cooperate and Play Fairly? Fine-Tuning Play

Playtime generally is a safe time during which transgressions are accepted by others, especially when one player is a youngster who is not yet a competitor for social status, food, or mates. Individuals must cooperate with one another when they play; they must negotiate agreements to play (Bekoff 1995). Fagen (1993, p. 192) noted that "Levels of cooperation in play of juvenile primates may exceed those predicted by simple evolutionary arguments." The highly cooperative nature of play has evolved in many other species (Fagen 1981; Bekoff 1972, 1995; Bekoff and Allen 1998; Power 2000; Burghardt 2002). Detailed studies of play in various species indicate that individuals trust others to maintain the rules of the game (Bekoff and Byers 1998). While there have been numerous discussions of cooperative behavior in animals (e.g., Axelrod 1984; Ridley 1996; Dugatkin 1997), none has considered social play—the requirement for coopera-

tion and reciprocity—and its possible role in the evolution of social morality, namely, behaving fairly (Bekoff 2001).

Individuals of different species appear to fine-tune ongoing play sequences to maintain a play mood and to prevent play from escalating into real aggression. Detailed analyses of films show that there are subtle and fleeting movements and rapid exchanges of eye contact that suggest that players are exchanging information on the run, from moment to moment, to make certain everything is all right—that this is still play. Why might they do this? Play in most species does not take up much time and energy (Bekoff and Byers 1998; Power 2000), and in some species only minimal amounts of social play during short windows of time early in development are necessary to produce socialized individuals. [For example, two 20-minute play sessions with another dog, twice a week, are sufficient for domestic dogs from 3 to 7 weeks of age (Scott and Fuller 1965).] Play appears to be very important in social, cognitive, and/or physical development, and may also be important for training youngsters for unexpected circumstances (Spinka et al. 2001). We know of no data concerning the actual benefits of social play in terms of survival and reproductive success. However, it generally is assumed that short-term and long-terms benefits vary from species to species, among different age groups, and between the sexes within a species. No matter what the functions of play may be, there seems to be little doubt that it has some benefits and that the absence of play can have devastating effects on social development (Power 2000; Burghardt 2002).

During early development there is a small time window when individuals can play without being responsible for their own well-being. This time period is generally referred to as the socialization period, for this is when species-typical social skills are learned most rapidly. It is important for individuals to engage in at least some play. All individuals need to play and there is a premium for playing fairly if one is to be able to play at all. If individuals do not play fairly, they may not be able to find willing play partners. As indicated, in coyotes, for example, youngsters are hesitant to play with an individual who does not play fairly or with an individual whom they fear (Bekoff 1977b). In many species, individuals also show play partner preferences, and it is possible that these preferences are based on the trust that individuals place in one another.

Social Play and Social Morality: Some Possible Connections between Structure and Function

Bekoff (2001) suggested that during social play, while individuals are having fun in a relatively safe environment, they learn ground rules that are acceptable to others—how hard they can bite, how roughly they can interact—and how to resolve conflicts. He argues that there is a premium on playing fairly (see Pellis, chapter 52 in this volume, for a discussion of the 50:50 rule in social play) and trusting others to do so as well. What could be a better atmosphere in which to learn social skills than during social play, where there are few penalties for transgressions? Individuals might also generalize codes of conduct learned in playing with specific individuals to other group members and to other situations, such as food sharing, hunting, and grooming (Dugatkin and Bekoff submitted).

To stimulate further comparative research on a wide array of species and the development of suitable models (Dugatkin and Bekoff submitted), Bekoff (2001) offered the hypothesis that social morality, in this case behaving fairly, is an adaptation that is shared by many mammals, not only by nonhuman and human primates. Behaving fairly evolved because it helped young animals acquire social (and other) skills needed as they matured into adults.

Group-living animals may provide insight into the evolution and expression of animal morality. Mech (1970) reported that the number of wolves who could live together in a coordinated pack was governed by the number of wolves with whom individuals could closely bond (social attraction factor) balanced against the number of individuals from whom an individual could tolerate competition (social competition factor). Codes of conduct, and consequentially packs, broke down when there were too many wolves. Whether pack structure was affected by individuals behaving fairly or unfairly is not known, but this would be a valuable topic for future research in wolves and other social animals.

In summary, we argue that mammalian social play is a useful behavioral phenotype on which to concentrate in order to learn more about the evolution and development of cognitive skills and perhaps social morality. There is strong selection for playing fairly because most if not all individuals benefit from adopting this behavioral strategy (and group stability may be also be fostered). Numerous mechanisms (play invitation signals, variations in the sequence of actions performed during play compared with other contexts, self-handicapping, role reversing) have evolved to facilitate the initiation and maintenance of social play in numerous mammals—to keep others engaged—so that an agreement to play fairly and the resulting benefits of doing so can be readily achieved.

Future comparative research that considers the nature and details of the social exchanges that are needed for animals to engage in play—reciprocity and cooperation—will undoubtedly produce data that bear on the questions raised in this brief essay. These are empirical questions for which there are few comparative data. Learning about the taxonomic distribution of animal morality involves answering many difficult questions. Perhaps it will turn out that the best explanation for existing data in some taxa is that some individuals do indeed on some occasions modify their behavior to play fairly.

Play may be a unique category of behavior in that asymmetries are tolerated more than they are in other social contexts. Play cannot occur if the individuals choose not to engage in the activity, and the equality (or symmetry) needed for play to continue makes it different from other forms of seemingly cooperative behavior (e.g., hunting, care giving, grooming, food sharing). This sort of egalitarianism is thought to be a precondition for the evolution of social morality in humans (Bekoff 2001).

All in all, our interdisciplinary collaboration has been a fruitful one in which each of us has brought to the table different skills and perspectives on common interests, namely, the evolution and development of cognitive skills in general and more specifically, how animals communicate their intentions to play and engage in the cooperative exchanges needed to maintain this activity. Social play is difficult to characterize but easily recognized, and its wide taxonomic distribution suggests that there is much potential for future comparative cognitive work, but that it will require more interdisciplinary collaboration.

References

Allen, C. (1992a). Mental content and evolutionary explanation. *Biology and Philosophy* 7: 1–12.

Allen, C. (1992b). Mental content. *British Journal for the Philosophy of Science* 43: 537–553.

Allen, C. (1999). Animal concepts revisited: The use of self-monitoring as an empirical approach. *Erkenntnis* 51: 33–40.

Allen, C. (in press). A tale of two froggies. *Canadian Journal of Philosophy*.

Allen, C. and Bekoff, M. (1994). Intentionality, social play, and definition. *Biology and Philosophy* 9: 63–74.

Allen, C. and Bekoff, M. (1995). Function, natural design, and animal behavior: Philosophical and ethological considerations. In *Perspectives in Ethology*. Vol. 11, *Behavioral Design*, N. S. Thompson, ed., pp. 1–47. New York: Plenum.

Allen, C. and Bekoff, M. (1997). *Species of Mind: The Philosophy and Biology of Cognitive Ethology*. Cambridge, Mass.: MIT Press.

Allen, C. and Hauser, M. D. (1991). Concept attribution in nonhuman animals: Theoretical and methodological problems in ascribing complex mental processes. *Philosophy of Science* 58: 221–240.

Allen, C. and Hauser, M. (1993). Communication and cognition: Is information the connection? *Philosophy of Science Association* 1992, 2: 81–91.

Allen, C. and Saidel, E. (1998). The evolution of reference. In *The Evolution of Mind*, D. Cummins and C. Allen, eds., pp. 183–203. New York: Oxford University Press.

Axelrod, R. (1984). *The Evolution of Cooperation*. New York: Basic Books.

Bateson, G. (1955). A theory of play and fantasy. *Psychiatric Research Reports A* 2: 39–51.

Bekoff, M. (1972). The development of social interaction, play, and metacommunication in mammals: An ethological perspective. *Quarterly Review of Biology* 47: 412–434.

Bekoff, M. (1977a). Social communication in canids: Evidence for the evolution of a stereotyped mammalian display. *Science* 197: 1097–1099.

Bekoff, M. (1977b). Mammalian dispersal and the ontogeny of individual behavioral phenotypes. *American Naturalist* 111: 715–732.

Bekoff, M. (1995). Play signals as punctuation: The structure of social play in canids. *Behaviour* 132: 419–429.

Bekoff, M. (2001). Social play behavior: Cooperation, fairness, trust, and the evolution of morality. *Journal of Consciousness Studies* 8: 81–90.

Bekoff, M. and Allen, C. (1992). Intentional icons: Towards an evolutionary cognitive ethology. *Ethology* 91: 1–16.

Bekoff, M. and Allen, C. (1998). Intentional communication and social play: How and why animals negotiate and agree to play. In *Animal Play: Evolutionary, Comparative, and Ecological Perspectives*, M. Bekoff and J. A. Byers, eds., pp. 97–114. New York: Cambridge University Press.

Bekoff, M. and Byers, J. A. (1981). A critical reanalysis of the ontogeny of mammalian social and locomotor play: An ethological hornet's nest. In *Behavioral Development: The Bielefeld Interdisciplinary Project*, K. Immelmann, G. W. Barlow, L. Petrinovich, and M. Main, eds., pp. 296–337. New York: Cambridge University Press.

Bekoff, M. and Byers, J. A. (eds.) (1998). *Animal Play: Evolutionary, Comparative, and Ecological Approaches.* New York: Cambridge University Press.

Burghardt, G. M. (in press). *The Genesis of Play: Testing the Limits.* Cambridge, Mass.: MIT Press.

Byrne, R. and Whiten, A. (1988). *Machiavellian Intelligence: Social Expertise and the Evolution of Intellect in Monkeys, Apes, and Humans.* New York: Oxford University Press.

Cheney, D. L. and Seyfarth, R. M. (1990). *How Monkeys See the World: Inside the Mind of Another Species.* Chicago: University of Chicago Press.

Dennett, D. C. (1987). *The Intentional Stance.* Cambridge, Mass.: MIT Press.

Dugatkin, L. A. (1997). *Cooperation Among Animals: An Evolutionary Perspective.* New York: Oxford University Press.

Dugatkin, L. A. and Bekoff, M. (submitted). The evolution of fairness: a game theory model.

Fagen, R. (1981). *Animal Play Behavior.* New York: Oxford University Press.

Fagen, R. (1993). Primate juveniles and primate play. In *Juvenile Primates: Life History, Development, and Behavior*, M. E. Pereira and L. A. Fairbanks, eds., pp. 183–196. New York: Oxford University Press.

Frith, C. D. and Frith, U. (1999). Interacting minds—a biological basis. *Science* 286: 1692–1695.

Humphrey, N. (1976). The social function of intellect. *Perspectives in Ethology* 2: 303–321.

Jolly, A. (1966). Lemur social behavior and primate intelligence. *Science* 153: 501–506.

Mech, L. D. (1970). *The Wolf.* Garden City, N.Y.: Doubleday.

Millikan, R. G. (1984). *Language, Thought, and Other Biological Categories.* Cambridge, Mass.: MIT Press.

Power, T. G. (2000). *Play and Exploration in Children and Animals.* Hillsdale, N.J.: Lawrence Erlbaum Associates.

Ridley, M. (1996). *The Origins of Virtue: Human Instincts and the Evolution of Cooperation.* New York: Viking.

Rosenberg, A. (1990). Is there an evolutionary biology of play? In *Interpretation and Explanation in the Study of Animal Behavior.* Vol. 1, *Interpretation, Intentionality, and Communication*, M. Bekoff and D. Jamieson, eds., pp. 180–196. Boulder, Col: Westview Press.

Scott, J. P. and Fuller, J. L. (1965). *Genetics and the Social Behavior of the Dog.* Chicago: University of Chicago Press.

Spinka, M., Newberry, R. C., and Bekoff, M. (2001). Mammalian play: Training for the unexpected. *Quarterly Review of Biology* 76: 141–168.

Watson, D. M. and Croft, D. B. (1996). Age-related differences in playfighting strategies of captive male red-necked wallabies (*Macropus rufogriseus banksianus*). *Ethology* 102: 33–346.

54 The Morals of Animal Minds

Lori Gruen

On January 13, 1962, the *London Daily Mirror* ran a story, "The Death of a Hero." The hero was a dog, Blackie, who tried to drag Ian Beech, the human infant with whom he lived, from a blazing fire that was consuming the Beech house. Blackie was unable to escape in time, and both he and Ian perished in the fire. The coroner remarked at the inquest that Blackie's teeth marks on Ian's shoulder were "gentle, gripping marks [indicating] that this dog made an attempt to get the body away from the fire." Ian was found just a foot away from Blackie's outstretched paws.

On August 16, 1996, at the Brookfield zoo near Chicago, a 3-year-old boy fell 18 feet onto the concrete floor of a gorilla enclosure. The child hit his head and was unconscious. Nervous onlookers and the child's parents panicked, believing the child was in danger from the gorillas. To their astonishment, a 7-year-old gorilla mother named Binti Jua, with her own baby on her back, gently picked up the child and carried him to a door within easy reach of zoo staff. The boy was taken to the hospital and recovered quickly.

In 1871, Darwin described a "case of a little American monkey":

Several years ago a keeper at the Zoological Gardens showed me some deep and scarcely healed wounds on the nape of his own neck, inflicted on him, whilst kneeling on the floor, by a fierce baboon. The little American monkey, who was a warm friend of this keeper, lived in the same compartment, and was dreadfully afraid of the great baboon. Nevertheless, as soon as he saw his friend in peril, he rushed to the rescue, and by screams and bites so distracted the baboon that the man was able to escape, after, as the surgeon thought, running great risk of his life. (Darwin 1871/1981)

Such anecdotes capture our attention perhaps because they show cross-species concern in an unusual direction. We humans care about animals, but it is remarkable when animals act as if they care about us. But does it make sense to describe these behaviors as motivated by what might be called moral concern? When animals behave in "heroic" or "generous" ways within or across species; when tolerance and assistance is offered to handicapped conspecifics; when animals behave as if they are experiencing guilt, shame, or embarrassment; when they build alliances, cooperate, and reconcile after conflicts; when they shun individuals who do not play fair; when they seek revenge or retribution for previous unacceptable behavior, can we usefully and meaningfully describe their actions as stemming from moral sentiments? (Bekoff 2001; de Waal 1989, 1996; Flack and de Waal 2000)

The systematic study of behaviors that might be described as morally motivated—behaviors that are empathetic, sympathetic, compassionate, trustworthy, shameful, vengeful, conciliatory; or that display a sense of friendship, loyalty, justice, or fairness—has just gotten under way. But interest in the topic is certainly not new and the idea that we might achieve a better understanding of morality by studying animal cognition has been suggested at least since Darwin's day. As he noted, his "investigation" into the development of mental capacities in humans and nonhumans "possesses, also, some independent interest, as an attempt to see how far the study of the lower animals throws light on one of the highest psychical faculties of man"—the moral sense (Darwin 1871/1981).

Recent studies being done to explain putatively moral behavior and the complex emotional and cognitive capacities on which such behavior depends, raise interesting and important questions relevant to those working in ethics. There are obvious questions about the boundaries of the moral community, i.e., who should matter, how they should matter, and why (DeGrazia 1996; Singer 1975; Van de Veer 1979; Varner 1998), as well as questions about human obligations to nonhumans within the moral community. For example, if animals feel sadness when they are separated from their kin, should we endeavor to keep families together in captivity? If anesthetized animals are being physically manipulated in laboratory experiments, should we be sure that their conspecifics do not witness these actions? What sort of cognitive enrichment should be provided for animals in captivity (Fouts 1986)?

Work that examines moral sentiments in nonhumans also raises less obvious but equally important questions about the conditions of moral agency and notions of practical reasoning. Because it has been presumed that humans are the only beings capable of moral sentiments and behavior, some ethical theorists believe they have license to theorize in ways that are disconnected from empirical studies in cognitive ethology (and even human psychology, but that is a topic for another day). In this brief discussion, I want to highlight the impact that cognitive ethology can have on debates about moral agency. I want to suggest that cognitive ethology can help reframe topics in ethical theory. Along the way I will identify dangerous ethical assumptions that cognitive ethologists should seek to avoid. Like Darwin, I believe there are exciting morals to be drawn from the study of animal minds.

One of the things that sets humans apart from other animals is our perennial efforts to establish our distinctiveness from them. Identifying which capacities are distinctively human, however, has not been easy. Showing how those capacities are morally important has been harder still. A number of candidate capacities and activities have been proposed over the years: using tools, maintaining family ties, generating culture, solving social problems, starting wars, having sex for pleasure, and using language are just a few. As it is turning out, while all of these capacities are to some degree cognitive, it looks as if none are distinctly human. Most of the ordinary activities that humans engage in have been observed, often in less elaborate form, in some nonhuman animal or other.

It has often been suggested that the morally important difference between humans and nonhumans is that humans are moral agents. Moral agency has been understood to mean a number of different things: the ability to have moral sentiments, the ability to engage in moral deliberation, the ability to make moral judgments, or the ability to act morally. These abilities in turn are thought to require different cognitive capacities. The notion of personhood has often been used to describe the capacity necessary for moral agency, and historically Kant is the most noted defender of this view. Kant maintained that "The fact that the human being can have the representation 'I' raises him infinitely above all the other beings on earth. By this he is a person.... that is, a being altogether different in rank and dignity from things, such as irrational animals, with which one may deal and dispose at one's discretion" (Kant 1798/1977; Wood 1998). More recently, some Kantian scholars have redescribed this distinctness as the capacity for reflective consciousness, a capacity that nonhumans supposedly lack:

A lower animal's attention is fixed on the world. Its perceptions are its beliefs and its desires are its will. It is engaged in conscious activities, but it is not conscious *of* them. That is, they are not the objects of its attention. But we human animals turn our attention on to our perceptions and desires themselves, on to our own mental activities, and we are conscious *of* them. That is why we can think *about* them. (Korsgaard 1996, p. 93)

Our reflective minds, on this view, allow moral agents to think about whether or not to act on their desires, or, to put it in popular philosophical terminology, moral agents have the capacity to determine whether particular desires should be elevated to reasons for action.

Philosophers are not the only ones trotting along this neo-Kantian path. For example, Marc Hauser has a hunch that nonhuman animals lack moral emotions or moral senses. He writes:

They lack the capacity for empathy, sympathy, shame, guilt, and loyalty. The reason for this emotional hole in their lives, I believe, is that they lack a fundamental mental tool: self-awareness; there is no evidence that they are actually aware of their own beliefs and desires. (Hauser 2000, p. 224)

This hunch, and the Kantian notion of personhood, rests on a number of problematic assumptions about moral agency. One is the idea that the reflective structure of human minds or self-awareness is either there or not. This obscures the possibility that there may be stages of cognitive moral development that lead to full-blown self-awareness in some humans, and it overlooks the cognitive continuity between human and nonhuman minds. As Darwin wrote:

[A]ny animal whatever, endowed with well-marked social instincts would inevitably acquire a moral sense or conscience, as soon as its intellectual powers had become as well, or nearly as well developed, as in man. For, firstly, the social instincts lead an animal to take pleasure in the society of its fellows, to feel a certain amount of sympathy with them, and to perform various services for them.... Secondly, as soon as the mental faculties had become highly developed, images of all past actions and motives would be incessantly passing through the brain of each individual.... Thirdly, after the power of language had been acquired, and the wishes of the community could be expressed, the common opinion how each member ought to act for the public good, would naturally become in a paramount degree the guide to action.... Lastly, habit in the individual would ultimately play a very important part in guiding the conduct of each member. (Darwin 1871/1981)

Darwin may not have been right about the particulars, but his suggestion that moral sentiments develop in stages seems quite plausible (and is substantiated by some psychological work on moral development in children; see Gilligan 1982; Kohlberg 1981).

Importantly, self-awareness is not the first stage. Consider one of your recent experiences of sympathy and ask yourself how much of a role self-awareness played in that experience. When you rush to assist an elderly woman whose leg has fallen between a train and the platform or when you jump into the street to prevent a car from hitting a stray dog, are you acting from sympathy and a desire to prevent the harm that will befall the woman or the dog, or are you acting for a reason that arose by self-reflectively testing your desire to determine whether it should be elevated to the status of a reason? This latter process certainly plays a role in our discussions and analyses of sympathetic experiences, but how much of a role does it play in the experience itself? [For a similar point about the role of language in thought, see Dretske (1993).]

Further, moral reflection may best be understood, not as a single, unified capacity, but rather as made up of a set of capacities: the capacity to have belieflike and desirelike states; the capacity to allow information to alter those states; the capacity to recognize conflicts in desires and to deliberate about solutions and to modify one's motivations so as to act on the solution; the capacity to sympathize or empathize with others; the capacity to weigh outcomes; the capacity to plan; the capacity to coordinate actions; the capacity to step back from all this believing, desiring, deliberating, planning, and coordinating and see it as one's own. In addition, reflective awareness itself comes in degrees. Reflective awareness manifests itself differently within members of our own species and at different times in the course of one's life; sometimes I am more reflective than other times and some humans are always more reflective than I am. Certainly it can be argued that the final suite of capacities, stepping back and endorsing one's believing, desiring, deliberating, etc., is what counts as the sine qua non of moral agency. But these arguments will be compelling only if they track empirical understandings of cognition and are based on defensible conceptions of moral motivation and behavior. Recent arguments about moral sentiments and moral agency in nonhumans by some philosophers and some ethologists do neither.

Consider the following claims: Moral agents must place value on moral emotions; moral agents must consider the beliefs, desires, and needs of others when planning action; and moral agents must understand the notions of duty and responsibility (Hauser 2000). Each of these assertions about what is required for moral agency rests on a particular type of normative theory that is focused on the intention of actors rather than on their ability to improve outcomes or the quality of the behavior. The adequacy of various types of normative theory is a matter of continuing philosophical debate. Cognitive ethologists or primatologists who simply pick up one type of normative theory (wittingly or not) and then uncritically adopt the conception of moral agency embedded in that theory are in danger of designing experiments that may not be helpful in understanding the moral cognition of nonhumans.

This brings me to a related concern at a different level of philosophical analysis. I worry that problems in the way most "moral psychology" is currently being done may also lead to the misformulation of the questions being asked by those studying moral sentiments in nonhuman animals. Let me briefly describe one popular view in moral psychology. To be a moral agent, one must act on reasons. Moral agency requires the ability to engage in practical reason, which has traditionally been explicated by reference to belief-desire psychology. Moral beliefs indicate what should be done or what the morally right thing to do is; moral desires motivate us to do that thing, and thus morally right actions are those based on the appropriate beliefs and desires. A problem emerges when our beliefs about what is right are in conflict with our desires—when we desire to do that which we believe to be wrong, or when we fail to desire to do that which we believe to be right. A moral agent will attempt to resolve this conflict (the attempt itself is enough for agency; one does not have to succeed to be considered a moral agent, although continual failure will mark one as immoral). This conflict can be resolved by an exercise in practical reasoning, either by reflecting on one's desires and altering them accordingly or by

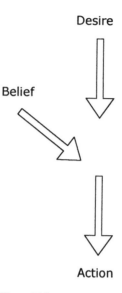

Figure 54.1

reflecting on one's beliefs about what is right and altering them accordingly.

This basic moral psychology, which posits simply beliefs and desires, has been at the heart of much debate among philosophers. Humeans, who accept something like the picture shown in figure 54.1, in which we start with desires and then add beliefs to guide the desires toward their object, are at odds with the Kantians, who accept something like the picture in figure 54.2, in which we start with beliefs and add a motivating desire that is an expression of respect for the moral law or a commitment to one's practical identity. However, both of these pictures seem limited and quite removed from much of the experiences and behaviors in which the complexity of moral sentiments is expressed.

The standard picture readily allows for the design and implementation of experiments that will determine whether nonhuman animals are moral agents as thus understood. However, the theorizing that lies behind these models is decidedly a prioristic and it seems quite possible,

Belief

Desire

Action

Figure 54.2

indeed likely, that because this model of agency simply posits certain moral concepts and categories and describes moral actions based on those posits, it may not accurately reflect actual moral agency at all.

We would do better, I believe, to find out what mental capacities are operating or available when we perform moral actions (and nonintentional mental capacities will probably play a role here) and then seek to understand their operation. Since the social psychological literature has provided interesting and useful challenges to certain philosophical understandings of folk psychology, studying animals' apparently moral behavior can help us obtain a better understanding of the cognitive mechanisms that are required for ethical behavior. Recent studies of animal cognition reveal that their mental capacities may be far more complicated and that moral belieflike states and moral desirelike states are much less homogeneous than one might think if one only looked at the armchair folk psychologizing that is done by ethicists discussing reasons for action.

Clearly, more conceptual work on what moral agency is, and what counts as evidence for it, as well as more empirical evidence from a variety of social species is needed before we can confidently assert that humans are moral agents and nonhumans are not. I believe cognitive ethologists interested in the moral sentiments of animals are in a position to lead the way, in the company of sympathetic moral philosophers, rather than following current fashions in philosophical ethics. Indeed, cognitive ethology can contribute greatly to work on naturalizing ethics if the conceptual dangers I have discussed are avoided.

One of the central intuitions at the base of controversies surrounding the legitimacy of cognitive ethology is the view that nonhuman animals do not really think and thus that cognitive ethologists and others who rely on their work are engaging in anthropomorphism, wishful thinking, or worse. My suggestion that cognitive ethology can reshape our thinking about the nature of ethical behavior will surely be met with even more extreme skepticism. Even if animals think, they cannot possibly think ethically. Much will depend on how we conceive of what it means to think ethically and what behaviors are considered ethical. Collaborations between those working in more naturalistic ethics and those working in cognitive ethology promise to create more controversy, but also to yield a deeper understanding of what it means to be moral.

Acknowledgments

I would like to thank the volume editors and Matthew Barrett and Valerie Tiberius for constructive comments and helpful discussions.

References

Bekoff, M. (ed.) (2000). *The Smile of a Dolphin: Remarkable Accounts of Animal Emotions*. New York: Random House/Discovery Books.

Bekoff, M. (2001). Social play behavior: Cooperation, fairness, trust and the evolution of morality. *Journal of Consciousness Studies* 8: 81–90.

Darwin, C. (1871/1981). *The Descent of Man, and Selection in Relation to Sex*. Princeton, N.J.: Princeton University Press.

DeGrazia, D. (1996). *Taking Animals Seriously: Mental Life and Moral Status*. New York: Cambridge University Press.

de Waal, F. B. M. (1989). *Peacemaking Among Primates*. Cambridge, Mass.: Harvard University Press.

de Waal, F. B. M. (1996). *Good Natured*. Cambridge, Mass.: Harvard University Press.

Dretske, F. (1993). The nature of thought. *Philosophical Studies* 70: 185–199.

Flack, J. C. and de Waal, F. B. M. (2000). Any animal whatever: Darwinian building blocks of morality in monkeys and apes. *Journal of Consciousness Studies* 7: 1–29.

Fouts, R. (1986). *Signs of Enrichment*. Videotape produced by Friends of Washoe, Central Washington University, Ellensberg, Wash.

Gilligan, C. (1982). *In a Different Voice*. Cambridge, Mass.: Harvard University Press.

Hauser, M. (2000). *Wild Minds*. pp. 250–252. New York: Henry Holt.

Kant, I. (1798/1977). *Anthropologie in pragmatischer Hinsicht* (translated as *Anthropology from a Pragmatic Point of View* by Victor Lyle Dowdell). Carbondale: Southern Illinois University Press.

Kohlberg, L. (1981). *The Philosophy of Moral Development*. Cambridge, Mass.: Harvard University Press.

Korsgaard, C. (1996). *Sources of Normativity*. Cambridge: Cambridge University Press.

Singer, P. (1975). *Animal Liberation*. New York: Avon Books.

Van de Veer, D. (1979). Interspecific justice. *Inquiry* 22: 55–70.

Varner, G. E. (1998). *In Nature's Interests? Interests, Animal Rights, and Environmental Ethics*. New York: Oxford University Press.

Wood, A. (1998). Kant on duties regarding non-rational nature. *Aristotelian Society Supplementary Volume*: 189–210.

55 Eye Gaze Information-Processing Theory: A Case Study in Primate Cognitive Neuroethology

Brian L. Keeley

Cognitive Neuroethology

The eyes of others represent an important social cue for humans and other animals, a cue that is often used to support inferences about the mental states of those whose eyes we observe. Is she *paying attention* to what I'm saying? Does he *know* that I was just smiling at his predicament? Does this person *find me attractive*? One important clue for answering such questions can be found in paying attention to where eye gaze has been directed. In essence, we use eyes to read the minds of others.

This is not a new or startling observation; as the sixteenth-century proverb goes, "The eyes are the window of the soul." This commonplace observation generates a number of interesting questions in a variety of different areas of inquiry. *Philosophy:* What do we really learn from observing eyes? Can we ever be certain that our mental attributions are correct? *Neurobiology:* What brain mechanisms are at play when we perceive eyes and draw inferences about them? *Ethology:* What are the relevant eye gaze behaviors and how do they relate to the life of the organism? For example, what is the relationship between eye gaze behavior and social hierarchy? Or, how is it used to facilitate turn taking during conversation? *Evolutionary and comparative biology:* How did this capacity evolve? How do human skills with eyes relate to those of other animals? Do canids or nonhuman apes, say, draw inferences from eye gaze similar to those that we do? Finally, all of these questions are related to the concerns of *cognitive ethology*: How do animals use the eye gaze behavior of other organisms to reason about their unseen mental attitudes? For lack of a better term or acronym, I refer to our scientific understanding of the capacity to process information about mental states from the eye gaze behavior of other organisms as eye gaze information-processing theory (EGIPT, for short).

Eye gaze phenomena come in a variety of forms (Argyle and Cook 1976). We *follow gaze* when we look at the eyes of another and then gaze in the direction they are looking, usually to see what they are looking at. *Mutual gaze* occurs when two individuals look each other in the eye; that is, they make "eye contact." There also seems to be an important difference between looking *at* another's eyes and looking *into* those eyes. (Consider the difference between a lover gazing into your eyes and your ophthalmologist doing the same thing.) *Gaze avoidance* is the behavior of avoiding eye contact with another. Some cultures single out certain eye gaze phenomena as socially significant, such as the phenomenon of the "evil eye." Finally, not all eye gaze reading requires making inferences about the mental states of others. It is possible to follow the gaze of another without also thinking about the other in mental terms; so we often need to be careful to rule out simpler, nonmental mechanisms when comparing the eye gaze perception of other animals with our own. Just because they seem to be doing the same thing that we are does not mean that they are in fact doing so.

My particular theoretical orientation is best described as cognitive neuroethology. I see cognitive neuroethology as the natural synthesis of the recent flowering of interest in the mental lives of animals represented by cognitive ethology with the traditional concerns of neuroethology. Neuroethology is the branch of biology that attempts to elucidate the evolutionary and neurobiological basis of naturally occurring animal behavior (Ewert 1980; Camhi 1984). In other words, neuroethologists first attempt to identify natural behaviors and perceptual capacities in organisms and then investigate (1) how the nervous systems of those animals mediate those behaviors and capacities and (2) how both the behavior and the brain mechanisms have evolved. ("Natural" in this context is meant to distinguish those behaviors that animals *in fact* perform in their native environments from those an organism can be induced to perform under artificial, laboratory conditions.)

Just as neuroethology was a natural outgrowth and addition to the science of ethology, cognitive neuroethology seems to me to be a natural outgrowth and addition to cognitive ethology. Also, just as neuroethology is not intended as a more fundamental replacement of ethology, neither should cognitive neuroethology be construed as eliminating the need for cognitive ethology. However, if successful, one would expect that discoveries concerning the neural basis of behavior should affect how cognitive ethologists characterize and understand the mental capacities of animals. Certainly, an influence in the opposite direction is also to be expected.

For the past several years I have been arguing that there is scientific and philosophical profit in taking a more neuroethological approach to the questions of cognitive science. I began by noting that a then-novel theoretical orientation in neuroethology—computational neuroethology—was successfully addressing questions about the evolution and neural basis of animal behavior within a framework mirroring that of cognitive science (Keeley 1997). From the perspective of philosophy of science, the structure is not identical. The primary difference is that neuroethology, by its nature, is much more concerned with evolution and comparative biology than traditional cognitive science has ever been. I argued that computational neuroethology's success suggests that cognitive science can be more comparative, and furthermore, that it should be (Keeley 2000a). [However, mentioning this in this volume is likely to be preaching to the choir; one of the features that separates cognitive ethology from cognitive science is the

former discipline's explicit embracing of the comparative method (e.g., Allen and Bekoff 1997; see also Keeley 1999b).] Central to the aforementioned "profit" of taking a more neuroethological approach to the issues of cognitive science is that it offers novel ways of dealing with long-standing philosophical conundrums in the philosophy of cognitive science, including those related to the indeterminacy of content and function (Keeley 1999a) and the alleged theoretical autonomy of psychology owing to the multiple realization of psychological states (Keeley 2000b).

One potential drawback to all the work just described is that it is based on the computational neuroethology of weakly electric fish. However, I argue that we can use the success of neuroethology in understanding the behavior of electric fish—exemplified in the work of the late Walter Heiligenberg (1991a,b)—as a philosophical model for a slightly reconfigured but potentially more successful cognitive science. But because fish are not the most cognitively interesting of creatures, this argument is at best one of principle. Even if correct, I am still left with the question of what this all would mean in practice; what exactly would a cognitive neuroethology of human intelligent behavior look like anyway? The goal of the work described here is to begin to put some flesh on the bones of my largely philosophically motivated proposal.

The State of Play in EGIPT

If any phenomenon were a currently strong contender for yielding to a cognitive neuroethological perspective, EGIPT would seem to be it. During the past decade, significant progress has been made in the relevant areas of study, although on the whole these investigations have been carried out in relative isolation from one another. However, as work in each of these areas continues to mature, it should be fruitful to combine their insights in order to come to a deeper and more general understanding of how human and nonhuman animals use eye gaze cues to read one another's minds. I am not alone in this assessment. In a recent, important review, N. J. Emery (2000) has brought together much of the research I describe here (although not the computational research; see the later discussion) in an attempt to sketch a series of interesting hypotheses about EGIPT. In the following sections I describe recent work in a number of areas brought together under the rubric of EGIPT: work on the developmental disorder of autism, the comparative biology and neurobiology of social cognition, and computational and robotic models of eye gaze-related phenomena.

Autism and Mindblindness

Autism is a pervasive developmental disorder of unknown etiology. In addition to some straightforward cognitive deficits affecting general intelligence, language skills, and attentional control, autistic subjects exhibit consistent and often severe problems with socialization. Compared with mental age-matched individuals, such as those with Down syndrome, autistic subjects have been shown to have difficulties reasoning about the mental states of others (Baron-Cohen et al. 1993). This difficulty in dealing with a theory of mind has led some theorists to propose that the core feature of autism is "mindblindness" (Baron-Cohen 1995; Frith and Happé 1999).

Autistic individuals also typically have an unusual relationship to eyes. Originally, gaze avoidance was considered a diagnostic feature of the disorder, although it is now thought that autistic individuals do not so much *avoid* gaze as restrict their sampling to quick glances at the eyes of others (O'Connor and Hermelin 1967; Volkmar and Mayes 1990). Some theorists, for example, Baron-Cohen (1995), interpret current data as showing that eye gaze detection in autism is unimpaired, meaning that the deficit is likely isolated in a putative theory of mind module,

which is receiving accurate information about eye gaze.

One intriguing possibility is that eye gaze stimuli adversely stimulate the autonomic nervous system in autism, perhaps inhibiting the normal development of a theory of mind. If so, any deficit in autism is likely to be much more low level and pervasive than a cortical theory of mind module. With this possibility in mind, I have been collaborating with William Hirstein and V. S. Ramachandran to encourage their exploration of this subject (Hirstein et al. in press).

The jury is still out with respect to the "autism as mindblindness" hypothesis. Nonetheless, it seems safe to say that there is something unusual about the autistic subject's perceptual relationship to eyes. This is an interesting observation in light of the current project for a historical reason. One of the final research projects of Niko Tinbergen—who shared a 1973 Nobel Prize for his role in creating the field of ethology—was the ethological study of autism (Tinbergen and Tinbergen 1983). In these studies, characterizing the behavior of autistic children in natural contexts was central, and chief among these behaviors were a variety of avoidance behaviors, including, not surprisingly, those related to gaze. The current project can be seen as a return to Tinbergen's final project, a promising subject that has been neglected since his passing.

The Comparative Perspective

The study of EGIPT in primates has been an explosive area of research in the past decade. It has long been known that the eyes of primate species vary; the color contrast between the sclera and pupil of the human eye is greater than that found in bonobos and chimpanzees, which in turn is greater than that found in other apes and monkeys. For this and other reasons, it seems reasonable to hypothesize that our eyes have evolved to be more easily read at a distance. What is more, the capacity to read gaze information seems to have evolved with the most recent common ancestor of apes and monkeys. While most primates can follow the gaze of others (Tomasello et al. 1998), lemurs—prosimians (i.e., the next closest relatives of primates)—cannot (Anderson and Mitchell 1999). At what point in evolution primates began inferring the mental states of others from their eyes is currently under debate. Some argue that this is a uniquely human adaptation and that even chimpanzees lack this skill (Povinelli and Eddy 1996). However, traditionally, apes have been thought to make such inferences, although the new debate has brought about a new series of more careful studies of exactly what apes do and do not understand about the mental significance of gaze (Hare et al. 2000, 2001).

EGIPT may be of central importance to the social cognition of primates, but it is clearly not restricted to them. For prey species, the visual cue of two black or colored circles in a horizontal orientation may be indicative of predation in the near future. Therefore it should not be surprising that a wide range of species, from fish to birds to mouse lemurs, have developed aversive responses to such stimuli (see Emery 2000 for a review). Animals as diverse as plovers (Ristau 1991) and black iguanas (Burger et al. 1992) have been shown to respond differently to approaching human experimenters who are looking directly at them and those who approach while looking in other directions. Even hognose snakes, which feign death when harassed by predators, use eye gaze cues when deciding whether to continue feigning death or whether it is safe to beat a retreat (Burghardt 1991; Burghardt and Greene 1988).

As a final example, Agnetta et al. (2000) report the interesting finding that while domesticated dogs are capable of using human gaze as a cue to perform tasks, such as finding hidden food, they do not spontaneously follow human gaze. This suggests potentially interesting questions about the EGIPT mechanisms of dogs

(and, by extension, other animals). Prior to this work, it might have been natural to think that an ability to draw inferences from gaze cues was built on top of a prior existing gaze-following mechanism. However, the situation may be more complicated than previously surmised. So suffice it to say, EGIPT offers the opportunity to compare the social cognition of humans with non-human primates and other animals that have evolved related capacities.

Neurobiology

What neural structures underlie the processing of information about gaze? Research on that workhorse of visual neurophysiology, the macaque monkey, has found cells that are selectively responsive to the face in several visual subareas of the temporal cortex. These include the lateral and ventral surfaces of the inferior temporal cortex and the upper bank, lower bank, and fundus of the superior temporal sulcus. Perrett and colleagues have found cells in the upper bank of the superior temporal sulcus that are selective for the orientation of facial stimuli. One obvious functional interpretation of this is that these cells help detect where faces are looking (Perrett et al. 1991, 1992; see also Hasselmo et al. 1989).

In humans, functional magnetic resonance imaging (fMRI) studies have identified the extrastriate cortex as a center of activity in the processing of facial images (Kanwisher et al. 1996a,b). Using both intracranial electrodes and fMRI, Puce and colleagues have discovered that "Faces primarily activated the fusiform gyrus bilaterally, and also activated the right occipito-temporal and inferior occipital sulci and a region of lateral cortex centered in the middle temporal gyrus" (Puce et al. 1996, p. 5205; see also Allison et al. 1994).

Of the areas that are implicated in the processing of faces, which are responsible for our being able to extract gaze information from a face? Both the amygdala and extrastriate cortex are suggested in separate studies. A study of D.R., a 51-year-old woman with a partial bilateral amygdalotomy, reported that she is impaired in verbally reporting gaze information (Young et al. 1996). (Since the amygdala is part of the limbic system, which in turn is intimately connected to the autonomic nervous system, findings such as these represent a potential bridge to current work on autism.) It has also been reported that results of evoked-response potential studies of the extrastriate cortex suggest the possible presence of an "eye detector" module (Allison et al. 1996; Bentin et al. 1996). So while we are far from a complete understanding of the neural basis of EGIPT, we have a firm and growing foundation of knowledge from which to test hypotheses generated from our understanding of autism and comparative studies.

Computational Work

In its early years, the field of artificial intelligence (AI) pointedly avoided building systems that were too close to the physical or biological nature of humans and other animals. All of this has changed in recent decades, as neural nets—based on a rough analogy to neurobiological systems—saw a resurgence of interest beginning in the 1980s. As we approached the millennium, a movement in AI started with the goal of exploring the possibility of "humanoid robots"; that is, robots with a roughly human form. This in turn has raised issues of EGIPT because these robots need to behave as humans do and understand human eye gaze behavior so that their human interlocutors can better understand the humanoid robots, and vice versa.

The vanguard of this new work is coming out of the Artificial Intelligence Laboratory at the Massachusetts Institute of Technology, particularly the work of Rod Brooks on "Cog" (Brooks et al. 1999) and Cynthia Breazeal on "Kismet" (Breazeal and Scassellati 1999, in press; Breazeal et al. 2000). While Cog is an attempt to build a general and sophisticated humanoid robot, Kismet is focused more on eye gaze behavior and communication via facial expressions. Breazeal and colleagues describe Kismet as

[A]n active vision head augmented with expressive facial features. Kismet is designed to receive and send human-like social cues to a caregiver, who can regulate its environment and shape its experiences as a parent would for a child. Kismet has three degrees of freedom to control gaze direction, three degrees of freedom to control its neck, and fifteen degrees of freedom in other expressive components of the face (such as ears and eyelids). To perceive its caregiver Kismet uses a microphone, worn by the caregiver, and four color CCD cameras. The positions of the neck and eyes are important both for expressive postures and for directing the cameras towards behaviorally relevant stimuli. (Breazeal et al. 2000, p. 2)

Kismet and other humanoid robots (see, for example, Mousset et al. 2000) are important testbeds of theories of eye gaze behavior and perception. The computational and robotic models of AI allow us to put into physical reality our theories of how biological and psychological phenomena seem to work and behave. This then allows us to test these theories and to discover more general theoretical principles of EGIPT that apply across species.

Conclusion

This project of synthesizing the recent work on EGIPT from a variety of scientific perspectives is still in its early stages, and in the short space allowed here, I can only begin to sketch the wealth of new discoveries that continue to appear. Since it is still "early days yet" for cognitive neuroethology and EGIPT, let me conclude by pointing to a number of questions that call for interdisciplinary interaction between the scientific areas I have discussed here:

What is the role of the autonomic nervous system in the eye gaze and facial perception of nonhuman animals? How does the emotional significance of eyes for humans compare with that of other animals who do (or do not) attend to eye gaze?

What differences in neural processing are there in humans between mere gaze following versus drawing mental inferences from gaze? If it is true that some animals follow gaze without attributing mental states to others, this raises the possibility that human mind reading is independent of gaze perception by itself. Is this indeed the case?

How do the deficits of autistic subjects compare with the natural capacities of nonhuman primates? If it is indeed the case that autism is mindblindness, what exactly is the relationship between EGIPT capacities in autistics and those of nonhuman animals that are similarly "mind blind"?

What general principles of EGIPT can be discovered in robotic and other computational models, and what does this tell us about the cognitive neuroethology of EGIPT in primates and other animals? Are there general principles that apply across phylogeny?

Finally, since my primary academic training is in philosophy, my hope is that in the end this work will shed light on that hoary chestnut of philosophical skepticism: the problem of other minds. If we can be said to mind read using the eyes of others, what are the epistemic limits of this process? How certain can we be of our mental attributions to others; to what illusions are we susceptible and under what conditions? There are many unanswered questions here, but there is also much promise of progress to be made and understanding to be had.

Acknowledgments

The work described here is supported by a grant to the author from the *McDonnell Project in Philosophy and the Neurosciences*. I thank my colleagues in that project for useful feedback, including Kathleen Akins, Tony Atkinson, Tori McGeer, Evan Thompson, and Pete Mandik. Thanks also to Bill Hirstein and to Eric Courschesne (and the members of his Laboratory for Research on the Neuroscience of Autism, San Diego) for help in the earliest stages of this project.

References

Agnetta, B., Hare, B., and Tomasello, M. (2000). Cues to food location that domestic dogs (*Canis familiaris*) of different ages do and do not use. *Animal Cognition* 3: 107–112.

Allen, C. and Bekoff, M. (1997). *Species of Mind: The Philosophy and Biology of Cognitive Ethology*. Cambridge, Mass.: MIT Press.

Allison, T., Ginter, H., McCarthy, G., Nobre, A. C., Puce, A., Luby, M., and Spencer, D. D. (1994). Face recognition in human extrastriate cortex. *Journal of Neurophysiology* 71: 821–825.

Allison, T., Lieberman, D., and McCarthy, G. (1996). Here's not looking at you kid: An electrophysiological study of a region of human extrastriate cortex sensitive to head and eye aversion. *Society for Neuroscience Abstracts* 22: 400.

Anderson, J. R. and Mitchell, R. W. (1999). Macaques but not lemurs co-orient visually with humans. *Folia Primatology* 70: 17–22.

Argyle, M. and Cook, M. (1976). *Gaze and Mutual Gaze*. Cambridge: Cambridge University Press.

Baron-Cohen, S. (1995). *Mindblindness: An essay on Autism and Theory of Mind*. Cambridge, Mass.: MIT Press.

Baron-Cohen, S., Tager-Flusberg, H., and Cohen, D. J. (eds.) (1993). *Understanding Other Minds: Perspectives from Autism*. Oxford: Oxford University Press.

Bentin, S., Allison, T., Puce, A., Perez, E., and McCarthy, G. (1996). Electrophysiological studies of face perception in humans. *Journal of Cognitive Neuroscience* 8: 551–565.

Breazeal, C. and Scassellati, B. (1999), "How to build robots that make friends and influence people." In *Proceedings of IEEE International Conference on Intelligent Robots and Systems (IROS99): Human and Environment Friendly Robots with High Intelligence*

Quotients, Oct 17–Oct 21, 1999, Kyongju, South Korea, IEEE, Piscataway, NJ, USA, pp. 858–863.

Breazeal, C. and Scassellati, B. (2000). Infant-like social interactions between a robot and a human caretaker. *Adaptive Behavior* 8: 49–74.

Breazeal, C., Edsinger, A., Fitzpatrick, P., and Scassellati, B. (2000). Social constraints on animate vision. In *IEEE Intelligent Systems and Their Applications*, v. 15. n. 4, Jul 2000, Piscataway, NJ, USA pp. 32–37.

Brooks, R. A., Breazeal, C., Marjanovic, M., Scassellati, B., and Williamson, M. W. (1999). The Cog project: Building a humanoid robot. In *Computation for Metaphors, Analogy and Agents*. Vol. 1562, *Springer Lecture Notes in Artificial Intelligence*, C. L. Nehaniv, ed., pp. 52–87. New York: Springer-Verlag.

Burger, J., Gochfield, M., and Murray, B. G. (1992). Risk discrimination of eye contact and directness of approach in black iguanas (*Ctenosaura similis*). *Journal of Comparative Psychology* 106: 97–101.

Burghardt, G. M. (1991). Cognitive ethology and critical anthropomorphism: A snake with two heads and hog-nose snakes that play dead. In *Cognitive Ethology: the Minds of Other Animals*. C. A. Ristau, eds., pp. 55–90. Hillsdale, N.J.: Lawrence Erlbaum Associates.

Burghardt, G. M. and Greene, H. W. (1988). Predator simulation and duration of death feigning in neonate hognose snakes. *Animal Behaviour* 36: 1842–1844.

Camhi, J. M. (1984). *Neuroethology: Nerve Cells and the Natural Behavior of Animals*. Sunderland, Mass.: Sinauer Associates.

Emery, N. J. (2000). The eyes have it: The neuroethology, function and evolution of social gaze. *Neuroscience and Biobehavioral Reviews* 24: 581–604.

Ewert, J.-P. (1980). *Neuroethology: An Introduction to the Neurophysiological Fundamentals of Behavior*. New York: Springer-Verlag.

Frith, U. and Happé, F. (1999). Theory of mind and self-consciousness: What is it like to be autistic? *Mind and Language* 14: 1–22.

Hare, B., Call, J., Agnetta, B., and Tomasello, M. (2000). Chimpanzees know what conspecifics do and do not see. *Animal Behaviour* 59: 771–785.

Hare, B., Call, J., and Tomasello, M. (2001). Do chimpanzees know what conspecifics know? *Animal Behaviour* 61: 139–151.

Hasselmo, M. E., Rolls, E. T., Baylis, G. C., and Nalwa, V. (1989). Object-centered encoding by face-selective neurons in the cortex in the superior temporal sulcus of the monkey. *Experimental Brain Research* 75: 417–429.

Heiligenberg, W. (1991a). *Neural Nets in Electric Fish*. Cambridge, Mass.: MIT Press.

Heiligenberg, W. (1991b). The jamming avoidance response of the electric fish, *Eigenmannia*: Computational rules and neuronal implementation. *Seminars in the Neurosciences* 3: 3–18.

Hirstein, W., Iversen, P., and Ramachandran, V. S. (in press). Autonomic responses of autistic children to people and objects. *Proceedings of the Royal Society of London*, Series B, vol. 268.

Kanwisher, N., Chun, M. M., and McDermott, J. (1996a). FMRI in individual subjects reveals loci in extrastriate cortex differentially sensitive to faces and objects. *Investigative Ophthalmology and Visual Science* 37: S193.

Kanwisher, N., Chun, M. M., McDermott, J., and Hamilton, R. (1996b). FMRI reveals distinct extrastriate loci sensitive for faces and objects. *Society for Neuroscience Abstracts*, 22: 1937.

Keeley, B. L. (1997). Cognitive Science as the Computational Neuroethology of Intelligent Behavior: Why Biological Facts Are Important for Explaining Cognition. Unpublished Ph.D. thesis, University of California, San Diego.

Keeley, B. L. (1999b). Review of C. Allen and M. Bekoff, *Species of Mind*. *Philosophical Psychology* 12: 543–546.

Keeley, B. L. (2000a). Neuroethology and the philosophy of cognitive science. *Philosophy of Science* 67 (proceedings): S404–S417.

Keeley, B. L. (2000b). Shocking lessons from electric fish: The theory and practice of multiple realization. *Philosophy of Science* 67: 444–465.

Mousset, E., Jabri, M., Carlile, S., and Sejnowski, T. (2000). Gaze-shifting in humans and humanoids. *Humanoids 2000*.

O'Connor, N. and Hermelin, B. (1967). The selective visual attention of autistic children. *Journal of Child Psychology and Psychiatry* 8: 167–179.

Perrett, D. I., Oram, M. W., Harries, M. H., Bevan, R., Hietanen, J. K., Benson, P. J., and Thomas, S. (1991). Viewer-centred and object-centred coding of heads in the macaque temporal cortex. *Experimental Brain Research* 86: 159–173.

Perrett, D. I., Hietanen, J. K., Oram, M. W., and Benson, P. J. (1992). Organization and functions of cells responsive to faces in the temporal cortex. *Philosophical Transactions of the Royal Society of London. B: Biological Sciences* 335: 23–30.

Povinelli, D. J. and Eddy, T. J. (1996). What young chimpanzees know about seeing. *Monographs of the Society for Research in Child Development* 61: 1–152.

Puce, A., Allison, T., Asgari, M., Gore, J. C., and McCarthy, G. (1996). Differential sensitivity of human visual cortex to faces, letterstrings, and textures: A functional magnetic resonance imaging study. *Journal of Neuroscience* 16: 5205–5215.

Ristau, C. A. (1991). Before mindreading: Attention, purposes and deception in birds? In *Natural Theories of Mind*, A. Whiten, ed., pp. 209–222. Oxford: Blackwell.

Tinbergen, Niko and Tinbergen, Elisabeth A. (1983). *"Autistic" Children: New Hope for a Cure*. London: George Allen and Unwin.

Tomasello, M., Call, J., and Hare B. (1998). Five primate species follow the visual gaze of conspecifics. *Animal Behaviour* 55: 1063–1069.

Volkmar, F. R. and Mayes, L. C. (1990). Gaze behavior in autism. *Development and Psychopathology* 2: 61–69.

Young, A. W., Hellawell, D. J., Van De Wal, C., and Johnson, M. (1996). Facial expression processing after amygdalotomy. *Neuropsychologia* 34: 31–39.

56 The Eyes, the Hand, and the Mind: Behavioral and Neurophysiological Aspects of Social Cognition

Vittorio Gallese, PierFrancesco Ferrari, Evelyne Kohler, and Leonardo Fogassi

Nevertheless, the difference in mind between man and the higher animals, great as it is, certainly is one of degree and not of kind.
—Darwin (1871, p. 193)

What most clearly distinguishes primates from other mammal species in terms of cognition concerns the social domain. Although it is not deeply understood, many authors have proposed that the uniqueness of the intelligence of primates has to be sought in their complex social world (see Humphrey 1976). Primates are not the only mammals displaying complex social systems, but it seems that the quality and complexity of their relationships within their social groups, compared with other taxa, are different.

In the evolution of primate cognition, the relationships between group members have played a fundamental role. As Tomasello and Call (1997) pointed out, nonhuman primates understand the quality of the relationships within their social group, not only in terms of kin and hierarchies, but also in terms of coalitions, friendship, and alliances. Primate cognition can be considered very different from that of other mammalian species because primates can categorize and understand third-party social relationships.

In the past few decades, an ever-growing literature has raised questions about the possibility that social behaviors of nonhuman primates are driven by intentions and that these animals understand the behavior of others as intentional. Most scientists would agree on the fact that monkeys and apes behave as if they possessed objectives and goals, although, contrary to humans, their awareness of purpose is not assumed. The fact that nonhuman primates may understand the behavior of conspecifics as goal related can have considerable benefits in an individual's life because it allows the individual to predict the actions of others.

The problem of intentionality in primates was almost simultaneously and independently raised by Humphrey (1980) and Premack and Woodruff (1978). The capacity to attribute mental states such as intentions, beliefs, and desires to others has been described as a theory of mind (Premack and Woodruff 1978). So far there is no firm evidence that nonhuman primates possess a theory of mind. For most authors, a theory of mind constitutes the mental Rubicon between humans and nonhuman primates.

One of the most influential models of a theory of mind is that proposed by Baron-Cohen (1994, 1995). According to this author, separate brain modules constitute a mind-reading system that is layered on multiple modularized levels of increasing complexity. The importance of such a model for the study of social cognition is twofold. First, it enables researchers to investigate several aspects of mind-reading abilities at both the behavioral and the brain level. Second, it allows an empirical evaluation of the cognitive stage reached by a given species, thus paving the way for comparative investigations. Even if something like a theory of mind really underpins mind-reading abilities in humans, this cognitive feature of the human mind must have evolved from a nonhuman ancestor who shared with the present primates—humans included—several cognitive features. In this context, one of the major tasks of cognitive neuroscience should be to investigate the behavioral and neural basis of intentional behavior. The behavioral study of social cognition of nonhuman primates and the examination of the neural mechanisms supporting it are therefore necessary for a thorough understanding of how the human mind evolved and how it works.

In this essay we review a series of investigations of the social behavior of macaque monkeys conducted at a behavioral and a neurophysiological level. First we summarize some recent behavioral experiments carried out in our laboratory to investigate the presence of gaze-following behavior in monkeys. Then we review neurophysiological data concerning the neural mechanisms that enable macaques to understand goal-directed actions performed by other individuals. We propose that such behavioral abilities and neural mechanisms corroborate the notion of cognitive continuity between human and nonhuman primates.

Gaze Following

A gaze-following response (GFR) is defined as the ability of one individual (X) to follow the direction of gaze of a second individual (Y) to a location in space (Emery et al. 1997). The ability to visually track the gaze direction of conspecifics to targets may have a considerable adaptive advantage because individuals can gain information about food sources, the social status of conspecifics, and the location of predators (Menzel and Halperin 1975; Whiten and Byrne 1988). It has been proposed that gaze perception plays a crucial role in social interactions (Whiten and Byrne 1988; Thomsen 1974; Tomasello et al. 1998).

It is well established that apes (chimpanzees and orangutans) are able to follow the gaze direction of conspecifics (Tomasello et al. 1998) and humans (Itakura 1996; Povinelli and Eddy 1996) using a combination of head and eye stimuli. Chimpanzees are also able to follow the gaze direction of a human experimenter by observing the eye direction alone, independently of head movement (Itakura 1996; Povinelli and Eddy 1996; Tomasello et al. 1999). Physiological studies have shown that in monkeys (Perrett et al. 1985, 1992), as in humans (Puce et al. 1998; Hoffman and Haxby 2000), there are neural correlates for detection of eye direction. There is little evidence at the behavioral level, however, of the presence and development of such abilities in monkeys (see Lorincz et al. 1999).

The aim of our study (see Ferrari et al. 2000) was to assess in juvenile and adult pig-tailed macaques (*Macaca nemestrina*) the capacity to use only eye cues to follow the gaze of an exper-

imenter. An experimenter presented biological stimuli (head, eye, and trunk movements) that were oriented in each presentation to one of four spatial directions (up, down, left, or right; see figure 56.1A). The stimuli were presented randomly to 11 monkeys of different ages who were free to move in their home cages. A nonbiological stimulus served as control. In order to judge a monkey's response to the movements of the experimenter, we recorded the direction of the first gaze movement after a stimulus visual engagement (see figure 56.1B).

The results (see figure 56.1C and D) showed that macaques, like humans, are able to determine the direction of another's gaze by using eye cues alone. Furthermore, also as in humans, gaze-following responses in macaques dramatically improve with age. Juvenile monkeys showed a marked difference in gaze-following behavior compared with adults; they were unable to determine the direction of another's gaze using eye cues alone. These results cannot be explained on the basis of differences in attentional factors because adults and juveniles devoted the same amount of time to interacting with and visually exploring the experimenter. In juveniles, the movement of the head and eyes together is the first feature that triggers a shift in a visual attention response, suggesting that in young macaques the combined movement of head and eyes provides more salient signals for the direction of another's gaze than the eyes alone.

This ontogenetic trend resembles that of humans; 3- to 6-month-old infants are able to follow the gaze of an adult by using a combination of head and eye cues, but it is not until 14–18 months that they are able to follow a gaze by using eye cues alone (Scaife and Bruner 1975; Butterworth and Jarret 1991; Moore and Corkum 1998). Thus, in humans this ability develops when children are still dependent on their parents. In monkeys, a gaze-following response based on head and eye cues develops between the second and the fourth year, a stage at which juveniles are weaned but still socially dependent on their mothers (Fa and Lindburg 1996; Smuts et al.

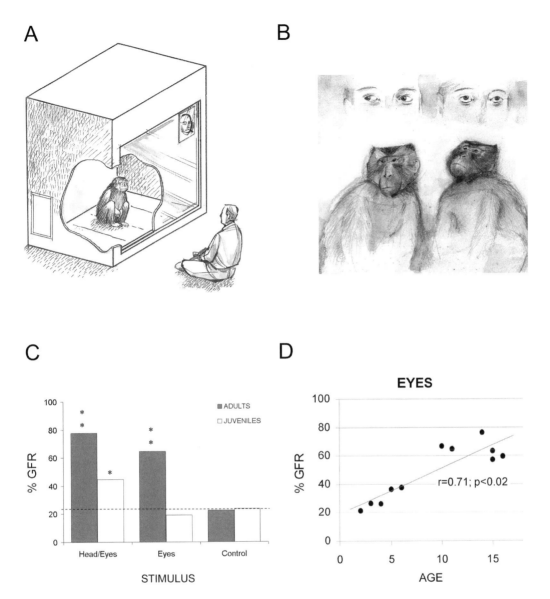

Figure 56.1

(A) Schematic illustration of the experimental setting. The experimenter was sitting facing the monkey at a 1.5-m distance. A camera placed behind the experimenter videotaped each session. A mirror placed on the front of the cage and facing the experimenter allowed the experimenter's gaze to be recorded. (B) Illustration of the results. The direction of the eyes as a stimulus presented by the experimenter (top) elicited a gaze-following response (GFR) in the adult macaque (bottom). (C) Percent of trials in which juvenile and adult monkeys showed a gaze-following response as a function of experimental condition. The data recorded for each monkey during two sessions are averaged and plotted together. Each bar represents the average of the percent of GFR in adults and juveniles. A single asterisk indicates the significance of the comparison between the experimental condition and the control ($p < 0.05$). A double asterisk indicates the significance of the comparison between the experimental condition and the control ($p < 0.05$) and between adults and juveniles ($p < 0.01$). The dashed line indicates the levels of expected probability by chance (25 percent; an animal's response could be oriented to one of the four possible spatial directions). (D) Percent of GFR to eyes alone as a function of age. (Modified from Ferrari et al. 2000.)

1987). However, according to our results, GFR based on eye cues alone seems to develop only at a later stage.

This finding makes it unlikely that the development of GFR is related to the maturation of visuomotor coordination and motor skills, which in monkeys are complete well before the end of the juvenile and adolescent period. It is possible therefore to hypothesize that the development of GFR could be related to the relatively long period preceding adulthood, during which young macaques have the opportunity to better explore the physical environment (Smuts et al. 1987; Janson and van Schaik 1993) and to learn the rules of their complex social world (Walters 1987). Indeed, it is well known that juveniles spend much of their nonfeeding time engaged in social play (for a thorough treatment of the relation between social play and cognition, see Allen and Bekoff 1997; Bekoff and Allen 1998). It has been proposed that through play, juveniles may establish dominance relationships and learn social and communication skills (Walters 1987).

Thus, it seems that in macaques the emergence of a gaze-following response based on head and eye cues could be linked to the processes of transition to adulthood, when individuals have to acquire social and cognitive skills that are crucial for their survival and reproduction. Juvenile monkeys are not able to orient their attention on the basis of eye cues alone. In general, gaze following is more frequent in adults than in juveniles. As in humans, however, in macaques such abilities dramatically improve with age, reinforcing the idea that the transition to a more complex social life is a key factor for the development of cognitive skills such as gaze-following behavior.

Understanding Action: The Role of Mirror Neurons

Observation of action appears to be important in order to build a meaningful account of conspecifics' behavior. How can individuals rec-ognize, and possibly understand, the actions performed by other conspecifics?

A possible neural correlate of the mechanism allowing understanding of action could be represented by a class of neurons—mirror neurons—that our group discovered in area F5 of the ventral premotor cortex of the macaque monkey (di Pellegrino et al. 1992; Gallese et al. 1996; Rizzolatti et al. 1996a). In a series of single-neuron recording experiments, we discovered that in a sector of the monkey ventral premotor cortex, area F5 (see Matelli et al. 1985) (see figure 56.2A), a particular set of neurons that are activated during the execution of purposeful, goal-related hand movements, such as grasping or holding or manipulating objects, also discharged when the monkey observed similar hand actions performed by another individual (see figure 56.2B). We designated these neurons "mirror neurons" (di Pellegrino et al. 1992; Gallese et al. 1996; Rizzolatti et al. 1996a). In order to be activated by visual stimuli, mirror neurons require an interaction between the agent (human being or a monkey) and the object of its action. Figure 56.2C1–C3 shows an example of the response of an F5 mirror neuron. The visual presentation of objects such as food items or objects at hand in the lab did not evoke any response. Similarly ineffective or showing very little effect in driving the neuron response were actions that achieved the same goal and looked similar to those performed by the experimenter's hand, but were made with tools such as pliers or pincers. Actions having emotional content, such as threatening gestures, were also ineffective.

Frequently, a strict congruence was found between an observed action that was effective in triggering the neuron and the executed action. In one-third of the recorded neurons, the effective observed and executed actions corresponded both in terms of the general action (e.g., grasping) and in the way in which that action was executed (e.g., precision grip). In the other two-thirds, only a general congruence was found (e.g., any kind of observed and executed grasping elicited the neurons' response). This latter class of mirror neurons is particularly interesting, because many of them appear to be used in generalizing across different ways of achieving the same goal, thus perhaps enabling a more abstract type of action coding.

Perrett and co-workers have described neurons buried within the superior temporal sulcus (STS) cortex that respond to the observation of complex actions, such as grasping or manipulating objects (for a review see Carey et al. 1997; Jellema and Perrett in press). These neurons, whose visual properties are in many respects similar to those of mirror neurons, could constitute the mirror neurons' source of visual information. The STS region, however, has no direct connection with area F5, but has links with the anterior part of the inferior parietal lobule (area PF or 7b, see figure 56.2A), which in turn is reciprocally connected with area F5 (Matelli et al. 1986; see also Rizzolatti et al. 1998). Area PF, or 7b, is located on the convexity of the inferior parietal lobule. Single-neuron studies showed that most of the PF neurons respond to passive somatosensory stimuli applied to the mouth, arm, leg, or chest (Leinonen and Nyman 1979; Leinonen et al. 1979; Hyvärinen 1981; Graziano and Gross 1995; Fogassi et al. 1998). A considerable number of neurons can be activated by both visual and somatosensory stimuli (bimodal neurons) (Leinonen et al. 1979; Graziano and Gross 1995). About one-third of PF neurons fire during mouth, arm, and hand goal-related movements (Leinonen et al. 1979).

Area PF, through its connections with the STS on one hand, and F5 on the other, could play the role of an intermediate step within a putative cortical network for understanding action, by feeding to the ventral premotor cortex visual information about an action received from the STS.

In a new series of experiments we decided therefore to better clarify the nature and the properties of such a cortical matching system in the monkey brain. The functional properties of area PF were studied by examining neuronal responses during monkeys' active movements and in response to somatosensory and visual stimuli. Visual stimuli included goal-related hand movements performed by the experimenter in front of the monkey. The results of this study showed that about one-third of the PF recorded neurons responded during both execution and observation of an action (Fogassi et al. 1998; Gallese et al. in press). Figure 56.2D1–D3 shows an example of a PF mirror neuron response. All PF mirror neurons responded to the observation of actions in which the experimenter's hand(s) interacted with objects. Similarly to the responses observed in F5, PF mirror neurons did not respond to presentation of an object or to actions performed using tools. Observed mimed actions evoked weaker responses, if any. What these experiments show is that the "mirror" system, matching observation of an action to its execution, is not a prerogative of the premotor cortex, but extends to the posterior parietal lobe as well.

On the basis of these findings, it appears that the sensorimotor integration process supported by the F5-PF fronto-parietal cortical network creates an internal copy of actions that is used to generate and control goal-related behaviors and to provide, at a preconceptual and prelinguistic level, a meaningful account of behaviors performed by other individuals.

Several studies that used different methodologies have demonstrated the existence of a similar matching system in humans (see Fadiga et al. 1995; Grafton et al. 1996; Rizzolatti et al. 1996b; Cochin et al. 1998; Decety et al. 1997; Hari et al. 1998; Iacoboni et al. 1999; Buccino et al. 2001). In particular, it is interesting to note that brain-imaging experiments in humans have shown that during observation of hand action, a cortical network composed of sectors of Broca's region, the STS region, and the posterior parietal cortex is activated (Grafton et al. 1996; Rizzolatti et al. 1996b; Decety et al. 1997; Decety and Grèzes 1999; Iacoboni et al. 1999; Buccino et al. 2001). Given the homology between monkey's area F5 and Broca's region in humans (for discussion see Matelli and Luppino 1997; Rizzolatti and Arbib 1998; Gallese 1999), it appears that a

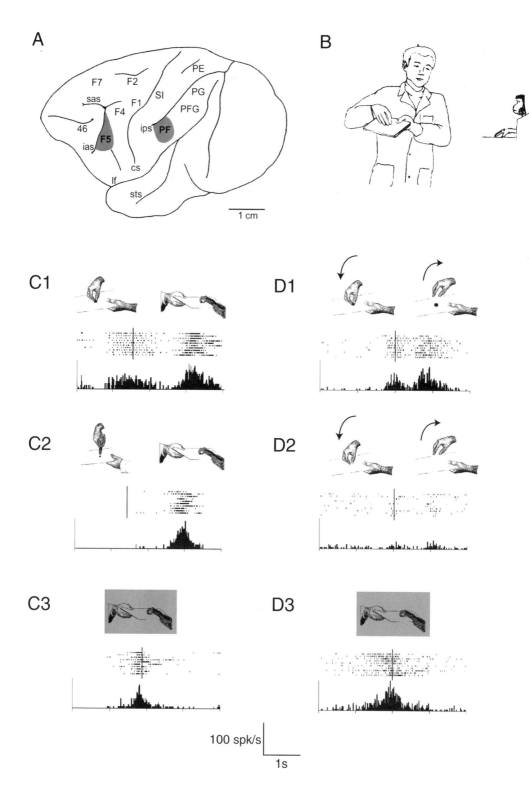

A

F7 F2 PE
sas F1 SI PG
46 F4 PFG
F5 ias ips PF
cs
lf
sts

1 cm

B

C1

C2

C3

D1

D2

D3

100 spk/s

1s

part of the human brain traditionally considered to be unique to our species nevertheless shares a similar functional mechanism with its nonhuman precursor area. In other words, Broca's region appears to be involved not only in speech control, but also, similarly to monkey's area F5, in a prelinguistic analysis of others' behavior. [For a discussion of the possible link between mirror neurons and the origin of language, see Fadiga and Gallese (1997), Rizzolatti and Arbib (1998), Gallese (1999).]

A recent brain-imaging study (Buccino et al. 2001) showed that when we observe goal-related behaviors executed with effectors as different as the mouth, the hand, or the foot, different specific sectors of our premotor cortex become active. These cortical sectors are the same sectors that are active when we perform the same actions. Whenever we look at someone performing an action, in addition to the activation of various visual areas, there is a concurrent activation of the motor circuits that are recruited when we ourselves perform that action. Although we do not overtly reproduce the observed action, nevertheless our motor system acts as if we were executing the same action that we are observing.

When a given action is planned, its expected motor consequences are forecast. This means that when we are going to execute a given action, we can also predict its consequences. Through a process of "equivalence" between what is acted and what is perceived, this information can also be used to predict the consequences of actions performed by others. According to this perspective, perceiving an action is equivalent to internally simulating it. This implicit, automatic, and unconscious process of motor simulation enables the observer to use his or her own resources to penetrate the world of the other without the need to explicitly theorize about it. The process of

Figure 56.2
(A) Lateral view of macaque monkey cerebral cortex showing frontal and parietal areas. Frontal agranular cortical areas (F_1, F_2, F_4, F_5, F_7) are classified according to Matelli et al. (1985; 1991). The posterior parietal areas (PE, PF, PFG, PG) are classified according to Von Bonin and Bailey (1947). Shaded areas indicate the cortical sectors where mirror neurons were recorded. Abbreviations: cs, central sulcus; ias, inferior arcuate sulcus; ips, intraparietal sulcus; lf, lateral fissure; SI, primary somatosensory area; sas, superior arcuate sulcus; sts, superior temporal sulcus. (B) Illustration of the experimental situation for testing the visual properties of mirror neurons. (Modified from di Pellegrino et al. 1992.) (C) Example of the visual and motor responses of an F5 mirror neuron. The behavioral situation during which the neural activity was recorded is illustrated schematically in the upper part of each panel. In the lower part, rasters and the relative peristimulus response histograms are shown. (C1) A tray with a piece of food placed on it was presented to the monkey; the experimenter grasped the food and then moved the tray with the food toward the monkey, which grasped it. A strong activation was present during observation of the experimenter's grasping movements and while the same action was performed by the monkey. Note that the neural discharge was absent when the food was only presented and moved toward the monkey. (C2) The same as C1, except that the experimenter grasped the food with pliers. Note the absence of response when the observed action was performed with a tool. (C3) The monkey grasped the food in the dark. Rasters and histograms are aligned (vertical bar) with the moment in which the experimenter (C1 and C2) or the monkey (C3) touched the food. Abscissae: time; ordinate: spikes/second; bin width: 20 ms. (Modified from Rizzolatti et al. 1996.) (D) Example of the visual and motor responses of a PF mirror neuron. (D1) A tray with a piece of food placed on it was presented to the monkey; the experimenter grasped the food and then released it, moving the hand away from the food. Note the strong response during the observation of the experimenter's grasping and releasing actions. The neuron did not respond during the presentation of the food on the tray. (D2) The same action as in D1 was mimed. Note that in this condition the neural response was virtually absent. (D3) The monkey grasped the food in the dark. Rasters and histograms are aligned (vertical bar) with the moment in which the monkey (D3) or the experimenter touched the food (D1) or the tray (D2). All other conventions are as in part C. (Modified from Gallese et al. 2002.)

simulating action automatically establishes a direct implicit link between agent and observer (see Gallese and Goldman 1998).

Action is therefore the most suitable a priori candidate principle that enables social bonds to be initially established. By an implicit process of simulating action when we observe other individuals acting, we can immediately recognize them as goal-directed agents like us, because a similar neural substrate is activated when we ourselves attempt to achieve the same goal by acting. In sum, we propose that observed behaviors acquire their full meaning only to the extent that they constitute one of the poles of the dynamic sensorimotor relation between agent and observer. Mirror neurons are the neural correlate of such a relation.

Mirror neurons are found in cortical regions endowed with motor properties because premotor neurons are able to establish relationships between expectancies and results. Thus, understanding action can be viewed as a subpersonally mediated function, that is, it relies on neural circuits involved in action control. In this respect, we can hypothesize a continuous path leading from action control to action representation (see Gallese 2000 a,b, 2001).

Conclusions

The results briefly summarized here suggest two things. First, humans share with nonhuman primates, not only particular types of social behavior (i.e., gaze following), but also similar developmental patterns leading to the full-blown acquisition of those same behaviors. Second, both humans and monkeys, when engaged in observing the actions of other individuals, display a similar pattern of covert action simulation, which is underpinned by the activation of a similar cortical network.

A link between gaze following and a theory of mind has been theorized by Baron-Cohen[1] (1995). According to this author, the perception of eye gaze is a crucial step to the development of a mind-reading system that allows individuals to understand not only what another individual is attending to but also what it is thinking about. Thus, the eyes can be considered as "windows" to the mind of others. Our results do not allow us to conclude that a theory of mind exists in monkeys. However, the fact that macaques, together with humans and great apes, display the ability to follow the gaze of others by using eye cues, suggests that essential mechanisms for the development of mind reading are shared by all these primate species.

The discovery that both monkeys and humans share a similar mechanism that matches the observation of an action with its execution is in our opinion an even stronger argument against the theses positing a sharp discontinuity between the cognitive toolkit employed by humans and by nonhuman primates in their social transactions. Most of the emphasis so far has been on clarifying the extent to which our species differs from other primates with respect to the use of propositional attitudes such as beliefs, intentions, and desires. According to this influential view, humans have a theory of mind, nonhuman primates do not, period. However, as pointed out by Allen and Bekoff (1997), this all-or-nothing approach to higher-order intentionality is strongly debatable.

When trying to account for our cognitive abilities, we tend to forget that these abilities are the result of a long evolutionary process. It is reasonable to hypothesize that this evolutionary process proceeded along a line of continuity (see Gallese and Goldman 1998; Gallese 2001). We think that it is worth taking a different heuristic approach; namely, establishing to what extent apparently different cognitive strategies may be underpinned by similar functional mechanisms. The empirical data briefly reviewed here represent, we think, a justification of this investigative approach. The data on mirror neurons and their human homolog seem to suggest that the ease with which we are capable of "mirroring" ourselves in the behavior of others and recognizing them as similar to us—in other words, our capacity to empathize with others—may rely on a series of matching mechanisms that we have just started to uncover.

Acknowledgments

We thank G. Braghiroli for the drawings in figure 56.1A and B. Our work is supported by Ministero dell'Università e della Ricerca Scientifica e tecnologica, by the Human Frontier Science Program, and by the Swiss National Science Foundation.

Note

1. Two cognitive components of Baron-Cohen's model of a theory of mind—the eye direction detector (EDD) and the shared attention mechanism (SAM) modules—are related to social gaze behavior. The EDD module supposedly detects the presence of eyes, computes the direction of gaze, and finally establishes the equivalence between the act of looking and the mental state of seeing. The SAM module would use information from the EDD to establish a link between the agent's and the observer's gaze behavior, thus enabling individuals to share the same mental state on a given object, action, or state of affairs. Both EDD and SAM components are characterized by behavioral and mental features. It could be argued that all behavioral aspects of both components could be underpinned by a matching system that enables the observer to map the agent's gaze behavior to his or her own gaze behavior. In other words, a putative gaze behavior mirror matching system could parsimoniously account for aspects of both the EDD and SAM modules.

References

Allen, C. and Bekoff, M. (1997). *Species of Mind.* Cambridge, Mass.: MIT Press.

Baron-Cohen, S. (1994). How to build a baby that can read minds: Cognitive mechanisms in mindreading. *Current Psychology of Cognition* 13: 513–552.

Baron-Cohen, S. (1995). *Mindblindness: An Essay on Autism and Theory of Mind.* Cambridge, Mass.: MIT Press.

Bekoff, M. and Allen, C. (1998). Intentional communication and social play: How and why animals negotiate and agree to play. In *Animal Play: Evolutionary, Comparative, and Ecological Perspectives*, M. Bekoff and J. A. Byers, eds., pp. 97–114. Cambridge: Cambridge University Press.

Buccino, G., Binkofski, F., Fink, G. R., Fadiga, L., Fogassi, L., Gallese, V., Seitz, R. J., Zilles, K., Rizzolatti, G., and Freund, H.-J. (2001). Action observation activates premotor and parietal areas in a somatotopic manner: An fMRI study. *European Journal of Neuroscience* 13: 400–404.

Butterworth, G. and Jarret, N. (1991). What minds have in common is space: Spatial mechanism serving joint visual attention in infancy. *British Journal of Developmental Psychology* 9: 55–72.

Carey, D. P., Perrett, D. I., and Oram, M. W. (1997). Recognizing, understanding and reproducing actions. In *Handbook of Neuropsychology*. Vol. 11, *Action and Cognition*. M. Jeannerod and J. Grafman, eds., pp. 111–130. Amsterdam: Elsevier Science.

Cochin, S., Barthelemy, C., Lejeune, B., Roux, S., and Martineau, J. (1998). Perception of motion and qEEG activity in human adults. *Electroencephalography and Clinical Neurophysiology* 107: 287–295.

Darwin, C. (1871). *The Descent of Man*. London: Murray.

Decety, J. and Grèzes, J. (1999). Neural mechanisms subserving the perception of human actions. *Trends in Cognitive Sciences* 3: 172–178.

Decety, J., Grèzes, J., Costes, N., Perani, D., Jeannerod, M., Procyk, E., Grassi, F., and Fazio, F. (1997). Brain activity during observation of actions. Influence of action content and subject's strategy. *Brain* 120: 1763–1777.

di Pellegrino, G., Fadiga, L., Fogassi, L., Gallese, V., and Rizzolatti, G. (1992). Understanding motor events: A neurophysiological study. *Experimental Brain Research* 91: 176–180.

Emery, N. J., Lorincz, E. N., Perrett, D. I., Oram, M. W., and Baker, C. I. (1997). Gaze-following and joint attention in rhesus monkeys (*Macaca mulatta*). *Journal of Comparative Psychology* 111: 286–293.

Fa, J. E. and Lindburg, D. G. (1996). *Evolution and Ecology of Macaque Societies*. Cambridge: Cambridge University Press.

Fadiga, L. and Gallese, V. (1997). Action representation and language in the brain. *Theoretical Linguistics* 23: 267–280.

Fadiga, L., Fogassi, L., Pavesi, G., and Rizzolatti, G. (1995). Motor facilitation during action observation: A magnetic stimulation study. *Journal of Neurophysiology* 73: 2608–2611.

Ferrari, P. F., Kohler, E., Fogassi, L., and Gallese, V. (2000). The ability to follow eye gaze and its emergence during development in macaque monkeys. *Proceedings of the National Academy of Sciences, U.S.A.* 97: 13997–14002.

Fogassi, L., Gallese, V., Fadiga, L., and Rizzolatti, G. (1998). Neurons responding to the sight of goal-directed hand/arm actions in the parietal area PF (7b) of the macaque monkey. *Society of Neuroscience Abstracts* 24: 257.5.

Gallese, V. (1999). From grasping to language: Mirror neurons and the origin of social communication. In *Towards a Science of Consciousness*, S. Hameroff, A. Kazniak, and D. Chalmers, eds., pp. 165–178. Cambridge, Mass.: MIT Press.

Gallese, V. (2000a). The acting subject: Towards the neural basis of social cognition. In *Neural Correlates of Consciousness. Empirical and Conceptual Questions*, T. Metzinger, ed., pp. 325–333. Cambridge, Mass.: MIT Press.

Gallese, V. (2000b). The inner sense of action: Agency and motor representations. *Journal of Consciousness Studies* 7: 23–40.

Gallese, V. (2001). From mirror neurons to empathy: The shared manifold hypothesis of intersubjectivity. *Journal of Consciousness Studies* 8: 33–50.

Gallese, V. and Goldman, A. (1998). Mirror neurons and the simulation theory of mind-reading. *Trends in Cognitive Sciences* 12: 493–501.

Gallese, V., Fadiga, L., Fogassi, L., and Rizzolatti, G. (1996). Action recognition in the premotor cortex. *Brain* 119: 593–609.

Gallese, V., Fogassi, L., Fadiga, L., and Rizzolatti, G. (in press). Action representation and the inferior parietal lobule. In *Attention and Performance XIX*, W. Prinz and B. Hommel, eds. Oxford, Oxford University Press.

Grafton, S. T., Arbib, M. A., Fadiga, L., and Rizzolatti, G. (1996). Localization of grasp representations in humans by PET: 2. Observation compared with imagination. *Experimental Brain Research* 112: 103–111.

Graziano, M. S. A. and Gross, C. G. (1995). The representation of extrapersonal space: A possible role for bimodal visual-tactile neurons. In *The Cognitive Neurosciences*, M. S. Gazzaniga, ed., pp. 1021–1034. Cambridge, Mass.: MIT Press.

Hari, R., Forss, N., Avikainen, S., Kirveskari, S., Salenius, S., and Rizzolatti, G. (1998). Activation of human primary motor cortex during action observation: A neuromagnetic study. *Proceedings of the National Academy of Sciences, U.S.A.* 95: 15061–15065.

Hoffman, E. A. and Haxby, J. V. (2000). Distinct representations of eye gaze and identity in the distributed human neural system for face perception. *Nature Neuroscience* 3: 80–84.

Humphrey, N. K. (1976). The social function of intellect. In *Growing Points in Ethology*, P. Bateson and R. A. Hinde, eds., pp. 303–321. Cambridge: Cambridge University Press.

Humphrey, N. K. (1980). Nature's psychologists. In *Consciousness and the Physical World*, B. D. Josephson and V. S. Ramachandran, eds., pp. 57–75. Oxford: Pergamon.

Hyvärinen, J. (1981). Regional distribution of functions in parietal association area 7 of the monkey. *Brain Research* 206: 287–303.

Iacoboni, M., Woods, R. P., Brass, M., Bekkering, H., Mazziotta, J. C., and Rizzolatti, G. (1999). Cortical mechanisms of human imitation. *Science* 286: 2526–2528.

Itakura, S. (1996). An exploratory study of gaze-monitoring in non-human primates. *Japanese Psychological Research* 38: 174–180.

Janson, C. H. and van Schaik, C. P. (1993). Ecological risk aversion in juvenile primates: Slow and steady wins the race. In *Juvenile Primates. Life History, Development, and Behavior*, M. E. Pereira and L. A. Fairbanks, eds., pp. 57–74. Oxford: Oxford University Press.

Jellema, T. and Perrett, D. I. (2002). Coding of visible and hidden actions. In *Attention and Performance XIX*, W. Prinz and B. Hommel, eds., pp. 356–380. Oxford: Oxford University Press.

Leinonen, L. and Nyman, G. (1979). Functional properties of cells in anterolateral part of area 7 associative face area of awake monkeys. *Experimental Brain Research* 34: 321–333.

Leinonen, L., Hyvärinen, J., Nyman, G., and Linnankoski, I. (1979). Functional properties of neurons in lateral part of associative area 7 in awake monkeys. *Experimental Brain Research* 34: 299–320.

Lorincz, E. N., Baker, C. I., and Perrett, D. I. (1999). Visual cues for attention following in rhesus monkeys. *Current Psychology of Cognition* 18: 973–1003.

Matelli, M. and Luppino, G. (1997). Functional anatomy of human motor cortical areas. In *Handbook of Neuropsychology*. Vol. 11, F. Boller and J. Grafman, eds., pp. 9–26. Amsterdam: Elsevier Science.

Matelli, M., Luppino, G., and Rizzolatti, G. (1985). Patterns of cytochrome oxidase activity in the frontal agranular cortex of the macaque monkey. *Behavioral Brain Research* 18: 125–137.

Matelli, M., Luppino, G., and Rizzolatti, G. (1991). Architecture of superior and mesial area 6 and of the adjacent cingulate cortex. *Journal of Comparative Neurology* 311: 445–462.

Matelli, M., Camarda, R., Glickstein, M., and Rizzolatti G. (1986). Afferent and efferent projections of the inferior area 6 in the macaque monkey. *Journal of Comparative Neurology* 251: 281–298.

Menzel, E. W. and Halperin, S. (1975). Purposive behavior as a basis for objective communication in chimpanzees. *Science* 189: 652–654.

Moore, C. and Corkum, V. (1998). Infant gaze-following based on eye direction. *British Journal of Psychology* 16: 495–503.

Perrett, D. I., Smith, P. A. J., Potter, D. D., Mistlin, A. J., Head, A. S., Milner, A. D., and Jeeves, M. A. (1985). Visual cells in the temporal cortex sensitive to face view and gaze direction. *Proceedings of the Royal Society of London B* 223: 293–317.

Perrett, D. I., Hietanen, J. K., Oram, M. W., and Benson, P. J. (1992). Organization and function of cells responsive to faces in the temporal cortex. *Philosophical Transactions of the Royal Society of London B* 335: 23–30.

Povinelli, D. J. and Eddy, T. J. (1996). Chimpanzees: Joint visual attention. *Psychological Science* 7: 129–135.

Premack, D. and Woodruff, G. (1978). Does the chimpanzee have a theory of mind? *Behavioral and Brain Sciences* 1: 515–526.

Puce, A., Allison, T., Bentin, S., Gore, J. C., and McCarthy, G. M. (1998). Temporal cortex activation in humans viewing eye and mouth movements. *Journal of Neuroscience* 18: 2188–2199.

Rizzolatti, G. and Arbib, M. A. (1998). Language within our grasp. *Trends in Neuroscience* 21: 188–194.

Rizzolatti, G., Fadiga, L., Gallese, V., and Fogassi, L. (1996a). Premotor cortex and the recognition of motor actions. *Cognitive Brain Research* 3: 131–141.

Rizzolatti, G., Fadiga, L., Matelli, M., Bettinardi, V., Paulesu, E., Perani, D., and Fazio, G. (1996b). Localization of grasp representations in humans by PET: 1. Observation versus execution. *Experimental Brain Research* 111: 246–252.

Rizzolatti, G., Luppino, G., and Matelli, M. (1998). The organization of the cortical motor system: New concepts. *Electroencephalography and Clinical Neurophysiology* 106: 283–296.

Scaife, M. and Bruner, J. S. (1975). The capacity of joint visual attention in the infant. *Nature* 253: 265–266.

Smuts, B. B., Cheney, D. L., Seyfarth, R. M., Wrangham, R. W., and Struhsaker T. T. (1987). *Primate Societies*. Chicago: University of Chicago Press.

Thomsen, C. E. (1974). Eye contact by non-human primates toward a human observer. *Animal Behaviour* 22: 144–149.

Tomasello, M. and Call, J. (1997). *Primate Cognition*. Oxford: Oxford University Press.

Tomasello, M., Call, J., and Hare, B. (1998). Five primate species follow the visual gaze of conspecifics. *Animal Behaviour* 55: 1063–1069.

Tomasello, M., Hare, B., and Agnetta, B. (1999). Chimpanzees, *Pan troglodytes*, follow gaze direction geometrically. *Animal Behaviour* 58: 769–777.

Von Bonin, G. and Bailey P. (1947). *The Neocortex of Macaca Mulatta*. Urbana: University of Illinois Press.

Walters, J. R. (1987). Transition to adulthood. In *Primate Societies*, B. B. Smuts, D. L. Cheney, R. M. Seyfarth, R. W. Wrangham, and T. T. Struhsaker, eds., pp. 358–368. Chicago: University of Chicago Press.

Whiten, A. and Byrne, R. W. (1988). *Machiavellian Intelligence*. Oxford: Oxford University Press.

Whiten, A. and Byrne, R. W. (1997). *Machiavellian Intelligence 2: Evaluations and Extensions*. Cambridge: Cambridge University Press.

Vigilance and Perception of Social Stimuli: Views from Ethology and Social Neuroscience

Adrian Treves and Diego Pizzagalli

To survive and protect their offspring, animals must detect threats before they suffer damage. This requires efficient information gathering, as well as rapid information processing. Studies of nonhuman primate behavior reveal that individuals direct frequent and time-consuming vigilance (information gathering by visual search of the environment beyond the immediate vicinity) toward members of the same species (conspecifics), particularly with increasing risk of aggressive competition. In addition, interactions with unfamiliar conspecifics are generally aversive.

In a complementary fashion, electrophysiological and neuroimaging research on the human brain provides independent lines of evidence that socially relevant stimuli are processed quickly (<200 ms) and by a phylogenetically ancient brain region. Hence, we propose that the primate brain is adapted to rapid and sensitive processing of information about conspecifics, which derives from vigilance directed to the dynamic interactions of associates. We use this link to illustrate the potential for fruitful collaboration between neuroscientists and ethologists, and to suggest improvements in current practices in both fields.

Interaction of Ethology and Neuroscience

Cognitive ethology has the potential to unite two heretofore separate biological disciplines: neuroscience (the study of brain–behavior relationships) and ethology (the study of animal behavior). Despite separate histories and different investigative methods, neuroscience and ethology often address related and complementary topics. Neuroscientists' concentration on brain function has advanced our understanding of the proximate mechanisms underlying behavior. Ethologists' comparative, functional approaches have elucidated ultimate, evolutionary explanations for behavior. Anyone interested in sensory-processing and information-gathering behavior would gain from understanding the brain functions of their subjects. Likewise, anyone unraveling the secrets of the brain should understand the evolutionary history and past environmental pressures that shaped the cognition and behavior of their subjects. Cognitive ethology can advance both disciplines because it places information on how brain and behavior interact in an evolutionary context that explains why they do so.

In this essay we illustrate how brain research can inform ethology and how in turn the study of animal behavior can inform neuroscience. We happened on this collaboration by chance, unaware of the similarity of our research questions. Our intuition tells us that many fruitful collaborations between neuroscientists and ethologists never occur because terminologies, techniques, and theories appear mutually unintelligible. This essay was designed to facilitate collaboration, using vigilance behavior to illustrate the utility of close communication between neuroscientists and ethologists. Our interdisciplinary approach, which could be useful in other domains of neuroscience and ethology, also suggests a modification of existing methods in both disciplines. Our focus is on the visual gathering and processing of social information related to members of the same species. Although many group-living species may show similar patterns of brain–behavior interactions involving several sensory channels, we concentrate here on visual cues used by primates.

Ethology Can Inform Neuroscience

Social neuroscientists often present stimuli to elicit changes in their subjects' brain activity. For humans, these stimuli are often images of the faces of strangers. In these studies, the null hypothesis is that strangers' faces are neutral stimuli, with pleasing or aversive properties being generated by different facial expressions. This may create a problem if strangers are inherently aversive. Today, seeing a stranger's face is commonplace for humans, yet our brains are the products of millions of years of evolution under different circumstances. For our ancestors, encounters with strangers were probably rare, owing to low population densities and territorial defense against outsiders, and they were possibly accompanied by strong emotions (e.g., anxiety, hostility, sexual interest). Hence, ethologists would caution neuroscientists to consider the evolutionary history of their subjects, particularly if they study phylogenetically older brain regions. Ethologists can clarify the socioecological relevance of candidate stimuli, based on their understanding of ancestral environments and selective pressures.

Neuroscience Can Inform Ethology

Ethologists study both the evolutionary origins of behavior and its consequences for current fitness. For example, variation in vigilance is considered to have fitness consequences when it correlates with risk or reduces the time allocated to other important activities. Consequently, ethologists have long assumed that individuals with low rates of vigilance perceive less risk, all else being equal. This assumption is invalid if different classes of targets of vigilance (e.g., associates versus escape routes) require glances of different durations irrespective of risk (Treves 2000a). Here neuroscientists can provide crucial information on processing speed for certain classes of target. Indeed, neuroscientists would caution ethologists that inferences about evolutionary origins of behavior—such as vigilance toward conspecifics—can be drawn with more confidence if associated stimuli receive priority in the information-processing flow or if a functionally specialized brain region is identified.

Nonhuman Primate Vigilance

Vigilance is defined differently in ethology than it is in neuroscience, where vigilance refers to

a brain state of receptivity to external stimuli directly associated with alertness. By contrast, ethologists studying a wide range of taxa usually define vigilance as looking up from foraging or simply as visual search of the environment beyond the immediate vicinity. As such, vigilance can be studied by observing an animal's eyes and its direction of gaze. Researchers have focused on measures of rate of vigilance in relation to variation in the risk of predation or environmental context (Elgar 1989). Unlike most other animals in the vigilance literature, primates also reveal the targets of their vigilance (Treves 2000a).

In the wild, nonhuman primates spend their waking hours in two forms of visual activity: inspection of close targets (e.g., nearby foods, grooming partners) and vigilance directed at more distant targets (e.g., travel routes). Among primates, a large proportion of vigilance is directed at associates within groups (Treves 2000a). This reflects in part the fact that competition with associates over resources and mates is frequent and may sometimes result in death (Dittus 1980; Treves 2000b). Indeed, vigilance increases with the intensity of competition from associates. In interspecific comparisons, species that live in more excitable, competitive groups monitor associates more frequently and for longer periods than do species that form calm, cohesive groups (Caine and Marra 1988; Treves 1999). Within species, female enemies in the same group receive more visual attention than do female allies (Watts 1998), and subordinates direct more glances to associates than do dominants (Keverne et al. 1978; Alberts 1994). When infants are born or begin to wander from their mother's reach, within-group vigilance increases significantly (Maestripieri 1993; Treves 1999, 2000a). In primates, therefore, the varied competitive and aggressive threats generated within groups of familiar conspecifics favor vigilance.

Nevertheless, threats posed by *unfamiliar* conspecifics are more likely to result in harm than those arising within groups (Bernstein 1969; Bernstein et al. 1974; Goodall 1986; Treves 1998,

2000b). Although responses to close encounters with unfamiliar conspecifics vary with sex and context, they are generally dangerous and stressful for all parties (Alberts et al. 1992). Accordingly, a higher risk of encountering unfamiliar conspecifics triggers increases in vigilance (Rose and Fedigan 1995; Steenbeck et al. 1999). In sum, for wild primates, unfamiliar conspecifics present many threats and few attractions. It is in light of this information that we should reconsider the use of strangers' faces as neutral stimuli in neuroscience, especially if phylogenetically older brain regions are involved.

Based on nonhuman primate vigilance toward conspecifics and responses to strangers, we propose that natural selection has favored individuals that are sensitive to rapid change in the social environment and those that interact cautiously with strangers. Modern humans are of course different from wild nonhuman primates. In many cultures, we are socialized and habituated to strangers from an early age. Culture and learning may override our evolutionary history. Yet, infants predictably go through a phase of aversion to strangers (Mangelsdorf 1992). If the human brain retains pathways that evolved millions of years ago, phylogenetically older brain regions may still produce aversive responses to strangers' faces, while more recently evolved regions may secondarily modulate aversion to strangers through social and cultural experience. We suggest therefore that experiments that present strangers' faces as stimuli should be expected to elicit a negative response at the outset.

Neuroscience and the Study of Social Perception

Social perception refers to the processing of information about conspecifics and the social environment. In humans, it can be studied by measuring electrical and physiological activation of the brain in response to stimuli. Typically, faces and direction of gaze are employed as stimuli because they convey considerable information about conspecifics' emotional and motivational state as well as the focus of their attention (Tomasello et al. 1999; Allison et al. 2000; Langton et al. 2000). Noninvasive electrophysiological studies have millisecond time resolution, allowing researchers to unravel the temporal dynamics and sequences of brain processes. However, they do not provide fine-grained spatial information about the brain regions involved. For this, neuroimaging techniques with their spatial resolution in the millimeter range can map brain regions involved in social perception, without, however, furnishing fine-grained temporal information. When information from the two techniques is combined, a more comprehensive view of human brain functions emerges.

The vigilance behavior of nonhuman primates suggests that information about the social environment has consequences for fitness; hence we would expect social perception to be rapid, sensitive, and dependent on functionally specialized brain regions. This idea is supported by pioneering work in nonhuman primates using invasive, single-unit recordings (Brothers et al. 1990; Perrett et al. 1992). In the next two sections we summarize studies of human social perception that extend this nonhuman work with independent lines of evidence demonstrating the involvement of (1) a preattentive response and (2) a phylogenetically ancient brain region in processing conspecific stimuli.

Human Electrophysiological Studies

Brain electrical activity can be monitored noninvasively by attaching electrodes to the scalp (electroencephalogram). Typically, differences in electrical potential are recorded from multiple scalp sites and sampled several hundred times per second. Changes in functional brain state, whether endogenous or induced by a task, can be measured. For instance, brain electrical activity (event-related potentials, ERPs) can be related directly to the presentation of a stimulus after background activity (noise) is eliminated.

Recently, ERPs have been used to study the time course of brain responses to a variety of socially relevant stimuli. For instance, strangers' faces elicited stronger responses than control stimuli—including heterospecific faces—as early as 170 ms after stimulus onset (for a review see McCarthy 2000). Also, stronger responses were elicited by an averted gaze than a gaze directed to the subject (Puce et al. 2000). Notably, even more subtle characteristics of faces, such as sex (145–185 ms: Mouchetant-Rostaing et al. 2000), likability (80–116 ms: Pizzagalli et al. 1999), expression (160 ms: Streit et al. 1999; 110 ms: Halgren et al. 2000), and attractiveness (170 ms: Halit et al. 2000) have been shown to be processed very rapidly. Thus, mechanisms exist that extract subtle information about conspecifics quickly and automatically (effortlessly and preattentively).

Neuroimaging Studies

Neuroimaging techniques (functional magnetic resonance imaging, fMRI, or positron emission tomography, PET) take advantage of fleeting increases in blood flow to the brain regions activated by a given stimulus. Such changes in regional cerebral blood flow and blood oxygenation can be measured with high spatial resolution to map human brain functions.

Neuroimaging has been employed recently to investigate brain regions underlying the processing of socially relevant information in humans (Davidson and Irwin 1999). Extrapolating from pioneering research in rodents that identified a subcortical (thalamo-amygdalar) pathway involved in rapid processing of fear-related stimuli (LeDoux 1996), several researchers have proposed that automatic and rapid processing of threat-related cues in humans may also be mediated by the amygdala[1] (Öhman 1993; LeDoux 1996; Whalen 1998).

Consistent with this view, human amygdalar activation has been reported during presentation of angry (e.g., Hariri et al. 2000) and fearful (e.g., Morris et al. 1996) faces, even when stimuli were presented below the level of conscious

awareness (Morris et al. 1999; Whalen et al. 1998). In agreement with our proposal that unfamiliar conspecifics may generally be aversive stimuli, amygdalar activation also followed the presentation of unfamiliar faces with neutral expression, whether of the subject's own race (Dubois et al. 1999) or another race (Phelps et al. 2000; Hart et al. 2000). Moreover, human patients with amygdalar lesions judged unfamiliar faces to be more approachable and trustworthy than did controls (Adolphs et al. 1998).

However, subsequent studies suggest that the human amygdala may play a broader role than simply responding to threat-related stimuli. For example, amygdalar activation increased when comparisons were made between biological and random motion (Bonda et al. 1996), direct and averted gaze (Kawashima et al. 1999), and images of friends and those of loved ones (Bartels and Zeki 2000). Clearly, not all of these stimuli conveyed threats, yet the amygdala was implicated. Notably, autistic individuals with deficits in social perception showed amygdalar dysfunction (Baron-Cohen et al. 1999). Collectively, animal and human studies suggest that one function of the amygdala is to process species-specific cues that predict biologically significant outcomes, based on either personal experience or evolved mechanisms (LeDoux 1996; Whalen 1998).

In summary, ERP studies reveal that the human brain performs social perception rapidly (<200 ms) and automatically, while fMRI and PET studies reveal a key role for an ancient brain region, the amygdala, in social perception. Although it is reasonable to infer a link between the two sets of studies, their radically different temporal resolutions preclude a direct connection so far. If a link is found between rapid processing and amygdalar processing of conspecific stimuli, it would represent evidence for a brain mechanism shaped by selective pressures imposed by the social environment. This does not contradict the idea that subcortical responses may prime the phylogenetically more recent cortical regions responsible for a slower, more complete, and richer response to the same stimuli. It is safe to assume that vigilance toward conspecifics involves both cortical and subcortical pathways (Morris et al. 1999; Hariri et al. 2000). Yet the involvement of the amygdala suggests to us that the primate brain contains an evolutionarily conserved pathway specialized to process conspecific stimuli with high priority.

Conclusion

In group-living primates, the social environment has shaped behavior and brain functions to coordinate and streamline the collection and processing of information about conspecifics. Both within-group vigilance and responses to unfamiliar conspecifics reflect selection imposed by allies, enemies, and strangers. Frequent, time-consuming, and sensitive vigilance toward conspecifics reflects the importance and priority of gathering socially relevant information. In turn, the primate brain shows adaptations to processing such socially relevant information, which is reflected by the involvement of a functionally specialized, ancient brain region and rapid extraction of subtle cues from conspecifics. Specifically, we hypothesize that the primate amygdala is adapted to simple, fast processing of socially relevant stimuli gathered by vigilance toward conspecifics—not only threat related but also other salient stimuli such as gaze direction and body movements. This integrated brain–behavior system would maximize fitness in environments where conspecifics hold the key to survival and reproduction.

Implications

If our proposal is correct, it has implications for neuroscientists and ethologists alike. Socially irrelevant stimuli should trigger a slower brain electrical response (using ERP) and less amygdalar activation (using neuroimaging) than so-cially relevant stimuli of comparable complexity. Primate ethologists should expect conspecific stimuli to have priority over other stimuli when individuals allocate vigilance effort. Moreover, the rapid processing of socially relevant stimuli suggests that glances to associates may be brief but sufficient to extract considerable information; hence the time spent in vigilance should not be equated with the importance of the target, especially during periods of social instability. A global recommendation for both neuroscientists and ethologists is to use paired stimuli (e.g., conspecific vocalizations versus heterospecific calls) carefully selected for their socioecological and evolutionary relevance. In this way, the findings from ethology and neuroscience may become mutually more interesting and intelligible.

Acknowledgments

A. Treves was supported by the Department of Psychology and the Graduate School of the University of Wisconsin—Madison and by a grant to C. T. Snowdon (National Institutes of Health grant MH 35,215). D. Pizzagalli was supported by grants from the Swiss National Research Foundation (81ZH-52864) and Holderbank-Stiftung zur Förderung der wissenschaftlichen Fortbildung. The authors wish to thank R. J. Davidson, W. Irwin, E. Nelson, A. J. Skolnick, and C. T. Snowdon for helpful comments.

Note

1. The amygdala is an almond-shaped brain region located in the depth of the medial temporal lobe. It is believed to be involved in threat processing in birds, mammals, and reptiles (LeDoux 1996).

References

Adolphs, R., Tranel, D., and Damasio, A. R. (1998). The human amygdala in social judgement. *Nature* 393: 470–474.

Alberts, S. C. (1994). Vigilance in young baboons: Effects of habitat, age, sex and maternal rank on glance rate. *Animal Behaviour* 47: 749–755.

Alberts, S. C., Sapolsky, R. M., and Altmann, J. (1992). Behavioral, endocrine and immunological correlates of immigration by an aggressive male into a natural primate group. *Hormones and Behavior* 26: 167–178.

Allison, T., Puce, A., and McCarthy G. (2000). Social perception from visual cues: Role of the STS region. *Trends in Cognitive Sciences* 4: 267–278.

Baron-Cohen, S., Ring, H. A., Wheelwright, S., Bullmore, E. T., Brammer, M. J., Simmons, A., and Williams, S. C. R. (1999). Social intelligence in the normal and autistic brain: An fMRI study. *European Journal of Neuroscience* 11: 1891–1898.

Bartels, A. and Zeki, S. (2000). The neural basis of romantic love. *NeuroReport* 11: 3829–3834.

Bernstein, I. S. (1969). Introductory techniques in the formation of pigtail monkey troops. *Folia Primatologica* 10: 1–19.

Bernstein, I. S., Gordon, T. P., and Rose, R. M. (1974). Factors influencing the expression of aggression during introductions to rhesus monkey groups. In *Primate Aggression, Territoriality and Xenophobia*. R. L. Holloway, ed., pp. 211–238. New York: Academic Press.

Bonda, E., Petrides, M., Ostry, D., and Evans, A. (1996). Specific involvement of human parietal systems and the amygdala in the perception of biological motion. *Journal of Neuroscience* 16: 3737–3744.

Brothers, L., Ring, B., and Kling, A. (1990). Response of neurons in the macaque amygdala to complex social stimuli. *Behavioural Brain Research* 41: 199–213.

Caine, N. G. and Marra, S. L. (1988). Vigilance and social organization in two species of primates. *Animal Behaviour* 36: 897–904.

Davidson, R. J. and Irwin, W. (1999). The functional neuroanatomy of emotion and affective style. *Trends in Cognitive Sciences* 3: 11–21.

Dittus, W. P. J. (1980). The social regulation of primate populations: A synthesis. In *The Macaques: Studies in Ecology, Behavior and Evolution*, D. G. Lindburg, ed., pp. 263–286. New York: Van Nostrand Reinhold.

Dubois, S., Roission, B., Schiltz, C., Bodart, J. M., Michel, C., Bruyer, R., and Crommelinck, M. (1999).

Effect of familiarity on the processing of human faces. *Neuroimage* 9: 278–289.

Elgar, M. A. (1989). Predator vigilance and group size in mammals and birds: A critical review of the empirical evidence. *Biological Review* 64: 13–33.

Goodall, J. (1986). *The Chimpanzees of Gombe*. Cambridge, Mass.: Harvard University Press.

Halgren, E., Raij, T., Marinkovic, K., Jousmaeki, V., and Hari, R. (2000). Cognitive response profile of the human fusiform face area as determined by MEG. *Cerebral Cortex* 10: 69–81.

Halit, H., de Haan, M., and Johnson, M. H. (2000). Modulation of event-related potentials by prototypical and atypical faces. *NeuroReport* 11: 1871–1875.

Hariri, A. R., Bookheimer, S. Y., and Mazziotta, J. C. (2000). Modulating emotional responses: Effects of a neocortical network on the limbic system. *NeuroReport* 11: 43–48.

Hart, A. J., Whalen, P. J., Shin, L. M., McInerney, S. C., Fischer, H., and Rauch, S. L. (2000). Differential response in the human amygdala to racial outgroup vs. ingroup face stimuli. *NeuroReport* 11: 2351–2355.

Kawashima, R., Sugiura, M., Kato, T., Nakamura, A., Hatano, K., Ito, K., Fukuda, H., Kojima, S., and Nakamura, K. (1999). The human amygdala plays an important role in gaze monitoring. *Brain* 122: 779–783.

Keverne, E. B., Leonard, R. A., Scruton, D. M., and Young, S. K. (1978). Visual monitoring in social groups of talapoin monkeys (*Miopithecus talapoin*). *Animal Behaviour* 26: 933–944.

Langton, S. R. H., Watt, R. J., and Bruce, V. (2000). Do the eyes have it? Cues to the direction of social attention. *Trends in Cognitive Sciences* 4: 50–59.

LeDoux, J. E. (1996). *The Emotional Brain*. New York: Simon and Schuster.

Maestripieri, D. (1993). Vigilance costs of allogrooming in macaque mothers. *American Naturalist* 141: 744–753.

Mangelsdorf, S. C. (1992). Developmental changes in infant–stranger interaction. *Infant Behavior and Development* 15: 191–208.

McCarthy, G. (2000). Physiological studies of face processing in humans. In *The New Cognitive Neurosciences*, 2nd ed., M. S. Gazzaniga, ed., pp. 393–409. Cambridge, Mass.: MIT Press.

Morris, J. S., Frith, C. D., Perrett, D. I., Rowland, D., Young, A. W., Calder, A. J., and Dolan, R. J. (1996).

A differential neural response in the human amygdala to fearful and happy facial expressions. *Nature* 383: 812–815.

Morris, J. S., Öhman, A., and Dolan, R. J. (1999). A subcortical pathway to the right amygdala mediating "unseen" fear. *Proceedings of the National Academy of Sciences U.S.A.* 96: 1680–1685.

Mouchetant-Rostaing, Y., Giard, M. H., Bentin, S., Aguera, P. E., and Pernier, J. (2000). Neurophysiological correlates of face gender processing in humans. *European Journal of Neuroscience* 12: 303–310.

Öhman, A. (1993). Fear and anxiety as emotional phenomena: Clinical phenomenology, evolutionary perspectives, and information-processing mechanisms. In *Handbook of Emotion*, M. Lewis and J. M. Haviland, eds., pp. 511–536. New York: Guilford Press.

Perrett, D. I., Hietanen, J. K., Oram, M. W., and Benson, P. J. (1992). Organization and functions of cells responsive to faces in the temporal cortex. *Philosophical Transactions of the Royal Society of London B* 335: 23–30.

Phelps, E. A., O'Connor, K. J., Cunningham, W. A., Funayama, E. S., Gatenby, J. C., Gore, J. C., and Banaji, M. R. (2000). Performance on indirect measures of race evaluation predicts amygdala activation. *Journal of Cognitive Neuroscience* 12: 729–738.

Pizzagalli, D., Regard, M., and Lehmann, D. (1999). Rapid emotional face processing in the human right and left brain hemispheres: An ERP study. *NeuroReport* 10: 2691–2698.

Puce, A., Smith, A., and Allison, T. (2000). ERPs evoked by viewing facial movements. *Cognitive Neuropsychology* 17: 221–239.

Rose, L. M. and Fedigan, L. M. (1995). Vigilance in white-faced capuchins, *Cebus capucinus*, in Costa Rica. *Animal Behaviour* 49: 63–70.

Steenbeck, R., Piek, R. C., van Buul, M., and van Hooff, J.A.R.A.M. (1999). Vigilance in wild Thomas's langur (*Presbytis thomasi*): The importance of infanticide risk. *Behavioral Ecology and Sociobiology* 45: 137–150.

Streit, M., Ioannides, A. A., Liu, L., Wolwer, W., Dammers, J., Gross, J., Gaebel, W., and Muller-Gartner, H. W. (1999). Neurophysiological correlates of the recognition of facial expressions of emotion as revealed by magnetoencephalography. *Cognitive Brain Research* 7: 481–491.

Tomasello, M., Hare, B., and Agnetta, B. (1999). Chimpanzees, *Pan troglodytes*, follow gaze direction geometrically. *Animal Behaviour* 58: 769–777.

Treves, A. (1998). Primate social systems: Conspecific threat and coercion-defense hypotheses. *Folia Primatologica* 69: 81–88.

Treves, A. (1999). Within-group vigilance in red colobus and redtail monkeys. *American Journal of Primatology* 48: 113–126.

Treves, A. (2000a). Theory and method in studies of vigilance and aggregation. *Animal Behaviour* 60: 711–722.

Treves, A. (2000b). Prevention of infanticide: The perspective of infant primates. In *Infanticide by Males and its Implications*, C. P. van Schaik and C. H. Janson, eds., pp. 223–238. Cambridge: Cambridge University Press.

Watts, D. P. (1998). A preliminary study of selective visual attention in female mountain gorillas (*Gorilla gorilla beringei*). *Primates* 39: 71–78.

Whalen, P. J. (1998). Fear, vigilance, and ambiguity: Initial neuroimaging studies of the human amygdala. *Current Directions in Psychological Science* 7: 177–188.

Whalen, P. J., Rauch, S. L., Etcoff, N. L., McInerney, S. C., Lee, M. B., and Jenike, M. A. (1998). Masked presentations of emotional facial expressions modulate amygdala activity without explicit knowledge. *Journal of Neuroscience* 18: 411–418.

Afterword: What Is It Like?

Donald R. Griffin

The chapters in this book demonstrate how rapidly our knowledge and understanding of animal cognition is expanding. One might even say that it is exploding. In the 1970s it seemed radical to suggest that there was such a thing as nonhuman cognition (Griffin 1976). However, even during the few months since I finished an expanded review of animal mentality (Griffin 2001), at least a dozen important and relevant publications have appeared.

One significant aspect of animal cognition has received relatively little attention, however. This is the question of subjective consciousness—what life is like for the animal itself. Most of the authors here implicitly or explicitly restrict their attention to information processing in animal nervous systems, although this information processing includes such activities as remembering and decision making that are ordinarily considered to involve conscious thinking. Thus an innocent reader might well conclude that none of the animals whose cognition is analyzed want, believe, hate, fear, or intend. Most of the authors would probably deny any intention to be dogmatically negative and take an agnostic position justified by a conviction that any such experiences are private and inaccessible to scientific investigation.

One reason for this strong aversion to consideration of conscious experience seems to be the assumption that consciousness is complex, sophisticated, and perhaps even ethereal. This view may, however, be misleading and needlessly restrictive. In our species, conscious experience obviously varies enormously in many dimensions, from extremely simple to profoundly complex and subtle. Certainly it is one aspect of cognition that is important in our species, and insofar as it occurs in others, it may well be important to the animals themselves. One approach that has been effective with other difficult problems in biology is to select simple examples and examine all available evidence bearing on their occurrence and nature. Familiar examples are the use of *Drosophila* in genetics and *Aplysia* ganglia in search of the biochemical correlates of learning.

If an animal experiences a simple and basic emotion such as pain or fear, this may well be subjectively real and important to the animal in question yet very simple, as pointed out by Marian Dawkins (1993). Rather than assuming that all consciousness must be similar to and as complex as ours, a sort of bottom-up approach to the challenge of identifying and analyzing nonhuman consciousness may make much more rapid progress by the tried-and-true method of starting with the simplest examples.

There is a significant difference between behavior that is or is not accompanied or influenced by conscious subjective experiences. We cannot fully understand particular animals until we know in which category they belong under various conditions. Insofar as they are conscious, we need to learn whatever we can about the content of their subjective experiences, what they feel or what they think about. It is therefore appropriate to inquire to what extent the animals whose cognition is discussed in this book are aware of their situation and whether they consciously choose what to do.

The behavioristic admonition that scientists cannot learn anything at all about nonhuman consciousness is rapidly becoming obsolete. For example, the search for neural correlates of consciousness has become one of the most active areas of neuroscience, as reviewed by Baars (1997), Crick and Koch (1998), Edelman and Tononi (2000), Taylor (1999), and Tononi and Edelman (1998). But determining the content of an animal's conscious awareness remains formidably difficult.

As Crist reminds us, Charles Darwin had no doubt that many animals are conscious of simple matters important in their lives. He became convinced by careful observation and experiments that earthworms probably experience very simple thoughts about the procedures by which they plug their burrows. The neural correlates of consciousness do not seem to entail specific features of gross neuroanatomy. Whatever basic processes lead to conscious awareness may therefore be present in a wide variety of animals.

Consciousness is obviously a highly heterogeneous phenomenon. We know it varies widely in our species, and there is every reason to expect even greater variability according to the physiological capabilities of various animals and the matters that are important in their lives.

The content of animals' consciousness is probably limited to matters of direct concern to them, and of course it does not include the more complex levels of human thought. However, simplicity of content does not mean total lack of subjective consciousness.

How can objective, verifiable, scientific data about animal consciousness be obtained? Similarity of neural structure and function is one type of evidence; and none of the proposed neural correlates of consciousness entail any element or process that is qualitatively unique to our species. Adaptive versatility in adjusting behavior to cope with novel challenges, for which neither evolution or learning provide specific instructions, constitutes suggestive evidence of simple conscious thinking about alternative behaviors available to an animal and their likely results. However, scientists reluctant to infer subjective consciousness in animals often claim that whatever an animal, or a human, does *might* be accomplished unconsciously.

With our human companions we avoid, or at least minimize, this uncertainty by relying on communication as evidence of their subjective thoughts and feelings. Such evidence stems from both verbal language and nonverbal body language. The versatility of animal communication suggests that the messages they exchange with others may also provide objective evidence about at least part of the content of their consciousness.

Communicating animals may be almost literally telling us what they are feeling or thinking. Here too, determined skeptics tend to imply that only human communication provides evidence of subjective thoughts and feelings. But evolutionary biologists are rightly suspicious of claims that some trait suddenly appears *de novo*, without any precursors. The difference between human consciousness and that of any animal is no doubt enormous, but this difference is probably one of degree rather than kind.

Total certainty is not attainable, even when we inquire about the thoughts and feelings of our human companions. Therefore cautious scientists have a strong tendency to avoid this question, and some insist that such questions are inappropriate for scientific inquiry (but see Bekoff 2000a,b). However, the question of whether some nonhuman animals have any conscious experiences is an important one, and science does not advance by ignoring significant challenges. Despite recent advances and continuing progress, we must deal with hypotheses that cannot yet be confirmed or rejected with total precision and certainty. Nevertheless, the tentative assumption that some animals experience simple levels of subjective awareness often enables us to make sense of their behavior. Like the once radical notion that the Earth revolves around the Sun, this approach "simplifies calculations."

References

Baars, B. J. (1997). *In the Theatre of Consciousness: The Workspace of the Mind.* New York: Oxford University Press.

Bekoff, M. (2000a). Animal emotions: Exploring passionate natures. *Bioscience* 50: 861–870.

Bekoff, M. (ed.) (2000b). *The Smile of a Dolphin: Remarkable Accounts of Animal Emotions.* New York: Random House/Discovery Books.

Crick, F. and Koch, C. (1998). Consciousness and neuroscience. *Cerebral Cortex* 8: 97–107.

Dawkins, M. S. (1993). *Through Our Eyes Only? The Search for Animal Consciousness.* Oxford: Blackwell.

Edelman, G. M. and Tononi, G. (2000). *A Universe of Consciousness: How Matter Becomes Imagination.* New York: Basic Books.

Griffin, D. R. (1976). *The Question of Animal Awareness: Evolutionary Continuity of Mental Experience.* New York: Rockefeller University Press.

Griffin, D. R. (2001). *Animal Minds, Beyond Cognition to Consciousness.* Chicago: University of Chicago Press.

Taylor, J. G. (1999). *The Race for Consciousness.* Cambridge, Mass.: MIT Press.

Tononi, G. and Edelman, G. M. (1998). Consciousness and complexity. *Science* 282: 1846–1851.

Index